MODERN HUMAN PHYSIOLOGY

Frederick D. Cornett
Pauline Gratz

HOLT, RINEHART AND WINSTON, PUBLISHERS
New York • Toronto • Mexico City • London • Sydney • Tokyo

FREDERICK D. CORNETT is a teacher of physiology and
Chairperson of the Science Department at Santa Monica
High School, Santa Monica, California

PAULINE GRATZ, Ed.D., is Professor Emeritus of Human
Ecology, School of Nursing, Duke University Medical Center,
Durham, North Carolina.

Cover design by Brian Molloy.

The large illustration on the cover shows a contour map of
the human head and neck. The smaller illustrations show
contour maps of a human nose (left), a single human
neuron (center), and the biting surface of a mandibular
premolar tooth (right).

Cover photographs: Dan McCoy/Rainbow (top); and Robin
Williams/Photo Researchers (bottom–left, center, and right).

Illustrations by Caliber Design Planning Inc.

Acknowledgments and photo credits appear on page x.
Human Body insert courtesy of W.B. Saunders Company.

ISBN: 0-03-005653-5
  123 071 98765

MODERN HUMAN PHYSIOLOGY CONSULTANTS AND CONTENT CRITICS

Consultants

Sue Falkenstine Beers
Biology Teacher
Westhill High School
Stamford, Connecticut

Sarah Sue Steed, Ph.D.
District Health Education
Coordinator
Albuquerque Curriculum
Center
Albuquerque, New Mexico

Content Critics

UNIT 1
Meyer Texon, M.D.
Associate Professor of Forensic
Medicine
New York University School of
Medicine
New York, New York

UNIT 2 AND UNIT 7
Henry Hermo, Jr., Ph.D.
Department of Biology and
Medical Laboratory Technology
Bronx Community College of
the City University of New York
New York, New York

UNIT 3
**Henry B. Peters,
O.D., M.A., D.O.S.**
School of Optometry
University of Alabama at
Birmingham
Birmingham, Alabama

UNIT 4
Patricia Long
Nutritionist and Science Writer
Albany, California

UNIT 5 AND UNIT 6
Betty Tevis, Ph.D.
Associate Professor and
Chairperson of the Health
Science Department
Mankato State University
Mankato, Minnesota

Kent Kalm, Ph.D.
Associate Professor
Physical Education Department
Mankato State University
Mankato, Minnesota

UNIT 8
Robert S. Bernstein, M.D.
Associate Professor of Clinical
Medicine
Columbia University College of
Physicians and Surgeons
New York, New York

Chief of the Endocrine Clinic
Saint Luke's Hospital
New York, New York

UNIT 9
Wilma Gladstone, M.D.
New York, New York

iii

PREFACE TO STUDENTS

The favorable response to previous editions of *MODERN HUMAN PHYSIOLOGY* is significant evidence of its value in the high school curriculum. In preparing this seventh edition the authors have maintained the systematic approach and continuity of topics that have proven effective for teachers and students who have used the text. The content has been updated to include the most recent advances in the various biological fields.

MODERN HUMAN PHYSIOLOGY is written for high school students who wish to learn about the structure and function of the body's systems in greater detail. Although a basic background in chemistry and biology, or health, may be helpful in understanding some concepts, courses in those subjects are not a prerequisite to learning the material in this text.

The authors recognize the importance of presenting basic facts that are relevant to the development of physiological principles. Thus the text contains anatomical illustrations and photos that help students to translate theoretical concepts into concrete descriptions. Technical words, printed in **boldface** and *italics*, are defined in the context of the discussion. In addition, marginal references serve as guides for learning new terms. Major concepts are clarified throughout the text by use of examples, practical applications, and laboratory experiences.

The material in *MODERN HUMAN PHYSIOLOGY* is organized into units and chapters. Every chapter begins with a list of student objectives. These help students direct their efforts while working on the chapter. Each chapter is organized into three sections: (1) The **anatomy** of the system identifies and relates the individual parts to the whole, (2) the **physiology** of the system provides an understanding of the function of the parts of the system, and (3) the **pathology** of the system relates a specific disorder to the abnormal functioning of the whole organism.

The end-of-chapter material contains a Summary for review and reinforcement of major concepts. Chapter questions, consisting of Vocabulary Review and Test Your Knowledge, allow students to evaluate their understanding of scientific terms and their comprehension of the material.

The Glossary in the back of the text contains a complete list of definitions of important biological terms. The Index is a valuable reference for quickly locating specific information. Two special features of *MODERN HUMAN PHYSIOLOGY* are the full-color insert on the human body, which helps students to identify the structures of different systems, and specially displayed features on people and topics of interest in the field of science.

EXPERIMENTS IN MODERN HUMAN PHYSIOLOGY contains laboratory activities that follow the sequence of the material in the text. The experiments are correlated to the anatomy and physiology sections within each chapter. Data sheets are included to allow students to record and analyze observations. Such laboratory experiences provide the background needed for understanding the chemical and physical phenomena that underlie the structure and function of the systems of the human body.

CONTENTS

Preface to Students   iv
Guide to Reading Science Material   ix

UNIT 1

The Organization of the Human Body   1

1 The Basic Plan   2
2 Principles of Chemistry   13
   Special Feature: Career Horizons   26
3 Cellular Structure and Function   29
4 Intercellular Organization   51
   Special Feature: New Frontiers   62
   Special Feature: Eyes of Science   66

UNIT 2

The Supporting Framework and Movement 67

5 Skeleton—The Framework   68
6 Muscle Tissue   96
7 The Skeletal Muscles   112

UNIT 3

Coordination and Control of the Body   129

8 Nervous Tissue   130
9 The Nervous System   145
   Special Feature: Seeing with
   Lasers and Ultrasound   168
10 The Eye and Vision   169
11 The Ears: Hearing and Equilibrium   190

UNIT 4

The Digestive System   205

12 Food and Nutrition   206
13 The Mouth, Pharynx, and Esophagus   228
14 The Stomach   241
15 The Intestines   251

UNIT 5

The Respiratory System   271

16 The Respiratory Organs   272
17 The Mechanics of Breathing   285
18 The Respiratory Gases and
    Barometric Physiology   299

UNIT 6

The Transport Systems   309

19 The Blood   310
20 The Heart   331
21 The Vascular System   346
22 The Lymph System   363

UNIT 7

The Regulatory Systems and Metabolism   373

23 The Skin and Its Appendages   374
24 Regulation of Body Temperature   386

25 Metabolism   395
26 The Kidney   408

UNIT 8

The Endocrine System   419

27 Endocrine Glands and
    the Function of Hormones   420
28 Specific Endocrine Glands   429
    Special Feature: Eminent Surgeon   450

UNIT 9

Reproduction and Heredity   451

29 Reproduction and Development   452
30 Human Genetics   468
    Glossary   486
    Index   497

GUIDE TO READING SCIENCE MATERIAL

New emphasis in learning and the reading process are a major concern in science education. Students who develop the necessary reading skills for comprehension of technical material will find it more interesting and easier to learn science. A common student complaint is in the inability to apply principles and perform operations in class and laboratory activities because of the difficulty in understanding the science content.

Every subject area has its own constellation of skills. The reading skills essential in science can be learned through reading materials required in the course. The following suggestions can be your guide to learn important science reading skills.

Instructional Techniques

Use the Textbook as a vehicle for learning and solving problems, as well as a resource for finding information. Read the preface to learn about the content organization and available student aids. **Survey** the table of contents, the glossary, and the index to find the sequence and arrangement of topics. **Preview** a chapter to note section headings, illustrations, marginal notes, summary, and exercises.

Studying Vocabulary Terms is an important component to understanding science. **Boldface** and *italics* are used to identify important terms. Word parts in the margin of the page can help to analyze a meaning. Words are defined within sentences by way of examples, illustrations, descriptions, and familiar relationships.

Identify Concepts using chapter section headings for the **main idea** and subheadings for **supporting** major concepts. The **sequence** of information is important in order to perform operations in activities and to **predict results** accurately. A group of common ideas, observations, or statements form a **generalization**.

Interpreting Data involves **locating** information on a graph, chart, table, or diagram that is used to **explain** or **support evidence** to verify concepts. **Compare** (similar) and **contrast** (different) information to **analyze** the results of experiments, to **understand relationships**, and to **apply facts** to new situations.

Following Directions is important in order to understand the purposes and problems involved in performing class activities and laboratory experiments. Carefully read or listen to instructions. Establish procedures in your own words. Ask **questions** to clarify steps in an investigation or the method used to observe results. **Record** all data on the laboratory sheet and **organize** information.

ACKNOWLEDGMENTS AND PHOTO CREDITS

Unit 1 p. 1 Paul Sutton/Duomo; p. 14 Granger Collection; p. 17 Courtesy of Memorial Sloan-Kettering Cancer Center; p. 18 Runk Schoenberger/Grant Heilman; p. 26 David York/Medichrome-Stock Shop; p. 31 C. Bruni; p. 32 D. W. Fawcett, M.D. from *The Cell—Its Organelles and Inclusion*, W. B. Saunders Co.; pp. 33, 34(l), 35(l) D. W. Fawcett, M.D.; p. 34(r) Lester V. Bergman & Assoc. Inc.; p. 35(r) Dr. Usha B. Raju, Henry Ford Hospital; p. 39 Phil Harrington/Peter Arnold; p. 46 Carolina Biological Supply Co.; pp. 47, 48 General Biological Supply House, Inc.; p. 53 from *Atlas of Microscopic Anatomy* by R. Bergman & A. Afifi, W. B. Saunders Co., Phila. © 1974; pp. 54, 55, 56, 59 from *Tissues and Organs, A Text-Atlas of Scanning Electron Microscopy* by Richard G. Kessel & Randy H. Kardon, W. H. Freeman & Co. © 1979; p. 57(l) Carolina Biological Supply Co.; p. 57(r) from *Atlas of Microscopic Anatomy* by R. Bergman & A. Afifi, W. B. Saunders Co., Phila. © 1974; p. 58 Courtesy of Memorial Sloan-Kettering Cancer Center; p. 59(r) Bausch & Lomb, Inc.; p. 60 Dr. Jerry Maynard, Dept. of Orthopedics, Univ. of Iowa; p. 61 (tl) General Biological Supply House Inc.; p. 61 (bl,br,tr) Fisher Scientific Co.; p. 62 Philips Medical Systems, Inc.; p. 66 (all) Marilyn Càldarolo, IBM-T. J. Watson Research Center.

Unit 2 p. 67 Sylvia Plachy/Archive Pix; p. 86 HRW by Russell Dian; p. 88 from *Atlas of Microscopic Anatomy* by R. Bergman & A. Afifi, W. B. Saunders Co., Phila. © 1974; p. 89(t) from *Tissues and Organs, A Text-Atlas of Scanning Electron Microscopy* by Richard G. Kessel & Randy H. Kardon, W. H. Freeman & Co. © 1979; p. 91 Dr. M. Meltzer; p. 92 Armed Forces Inst. of Pathology; p. 97(t) Ida Wyman; p. 97(c) Walter Dawn; p. 97(b) Courtesy of Carolina Biological Supply Co.; p. 98(t) E. Leita Inc.; p. 98(c) Walter Dawn; pp. 98(b), 99 Dr. H. E. Huxley; p. 106 Taurus Photo; p. 107 HRW by Richard Haynes.

Unit 3 p. 129 Peter Angelo Simon/Photo Researchers; p. 134 from *Tissues and Organs, A Text-Atlas of Scanning Electron Microscopy* by Richard G. Kessel & Randy H. Kardon, W. H. Freeman & Co. © 1979; p. 140 Dr. A. R. Lieberman of Univ. College, London; p. 163 from the "Dreamstage Catalogue," © 1977 by J. Allan Hobson & Hoffman-La Roche Inc.; p. 168(t) Sensory Aids Corporation; p. 168(b) Nurion Industries; p. 174 HRW by Russell Dian; p. 175 Rotker/Taurus Photo; p. 177 HRW by Richard Haynes; p. 178(r,l) David Parker/Photo Researchers; p. 195(tr) from *Tissues and Organs, A Text-Atlas of Scanning Electron Microscopy* by Richard G. Kessel & Randy H. Kardon, W. H. Freeman & Co. © 1979; p. 200 C. G. Conn, Ltd.

Unit 4 p. 205 HRW by Richard Haynes; p. 218 HRW by Russell Dian; p. 232 Len Barbiero/Medichrome-Stock Shop; pp. 233, 236 Structural Biology, Stamford Medical School/BPS; pp. 242, 248 Jerry E. Harrity; p. 243 John H. Nothrop; p. 256 Armed Forces Inst. of Pathology; p. 258 Lester V. Bergman & Assoc.

Unit 5 p. 271 Douglas Kirkland/Sygma; p. 281 Rotker/Taurus Photo; p. 292 Stephenson Corp.

Unit 6 p. 309 Richard Laird/Freelance Photographers Guild; p. 311 NIH, National Heart, Lung & Blood Inst.; p. 314 Warren Rosenberg; p. 315 Richard J. Feldman, NIH; p. 320 Tony Brain/Peter Arnold; p. 327 NIH, National Heart, Lung & Blood Inst.; p. 332 Lester V. Bergman & Assoc.; p. 364 Dr. Daniel R. Alonso; p. 365 General Biological Supply House Inc.

Unit 7 p. 373 Tannenbaum/Sygma; p. 377 Eastman Kodak Co., Industrial; p. 379 Manfred Kage/Peter Arnold; p. 399 Gilson Int./Gilson Medical Electronics, Inc.

Unit 8 p. 419 Jim Anderson/Woodfin Camp; pp. 434,435 Lester V. Bergman & Assoc. Inc.; p. 444 UPI; p. 445(tl) Liu Hueng Shing/Woodfin Camp; p 445(tr) Omikron/Photo Researchers; p. 450 M. & M. Smith/Schomburg Center for Research in Black Culture, NYPL.

Unit 9 p. 451 Hella Hammid/Photo Researchers; p. 457 (t) Ida Wyman; p. 465 Courtesy of the Cleveland Health Museum; p. 469 Theodore T. Puck; p. 470 The Jackson Lab.

UNIT 1

THE ORGANIZATION OF THE HUMAN BODY

CHAPTER 1

THE BASIC PLAN

Objectives

A. Identify some of the important fields of study that comprise the biological sciences.
B. Describe the various levels of structural organization within the human body
C. Define basic directional terms used in anatomy and physiology
D. Identify the various body cavities
E. List the nine body systems

THE BIOLOGICAL SCIENCES

Throughout the ages humans have had a natural curiosity about the universe—the heavens, the earth, and the mechanisms of life. The biological sciences cover a broad spectrum in the study of living things. In recent years biologists, chemists, and physicists have made discoveries that have greatly expanded our understanding of living things.

An Orientation to Anatomy and Physiology

Biology is fundamental to all the life sciences. It encompasses a vast collection of facts and theories about living organisms. So complex is this science that it is divided into two major categories—*botany*, the study of plants, and *zoology*, the study of animals.

Morphology deals with the form and structure of plants and animals. To learn how the human body operates, you need a clear understanding of its architecture (anatomy).

Anatomy is a subdivision of morphology. It is primarily concerned with the structures of cells and organs and their relationship to the whole organism. The subject matter of anatomy is further divided according to those structures that can be seen and identified with the naked eye (*gross anatomy* or *macroanatomy*) and those visible only with a microscope (*histology* and *cytology*).

macro = large
scope = observe

Physiology is concerned with how an organism functions. Human physiology probes into all of the chemical and physical mechanisms

2

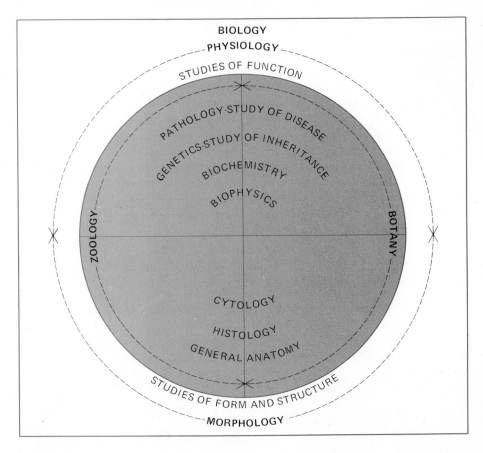

Figure 1-1 The relationship of some of the main subdivisions of the biological sciences.

that are characteristic of living matter within the human body. Not only is it concerned with what happens but also with how and why it happens. Like anatomy, physiology has become highly specialized. Some examples of whole new fields of study are: *mammalian physiology*, which deals with the organ systems of mammals; *comparative physiology*, which studies the life processes and dysfunctions of different kinds of animals; and *environmental physiology*, which focuses on the natural environment and how certain conditions affect the human body.

Compared to anatomy, physiology is a relatively young science and relies on many other fields of knowledge. One such field is *biochemistry*, the study of the chemistry of organisms and their life processes. We can also include *psychology*, which deals with the study of the mind, human emotions, and behavior. These psychological factors frequently have a profound effect on the physiological well-being of the individual.

In our discussion of anatomy and physiology, it is also important to study the causes and effects of bodily dysfunction or disease as it relates to the physiology of the human body. This science is called *pathology* and has contributed much information to the field of medicine.

One additional division of the biological sciences that is essential to understanding the human body is **genetics.** Genetics explains the manner in which various physical traits (such as eye color) are transmitted from one generation to the next. Patterns of inheritance determine not only physical features but physiological traits as well. Hereditary traits are transmitted by the genetic material that is contained in the *chromosomes* of the cells. Research scientists have been able to reconstruct hereditary information in *genetic engineering*. These advances can prove to be beneficial but also can involve serious risks.

Many other subjects exist that are important in understanding life processes. As you study physiology, you will become increasingly aware of the need for more detailed and quantitative information. The human body reveals an unending complexity of structure as well as a diversity of physical and chemical activity.

PLAN OF THE HUMAN BODY

As in most areas of study, anatomy has a language all its own. Specific terms are needed to describe positions, surfaces, and anatomical relationships of parts. Several anatomical reference systems are illustrated and defined on pages 5–9 and apply to nearly all anatomical descriptions. They include levels of structural organization, directional terms, planes of sections, and body cavities.

Levels of Structural Organization

There are several levels of structural organization that are associated with each other, ranging from the simple to the more complex. The first level of organization, the *chemical level*, is common for all living organisms, both plants and animals. Different kinds of atoms join together in various ways to form compounds from which all living matter is made (Fig. 1-2). The next ascending level of organization is the *cellular level*. The cell is the structural and functional unit of all living things. Both the structure and function of the cell will be discussed in detail in Chapter 3.

Specialization of cells.   In very early stages of the embryonic development of a young organism, cells begin to change in appearance. This process is called **differentiation;** that is, groups of cells having a common origin become specialized for certain physiological functions. They also develop structural characteristics that are unique to their group. Cells with a common origin, appearance, and function form the next level of organization, called the *tissue level*. There are four types of tissues in the human body: *epithelial tissue* covers and protects surfaces; *connective tissue* joins various parts together and provides support; *muscular tissue* allows movement; and *nervous tissue* responds to environmental stimuli and coordinates bodily activity. The isolated cells

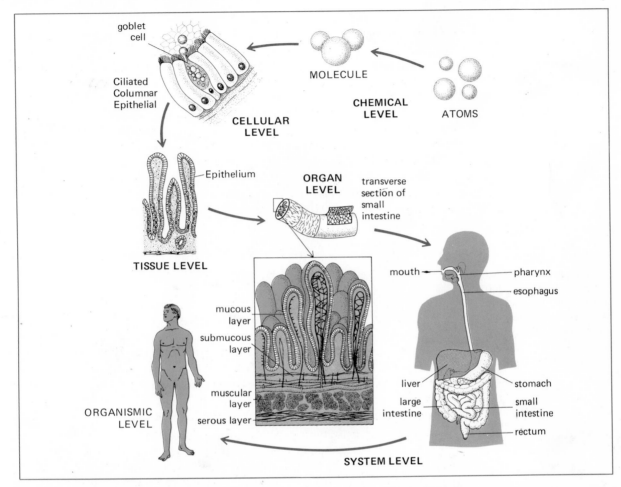

Figure 1-2 Structural levels of organization as seen in the human body.

shown in Fig. 1-2, when grouped together, form a type of epithelium that lines the small intestine. Some of these cells produce mucus, which mixes with nutrients, to permit their easy passage through the small intestine. Other cells of this tissue produce fluids to prepare food materials for absorption by the small intestine.

Organ systems.   When different kinds of tissues are joined together, they form another level of organization—the *organ level*. An organ is a differentiated part of an organism adapted for a specific function. The heart, brain, liver, stomach, and eye are all examples of organs, each having its own function. As seen in Fig. 1-2, the wall of the small intestine (an organ) shows a number of different kinds of tissues. The epithelial lining produces mucus and digestive fluids and prepares certain-sized molecules for absorption. The submucous layer is largely connective tissue, which supports blood vessels and nerves. The muscular layer provides for movement, and the outer serous layer serves to lubricate the outside walls of the small intestine.

Table 1-1   Terms of Direction

Term	Definition	Example
Anterior or ventral	Toward the front of the body or body part.	The sternum is *anterior* to the heart.
Posterior or dorsal	Toward the back of body or body part.	The spinal cord is *posterior* to the trachea.
Superior or cranial	Uppermost part of a structure or above.	The head is *superior* to the chest.
Inferior	Lowermost part of a structure or below.	The stomach is *inferior* to the heart.
Medial	Nearest the midline of the body.	The big toe is on the *medial* side of the foot.
Lateral	Away from midline, to the side.	One ear is on each *lateral* aspect of the head.
Proximal	Nearest the attachment of an appendage.	The knee is *proximal* to the ankle.
Distal	Away from the point of attachment.	The wrist is *distal* to the elbow.
Parietal	Pertains to the outer layer or wall.	The *parietal* pleura lines the chest cavity.
Visceral	Pertains to the organs within a body cavity.	The *visceral* pleura covers the lung.

Figure 1-3 The anatomic position, anterior and posterior view, of the human body to show descriptive terms, general regions, and the common terms of various body structures, with the anatomical terms in parentheses.

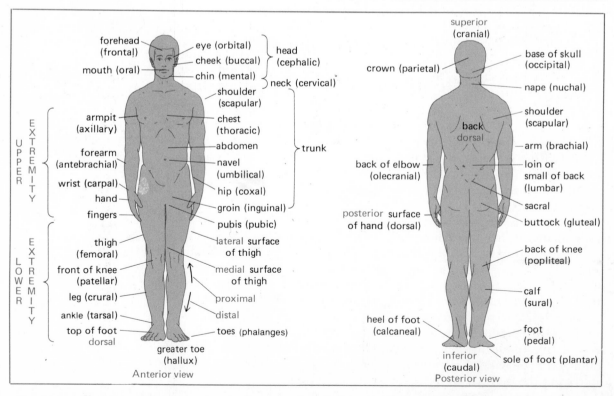

When groups of organs join together, they form a still higher level of organization—the *organ system*. In such a system, the organs act together to perform a highly complex but specialized function. For example, all of the mechanical and chemical processes of digestion are carried out through the coordinated activities of the digestive system.

ANATOMICAL REFERENCE SYSTEMS

Directional Terminology

Directional terms describe the exact location of certain structures of the body, in the anatomical position. (In this position the individual is standing erect with head and palms facing forward.) It is important that you become familiar with these words in order to avoid complicated descriptions. Many of the directional terms that are defined in Table 1-1 may be located as labeled on Fig. 1-3.

Planes of Sections

Humans, like all vertebrates (animals with backbones), have a type of *body symmetry* that is known as **bilateral**. This means that the body can be bisected vertically into equal right and left segments along a **midsagittal plane**. A **parasagittal plane** also runs vertically, but it divides the body into unequal portions. If a vertical plane is drawn through the human body at right angles to the midsagittal plane, the body is divided into anterior (ventral) and posterior (dorsal) portions along a **frontal** or **coronal plane**. When the body is sectioned in a horizontal direction (at right angles to the sagittal and frontal) it is then divided into upper (superior) and lower (inferior) portions along a **transverse plane**. When you examine an organ that has been sectioned, it is important to know how the sections were made so that you can understand the structural relationship of one part to another. Fig. 1-4 shows various sections through the human body.

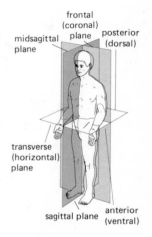

Figure 1-4 **The human body showing the fundamental planes of reference.**

Body Cavities

Nearly all of the internal organs of the human body lie within two major spaces, the **dorsal cavity** and the **ventral cavity.** These can be distinguished easily if the human body is viewed from the side of a midsagittal section. (See Fig. 1-5.) The dorsal cavity is found within the skull and vertebral column. It is divided into a *cranial cavity*, which contains the brain, and *spinal cavity*, which contains the spinal cord.

The larger of the two major spaces is the ventral cavity. A dome-shaped muscular partition, the *diaphragm*, divides this cavity into the superior *thoracic cavity* and the inferior *abdominal cavity*. The thoracic cavity contains the lungs and their air passages, the esophagus, and the heart and major blood vessels.

Inferior to the diaphragm lies the abdominal cavity. The organs that occupy this space include most of the digestive organs as well as the kidneys and spleen. The lowest portion of the abdominal cavity is often referred to as the *pelvic cavity* and contains the urinary bladder, rectum, and certain reproductive organs. The cavities and their organs, frequently called **viscera**, are illustrated in Fig. 1-5 and Fig. 1-6.

A systematic approach to physiology.   The grouping of organs into systems is a logical way to study the structure and function of the human body. The chapters that follow present human physiology using a systematic approach. Each organ system is responsible for maintaining and adjusting its activities to provide the proper internal environment for every cell.

The **skeletal system** is composed of bones, cartilage, and joints, which connect one bone to the next. The skeleton provides the body with a supporting framework that protects vital organs. It also serves as a place for the attachment of muscles, permitting a wide range of movement.

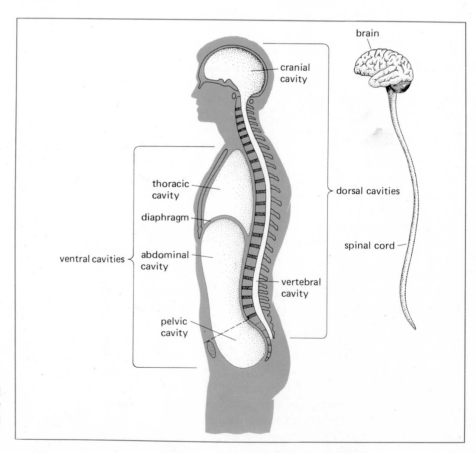

Figure 1-5 **Midsagittal section of the body showing the dorsal and ventral body cavities.**

The **muscular system** consists of skeletal or striated muscle, and is responsible for movement. Smooth muscle is found in the internal organs, as in the stomach and in the intestine. Cardiac muscle is only found in the heart.

The **nervous system** includes the brain, the spinal cord, nerves, and special sense organs that enable us to see, hear, taste, smell, and touch. This highly integrated system makes us aware of environmental changes (stimuli). It also enables us to react to these changes, to think, and to coordinate the activities of all of the other systems.

The **digestive system** consists of a long tubular passageway starting with the lips and ending at the anus. Food is ingested into the mouth and then is broken down by physical and chemical processes for absorption through the intestinal wall. The large intestine absorbs excess water and minerals and eliminates undigested materials. Numerous

Figure 1-6 **Anterior view of the organs of the body cavities.**

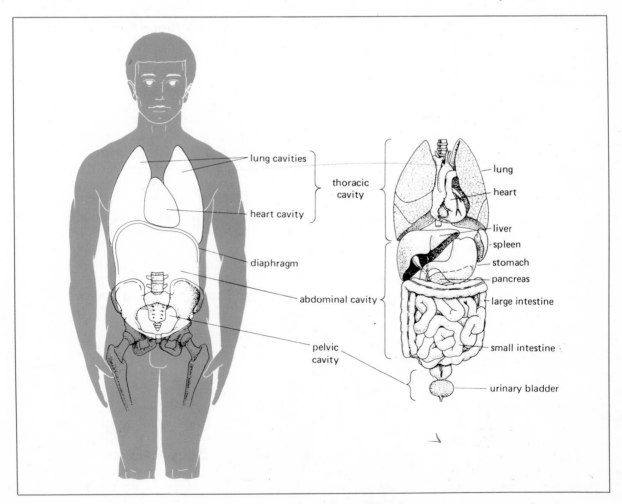

accessory structures, including the salivary glands, pancreas, liver, and gall bladder, assist in the process of digestion.

The **respiratory system** provides the body with a continuous supply of oxygen needed for cellular activity and rids the body of carbon dioxide. The organs responsible for this activity include the nose, pharynx, larynx (voice box), trachea (windpipe), bronchi, and lungs.

The **circulatory system** consists of an extensive network of vessels that distributes blood throughout the body. The two divisions are the *cardiovascular system* (heart, blood vessels, and blood) and the *lymphatic system* (lymph vessels, lymph nodes, and lymph). The transport of oxygen, absorbed nutrients, and hormones and the elimination of waste products are but a few of this system's functions.

The **urinary system** includes the kidneys, two ureters, a urinary bladder, and the urethra. Together these organs function to eliminate excess water, various salts, foreign substances, and cellular waste products from the body in the form of urine. The urinary system is also largely responsible for maintaining constant conditions in the body's internal environment.

The **endocrine system** consists of *ductless glands* that secrete hormones directly into the bloodstream to target organs throughout the body. The ductless glands include the pituitary, pineal, thyroid, parathyroids, adrenals, and portions of the pancreas, ovaries, and testes. These glands along with the nervous system coordinate many of the body's physiological activities. The secretions of the endocrine glands regulate the body's internal processes, whereas the nervous system regulates the body's responses to changes in the external environment.

The **reproductive system** in the female includes the ovaries, which produce eggs or ova; the oviducts or fallopian tubes, which transport the ova to the uterus or womb, and the vagina and external genitalia. In the male, the reproductive organs include the testes, sperm-storing tubules (epididymis), sperm ducts (vas deferens), various glands (seminal vesicles and prostate), and the penis and urethra. These two distinct systems are uniquely designed to perpetuate the species. In addition, their respective hormones influence maturation and development.

SUMMARY

The study of all living matter comes under the broad heading of the biological sciences. Morphology is concerned with the structure and form of living organisms. Anatomy is a branch of morphology that studies the form and structure of organisms.

Physiology delves into the physical and chemical functions of living organisms. Subdivisions include: human physiology, mammalian physiology, comparative physiology, and environmental physiology. Genetics studies the manner in which the characteristics or traits of an organism are transmitted from generation to generation. The human body is organized into several structural levels ranging from the chemical level to the higher level of organ systems. Anatomical reference systems define and describe location, position, and relationship of one part to another in the body.

VOCABULARY REVIEW

Match the statement in the left column with the word(s) in the right column. Place the letter of your choice on your answer paper. *Do not write in this book.*

1. This science studies the causes and effects of bodily dysfunctions.
2. The process by which cells develop structural characteristics for specialization.
3. A group of tissues that performs one or more specific functions.
4. A branch of the biological sciences that studies the form and structure of plants and animals.
5. The plane that divides the body into equal right and left segments.
6. Pertains to the organs within a body cavity.
7. The skull and vertebral column are found within this cavity.
8. The lowest portion of the abdominal cavity.
9. A subdivision of physiology.
10. Uppermost part of a structure.

a. superior
b. system
c. botany
d. organ
e. sagittal
f. differentiation
g. morphology
h. midsagittal
i. pathology
j. visceral
k. dorsal cavity
l. pelvic cavity
m. mammalian physiology

TEST YOUR KNOWLEDGE

Group A

Are the following statements *true* or *false?* If the statement is correct, mark *true* on your answer sheet. If the statement is *false*, change the word(s) in the statement to make it correct.

1. The science of zoology is fundamental to the study of all living things.
2. You need to have an understanding of anatomy when you study human physiology.
3. Physiology is probably the oldest of all known sciences.
4. The structural and functional units of all living organisms are cells.

5. Cells with a common origin, appearance, and function form tissues.
6. Different kinds of tissues joined together to perform a specific function(s) are called organelles.
7. The abdomen is posterior to the back in the human being.
8. There are seven systems in the human body.
9. A sagittal plane divides the body into anterior and posterior portions.
10. Physiology is mainly concerned with the diseases of the body.

Group B

On your answer sheet, write the letter of the word that correctly completes the statement.

1. The type of symmetry shown by the human body is   (a) radial   (b) bilateral   (c) asymmetrical   (d) unilateral.
2. The chest region of the human body is called the   (a) abdomen   (b) pelvis   (c) cervix   (d) thorax.
3. A closely linked subdivision in the study of anatomy is   (a) chemistry   (b) morphology   (c) histology   (d) sociology.
4. The science primarily concerned with the study of disease is   (a) pathology   (b) embryology   (c) cytology   (d) physiology.
5. Physiology is mainly concerned with (a) growth of the body   (b) functions of the body   (c) structure of the body   (d) diseases of the body.
6. The plane that divides the body into superior and inferior portions is the   (a) sagittal   (b) transverse   (c) coronal   (d) midsagittal.
7. The cavity occupied by the brain is the   (a) spinal   (b) thoracic   (c) cranial   (d) pelvic.
8. The branch of the biological sciences that is primarily structural is   (a) morphology   (b) physiology   (c) pathology   (d) botany.
9. The end of a limb that lies toward the main part of the body is called   (a) distal   (b) dorsal   (c) proximal   (d) ventral.
10. The surface of an appendage that is toward the outside of the body is said to be   (a) medial   (b) lateral   (c) anterior   (d) posterior.

Write the word or words on your answer sheet that will correctly complete the sentence.

11. The heart, brain, and stomach are all examples of _____ that carry on their own respective functions.
12. A _____ plane runs vertically and divides the body into unequal right and left portions.
13. The _____ system consists of a long tubular passageway starting with the lips and ending at the anus.
14. The study of the mind, human emotions, and behavior is called _____.
15. The science that is concerned with how an organism functions is called _____.

CHAPTER 2

PRINCIPLES OF CHEMISTRY

Objectives

A. Define the word *matter* and list the three general classes of matter
B. List the three basic subatomic particles and describe the characteristics of each
C. Explain the meaning of the terms *atomic number* and *atomic mass*
D. Explain the difference between ionic bonds and covalent bonds
E. Define the term *organic compound* and list four examples
F. Compare different types of energy
G. List four categories of chemical reactions and give examples
H. Explain the differences between acids and bases

BASIC CHEMICAL ORGANIZATION

The cell has been described as the basic structural unit of living things. **Protoplasm** is a chemically complex substance made up of the components of matter, **elements** and **compounds.** Every cell is capable of carrying on the vital life functions, but the ability to do so depends upon the composition of protoplasm. To understand the physiology of the cell, it is important to study the most basic components of matter.

The Nature of Matter

In scientific terms, *matter is anything that occupies space and has mass.* Matter can be divided into three general classes: elements, compounds, and mixtures. When two or more elements combine in a chemical reaction, a compound is formed. A mixture is a combination of two or more substances, each of which retains its own characteristic properties. An **atom** is the smallest unit of an element that can exist alone or in combination with other elements. The compound sodium chloride (ordinary table salt) is formed when atoms of the element sodium chemically combine with atoms of the element chlorine. Atoms themselves cannot be divided in a chemical reaction. It is only as a result of nuclear reactions, such as fusion or fission, that atoms can be broken down into smaller particles. Matter may exist in three physical forms: *solid, liquid,* or *gas*. In the body, all three phases are present: the skeleton is a solid, the blood is a liquid, and the oxygen in our lungs is a gas.

Figure 2-1 In 1897 J. J. Thompson discovered the electron when using a glass tube such as this.

The Structure of Atoms

Though individual atoms are exceedingly small, they are quite complex. In 1897, Sir J. J. Thomson at the Cavendish Laboratory in Cambridge, England, experimenting with a special glass tube (Fig. 2-1), discovered the first of a number of *subatomic particles.* Through his investigations, Thomson concluded that the mass of this particle was 1,860 times less than the smallest known atom—hydrogen. This particle became known as an **electron.** We can see the effect of a beam of electrons when it strikes the surface of the screen in the electron microscope. In addition to the electron, numerous other subatomic particles have since been discovered. Two of these are important to our study of physiology—the **proton** and the **neutron.**

Subatomic particles can be distinguished on the basis of their location in the atom, their mass (weight), and their electric charge. The extremely dense center of the atom is called the **nucleus.** The nucleus is composed of clusters of protons that have a positive charge, and neutrons that carry no electric charge. Since the weight of the electrons is insignificant, the actual weight of the atom is concentrated in the nucleus. Both its protons and neutrons have a definite mass that is expressed as the **atomic mass** of the atom. The negatively charged electrons move around the nucleus. A neutral atom has an equal number of protons and electrons. The **atomic number** of an atom refers to the number of protons in the nucleus. All atoms of a particular element have the same atomic number. Atoms of the element *hydrogen* have atomic number 1; *helium*, atomic number 2; and *lithium*, atomic number 3. There are additional elements, each with its characteristic number of protons, its own name, and structure. At present, there are 107

Chemical symbol ($_z$X^a)
   X = *element; a = at. weight; z = at. number*

known elements. Element 107 (as yet, unnamed) has 107 protons in its nucleus. Each element is assigned a symbol, the scientists' shorthand, which represents one atom of the element. The first letter or the first two letters of the element's name is usually used as the symbol. For example, **H** stands for the element hydrogen, **He** for the element helium, **C** for the element carbon (six protons in the nucleus), and **O** for the element oxygen (eight protons in the nucleus). A basic understanding of the chemical behavior of the atoms of elements is important because the body and its processes are based on the same chemical principles.

Table 2-1
Chemical Elements Found in Animals

Element	Symbol	Weight by Percentage
Oxygen	O	62
Carbon	C	20
Hydrogen	H	10
Nitrogen	N	3.0
Calcium	Ca	2.50
Phosphorus	P	1.14
Chlorine	Cl	0.16
Sulfur	S	0.14
Potassium	K	0.11
Sodium	Na	0.10
Magnesium	Mg	0.07
Iodine	I	0.01
Iron	Fe	0.01

Trace Elements

Copper	Cu	
Manganese	Mn	
Molybdenum	Mo	needed in very small amounts
Cobalt	Co	
Boron	B	
Zinc	Zn	

Atomic energy levels.   The electrons move around the nucleus at very high speeds in a number of definite regions called *shells* or *energy levels*. Scientists describe the atom as having a spherical *electron cloud* appearance, as shown in Fig. 2-2. The chemical activity of any specific atom, that is, the way in which one atom reacts with other atoms, is determined by the number and arrangement of electrons in its outermost shell. Each shell has a definite number of electrons it can hold. When this number of electrons occupies the shell, the atom is considered "*shell-complete.*" The energy level nearest the nucleus never holds more than two electrons, no matter how large the atom. This shell can be referred to as the first energy level. The second shell or energy level

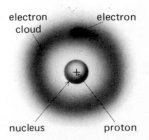

Figure 2-2 **Model of the hydrogen atom.**

holds a maximum of eight electrons. The third energy level can hold up to eight electrons. However, if there is a fourth energy level, the third energy level can hold a maximum of 18 electrons. Although the third and successive shells can hold more than eight electrons, they seem to be more stable with only eight electrons. Atoms of the elements helium, neon, argon, xenon, and krypton have complete outer shells and do not attract or transfer electrons. The atoms of these elements are very stable and therefore are called *inert elements*.

Except for those of the inert gases, the atoms of the 92 different kinds of naturally occurring elements, as well as the fourteen artificially made elements, have incomplete outer shells. Such atoms are unstable. They will try to share electrons with or transfer electrons to other atoms so that each can become shell-complete and attain greater stability. For example, the element chlorine (Cl) has the atomic number 17. This indicates that there are 17 protons in the nucleus and 17 electrons orbiting the nucleus. Two electrons occupy the first shell, eight the second shell, and seven the third shell. The outermost shell needs one electron to make a complete shell of eight electrons. Thus, chlorine tends to attract one electron from another atom to complete the third energy level. The element sodium (Na) has the atomic number 11. An atom of sodium has 11 protons and 11 electrons. What is the configuration of its electrons? There are two electrons in the first shell, eight in the second shell, and only one in the third shell. The sodium atom gives up the single electron in the third shell. When a chlorine atom accepts this electron, the two atoms unite, both forming a stable configuration in their outermost shells. (See Fig. 2-3.)

Isotopes.   Although all the atoms of an element have the same atomic number, they can differ in atomic mass. For example, some atoms of hydrogen contain one proton and no neutrons. This is the most common form of hydrogen, known as *protium*. Another form contains one neu-

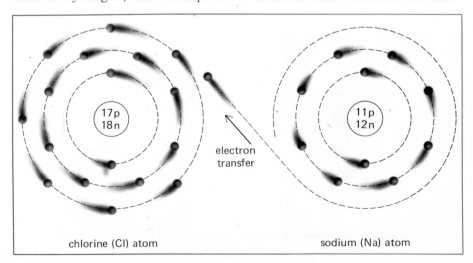

Figure 2-3 Sodium chloride is formed by the transfer of a single electron in the outer shell of the sodium atom to the outer shell of the chlorine atom.

17p
18n

electron transfer

11p
12n

chlorine (Cl) atom

sodium (Na) atom

tron in addition to one proton. This form is called *deuterium*. Finally, a rare form of hydrogen, called *tritium*, contains two neutrons and one proton. These three atoms of the same element have the same atomic number, but a different number of neutrons in their nuclei. This results in the three atoms having different atomic masses. Atoms of the same element that have the same atomic number but different atomic masses are called **isotopes.**

Isotopes may occur naturally or be produced artificially. The atomic weight of an element is the average mass of the atoms of all the naturally occurring isotopes of that element. The number of isotopes varies among different elements. Carbon has three isotopes, of which two are stable (C^{12}, C^{13}) and one is radioactive (C^{14}). The nuclei of such radioactive isotopes decay or disintegrate, releasing radioactive particles or rays. Such atoms are said to be naturally radioactive. It is also possible to artificially make some atoms radioactive. In a nuclear reactor atomic particles are used to bombard the nuclei of atoms to produce these radioactive isotopes.

Radioactive materials have become invaluable tools in the treatment and diagnosis of illness. When introduced into the living body, radioactive particles can be traced by special instruments. The amount of radiation can be picked up and recorded in body organs. Some radioactive isotopes are used to study rates of metabolic processes in the body. For example, radioactive iodine is used to obtain important information on thyroid function. Radioactive iron can indicate the iron reserves in red blood cells.

A picture or image of an organ can be produced by instruments called *scanners* or *gamma ray* cameras. They show the distribution of radioactivity within a certain body organ. These are diagnostic tools which provide important information about the organ's function and structure. Radioactive compounds can be used to treat certain diseases by destroying affected tissues.

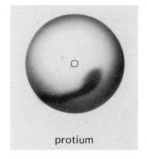

protium

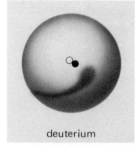

deuterium

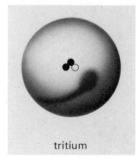

tritium

Figure 2-4 **Models of the isotopes of hydrogen.**

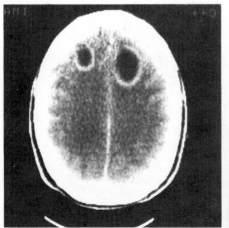

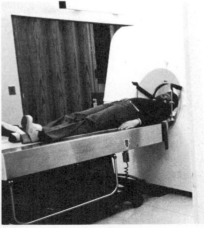

Figure 2-5 **The advances in medical technology can offer possible cures and prevention of diseases. Here a brainscanner is used to diagnose brain disorders.**

Chemical Bonding

The transfer of electrons or the sharing of electrons between two atoms establishes a **chemical bond** between the atoms. In the previously discussed reactions of sodium with chlorine, the chlorine atom gained an electron. The chlorine atom is no longer a neutral atom with equal numbers of protons and electrons. It now has an *excess* of electrons and carries a negative charge ($-$). The sodium atom transferred an electron to chlorine. In turn, the sodium atom now has a *deficiency* of one electron and carries a positive charge ($+$). Each of these charged atoms is known as an **ion**.

Ionic bonds.   When two or more atoms combine chemically, they form a new substance called a **compound.** Sodium and chlorine unite because of the transfer of electrons between the sodium and chlorine. This union forms the compound sodium chloride (NaCl). In such compounds, chemical bonds hold the oppositely charged ions together. The positively charged ions are called *cations*. With the exception of hydrogen, cations are metals. Some familiar examples are potassium (K^+), magnesium (Mg^{++}), calcium (Ca^{++}), and sodium (Na^+). The negatively charged ions are called *anions* and are nonmetals such as oxygen (O^-), fluorine (F^-), and chlorine (Cl^-). When metals react with nonmetals, the electrical attraction between the ions forms **ionic bonds.**

Covalent bonds.   Compounds can also be formed by the sharing of electrons. This sharing forms a **covalent bond.** When atoms of nonmetals combine, they do so by sharing equal numbers of electrons rather than by losing or gaining them. For example, two atoms of hydrogen, each having one electron, will combine with one another to form a **molecule** of hydrogen gas (H_2). Each atom of the pair shares its electron with the other in order to complete their outer shells. The shared pair of electrons orbit the nuclei of both atoms. When atoms share one electron, they form a *single* covalent bond.

Figure 2-6 (a) Diagram shows the arrangement of the sodium and chloride ions. (b) Sodium chloride crystals seen in a photograph.

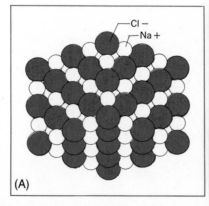

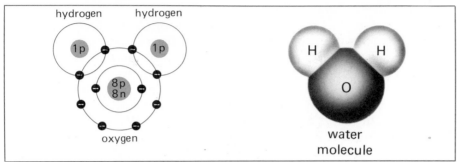

Two pairs of electrons (four electrons) are shared by two atoms of oxygen in the formation of oxygen gas (O_2). (See Fig. 2-8.) This is termed a *double* covalent bond. Each hydrogen atom has the same number of protons in the nucleus, and, therefore, the same attraction for the shared electrons. Two or more different elements may form compounds by means of covalent bonding. In the formation of water (H_2O), two atoms of hydrogen share electrons with one atom of oxygen. Within the individual bonds of the water molecule, the electrons being shared between each hydrogen atom and the oxygen atom have a greater attraction to the more positive nucleus of the oxygen atom. As a result, there is an unequal sharing of electrons in a molecule of water. These bonds produce a polarity that causes the asymmetrical shape of the water molecule, as seen in Fig. 2-7.

CHEMICAL COMPOUNDS

A chemical compound, like each chemical element, has a definite set of properties that differ from those of its component elements. Each compound has characteristics that relate to the proportions and positions of the atoms in the different elements. A **molecule** is the basic structural unit of a compound and retains the same properties of the compound. A molecular formula indicates the kinds and number of atoms in a molecule of a compound.

The number and arrangement of the atoms in a specific compound are always the same. For example, the *molecular formula* for water is H_2O. This means two atoms of hydrogen always combine with one atom of oxygen to form water. Two simple sugars that play important roles in our body's chemistry also illustrate this principle. *Glucose*, a very common sugar, is composed of 6 atoms of carbon, 12 of hydrogen, and 6 of oxygen ($C_6H_{12}O_6$). Within this molecule, the atoms have a definite arrangement (Fig. 2-9). If these same atoms are arranged in a slightly different manner, another type of sugar is formed. This sugar is called *fructose*. Its molecular formula is also written as $C_6H_{12}O_6$. Since these two sugars are so nearly alike, ordinary chemical tests will not distinguish between them. However, if each is exposed to a beam of polarized light, the glucose will rotate the light to the right, the fructose to the left. Glucose is often known as *dextrose* and fructose as *levulose*.

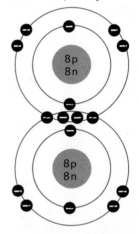

Figure 2-7 A molecule of water. One atom of oxygen is covalently bonded to two atoms of hydrogen. The molecule is bent at an angle between the two bonds due to the polarity.

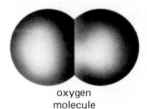

oxygen molecule

Figure 2-8 A double covalent bond between two oxygen atoms. The atoms share two pairs of electrons so that each becomes "shell-complete."

dexter (Lat.) = right

laevus (Lat.) = left

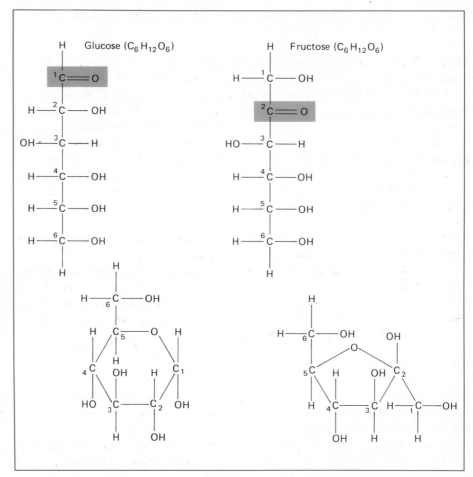

Figure 2-9 Glucose and fructose portrayed in different ways. The "ring" formula gives a better idea of how the atoms are arranged in space. (By convention, the carbon atoms are numbered.) Both compounds have the same general formula, but the arrangement of the atoms in the molecules is different.

Organic Compounds

Organic compounds were originally derived only from living organisms, hence the name "organic." Carbon is the key element in organic compounds. It combines with oxygen, hydrogen, and frequently nitrogen and sulfur to form various compounds that act as the building blocks of living matter. The main categories of organic compounds in living things are *carbohydrates*, *lipids* (fats), *proteins*, and *nucleic acids*. Many organic compounds are now produced industrially.

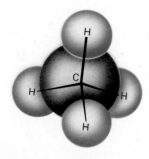

Figure 2-10 The carbon atoms form complex organic molecules.

Carbohydrates.   These compounds have the general formula $C_n (H_2O)_n$. Sugars and starches are carbohydrates. Although these substances account for only 1 percent of body weight, they are among the principal sources of chemical energy used for cellular activities. *Glucose* is the most common simple sugar found in living things. It can be stored as is or as animal starch (**glycogen**).

Lipids.   This group of compounds consists of fats, oils, waxes, and steroids. They are all insoluble in water. The steroids are large molecules with hydrogen and oxygen atoms joined to four rings of carbon atoms. Such important compounds as cholesterol and certain hormones are steroids.

   Other lipids are made up of fatty acids joined to compounds such as glycerol or proteins. Many are important constituents of cell membranes. Because lipids are insoluble in water, they are able to form barriers between the contents of the cell and the environment of the cell, which are both aqueous.

Proteins.   Protein molecules are made up of long chains of amino acids. There are twenty-two naturally occurring amino acids. The chains may be straight or branched, but for any one protein the types of amino acids, their sequence in the chain, and the shape of the molecule are very specific. Proteins have several functions in cells. They are the cell's basic building material and are found in its membranes, cytoplasm, and organelles. Other proteins known as *enzymes* regulate the chemical activities of cells. For instance, the chemical reactions that split glucose into smaller molecules are regulated by enzymes.

Nucleic acids.   There are two different nucleic acids. **Deoxyribonucleic acid (DNA)** is found in the nucleus and contains the hereditary information. It is a double-stranded molecule containing carbon, hydrogen, oxygen, nitrogen, and phosphorus. **Ribonucleic acid (RNA)** is similar in many ways to DNA but is a single-stranded molecule. RNA directs the synthesis of proteins in the cytoplasm.

deoxy = deprived of oxygen

ENERGY FOR LIFE FUNCTIONS

   Whether substances are changed physically or chemically, energy is always required. How is energy defined? Energy is a concept that is difficult to define, but easier to identify in terms of what it *does* or what it is *capable of doing*. Energy is the ability to do work. Energy that is stored is called **potential energy**; energy of motion is called **kinetic energy.** Heat, light, and electricity are examples of kinetic energy. Food that is stored in the body for future use is an example of potential energy. Potential energy and kinetic energy can be easily interchanged. Whenever bonds between atoms are chemically formed or broken, some form of energy is exchanged. When two atoms unite by transferring or sharing electrons, energy is usually required. When a bond is broken, energy is usually released.

Energy Exchange

A chemical reaction is always accompanied by the absorption or release of energy. Chemical reactions, or the processes by which chemical changes occur, can be grouped into four general categories:

(1) When two or more substances combine to form a new, more complex substance, the reaction is called **synthesis.** For example, atom A combines with atom B, to produce a new product—AB.

$$A + B \longrightarrow AB$$

In living organisms, synthesis occurs when small molecules combine to form large molecules such as proteins, fats, and carbohydrates. Biological synthesis is called **anabolism** and always requires a supply of energy.

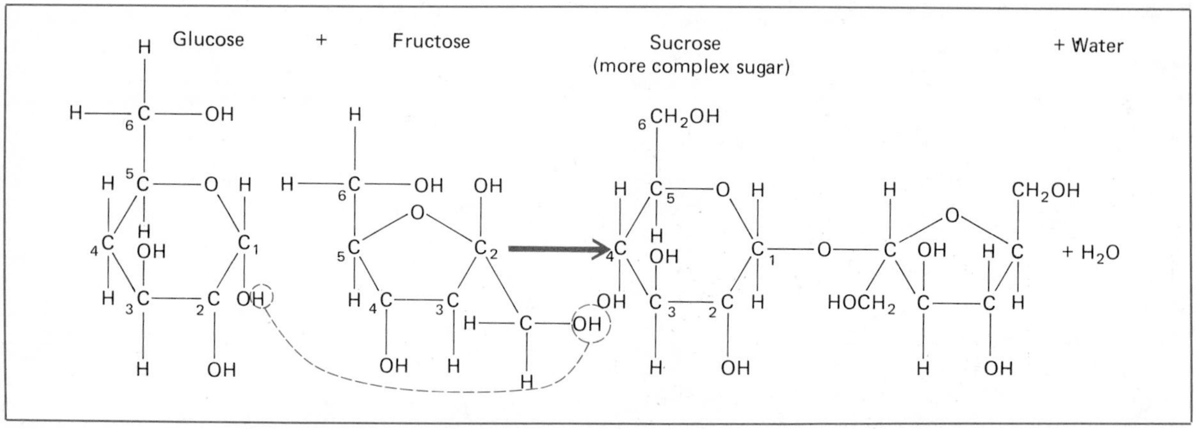

Figure 2-11 When glucose combines with fructose, it produces a more complex sugar plus water.

(2) The reverse of synthesis is **decomposition.** Complex substances are broken down into two or more simpler substances.

$$AB \longrightarrow A + B$$

hydro = water

lysis = breaking

The most common biological decomposition is called **hydrolysis.** In this reaction a water molecule, when added to a more complex molecule, causes the larger molecule to split into smaller molecules. Carbohydrates, fats, and proteins are digested in this way. Decomposition in living things is called **catabolism**, which is the opposite of anabolism, and always involves the release of energy.

(3) When one substance replaces another substance in a compound, a reaction known as *single replacement* (or displacement) takes place. In single replacement, an uncombined metal replaces the metal in the compound, or an uncombined nonmetal replaces the nonmetal in the compound.

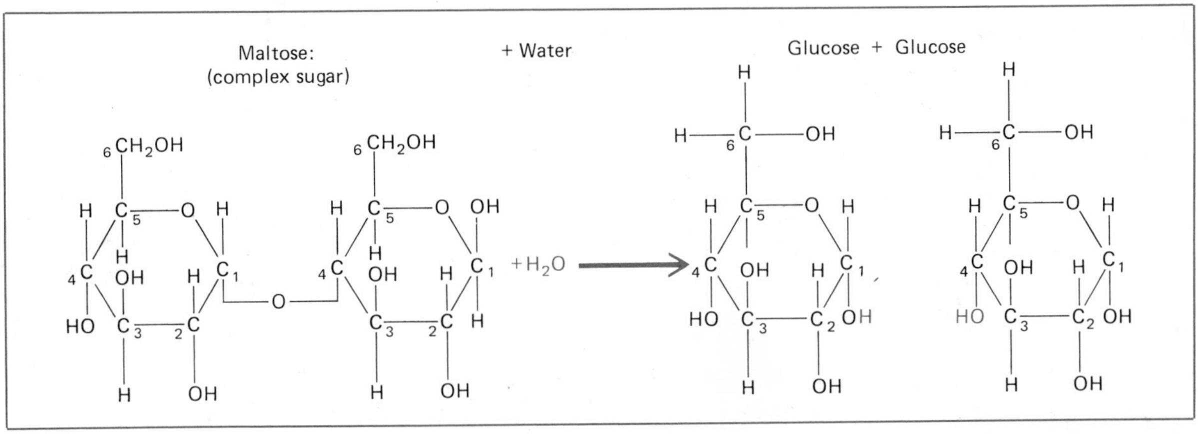

Maltose: (complex sugar) + Water → Glucose + Glucose

Figure 2-12 **The addition of water splits the maltose molecule into the more simple sugar— glucose.**

metal compound

$$A + BC \longrightarrow AC + B$$

metal .... metal

nonmetal compound

$$D + BC \longrightarrow DB + C$$

nonmetal .... nonmetal

(4) In *double replacement* reactions the positive and negative portions of the two reacting compounds are interchanged.

$$AB + CD \longrightarrow AD + CB$$

The reaction between an acid and a base to form water and salt is an example of a double replacement reaction. If a base such as sodium hydroxide combines with hydrochloric acid, the reaction produces sodium chloride (table salt) and water.

sodium hydroxide ..... hydrochloric acid

$$NaOH + HCl \longrightarrow NaCl + H_2O$$

sodium chloride ... water

Oxidation-reduction reactions. Oxidation-reduction reactions are chemical reactions that supply the energy for life functions in the body. *Oxidation* reactions are reactions in which there is a release of hydrogen atoms from a compound or the addition of oxygen. Either process involves the loss of electrons and the release of energy. Oxidation is always coupled with **reduction.** During oxidation electrons are removed from a molecule while in reduction these same electrons are gained by another molecule.

Food molecules are the source of energy for living organisms. When sugar (reducing agent) is oxidized, it will lose electrons. Oxygen (oxidizing agent) is the substance that will acquire the electrons. Carbon dioxide will be the product of the oxidation, and water will be the product of the reduction. This reaction is shown below:

$$\underset{\text{Sugar}}{C_6H_{12}O_6} + \underset{\text{Oxygen}}{O_2} \longrightarrow \underset{\text{Carbon dioxide}}{CO_2} + \underset{\text{Water}}{H_2O} + \text{energy}$$

Cellular respiration is the process by which energy is obtained from the breaking down of molecules. This occurs through a series of cellular reactions. The above reaction is a simplified form of cellular respiration.

The Hydrogen Ion Concentration (pH)

Substances that are acidic, such as lemon juice, have a sour flavor. Basic substances, such as baking soda, have a bitter taste. What are the chemical properties that determine the difference between the two? When an acid *dissociates*, or breaks apart in a solution, hydrogen ions (H^+) are released. When bases dissociate, they release hydroxyl ions (OH^-) in the solution. The greater the number of H^+ ions that dissociate in a solution, the stronger the acid. The strength of a base is also directly proportional to the number of OH^- ions that dissociate in a solution.

Figure 2-13 The pH scale. If a solution has a pH of 5.0, what is the concentration of hydrogen ions?

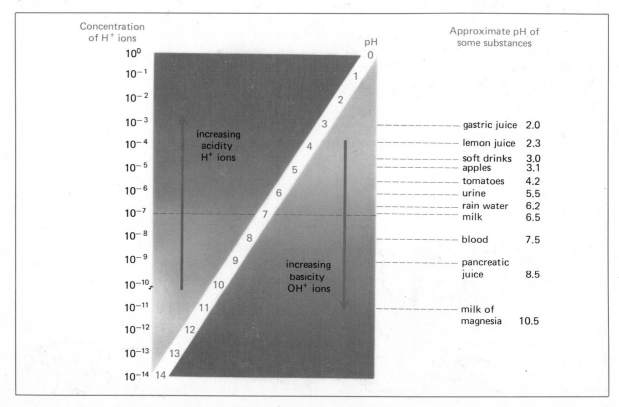

Salts (a class of compounds) have neither properties of an acid or base. A salt yields neither H^+ nor OH^- ions when it dissociates. If an acid and a base combine, they neutralize each other, forming a salt and water. When pure water dissociates it yields an equal number of H^+ and OH^- ions and is therefore neutral.

In 1909, a Danish biochemist developed a scale to measure the H^+ concentration in a solution. The **pH** scale runs from 0 to 14, with 7.0 as the neutral point. At this point there are equal numbers of H^+ and OH^- ions in a solution. Any number below 7 indicates an acid with increasing acidity (H^+ ions) toward the lower numbers. In like manner, any number above 7 indicates a base with progressively more OH^- ions toward the higher numbers. The concentration of H^+ ions in pure water is 1×10^{-7} or 0.0000001. To convert this value to pH, the negative exponent -7 is changed into a positive number 7. Thus, a solution with a concentration of hydrogen ions of 1×10^{-8} would have a pH of 8. The complete pH scale is shown in Fig. 2-13. The cells of the body are extremely sensitive to changes in pH. The fluids that surround these cells must maintain a constant acid-base balance. The tendency of the body to maintain constant conditions in this internal environment is called **homeostasis.** All the organs work together to accomplish this.

pH = hydrogen ion concentration

Electrolytes in solutions.   The cells of the body and the fluid surrounding the cells are composed of approximately 80 percent water. Water is considered to be a most versatile ***solvent,*** since many different substances dissolve completely in water. The substance being dissolved in the water is called the ***solute.*** Both components make up a ***solution.***

A solution containing ions is capable of conducting an electric current. In a solution of salt and water, the salt dissociates into positive

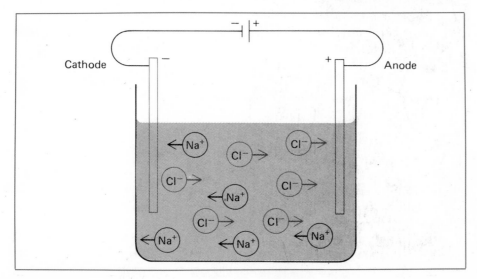

Figure 2-14 Ionized solutions conduct an electric current because of the movement of charged particles.

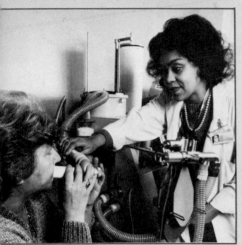

Respiratory therapists, also called inhalation therapists, work with people who have cardiorespiratory problems. They are also called on to care for patients who have undergone surgery because anesthesia depresses respiration. The therapists always work under the supervision of a physician. In their treatment of patients they use respirators and positive pressure breathing machines which they are trained to keep in good repair. They also show patients and their families how to use the equipment.

Training in this field is offered to high school graduates at hospitals, medical schools, colleges and universities, vocational-technical schools, and in the armed forces. Programs vary in length from 1 to 4 years. The shorter programs award certificates; the longer ones lead to associate or bachelor's degrees. Some of the subjects studied are anatomy and physiology, chemistry, physics, microbiology, mathematics, clinical testing, and equipment procedures.

Respiratory therapy workers should enjoy working with people and be sensitive to the physical and psychological needs of patients. A person in this field must also pay attention to details and have some mechanical ability.

and negative ions. If a source of current is introduced into such a solution, the negatively charged ions (Cl^-) will be attracted to the positive electrode. The positively charged ions (Na^+) will move to the negative electrode. This flow of ions produces an electric current. Compounds that ionize when dissolved are called **electrolytes.**

Ions of potassium, magnesium, sodium, and chloride are important in certain cellular activities. For example, potassium ions found along the cell membrane allow transmission of electrochemical impulses in nerve and muscle fibers. Other ions within the protoplasm of the cell regulate the activity of enzymes necessary for cellular respiration.

Summary

Atoms are composed of subatomic particles: electrons, protons, and neutrons. The number of protons of an atom is equal to the atomic number of the atom. A substance with two or more different elements is called a compound. Ionic chemical bonds occur when there is a transfer of electrons between atoms. Covalent chemical bonds occur between nonmetals by sharing pairs of electrons in the outer shell. In oxidation-reduction reactions electrons are lost or gained, resulting in the release of energy. Acids dissociate into H^+ ions and bases dissociate into OH^- ions. Compounds that ionize in water are called electrolytes.

Vocabulary Review

Match the statement in the left column with the word(s) in the right column. Place the letter of your choice on your answer paper. *Do not write in this book.*

1. Formed when two or more atoms unite by sharing or transferring electrons.
2. The number of positively charged particles in the nucleus of the atom.
3. A symbol for the value of hydrogen ion concentration of a solution.
4. The basic unit of structure of all matter.
5. Elements chemically inactive due to complete outer shells.
6. Atoms that occur in several different forms due to varying numbers of neutrons.
7. The splitting of complex molecules by the addition of a water molecule.
8. A positively charged particle of an atom.
9. Energy of motion.
10. Atoms that release particles or rays.
11. A positively or negatively charged atom.
12. Characteristic of reactions in which smaller molecules combine to form larger molecules.
13. A subatomic particle that has no electric charge and may be found in the nucleus of an atom.
14. The negatively charged particle of an atom.
15. The effort by the body to maintain a constant internal environment.

a. atom
b. atomic number
c. hydrolysis
d. electron
e. homeostasis
f. inert element
g. ion
h. isotope
i. kinetic energy
j. molecule
k. neutron
l. pH
m. element
n. anabolism
o. proton
p. radioactive atom

Test Your Knowledge

Group A

On your answer sheet, write the letter of the word that correctly completes the statement.

1. Energy of motion is said to be   (a) potential   (b) kinetic   (c) mechanical   (d) dynamic.

2. Nearly all the weight of an atom is concentrated in its   (a) protons
   (b) neutrons   (c) electrons   (d) nucleus.
3. The number of protons in an atom determines its   (a) atomic
   weight   (b) atomic number   (c) atomic mass   (d) atomic energy.
4. The chemical activity of an element is determined by its
   (a) nucleus   (b) protons   (c) outer-shell electrons   (d) neutrons.
5. Proteins are chains of   (a) carbohydrates   (b) glucose   (c) amino
   acids   (d) lipids.
6. A substance made up of atoms that all have the same number of
   protons is a(n)   (a) element   (b) compound   (c) molecule
   (d) isotope.
7. A compound   (a) contains only one kind of atom   (b) cannot be
   broken down to smaller particles   (c) may have a variety of positions for its elements   (d) is formed when two or more elements
   are bonded together.
8. An example of potential energy is   (a) sugar   (b) a ball in
   flight   (c) light from the sun   (d) a burning log.
9. An example of decomposition is   (a) A + B + energy $\longrightarrow$ AB
   (b) A + B $\longrightarrow$ AB   (c) AB $\longrightarrow$ A + B + energy   (d) AB + CD
   $\longrightarrow$ AC + BD
10. A fluid with pH greater than 9.5 is   (a) water   (b) basic
    (c) acidic   (d) none of these.

Group B

Write the word or words on your answer sheet that will correctly complete the sentence.

1. The atom is composed of numerous small structures, collectively
   called _____ particles.
2. Depending on the number of neutrons present in the nucleus, an
   element may exist in several different forms called _____.
3. A charged atom is called an _____ and is formed by the transfer or acceptance of electrons.
4. Atoms and molecules are electrically _____.
5. Acids dissociate in water to form _____ while bases dissociate
   to form _____. The neutral point on the pH scale is
   _____.
6. Explain two processes that involve decomposition. Gives examples
   of each.
7. Describe by using a diagram the covalent bonding of a water molecule.
8. Explain the chemical behavior of radioactive isotopes and some of
   their uses in medical science.

CHAPTER 3

CELLULAR STRUCTURE AND FUNCTION

Objectives

A. Describe basic cellular organization

B. Describe important functions of cellular organelles

C. Explain the processes of diffusion, facultative diffusion, osmosis, filtration, dialysis, and active transport

D. List the general energy changes in cellular metabolism

E. Describe the interaction of mRNA, tRNA, and rRNA in the synthesis of proteins

F. Outline the stages of mitosis

THE STRUCTURES OF CELLS

The cell is so small and so simple in appearance when viewed with a light microscope that it is difficult to conceive that each cell is a living entity unto itself. Cells are the smallest structures capable of carrying on all vital life functions of living organisms. Some 75 trillion of them make up the human body. Cells appear to be simple structures, but are highly complex in design and in the specialized functions they perform. The physiology of the human body is derived from the complex functions of the cells which compose tissues and organs.

Almost all human cells are microscopic in size. Their diameters range from 0.008 millimeter in red blood cells to 0.3 millimeter in the female sex cell, the ovum. The period at the end of this sentence measures about 0.1 millimeter, roughly 13 times as large as the body's smallest cells and one-third the size of the largest ovum.

Ideas about cell structure have changed considerably over the years. Early biologists saw cells as simple membranous sacs containing fluid and a few floating particles. Today's biologists know that cells are infinitely more complex than this. They know that cells consist of a surface membrane (the plasma membrane), a relatively large centrally located structure (the nucleus), and between the two a semifluid substance (the cytoplasm). Within the cytoplasm lie intricate arrangements of fine fibers and hundreds or even thousands of miniscule distinct structures called organelles.

The traditional concept of the cell has changed in the last two decades. New tools and methods of investigation, such as the electron microscope, have permitted the detailed study of the structures of cells. Advances in biochemical techniques have been helpful in identifying cellular components and their chemical composition. As a result, the structure, as well as functions, of many intracellular components can be accurately described.

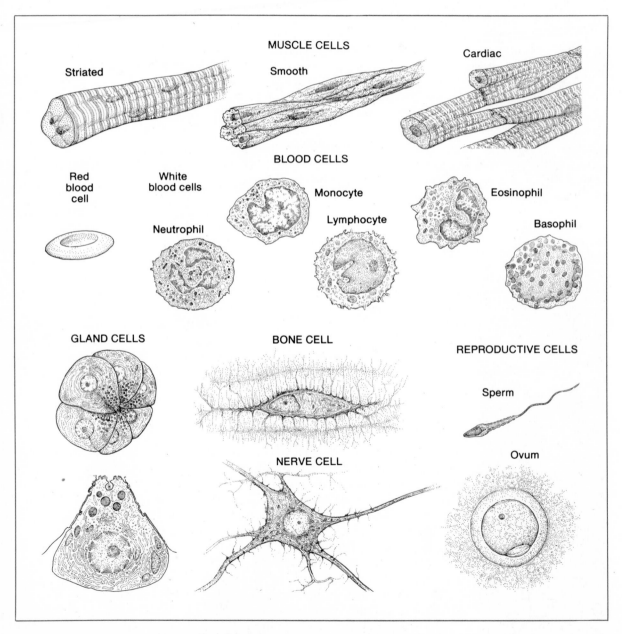

Figure 3-1 Different types of cells.

Variations among Cells

All cells have certain basic components that enable them to carry out life processes. In addition, each cell contains structures that allow it to make a specialized contribution to the functioning of the body.

There are several hundred basic types of cells in the human body. The sizes and shapes of cells vary and are usually characteristic of the particular function they perform. The human red blood cell is 0.008 millimeter in diameter and is disc-shaped in appearance. These cells are adapted to transport and distribute oxygen to the cells in the body. Voluntary muscle cells have the ability to provide movement and produce the required actions. Nerve cells may in some cases have long projections (axons) that extend for as long as 1 meter. They are adept at transmitting impulses through long distances. Examine Fig. 3-1 and note the different cells in the human body.

The structures common to all cells are the outer limiting membrane, the cytoplasm, and smaller structures (organelles) inside the cytoplasm. Almost all cells also contain a nucleus. The diagram in Fig. 3-2 shows what a cell looks like under an electron microscope. All of the cellular structures shown here are found in cells; however, different cells may lack one or more of these structures.

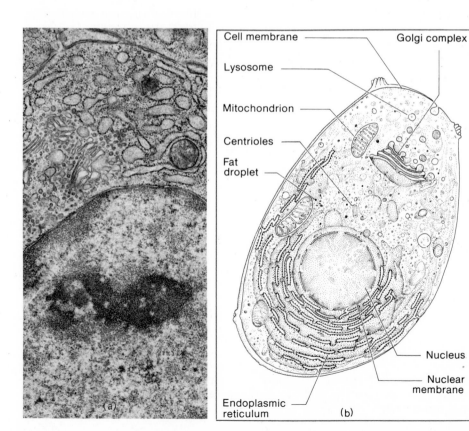

Figure 3-2 The cell as seen through an electron microscope: (a) photograph and (b) diagram.

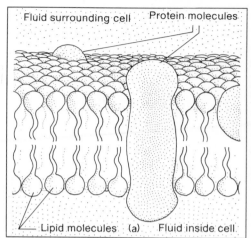

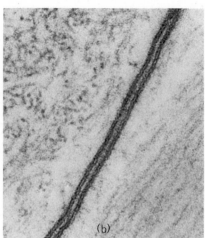

Figure 3-3 (a) The plasma membrane is made of a double layer of lipid molecules embedded with protein molecules. (b) Electron micrograph showing the plasma membranes of two adjacent cells. The dark lines are the protein layers.

Cellular Membranes

The most conspicuous structural feature of a cell is its membranes. They form both the outer boundary of the cell itself and also the boundaries or walls of most of the cell's internal parts. The membrane that constitutes the surface boundary, or outline of the cell, is called the **plasma membrane,** or more descriptively, the **cytoplasmic membrane.** The internal structures of the cell are also covered by membranes. The intracellular membranes include the *nuclear membrane,* and the membranes of the *endoplasmic reticulum, Golgi complex, mitochondria,* and *lysosomes.* The cytoplasmic membrane and the internal membranes have similar physical and chemical properties. Both are composed largely of lipid and protein molecules. Most scientists today postulate that the cell membranes consist of two layers of phospholipid molecules in which the protein molecules are embedded (Fig. 3-3).

Some protein molecules are found in an organized pattern. Others are randomly distributed throughout the membrane. The protein and lipid molecules which make up the membranes actually move about, forming the fluid framework for cell membranes.

Function of cell membranes.   Cell membranes perform many functions essential for the survival of the cell. The cytoplasmic membrane serves as the boundary between a cell's internal fluid and its external fluid environment. It maintains the integrity of the cell and preserves the arrangement of its internal structures. The membrane has a total thickness between 75 and 100 angstroms. If a cell's cytoplasmic membrane becomes torn, the cell's contents leak out and the cell dies. Another function of the cytoplasmic membrane is to control the flow of substances into and out of the cell. Such membranes are said to be **selectively permeable.** Some substances are entirely excluded while others are allowed to pass through. Communication is another function of the cytoplasmic membrane. For example, the surface proteins in some cell membranes serve as receptors for chemical messages. Some *hormones* and other substances called *neurotransmitters* bind to these surface proteins and transmit their message to the cells. Other cytoplasmic membrane proteins present on lymphocytes (white blood cells) function as *antibodies.* They combine with potentially harmful foreign proteins called antigens to produce an *immune response.* Many of the proteins that are an integral part of internal cell membranes are *enzymes.*

Cytoplasm and Organelles

When cells are viewed under the light microscope, everything inside the plasma membrane except the nucleus appears as a slightly granular semifluid. This material is called **cytoplasm.** The electron microscope reveals that the cytoplasm has a number of distinct structures called **organelles.** These structures play specific roles both in the life processes of the cell and, in turn, the organism as a whole.

cyte = cell

Mitochondria.   *Mitochondria* number among the few cytoplasmic organelles visible under the light microscope. They are small rod-shaped granules in the cytoplasm. A diagrammatic cutaway drawing of an electron micrograph of one mitochondrion appears in Fig. 3-4. It appears as a fluid-filled vesicle, bounded by a double membrane. The inner membrane displays numerous infoldings called *cristae.* Both membranes consist of alternating layers of protein molecules and phospholipids. Mitochondria have often been called the "power plants" of the cell. They are responsible for cellular respiration, during which energy is released for cellular activities.

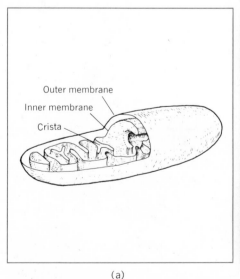

(a)

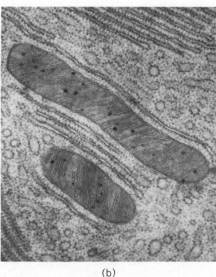

(b)

Figure 3-4 (a) Diagram and (b) electron micrograph showing structure of mitochondria.

Endoplasmic reticulum.   Throughout the cytoplasm there is a network of canals known as the *endoplasmic reticulum.* Some of these canals appear to form an intricate connecting link between the plasma membrane and the membrane that surrounds the nucleus of the cell. The endoplasmic reticulum plays an important role in protein synthesis and in transportation of cellular substances.

Ribosomes.   On the outer surface of some of the canals that form the endoplasmic reticulum, there are numerous small granules. The granules are called *ribosomes.* (See Fig. 3-5.) These particles contain a large amount of *ribonucleic acid (RNA).* In addition to their RNA content, ribosomes possess enzymes which aid in the production or synthesis of various kinds of proteins necessary for cellular life.

Golgi apparatus.   A network of cytoplasmic structures seen as a stack of flattened sacs or vesicles is known as the *Golgi apparatus.* These vesicles are surrounded by smooth membranes. This specialized complex is often found near the nucleus of the cell. It seems that they may be related to the secretory activities of the cell. Several research teams have presented evidence that the Golgi sacs synthesize large carbohydrate molecules. These carbohydrates combine with proteins to form chemical compounds called *glycoproteins.* The Golgi apparatus seems to be the main agency for building glycoproteins. (See Fig. 3-6.)

The centrosome.   This structure lies quite close to the nucleus and Golgi apparatus. The centrosome contains a pair of *centrioles.* Through a light microscope they appear as a pair of cylinders situated at right angles to each other. The wall of each cylinder is composed of nine double rods or tubules arranged around an open center. The centrioles play a major role in the division of the cell.

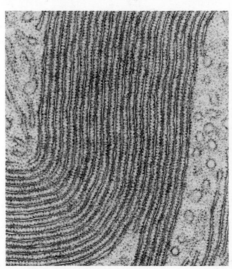 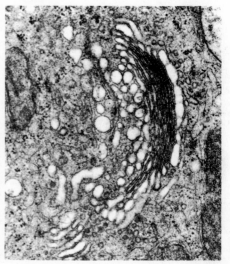

Figure 3-5 (left) Electron micrograph of endoplasmic reticulum lined with ribosomes.

Figure 3-6 (right) Electron micrograph of Golgi apparatus (29,520 X).

Lysosomes.   The *lysosomes* are membrane-enclosed vesicles within the cytoplasm. They vary in shape and size. Lysosomes are thought to contain enzymes that are capable of destroying all the main components of cells. The enzymes within the lysosomes can dispose of bacteria or other foreign material that enter a cell. Consequently, lysosomes are sometimes called "digestive bags" or "cellular garbage disposal units." White blood cells serve as scavenger cells for the body. The cytoplasm of these cells contain a great many lysosomes. When a cell dies, the enzymes within the lysosomes digest the rest of the cell.

The Nucleus

The *nucleus* of a cell has two vital functions. It controls and regulates the metabolic activities of the cell. It is also essential in the processes of cell division and heredity. The protoplasm within the nucleus is called **nucleoplasm.** The nucleus is enclosed by nuclear membrane. This thin, double-layered membrane behaves in a manner similar to that of the plasma membrane. The nuclear membrane has large porelike openings which connect with the canals of the endoplasmic reticulum. The nucleus manufactures the nucleic acids needed for protein synthesis and contains the genetic material called **chromatin.** The chromatin material appears as a dark mass. From a chemical standpoint, chromatin is composed of a type of protein known as a *nucleoprotein.* Nucleoproteins are combinations of proteins and nucleic acids. The nucleic acid found in chromatin is *deoxyribonucleic acid (DNA).* We shall have more to say about DNA in the chapters that follow.

chroma (Gk.) = color

When a cell is not dividing, the chromatin is usually in the form of long threads so thin and interwoven that they give a netlike appearance to the contents of the nucleus. However, when a cell is in the process of dividing, this network becomes distinct, forming thick, rodlike bodies called **chromosomes.**

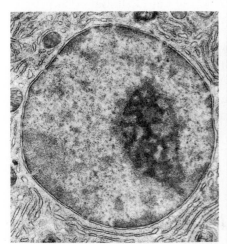 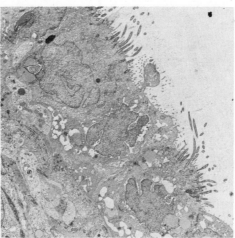

Figure 3-7 (left) Electron micrograph of the nucleus. Note the nucleolus as the dark mass with the nucleus and the pores in the double nuclear membrane (approx. 8,000 X).

Figure 3-8 (right) Tiny protoplasmic projections on the surface of epithelial cells increase surface area.

The nucleolus.   The *nucleolus* is found in the nuclei of many cells that are not in an actively dividing state. It appears as a dark-staining area, as shown in Fig. 3-7. The nucleolus disappears when cell division begins and reappears when division is complete. It has no membrane and is rich in proteins and ribosomal RNA. The nucleolus may be the site where ribosomes are assembled and stored.

Special Cell Structures

Microtubules, microvilli, cilia, and *flagella* are structures characteristic of certain types of cells.

Microtubules.   These are hollow tubes that can combine to form certain structures in the cell as needed. For example, they form the apparatus that pulls chromosomes away from each other in cell division (this will be described later). They also form cilia, flagella, and centrioles. Both cilia and flagella are composed of microtubules arranged so that there are nine pairs of microtubules in the periphery surrounding a single pair of microtubules in the center. Centrioles are made up of nine groups of three microtubules each, surrounding a hollow core.

Microvilli.   These are special structures of epithelial cells of absorbing surfaces such as the lining of the intestines. Microvilli consist of projections of cytoplasm and cytoplasmic membrane. They resemble tiny fingers crowded close together. A single microvillus measures about 0.5 millimeter long and 0.1 millimeter or less in diameter. One cell may have hundreds of these projections, thereby greatly increasing the surface area of the cell.

Cilia.   Among cells possessing cilia are the epithelial cells of the mucous membrane that lines some parts of the respiratory tract. One cell may have a hundred or more cilia. They move simultaneously to propel a fluid in one direction over the surface of the cell. Ciliated cells propel mucus upward in the respiratory tract.

Flagella.   Flagella have the same basic structure as cilia, but they are longer and fewer in number. A flagellum is a hairlike projection stemming from the surface of a cell. Its function is to provide movement. Each male sex cell, for example, has a flagellum that propels the spermatozoan forward in its fluid environment.

CELL FUNCTIONS

Every cell must carry on all the processes of life. In addition, most cells in the body perform some specialized function. For example, muscle cells provide movement, nerve cells communicate, and red blood cells transport oxygen.

Transport Mechanisms Through Cell Membranes

Heavy traffic moves continuously in both directions through cell membranes. Streaming in and out of all cells in endless procession are molecules of water, nutrients, gases, wastes, and different kinds of ions. The mass transportation of substances across a cell membrane can be classified as **physical** (passive) or **physiological** (active) processes. The main distinction between these two kinds of processes is in the source of the energy that does the work of moving substances through the membrane. A transport system that requires energy from chemical reactions within a cell is known as an active or physiological process. If the energy is from a source other than the cell's chemical reactions, transport is then a passive or physical process. Another way to understand transport systems is to remember that active processes can move substances *only* through living cell membranes. Passive processes can move materials through living or dead cell membranes.

Physical Processes

Physical transport processes constitute the net movement of ions or molecules that pass through a membrane. The movement of these particles is in a "downhill" direction or gradient. A gradient is the degree of difference between two divided areas, and is determined by the concentrations of the particles on each side of the membrane. These processes include **diffusion, osmosis, filtration,** and **dialysis.**

Diffusion.   Diffusion means the scattering or spreading out of particles. Small particles such as molecules and ions move continuously and rapidly in all directions. In Fig. 3-9, a 10% sodium chloride (NaCl) solution is separated from a 20% NaCl solution by a membrane. The membrane

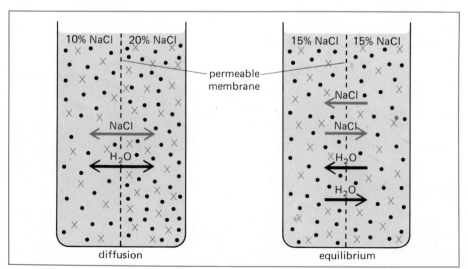

Figure 3-9 Diffusion is one basis for the intake of digestion products by cells lining the intestines.

is permeable to both NaCl and to water. This means that both these substances can and do pass through the pores of the membrane in either direction. Inevitably, some NaCl and water molecules collide with each other and with the membrane. Because they are more numerous in the 20% solution, more NaCl particles move through the membrane from there into the 10% solution on the opposite side. *Net diffusion* occurs down a *concentration gradient,* or from an area of higher concentration to an area of lower concentration. Diffusion of NaCl and water continues even after equilibrium has been achieved. At equilibrium there is no net movement, but diffusion takes place in both directions through the membrane. Similar examples of this passive mechanism occurs in the body. Oxygen and carbon dioxide are exchanged in the lungs and in the tissue cells by diffusion.

Osmosis.   Osmosis is the diffusion of a solvent, usually water, through a *selectively permeable membrane* (Fig. 3-10). A selectively permeable membrane is one that is not equally permeable to all solutes in a solution. It permits some solutes to diffuse through it but totally prevents the diffusion of others. Consequently, the selectively permeable membrane maintains a concentration gradient of the solute that cannot freely diffuse through the membrane. Osmosis is of great importance because it is the process by which water moves between the various fluid compartments of the body.

In osmosis, the water diffuses freely through the selectively permeable membrane, whereas the dissolved solute does not. If a 20% glucose solution is separated from pure water by a selectively permeable membrane, only the solvent (water) will penetrate the membrane. In this case, the water concentration is greater in the pure water than it is in the 20% glucose solution. Therefore, net diffusion of water takes place from the pure water into the 20% glucose solution. Eventually, the gradient and thus the rate of osmosis both decline. There is then no

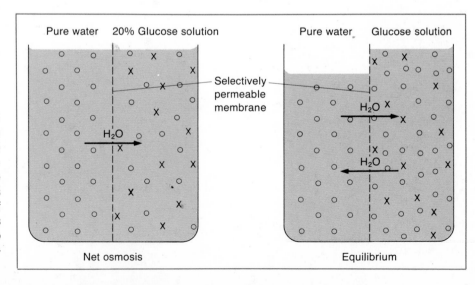

Figure 3-10 Water diffuses into the glucose solution at a faster rate than it moves in the opposite direction until the osmotic pressure is equal on both sides of the membrane. At this point, net diffusion is zero and equilibrium is attained.

further net movement of water and an equilibrium is established. Note that osmosis of water into the glucose solution increases the volume of the solution. At the same time the volume of pure water decreases. The increase in volume in the glucose solution causes it to exert a pressure called *osmotic pressure*. The higher the concentration of solute in a solution, the higher the osmotic pressure.

A solution that has the same osmotic pressure as the fluid inside a cell is said to be an *isotonic* solution. Since the osmotic pressures are equal, cells in isotonic solutions neither gain nor lose water. Solutions with a lower osmotic pressure than the fluid inside a cell are called *hypotonic* solutions. A cell placed in a hypotonic solution will take in water and swell. When cells are placed in a solution (such as seawater), which has a higher osmotic pressure than the cells, water will leave the cells and their contents will shrink. Such a solution is called a *hypertonic* solution. Fig. 3-12 shows red blood cells that have been placed in a hypertonic solution.

iso = same
tonic = strength

hypo = beneath
hyper = above

Filtration.   Filtration is the net movement of fluid and solutes through a membrane caused by a mechanical-pressure gradient across that membrane. A greater concentration of molecules on one side of a membrane creates a *filtration pressure*. Filtration pressure is the force of a fluid pushing against a surface. Filtration is a major mechanism for moving substances through the walls of the blood capillaries, especially in the kidneys (Fig. 3-11).

Dialysis.   Dialysis is a method of separating solutes by taking advantage of the differences in their abilities to pass through a semipermeable membrane. The net movement of particles across the membrane can be controlled by changing concentration gradients. In effect, this is the principle of the artificial kidney. Waste products in the blood pass through the dialysis tubing to the surrounding solution, where their concentration is low.

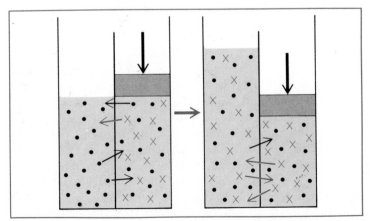

Figure 3-11 Filtration is the introduction of a mechanical force (↓) in transport of solute (x) and solvent (o) through a membrane.

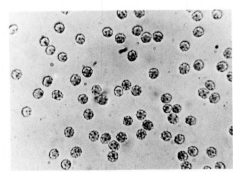

Figure 3-12 Micrograph of red blood cells in a hypertonic solution, showing shrinkage of the cells.

Facilitated diffusion.   In facilitated diffusion, molecules can move more rapidly than they ordinarily would from a high to a low concentration with the aid of carrier molecules in the cell membrane. This movement does not require the expenditure of cellular energy. The molecule to be moved across the membrane combines with the carrier molecule on one side of the membrane. The carrier moves to the other side of the membrane and releases the molecule. Carrier molecules, usually proteins, are generally specific for one substance. The most important example of facilitated diffusion in the body is the movement of glucose from the blood across the membranes of most cells in the body.

Physiological Processes

Active transport.   Movement of molecules and ions against a concentration gradient, from lower to higher concentrations, requires the use of energy obtained from ATP. This type of transport is termed *active transport*. As in facilitated diffusion, carrier molecules are used. A molecule or ion to be transported through a membrane binds to a specific carrier molecule on one side of the membrane. The bound carrier rotates rapidly in the membrane and moves the bound substance to the opposite side of the membrane, where the substance is released. Active transport carriers are often referred to as "pumps." While some of these systems transport only one molecule or ion at a time, others exchange one ion for another. The most important of these latter type carriers is the Na/K pump.

Metabolic Functions

Metabolism includes all the chemical reactions that occur in the body. In some biochemical reactions large molecules are broken down into smaller ones with the release of energy. These reactions are known as **catabolic reactions**. In other biochemical reactions, small molecules are united to form larger ones and, in so doing, consume energy. These are known as **anabolic reactions**. Catabolism furnishes the energy necessary for all cellular activity. Some of this energy is dissipated as heat. Anabolism utilizes energy to synthesize products for growth, repair, and reproduction. Metabolism includes the sum total of all the reactions of catabolism and anabolism. The diagram in Fig. 3-13 summarizes the primary activities in a "protein cycle." Each step includes an explanation of what is taking place and names the organelle involved.

cata = down

ana = up

Metabolism and energy exchange.   Without food no living organism can maintain life. In animal cells, food is the ultimate source of energy. Since cells cannot create matter, food is the only source that can be used to maintain or produce new protoplasm. In Step 1 (Fig. 3-13), the principal organic substances are taken into the cell by **absorption.** This is the process by which organic substances pass into the cells of the body. These organic substances are the end products of carbohydrate digestion (mainly glucose), fat digestion (glycerol and fatty acids), and of protein digestion (various amino acids). Most absorption takes place by the diffusion of materials through the plasma membrane. However, the larger molecules cannot pass through. Electron micrographs show that the plasma membrane forms an indentation that engulfs these large molecules. Taking materials into the cell by this process is an example of pinocytosis.

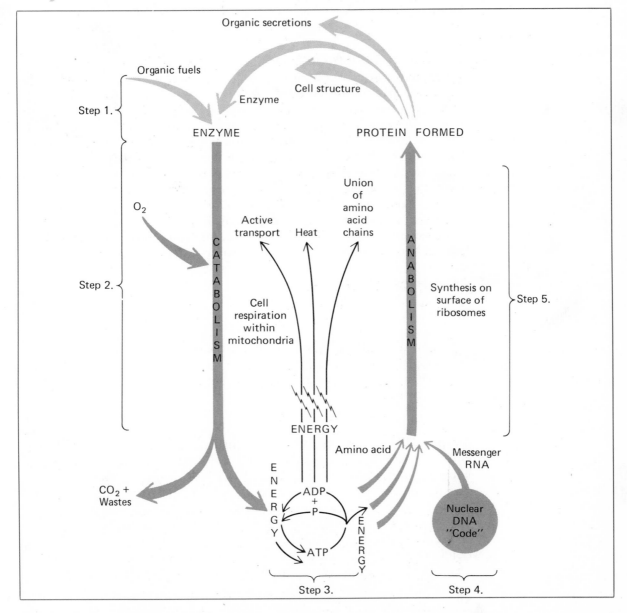

Figure 3-13 Chart showing how the different metabolic processes are related.

Catabolism, Step 2, is mainly concerned with the oxidation of organic molecules, primarily glucose. Chemical bonds are severed and large amounts of energy are released. This reaction is known as *cellular respiration*. While many steps are involved in the oxidation of glucose, it can be simplified in the following equation:

$$C_6H_{12}O_6 + 6O_2 \xrightarrow{\text{enzymes}} 6CO_2 + 6H_2O + \text{energy}$$

$$\underbrace{C_6H_{12}O_6}_{\text{glucose}} \; \underbrace{6O_2}_{\text{oxygen}} \qquad \underbrace{6CO_2}_{\substack{\text{carbon} \\ \text{dioxide}}} \; \underbrace{6H_2O}_{\text{water}}$$

Note that in this form of oxidation, oxygen must be supplied and the end products are carbon dioxide, water, and energy. As carbon dioxide and water accumulate, they are removed as wastes.

The step-by-step breakdown of organic molecules is made possible by the availability of the needed enzymes. An *enzyme* is a type of protein that acts as a *catalyst*. A catalyst is an agent that affects the speed of a chemical reaction without itself being permanently changed by the reaction. Metabolic processes are controlled by enzymes, with at least one enzyme needed at each step of a reaction.

There are thousands of chemical reactions that occur within cells. Virtually each cell may contain several thousand different enzymes. In certain chemical reactions, enzymes require the assistance of other substances known as **coenzymes** (enzyme activators) to complete the chemical change. Many of the vitamins, especially the B-complex group, act as coenzymes in cell metabolism.

The energy resulting from the catabolic reactions provides the energy for the anabolic reactions. Some of the energy is lost as heat which the cells cannot use.

Most of the cellular activities mentioned in Step 2 occur in the mitochondria of the cell. Small particles located on the inner folds (cristae) are believed to be responsible for carrying out the oxidation reactions that supply energy.

ATP and ADP.   Step 3 shows how the utilization of energy within living cells involves a universal intracellular carrier of chemical energy known as **adenosine triphosphate (ATP).** ATP is able to transfer energy because of the presence of high-energy phosphate bonds (P). Two of the three phosphate groups in ATP are attached to adenosine by high-energy bonds (Fig. 3-14). When some part of the cell needs energy, it is immediately made available by splitting the outermost high-energy bond. This removes one of the phosphate units, and *adenosine triphosphate* (ATP) becomes *adenosine diphosphate* (ADP).

$$ATP \xrightarrow{\text{breakdown}} ADP + \text{phosphate} + \text{energy}$$

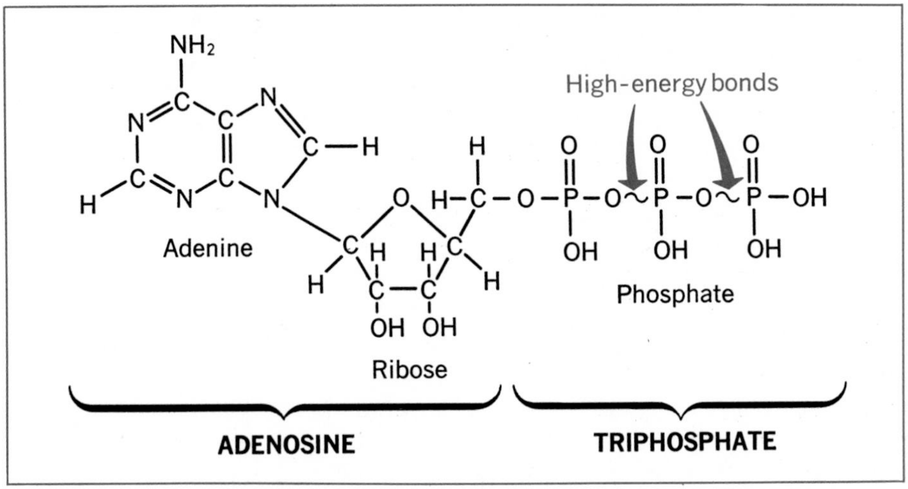

Figure 3-14 One of the most important substances in all living cells is adenosine triphosphate (ATP). Note the high-energy bonds that transmit all of the energy. This bond has a special symbol ($\sim$).

As ATP is used up by the energy requirements of the cell, ADP accumulates. Certain reactions in the cell release energy, and some of this energy is used to convert ADP back to ATP. The reverse reaction can be written as follows:

$$ADP + P + energy \longrightarrow ATP$$

One of the sources for the chemical energy required for the above reaction is the breakdown of sugar (glucose) described in Step 2. This involves 14 different steps, each step requiring a different enzyme. These reactions provide the energy necessary for all metabolic processes, particularly the synthesis of protein.

Protein synthesis. *Protein synthesis* is an anabolic process. Protein molecules are composed of long chains of subunits called amino acids. There are about 22 different naturally occuring amino acids that are important in protein synthesis. Each protein is made up of a specific sequence of amino acids. The chain of amino acids may be straight or branched. The shape and length of the chain, and the sequence of amino acids, determine the kind of protein that is produced. The sequence of amino acids varies for each different protein. Therefore, the number of possible kinds of protein is tremendous. Some proteins are used for cell reproduction, growth, and repair. Others make up the thousands of different enzymes found inside and outside the cell. Still others are hormones—the products of various endocrine glands. How is the cell able to produce the kind of proteins it needs?

The basic instructions for producing proteins are coded in DNA molecules. A molecule of DNA resembles a very long, twisted ladder. This shape, with two strands turning around each other, is known as a ***double helix***. The sides of the ladder are formed by alternating molecules of deoxyribose (a five-carbon sugar) and phosphate groups. The "rungs" are pairs of nitrogen-containing bases. A combination of a sugar, a phosphate, and a nitrogen-containing base is called a ***nucleotide***. There are four nitrogen-containing bases found in DNA: adenine (A), thymine (T), cytosine (C), and guanine (G). It was found that in the "rungs," A always paired with T, and G with C.

Each strand of a DNA molecule serves as a template for the formation of a complementary strand of **messenger ribonucleic acid (mRNA).** RNA contains ribose instead of deoxyribose, and uracil (U) instead of thymine. When the mRNA strand is formed, it moves out of the nucleus and into the cytoplasm of the cell. A different type of RNA, **ribosomal RNA (rRNA)**, is found in the ribosomes along the endoplasmic reticulum. An mRNA strand lines up on a ribosome to begin the assembly of a protein molecule. A third kind of RNA, **transfer RNA (tRNA)**, picks up amino acids in the cytoplasm and brings them to the ribosome. At a specific site on the tRNA molecule there is a sequence of three bases known as an *anticodon*. The kind of amino acid bound to the RNA depends on the sequence in the anticodon. The tRNA, carrying its amino acid, matches its anticodon with the proper base sequence on the mRNA on the ribosome. Again, A pairs with U, and G with C. As the tRNA molecules fit into position, the amino acids they are carrying join to form the chain of the protein molecule encoded in the DNA.

The new protein molecule, whose amino acid sequence has been determined by DNA, now can perform its specific function. Possibly it will become part of a cytoplasmic organelle or a cell membrane, or act as an enzyme or hormone. (Protein synthesis is discussed in further detail in Chapter 30.)

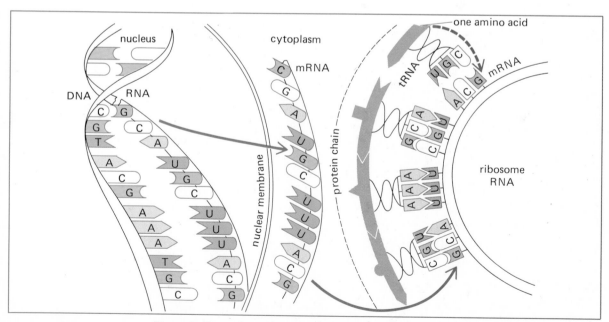

Figure 3-15 **A single strand of DNA forms a strand of RNA. Messenger RNA moves to a ribosome. As the amino acids are laid down, a protein is synthesized.**

Irritability

Another characteristic of protoplasm is its ability to respond to environmental changes. A single-celled animal such as the amoeba is quite sensitive to differences in its surroundings even though it lacks a definite nervous system or specialized sense organs. This is an example of how a cell, even without benefit of highly differentiated parts, can respond to variations in its environment. An increase in the size of an organism requires highly specialized organs which are capable of receiving and responding to the stimuli from outside the body. Other parts within the body must be able to transmit these messages to distant regions. A nervous system is an essential part of such an animal's organization. The human organism has the most highly specialized nervous system of all living organisms.

Growth of Cells and Reproduction

The growth of a living organism is dependent on two factors concerning the cells. There is a limit to the size or volume of individual cells. The increase in the number of cells depends upon the types of cells and where they are found in the body. Cells that have become specialized divide less frequently.

Individual cells grow in volume as a result of a process known as **assimilation.** During this process the cell takes in digested food substances to make new protoplasm. The cell forms more complex substances from the simpler ones, as in the synthesis of proteins from amino acids.

The rate of normal division of individual cells is a direct result of their capacity for growth. In the human body, cells divide at varying rates of speed, depending on their functions. The result is that some parts of the body may grow very rapidly for a period of time and then slow down. Other parts grow slowly or not at all once they have developed. Good examples of how different kinds of cells grow and divide at varying rates of speed are the cells of the middle and inner ear, and those in the outer layers of the skin. The tiny bones of the ear and the hearing apparatus have reached their maximum size some time before birth and do not grow after that time. The outer layers of the skin, however, are in a constant state of active growth. These cells are constantly shed and replaced by new cells formed by the underlying skin layers.

The continuation of any species of living organisms is dependent on the ability to produce others of a like nature. One-celled animals simply divide the material of the cell into two halves. Each half then develops into a new individual. The complexity of the higher animals has necessitated the development of specialized cells (sex cells) for reproduction. These cells contain all the information necessary for the development of a new individual and insure the transmission of hereditary characteristics from one generation to the next.

mito (Gk.) = thread

Cell division.   The process of cell division results in the development of the body as a whole. The most common process by which cells divide is called **mitosis.** During this process, the cell passes through a series of phases or stages which result in the formation of two new cells. Each of the two new cells contains an equal quantity and quality of chromatin material. Thus all the information encoded on the DNA is passed on to all new cells. The process of mitosis is continuous and usually occurs in a short period of time. The various stages through which the nucleus passes in its division have been grouped into four principal phases.

Interphase (Fig. 3-16): This stage is not in any of the phases of mitosis but represents that growth period in the cell's life when it is not actively dividing. However, the nucleus, during the latter part of this period, is duplicating the DNA molecules in preparation for cell division. This self-duplication of the DNA molecules is referred to as **replication** and results in an additional set of chromosomes.

Prophase (Fig. 3-17): One of the first indications that a cell is about to divide is that the nucleus changes in appearance. The chromatin material (now doubled) condenses into visible chromosomes. The nuclear membrane and the nucleolus disappear. The two pairs of *centrioles* move slowly away from each other toward opposite poles of the cell. As the centrioles separate, lines of granules, *the spindle,* appear between them. Around each centriole similar fibers, called asters, radiate into the cytoplasm. If a fine-hooked needle is thrust into the cell, the spindle can be stretched and drawn out of position, indicating its fiberlike consistency. For many years the chromosomes were thought to be single structures, but recent advances in the techniques of staining cell parts show that they are not. If special stains are used, these structures can be seen to consist of two daughter chromosomes, each one called a **chromatid.** The chromatids are joined to each other at a single point by a very small piece of chromatin substance, the *centromere.*

Metaphase: Metaphase is characterized by the separation of the chromatids. Up to this time the chromatids have been attached to each other by the centromere. The centrioles, linked to the spindle, move to

Figure 3-16 **Interphase** (resting stage). The individual chromosomes are not yet visible.

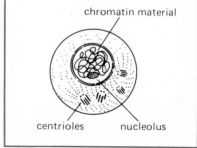

chromatin material

centrioles   nucleolus

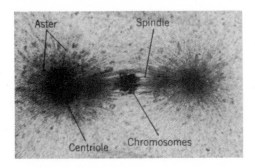

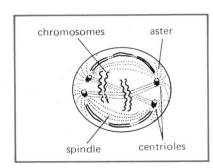

Figure 3-17 **Prophase.** The nuclear membrane and nucleolus disappear. The centrioles move to opposite poles of the cell, each one the center of an aster. The chromosomes and spindle appear.

opposite poles of the cell. The chromosomes become neatly arranged along the center (equator) of the cell.

Anaphase: Following their separation, the chromatids move toward opposite poles of the cell. Just how this movement is accomplished is not clear, but the spindle fibers appear to attach to the centromere of the chromatids and contract, drawing the two members of the chromatid pair to opposite poles. The centrioles now divide and an indentation appears in the cell membrane at a point in the middle of the cell. This indentation begins to divide the cytoplasm of the cell in half.

Telophase: The final stage of mitosis completes the division of the nuclei. This is usually followed by *cytokinesis* as new cell membranes completely divide the old cell into two new and separate cells. The chromatin material of each daughter cell then rapidly goes through a thread-like stage and eventually assumes the network appearance typical of interphase. The nuclear membrane forms around the chromatin of each new cell, and the nucleoli reappear. The two new cells are identical to one another and with the parent cell.

CELL PATHOLOGY

Tissues are formed by cell divisions. When the proper number of cells have been formed to meet the requirements of the individual, cell division stops. In the majority of tissues, replacement of lost or injured cells is regulated by different controlling mechanisms. Occasionally, the controlling factors cease for some unknown reason, and some of the cells in a tissue may begin to divide rapidly and in an abnormal manner. When this occurs, the abnormal cells invade areas occupied by normal cells and upset their metabolic activities. These invading cells form a cancerous tumor.

Cancer

Cancer can occur in any part of the body. It results from a change in certain cells in which the normal growth limits are out of control. Researchers have found that the genetic make-up of cancer cells is differ-

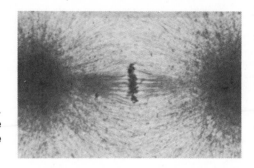
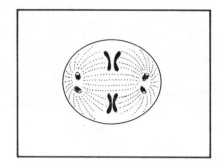

Figure 3-18 **Metaphase.** The chromosomes are arranged along the equator of the cell.

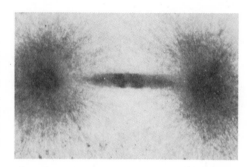
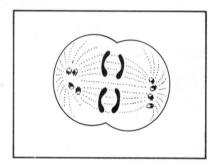

Figure 3-19 **Anaphase.** The chromatids separate and move toward opposite poles of the cell. As they move farther apart, the spindle begins to break down and the centrioles now divide.

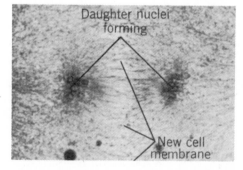
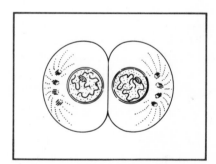

Figure 3-20 **Telophase.** A new cell membrane appears and two distinct nuclei re-form.

ent from that of normal cells. Scientists are continuing to investigate the mechanisms that cause the abnormal growth of cells. Cancerous cells compete with the normal cells for nutrients and cause, eventually the destruction of normal cells.

SUMMARY

The size and structure of cells are usually indicative of their particular function. Cells contain various organelles for specific cellular activities. The organization of the lipid and protein molecules in the plasma membrane determines the kind of substances that go into and out of the cell. The ribosomes, located along the endoplasmic reticulum, are important to the production of proteins. The mitochondria are called the "power plants" of the cell. The nucleus of the cell controls and regulates all the metabolic activities of the cell. DNA and RNA are nucleoproteins that play a major role in the synthesis of proteins and transmission of hereditary material. Transportation of substances across a cell membrane can be classified as passive or active processes. Diffusion and osmosis occur down a concentration gradient. Phagocytosis and pinocytosis are examples of active transport mechanisms. Protein synthesis involves both catabolic (breaking-down) and anabolic (building-up) activities. Enzymes are proteins which control metabolic processes. Assimilation involves the growth of cells, while mitosis is the process of forming new cells.

VOCABULARY REVIEW

Match the statement in the left column with the word(s) in the right column. Place the letter of your choice on your answer paper. *Do not write in this book.*

1. An organic catalyst.
2. A maze of small canals that appears to connect the surface of a cell to its nucleus.
3. The sum total of all chemical activity in a living organism.
4. The ability to respond to environmental stimuli.
5. That part of metabolism that results in the liberation of energy.
6. All the minute structures found in cytoplasm.
7. Movement of substances through a cell membrane against a concentration gradient.
8. The outer boundary of all animal cells.
9. That part of metabolism that uses energy in the formation of new protoplasm.
10. A cytoplasmic organelle which is considered the "power plant" of the cell.
11. The oxidation of organic molecules with the liberation of chemically bound energy.
12. Networks of nucleoproteins found in the nucleus of nondividing cells.
13. Formation of a small pocket around a liquid substance by the cell membrane.

a. anabolism
b. catabolism
c. cellular respiration
d. chromatin
e. cytoplasmic organelles
f. endoplasmic reticulum
g. enzyme
h. irritability
i. metabolism
j. cytoplasm
k. mitochondrion
l. ribosomes
m. plasma membrane
n. phagocytosis
o. active transport
p. pinocytosis

TEST YOUR KNOWLEDGE

Group A

Write the letter of the word(s) that correctly completes the statement. *Do not write in this book.*

1. Activities of the cell are directed and controlled by the   (a) Golgi apparatus   (b) cytoplasm   (c) nucleus   (d) polar body.
2. Mitochondria, Golgi apparatus, and centrosomes are called   (a) inclusions   (b) organelles   (c) nucleus   (d) ribosomes.
3. The nucleus of a cell contains   (a) ribosomes   (b) mitochondria   (c) chromosomes   (d) centrosomes.
4. An outer covering found on all cells is the   (a) nuclear membrane   (b) plasma membrane   (c) permeable membrane   (d) cell wall.
5. The cytoplasm of a cell normally contains   (a) endoplasmic reticulum   (b) cellulose   (c) DNA   (d) chromatin granules.
6. A structure which contains enzymes of a digestive nature is the   (a) Golgi apparatus   (b) ribosome   (c) lysosome   (d) mitochondrion.
7. A process in which molecules move more rapidly than they ordinarily would from an area of high concentration to one of low concentration is called   (a) facilitated diffusion   (b) dialysis   (c) active transport   (d) osmosis.
8. During the interphase DNA is found mainly in the   (a) nucleus   (b) cytoplasm   (c) endoplasm   (d) nucleolus.
9. Oxidation reactions within a cell occur chiefly in the   (a) mitochondria   (b) centrioles   (c) ribosomes   (d) vacuoles.
10. A compound that is formed in the nucleus and moves to the cytoplasm to take part in protein synthesis is   (a) tRNA   (b) DNA   (c) ribosomal RNA   (d) mRNA.

Group B

1. Briefly discuss the functions of the cell.
2. Outline the structure and function of the various organelles that make up the cytoplasm and nucleus of a cell.
3. Briefly define each of the following: irritability, metabolism, growth of cells, and reproduction.
4. Summarize the various steps in a "protein cycle."
5. What part do organic catalysts play in cellular activity? Define coenzymes and their function.
6. Describe the various stages of the nucleus of a cell as it goes through the process of mitosis. Stress the part played by the chromatin.

CHAPTER 4

INTERCELLULAR ORGANIZATION

Objectives

A. Define the terms *tissue* and *organ*
B. Differentiate between the structure of various kinds of tissue
C. Explain the function of each of the various kinds of tissue
D. Describe the common characteristics of the four basic tissues

TYPES OF TISSUES

The cells of the body are arranged into various groups having similar structural and functional characteristics. Such groups of cells are called **tissues**. Each kind of tissue performs one or more specialized functions for the body as a whole. In addition, tissue cells carry on all the essential activities necessary for their own survival.

There are four primary kinds of tissues: **epithelial, connective, nervous,** and **muscle.** Epithelial tissue is important in the passive and active transport of substances into and out of the body tissues, and equally important is its role as a protective tissue. Connective tissue, as its name implies, connects other tissues and forms the components of the intercellular material. Muscle tissue has the property of **contractility,** or being able to change shape. This special feature enables the action of muscles to produce movement of the body's parts. Nervous tissue specializes in the ability to respond to certain changes in the external or the internal environment. The appropriate activity is then communicated to the different parts of the body.

The primary tissues are organized into organs such as the heart, lungs, and stomach. Several of the tissues combine to form most of the organs of the body. The variation in the kinds of tissues and the manner in which they are organized determine the activities of a particular organ in the body.

General Characteristics of Primary Tissues

epi = on, upon

Epithelium is the name given to those tissues that cover surfaces and line the internal structures and cavities of the body. Their cells are close together with little intercellular material between them. Epithelial tissue can be arranged in one or more layers. A single layer is capable of transporting materials. Two or more layers are more suited for protection against the invasion of bacteria, or serve as a buffer against mechanical injury. The size and shape of epithelial cells are also related to the functions they perform. Cells particularly adapted for secreting fluids are found lining the ducts of many glands. Certain other cells may be found lining the abdominal cavity or the walls of the blood vessels. Another type of epithelium contains tiny projections on the surfaces of the cells. The most common examples are cilia for movement and microvilli for absorption of substances.

Since there are no blood vessels present in epithelial layers, nutrients reach the cells by diffusion from nearby capillaries in the underlying tissues. One characteristic common to all types of epithelial tissue is its ability to replace cells at a high rate.

Connective tissue consists of various types of cells, fibers, and a mixture of proteins known as the intercellular material. These components may be organized in numerous ways and their particular arrangement determines the different types of connective tissue. In addition, each component contributes to a number of important functions. For example, the types of connective tissue that contain fibers provide support and flexibility and bind structures together. Such tissues are commonly found in the muscles, the tendons, the ligaments, and in the walls of blood vessels. The cellular components of connective tissue perform important functions for transport and storage of materials and for protection against foreign substances. Connective tissue cells are found in many places such as in bone, cartilage, fat (adipose), blood, and lymph tissue.

The intercellular substance, called *matrix*, may exit as a fluid, semifluid, gelatinous, or solid substance. Because of the unique properties of this component, connective tissue can serve as a medium to transport and store many soluble substances, a protective lining between and around structures, and a medium filling the spaces between the other tissue cells. Connective tissue, in its more solid form, can be found in bone and cartilage and provides for support and for weight-bearing. Finally, connective tissue has the capacity for regenerating and repairing damaged areas with new fibers and intercellular material.

Nervous tissue is specially adapted to collect and transmit information by means of different types of nerve cells. These cells vary in structure and in function. Some of the cells are sensitive to external changes and therefore are capable of receiving stimuli. The information is then relayed through nerve impulses to a center in the body. Other types of nerve cells are responsible for carrying out the proper responses. In this way, the activities of organ systems are regulated to maintain a constant internal environment.

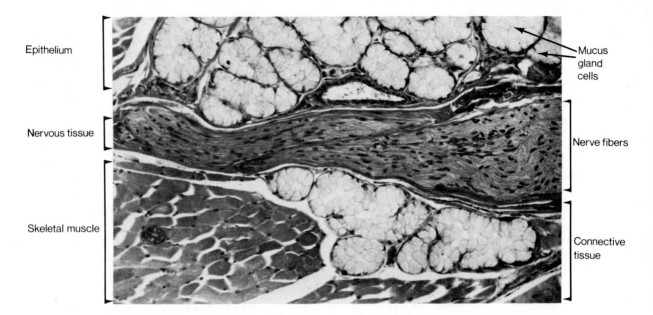

Epithelium

Mucus gland cells

Nervous tissue

Nerve fibers

Skeletal muscle

Connective tissue

Figure 4-1 The four basic tissues in one section.

Muscle tissue accounts for 40 to 50 percent of the body weight. The movements of the body organs are due to the action of the muscle cells. There are certain types of movement that are so slight that they are only visible with the aid of a microscope. The rate and control of movement of any part of the body depends upon the kind of cells that make up the muscle tissue. The individual muscle cells differ in size, shape, and function, including some of their physiological processes. The function of some muscle cells is to move the parts of the body skeleton. Another type of muscle cells, found only in the heart, enables the heart to pump blood to all the parts of the body. Still a third type is responsible for the internal activities of the organs of the body.

Organization of Tissues

An *organ* can be defined as a group of tissues that function together in order to perform some vital activity. The tissues in an organ are not necessarily all alike in structure or function, but by the coordination of their individual activities they form a distinct part of the body.

The hand may be taken as an example of an organ. The collection of different tissues in the hand work simultaneously to carry out important functions. The outside of the hand is covered by a highly complex type of stratified epithelium. In certain areas this tissue folds into ridges (the fingerprints). These ridges produce a slightly roughened surface to aid in grasping. Within the skin and throughout the entire hand are blood vessels that carry the blood to the tissues, supplying them with food and oxygen. In the walls of the blood vessels are smooth muscle fibers, which produce involuntary muscular contractions. The fingers contain striated muscle tissue for more precise movements. Bone and cartilage tissue serve as support of the hand. Adipose tissue in the palm of the hand absorbs shock. All of these various types of tissues are held together by bands of connective tissue. This is an example of a group of tissues functioning as a coordinated organ.

CLASSIFICATION OF PRIMARY TISSUES

The fundamental properties of tissues previously discussed are related to the particular activities each performs. Although the cells of one tissue do not resemble the cells of another type, there are a number of dissimilarities between cells of the same primary tissue. The various cells within the primary types of tissue are each responsible for some particular aspect of the internal environment that is necessary for the survival of all cells.

Epithelial Tissue

Simple squamous epithelium is composed of cells that are flat and slightly irregular in outline—almost scalelike. All tissues composed of such cells are extremely thin, the average depth being approximately 0.0025 millimeter. Simple squamous epithelium is found in the lining of blood vessels, the heart (pericardium), and in the covering of the lungs (pleurae). The cells around the heart and lungs secrete fluid (serous fluid) to reduce the friction between the actively moving organ and its covering. The fluid secreted by the cells covering the heart is called the *pericardial fluid.* The secretion found between the layers of the pleura is known as the *pleural fluid.*

The layer of cells forming the thin inner walls of all blood vessels is a specialized type of squamous epithelium called *endothelium.* These cells are also flat and irregular in outline. They resemble the shape of the epithelial cells of the inside lining of the cheeks. An outstanding characteristic of these cells is that they are held together by a cement-like substance that changes in permeability as the situation demands.

 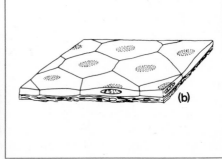

Figure 4-2 Simple squamous epithelium. (a) Electron micrograph shows surface folds often seen in the cells and the dome-shaped area. (b) Diagram of cells as they appear from above.

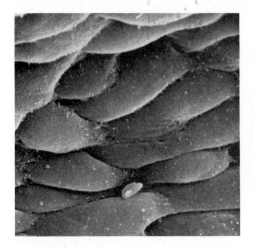

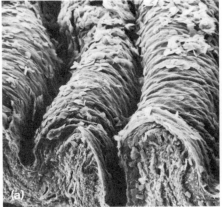

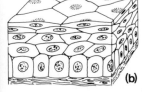

Figure 4-3 **An electron micrograph of endothelium tissue of a blood vessel.**

Figure 4-4 (a) Stratified squamous epithelium of a higher magnification shows the actual folds on the cell surfaces. (b) Diagram shows the organization of the cells.

The walls of the smallest blood vessels, the capillaries, are composed of endothelial cells. Thus, the diffusion of substances through the capillary walls to the surrounding tissues is regulated. The endothelial cells also line the walls of the arteries and veins, and the cavities between bones at the joints. *Endothelial tissue* will be further discussed in the chapters concerning the circulatory system.

Stratified squamous epithelium is a more complex type of epithelial tissue. Its primary function is the protection of some body surfaces that may be subject to mechanical injury. This tissue is composed of several layers of cells, which gives it added strength. The lowest layer is composed of cells that are tall and cylindrical in shape. Above this layer are irregularly shaped cells which gradually transform into flat cells that cover the surface. In the skin, for example, the outermost cells may eventually become nonliving because their protoplasm is replaced by a harder and more resistant material (keratin), which protects the more delicate lower layers. Under a microscope (Fig. 4-4), the general form of the cells lining the inside of the cheek appears irregular, with gently folded surfaces. Such cells lack the keratin because they are not subjected to the same degree of mechanical abrasion.

The **simple cuboidal epithelial** cells shown in Fig. 4-5 are equal in height and width. The cells are not actually cube-shaped. The name has been applied to them because of the way they appear through a cross section of the cell. Cuboidal epithelium is commonly found in the secretory portions of glands, or in the tubules of the kidneys and the tissue covering the ovary.

Simple columnar epithelium is composed of cells that are much longer than they are broad. They are packed close together to form a protective covering for the inner surface of an organ. The lining of the digestive tract is composed of cells of this type. Columnar epithelial

Figure 4-5 **Simple cuboidal epithelial cells consist of a single layer of cells.**

Figure 4-6 Diagram of simple columnar epithelium.

cells are able to produce secretions which aid in digestion. In some tissues, cells of the columnar type, called *goblet cells,* produce mucous secretions which lubricate the epithelial surface.

Ciliated epithelium cells are a type of modified columnar epithelium. These cells have small hairlike projections called **cilia.** The cilia are capable of moving small particles of debris or individual cells along a surface or through a tube. By the continuous movement of the cilia, particles of dust and dirt are moved up the trachea (windpipe) from the lungs. Cilia also aid the movement of the eggs from the ovary through the oviducts, and the sperm from the testis through the seminal tubules.

Connective Tissue

Loose connective tissue is composed of a relatively few inconspicuous cells in a semifluid matrix. Found scattered throughout the matrix are bundles of flexible but strong white fibers composed of a protein called **collagen.** Also present are single interlacing fibers of great elasticity. They are composed of the protein **elastin.** The spaces between the fibers are filled with tissue fluid. Loose connective tissue is found within the dermis of the skin and in the subcutaneous layer along with fat cells. It surrounds various organs and supports the nerves of the body and the network of blood vessels that bring nutrients and carry away wastes from various structures. (See Fig. 4-8.)

Adipose tissue or fat tissue is quite widely distributed throughout the body. The cells that make up this type of tissue are relatively large and have a single vacuole containing a droplet of fat (Fig. 4-9). Fat accounts for a large percentage of the total body weight; in males it is approximately 18 percent of the total weight and in females about 28 percent. Fat plays three important roles in the body. It serves as a reserve supply of energy-producing materials. The fat stored in the cells can be oxidized and used as food. It also serves as padding to absorb the jolts and

Figure 4-7 Ciliated epithelium found in the lining of the trachea.

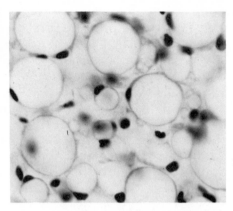

Figure 4-8 Photomicrograph of loose connective tissue. Note the cobweblike white fibers and the large amount of intercellular space.

Figure 4-9 The fat globules appear as empty spaces because the fat has been dissolved out in preparing the slide.

jars to which the body is being constantly subjected. Fat tissue is deposited around the eyes, in the palms of the hands, in the soles of the feet, and between the joints. The third function of fat is to serve as an insulator. Deposits of fat beneath the skin help to maintain the normal body temperature by preventing the loss of heat generated within the body tissues.

Liquid or circulating tissue is necessary, for all cells of the body must be supplied with food and oxygen and the means for removal of waste products. To accomplish this, there is a transportation system which contains the liquid connective tissue, commonly known as **blood.**

The blood is composed of two distinct parts: a fluid portion, the plasma, which forms the matrix and solid elements, the corpuscles. The particular characteristics of each of these will be dealt with in later chapters. The solid components of the blood can be divided into three main types of cells. These are the red corpuscles (*erythrocytes*), the white corpuscles (*leukocytes*), and the platelets (*thrombocytes*). The fluid that is present in the cells, and also bathes the cells and surrounding tissues, is formed from the plasma.

Fibrous connective tissue is a dense tissue sometimes referred to as *white fibrous tissue* because it contains closely packed white collagen fibers, as seen in Fig. 4-11. These unyielding fibers run parallel, giving strength to the length of the tissue. Although these bundles of fibers are flexible, they have a limited amount of elasticity. Fibrous connective tissue forms the major part of some ligaments, which hold bones firmly in place at the various joints. Tendons are also made of this tissue and serve to attach the skeletal muscles to bones of the body. Sheets of fibrous connective tissue cover the muscles and keep the bundles of muscle in place.

Cartilage, commonly called gristle, is a more flexible yet firm material. In the early stages of development before birth, it forms the model for the future bones of the body. In the adult, cartilage is present in those parts where flexibility is a desired condition. Cartilage attaches

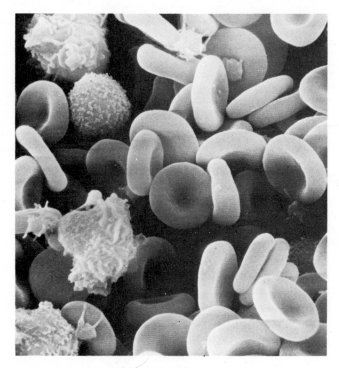

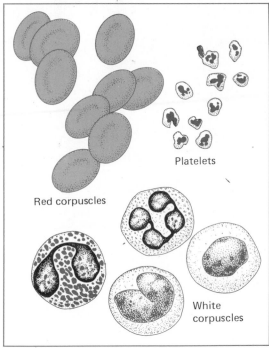

Platelets

Red corpuscles

White corpuscles

Figure 4-10 Blood as seen through a microscope and its component parts as shown in a diagram.

the ribs to the sternum (breastbone), which permits the rib cage to move during breathing. It is also present in the tip of the nose and in the external ear. Cartilage is composed of a nonliving material, the *matrix*, in which are scattered groups of cartilage cells. These cells may be in groups of two, four, etc., each group surrounded by a capsule of semitransparent material. Microscopically (Fig. 4-12), it is possible to distinguish several different types of cartilage. Two of the more common forms are the **hyaline cartilage,** with a uniformly clear matrix, and the **elastic cartilage,** in which there are strands of denser substances between the cells.

Bone or osseous tissue forms the framework of the body and consists of two types of supporting tissues: cartilage and bone. Although other types of tissue play some role in the general support of the body, these two are the principal ones, since they make up the skeleton. Most of the bones of the skeleton first appear as cartilage structures, although in some instances special membranes form the bases into which bone cells migrate. The process of bone formation (ossification) by the replacement of cartilage, or of membranes, proceeds slowly throughout life.

In a very young child the skeleton is characterized by the presence of a relatively large number of living cartilage cells and their products. In an elderly person most of the cartilage has been replaced by nonliving calcium and phosphorous salts that have been laid down by the bone-forming cells. Thus, a young person has quite flexible bones, but those of an older individual are brittle because they contain more of the inorganic salts. Bone tissue is composed of two general types of cells: one that forms the bony material, the **osteoblasts,** and another that reabsorbs it, the **osteoclasts.** By the action of these two opposing types, the skeleton undergoes continual changes that result in the building up of new bone and the sculpturing and reabsorbing of old material. The formation and fine structure of bone is discussed in greater detail in Chapter 5.

osteo = bone

Nervous Tissue

The basic unit of structure in all nervous tissue is the nerve cell, or **neuron.** This type of cell shows many variations in form and in size. Its distinct location is associated with its particular function. In the brain, for example, the cells may be relatively short but finely branched, while in other parts of the body some of the extensions of the nerve cells may be several feet long. Nerve cells are among the most specialized of all cells. Once they are destroyed, however, they cannot be replaced. The details of the structure of the neuron are found in Chapter 8.

Neurons can be classified into three different groups according to their individual functions. The **sensory neurons** receive impulses from the sense organs, the **motor neurons** carry impulses to muscles, and **associative neurons** relay impulses from the sensory to motor neurons.

Figure 4-11 Micrograph of the dense network of collagen fibers, which provide resistance to stress.

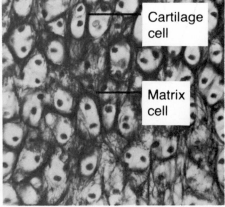

Cartilage cell

Matrix cell

Figure 4-12 Photomicrograph of elastic cartilage showing the relation of the cartilage cells to the matrix.

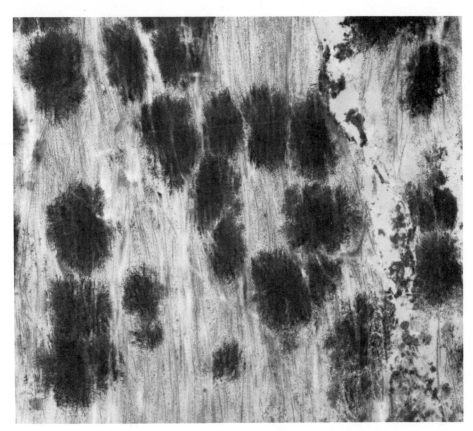

Figure 4-13 **An electron micrograph from an area of bone matrix that is in the process of ossification.**

Muscle Tissue

The majority of **striated muscles** require conscious effort to make them contract. They are therefore often called **voluntary muscles.** Under the microscope the cells making up the striated tissue appear crossed by very delicate lines. Figure 4-14c is typical of muscles that are attached to the skeleton of an animal. Each individual line is known as a **striation.** In human muscles, these cells may vary from 1 millimeter to over 40 millimeters in length and from 10 to 40 microns in width (1 micron = 0.001 millimeter). Muscle cells are relatively large and very active. They require several centers (nuclei) to regulate their activities. It is not unusual, therefore, to find cells containing 20 or more nuclei, which lie in the cytoplasm outside the main body of the cell. The cells are surrounded by a very thin membrane, the **sarcolemma.** Internally, the muscle cell is made up of many small fibers called **myofibrils.** The myofibrils lie parallel to each other and run lengthwise in the cell. These in turn are composed of still finer **myofilaments.** When a muscle contracts, the myofilaments slide past one another as a result of the chemical energy released when various forms of stimulation take place. More will be said about these changes in Chapter 6.

*myo = muscle
fibra = fiber or
filament*

Smooth muscle (visceral) tissue is called **involuntary muscle** since its action, unlike that of striated muscle, is not consciously controlled. Striated muscle cells may be over 40 millimeters in length, whereas the length of the smooth muscle cells is measured in thousandths of a millimeter. Their appearance is also quite different because they lack the cross-markings which are characteristic of striated muscles. Another difference is that each smooth muscle has a single nucleus which lies approximately in the center of the cell. Smooth muscle cells are found in the walls of the organs of the digestive system. Their slow movements pass food along the canal during the process of digestion. They are also present in the walls of arteries and veins. The contraction and relaxation of the muscle cells control the flow of blood to the various parts of the body. (See Fig. 4-14b.)

Cardiac muscle contracts in a rhythmic pattern. The word *cardiac* always refers to the heart, which is the only place where cardiac muscle tissue is found. The cells of the heart are unlike those of either smooth or skeletal muscles. They are greatly branched and join each other to form a protoplasmic network. The muscle cells show some striation, but it is not as distinct as that found on skeletal muscles. Also, discs are present which cross the muscle cells at more or less regular intervals. The nuclei are found in the middle of the cells. Functionally, the unique characteristic of cardiac tissue is its ability to contract without being stimulated by a nerve-bone impulse. If the heart is removed from a freshly killed animal, it will continue to beat. Individual cells isolated from fresh heart tissue also contract spontaneously. (See Fig. 4-14d.)

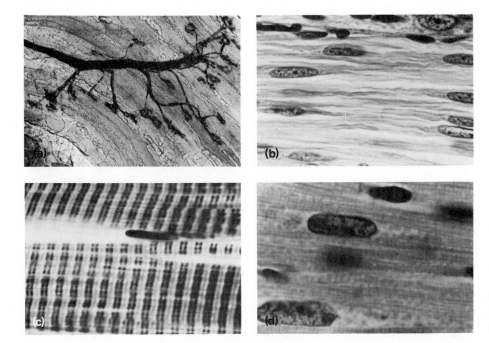

Figure 4-14 (a) An example of a neuron that transmits impulses to different parts of the body. Photomicrographs of the three types of muscle tissue: (b) smooth, (c) skeletal, and (d) cardiac.

NEW FRONTIERS

Doctors have a new technique called nuclear magnetic resonance imaging (NMRI) that allows them to see into the body. It does not carry the risks of surgery or exposure to X rays.

NMRI uses a strong magnetic field and a pulse of radio waves. The patient is placed in a large doughnut-shaped magnet which lines up the nuclei of certain atoms. The atom most commonly used is the H-1 found in water, which is abundant in nonbony tissues. The strong field from the magnet (3,000 to 28,000 times stronger than the earth's magnetic field) aligns the nuclei. The radio pulse disturbs the alignment, causing some nuclei to absorb energy. When the radio pulse is turned off, those nuclei that have absorbed energy release it as radio waves. These waves are measured by a computer which forms the image. For reasons not yet understood, different tissues react in specific ways to this technique, and differences in types of tissue become apparent in the computer image. It is also possible to demonstrate the presence of diseased or tumorous tissue because such tissue seems to react differently from normal tissue. Because of the strong magnetic field, NMRI cannot be used on people with heart pacemakers or metal replacement joints. The device must also be shielded so that it does not attract metal objects or interfere with the working of other devices that also depend upon magnets.

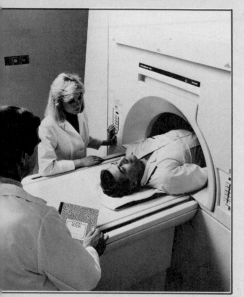

TISSUE PATHOLOGY

Growth disturbance is used as a general term to include diseases that involve an increase in tissue mass because of an excessive division or growth of a particular group of cells. **Neoplasms** are growths of a particular group of cells. Neoplasms can occur from a mutation of a single body cell. The cell multiplies to form a tissue mass. However, this has not been proven for all neoplasms.

Types of Neoplasms

Most neoplasms can be classified as **benign** or **malignant.** Benign neoplasms are single masses of cells that remain localized at their site of origin and limited in their growth. Malignant neoplasms are defined by their potential to *metastasize,* or invade other tissues at some point in their life span. *Invasion* refers to direct extension of neoplastic cells to a new site by way of the blood or lymph tissue. A growth of cells will appear at the new site. The term *cancer* is a synonym for malignant neoplasm. The term *tumor* is used in reference to any growth disturbance, whether benign or malignant.

SUMMARY

Type of Primary Tissue	Organization of Tissue	Function
	Simple squamous epithelial—cells are flat and thin; found lining the blood vessels, the heart, and the lungs.	Secretes serous fluid to reduce friction between membranes.
	Endothelial epithelial—cells are flat and irregular; found in the walls of the blood vessels and between the joints.	Regulates diffusion of materials.
	Stratified squamous epithelial—cells can be tall and cylindrical or flat, form layers; found covering tissues or organs.	Protects organs against injury.
Epithelium	*Simple cuboidal epithelial*—cells almost square-shaped; found in secretory glands, kidney tubules, and covering the ovaries.	Secretes certain tissue fluids.
	Simple columnar epithelial—cells are tall and packed together; found lining the digestive tract.	Protects organs, secretes mucus and enzymes.
	Ciliated epithelial—cells are modified columnar, contain hairlike projections; found in the trachea and reproductive organs.	Cilia aid in the movement of particles.
	Loose connective—few cells and many fibers in a semifluid matrix; found directly beneath the epithelium of the skin.	Supports and surrounds organs.
	Adipose (fat)—cells with large vacuoles; found throughout the body.	Insulates and protects the body; reserve supply for energy.
	Liquid tissue (blood)—contains plasma and different kinds of corpuscles; part of the circulatory system.	Transports material throughout the body.
Connective	*Fibrous connective*—densely packed white fibers; found in ligaments, tendons and covering for muscles.	Supports and binds joints together.
	Cartilage—cells within a matrix; found in early stages of skeletal development, and in nose and ears throughout life.	Provides flexibility.

	Bone—two general types of cells, osteoblasts and osteoclasts; found in the skeleton of the body.	Provides a supportive framework.
Nerve	*Nerve cell* (*neuron*)—neurons vary in form and size; types are sensory, motor, associative; found in the brain and spinal cord.	Provides communication for the parts of the body.
Muscle	*Striated muscle* (*voluntary*)—cells are striated and contain many nuclei; found throughout the skeleton.	Provides movement of skeletal muscles.
	Smooth muscle (*involuntary*)—cells lack striations; found in the walls of the digestive tract and blood vessels.	Controls movement of internal organs.
	Cardiac muscle—cells show some striations; found only in the heart.	Controls the beating of the heart.

VOCABULARY REVIEW

Match the statement in left column with the correct word(s) in the right column. Place the letter of your choice on the answer paper. *Do not write in this book.*

1. A group of cells having the same origin and function.
2. A group of different tissues having a common function.
3. A tissue in which fat is stored.
4. Connective tissue which joins the bones together at a joint.
5. A type of connective tissue that provides flexibility.
6. Groups of different organs working together with a common function.
7. The inner lining of all blood vessels.
8. A lubricating secretion.
9. The intercellular material in connective tissue.
10. A protein found in nonelastic connective tissue fibers.
11. Small hairlike projections on certain cells.

a. adipose
b. cilia
c. fibrous
d. endothelium
e. simple squamous epithelium
f. cartilage
g. matrix
h. mucus
i. organ
j. system
k. collagen
l. tissue
m. neuron

TEST YOUR KNOWLEDGE

Group A

On your answer sheet, write the letter that correctly completes the statement.

1. Cartilage   (a) can be replaced by bone during ossification   (b) binds muscles to bones   (c) is part of epithelial tissue   (d) covers muscles.
2. A group of cells similar in structure and function is a(n)   (a) organ   (b) system   (c) tissue   (d) organism.
3. Smooth muscle tissue   (a) is attached to the skeleton   (b) contains many nuclei in each cell   (c) forms the walls of the heart   (d) is in the walls of veins.
4. Adipose tissue is   (a) a storage tissue   (b) a muscle tissue   (c) held together by cartilage   (d) an epithelial tissue.
5. A leukocyte is a type of   (a) muscle cell   (b) blood cell   (c) receptor   (d) cellular inclusion.
6. Epithelial cells   (a) line body cavities   (b) cover bone surfaces   (c) form connective tissues   (d) have many nuclei in a single cell.
7. Voluntary muscle cells are   (a) smooth   (b) striated   (c) ciliated   (d) branched.
8. Cardiac muscle is found in the   (a) stomach   (b) intestine   (c) heart   (d) veins.
9. Red blood corpuscles are called   (a) lymphocytes   (b) erythrocytes   (c) thrombocytes   (d) leukocytes.
10. One type of bone cell is a(n)   (a) osteocyte   (b) neuron   (c) thrombocyte   (d) goblet cell.

Group B

1. Briefly explain the importance of each type of primary tissue in the human body.
2. Describe the various types of cells that are found in each of the four types of tissues.
3. What significant differences are there beween epithelial and connective tissues?
4. List the activity of each of the following: osteoblasts and osteoclasts.
5. Classify the following as a tissue or an organ: skin, lining of the stomach, fingernails, bone, and heart. State the basis for your answer in each case.
6. What are some of the differences between voluntary muscle, involuntary muscle, and cardiac muscle. (Hint: refer to structure and function.)

EYES OF SCIENCE

X-ray microscopy bridges the gap between light microscopes and electron microscopes. Light microscopes can be used to view living cells, but there is a limit to their power of magnification (about 2,000 times). Electron microscopes have much greater power of magnification, up to 200,000 times, but cannot be used to view living cells because the object to be viewed must be dried and placed in a vacuum chamber. The X-ray microscope has the magnification power of the electron miscroscope, and it can be used to view living cells. It uses a flash of long-wavelength (soft) X rays of 100-nanosecond duration and an X-ray-sensitive material called the **X-ray resist**. The **X-ray source**, called LEXIS, is an imploding gas jet plasma. With the use of such a short flash of X rays an image of a living specimen can be made before the specimen is destroyed by the X rays. A scanning electron microscope is used to view the developed X-ray resist.

This technique has been used to view living human platelets. The images produced reveal details not seen before in dried platelets. Flash X-ray microscopy should be able to provide detailed stop-motion viewing of the interactions of many types of cells.

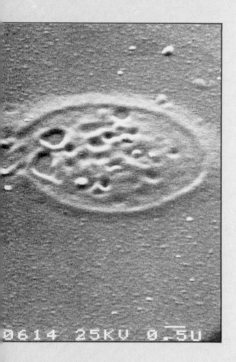

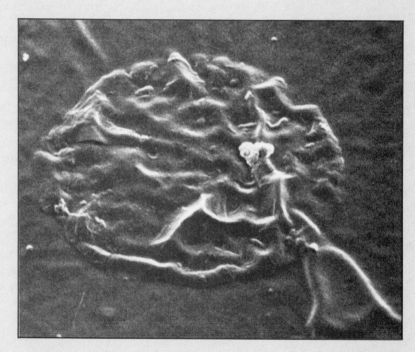

UNIT 2

THE SUPPORTING FRAMEWORK AND MOVEMENT

CHAPTER 5

SKELETON—THE FRAMEWORK

Objectives

A. List five functions of the human skeleton

B. Identify the structural parts of a typical long bone

C. Distinguish between the axial division and the appendicular division of the skeleton and name the major bones in each

D. Identify the three classes of joints, give examples, and describe the action of each

E. Explain the processes involved in the development and maintenance of bones

F. Identify some skeletal disorders and types of fractures

ANATOMY OF THE SKELETON

Living organisms must have support to retain their body shape and form. This is true for microscopic one-celled organisms with delicate cell membranes as well as for large multicellular plants and animals. Among the animals, means of support range from the chitinous **exoskeletons**, or outside skeletons, of insects and crustaceans to the **endoskeletons**, or internal skeletons, of the vertebrates. The chitin of exoskeletons is a polysaccharide. The bone of internal skeletons is very different. The bones of the human skeleton perform numerous functions: (1) As a framework or support for the body, all of the internal organs are directly or indirectly connected to and supported by the skeleton. (2) The skeletal muscles that are attached to it provide for movement such as walking, running, or jumping. (3) The bones serve as protection for the heart, lungs, spinal cord, and brain.

The bones of the skeleton have other important functions that are less well known. For example: (4) The bone marrow is the site for the formation of many blood cells. (5) The bone tissue stores mineral salts that are essential for maintaining metabolic functions.

The skeletal system is a living, dynamic organ system that is constantly being remodeled. Continual bone formation and reabsorption occur throughout life. First, rapid bone growth occurs during the early years, and then during middle life a balance is reached. Finally, there is more reabsorption than formation, and the total bone mass slowly decreases. The constant exchange of materials that takes place during bone remodeling is an example of homeostasis.

ex = out
endo = in, within

Structure of Bone

The 206 individual bones that make up the human skeleton vary greatly in size and shape. There are four principal types of bones: **long, short, flat,** and **irregular.** Long bones are found in the arms and legs of the body. The bones of the wrist and ankle are examples of short bones. The shoulder blades and the cranium bones are among the flat bones. Irregular bones have various shapes and sizes and include the vertebrae. There are also some small round nodules that develop in a tendon. These bones are called **sesamoid** bones. The **patella** (kneecap) is an example of such a bone.

A long bone has a central shaft, the **diaphysis,** each end of which is called the **epiphysis.** Covering the bone is the **periosteum,** a thick, double-layered membrane. The outer layer is connective tissue containing blood vessels, lymphatic vessels, and nerves. The inner layer contains elastic fibers, blood vessels, and bone-forming cells. The growth, development, and repair of bone is initiated from this membrane. Hyaline cartilage, which covers the ends of bones, forms the **articular surface** (place of contact) between bones.

Bone tissue.   As in any connective tissue, bone, known as **osseous tissue,** contains few cells but has an abundance of intercellular material. The intercellular substance of bone consists of deposits of a solid matrix rich in mineral salts. These minerals, largely *calcium* and *phosphorus*, give bone its hardness and rigidity. The major salt content of bone is called *hydroxyapatite*, $(Ca_3(PO_4)_2)_3 \cdot Ca(OH)_2$. This substance forms a crystalline structure. The other components of bone material—the cells, fibers, and ground substance—are mostly protein. They reinforce the tissue and provide a certain amount of flexibility. There are two types of bone tissue: **spongy** or **cancellous** bone and **compact** or **dense** bone. Spongy bone is found in the interior of a bone. It is porous and contains more blood vessels than does the dense material of compact bone. It is thicker in the diaphysis (shaft) of a long bone. The epiphyses (ends) are composed mostly of cancellous bone. Within the diaphysis is a long **medullary canal** containing **bone marrow.** Most bone contains **yellow marrow,** composed of blood vessels, nerve cells, and fat cells. **Red marrow** is found in the ends of long bones in the cancellous tissue. There are fat cells in red bone marrow and blood cells in various stages of development. Lining the medullary canal is a thin membrane called the **endosteum.** Short and irregular bones are filled with cancellous tissue and covered with a thin layer of compact bone. Flat bones consist of two plates of compact bone with a layer of cancellous bone in between. Refer to Fig. 5-2.

Surface features of bone.   The surfaces of various bones are different. Some have **projections, depressions,** or even **holes.** These surface features help to provide for the attachment of muscles, to form articulations (joints), and to serve as passageways for blood vessels and nerves.

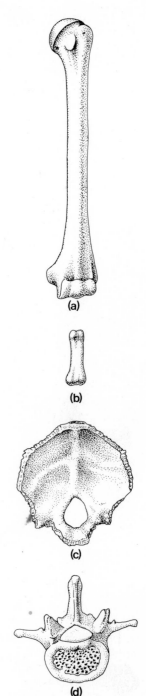

(a)

(b)

(c)

(d)

Figure 5-1 **Types of bones. (a) Long bone. (b) Short bone. (c) Flat bone. (d) Irregular bone.**

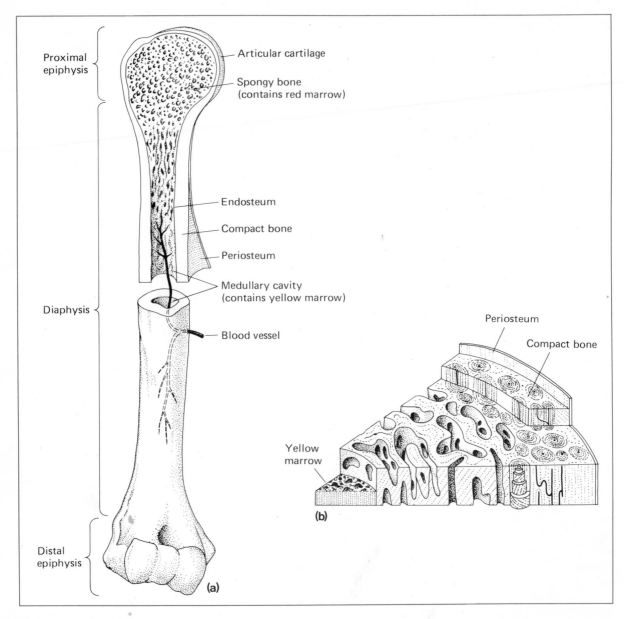

Figure 5-2 (a) A section through the humerus to show the internal and external features. (b) A cross section of bone.

Projections on the surface of bones are called **processes** and may include:

head—a large rounded articular surface.

neck—the narrow part of a bone that is between the head and shaft.

spine—a sharp slender process as seen on the back of the shoulder blade (scapula).

condyle—a rounded knucklelike process located where the bone articulates with another bone.

crest—a ridge of bone that is unusually narrow.
trochanter—a large projection for the attachment of muscles.

Depressions and openings on the surface of bones include:

fossa—a shallow depression in the surface of a bone.
foramen—a hole in a bone allowing passage of blood vessels and nerves.
    The largest foramen is at the base of the skull through which the
    spinal cord passes and is called the *foramen magnum.*
meatus—a long tubelike canal in a bone as seen in the skull at the ear.
sinus—usually an air-filled cavity in a bone.

Divisions of the Skeleton

The bones of the skeleton can be grouped into two divisions: the **axial** skeleton and the **appendicular** skeleton. The axial skeleton includes the bones of the **skull, vertebral column,** and **thorax.** The appendicular skeleton consists of the appendages and those structures by which they are attached to the axial skeleton.

The axial skeleton.   The axial skeleton has 80 bones including those of the **skull,** the **vertebrae,** the **sternum** (breastbone), and the **ribs.** The skull can be divided into the **cranium** and the **facial bones.** The bones of the cranium house the brain. They are mostly flat bones. The facial bones are irregularly shaped bones. They support and protect the mouth, nose, eyes, and ears.

The eight bones of the cranium consist of four single bones—the **frontal, occipital, sphenoid, ethmoid**—and two sets of paired bones—the **parietal** and **temporal.** The *frontal bone* forms the forehead and the upper portion of each *orbit* (eye socket). The *occipital bone* forms part of the base of the skull. This bone is found in back of the cranium. The spinal cord passes through a hole in the occipital bone called the *foramen magnum.* Between the frontal and occipital bones are the two *parietal bones.* These bones form part of the sides and top of the cranium. The *sphenoid,* the central bone of the skull, forms part of the anterior of the cranium. It resembles the shape of a bat and is composed of a body and two sets of wings. The wings are processes known as the *greater* and *lesser* wings. On the upper part of the body is a fossa which is saddle-shaped, called the **sella turcica.** This is where the pituitary gland is located. There are also several openings, or foramina, in the sphenoid for the passive of nerves. The *ethmoid bone* is an irregular bone. It forms most of the upper nasal region and joins with a bone of the face to form the **nasal septum.** The two *temporal bones* are below the parietal bones on either side. The temporal bones form part of the sides and base of the cranium. Each temporal bone contains the middle and inner portion of the hearing mechanism. The opening on the lateral surface of the temporal bone is the **external auditory meatus.** Below and posterior to the meatus is the rounded **mastoid process.** The mastoid can be felt by placing the hand behind the ear. Part of the temporal bone extends out to form the **zygomatic arch.** (See Fig. 5-4.)

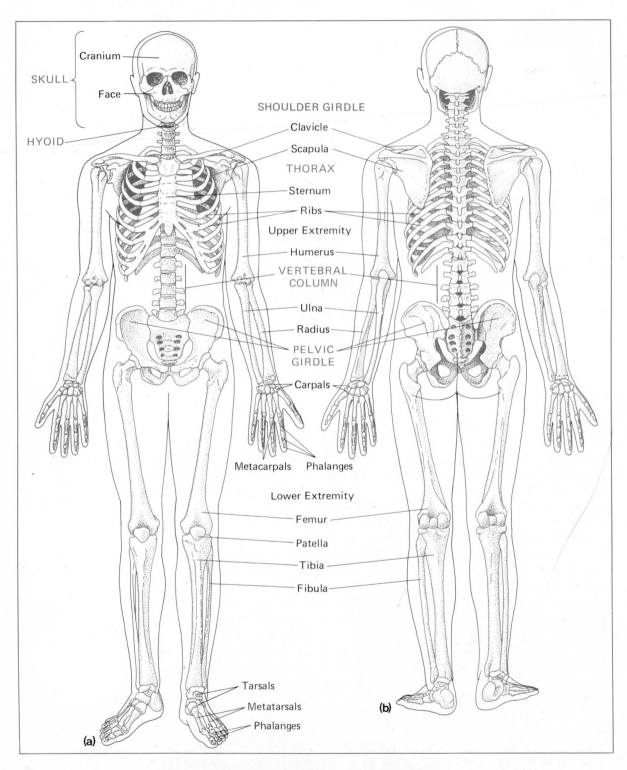

Figure 5-3 Divisions of the skeletal system. (a) Anterior view. (b) Posterior view.

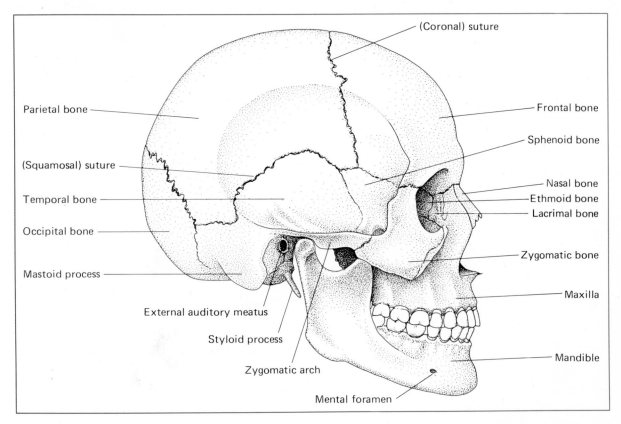

Figure 5-4 **Lateral view of the skull.**

The 14 facial bones of the skull are all paired bones except the **mandible** and the **vomer.** The muscles attached to the facial bones allow movement of the mouth and different facial expressions. The **zygomatic bones,** or cheek bones, join with a temporal bone to complete the zygomatic arches. The **maxillae** (upper jawbones) join to form the hard palate in the roof of the mouth. These bones also form part of the floor of the orbits and the lateral walls of the nasal cavity. Some facial features are composed of both bone and cartilage. For example, in the nasal cavity the *nasal bones* form the upper part of the bridge of the nose. Cartilage forms the lower part of the nasal framework. The partition between the two nasal cavities (septum) is formed by the *vomer.* The vomer is a thin, single bone shaped like a plowshare. Along the lateral walls of the nasal cavity are two scroll-shaped bones called the **inferior nasal conchae.** These are similar to the middle and superior nasal conchae which are part of the *ethmoid bone* (a cranial bone). Posterior and lateral to the nasal bones are **lacrimal bones.** These are the smallest bones of the face. They help to form the medial wall of each orbital cavity (eye socket). Each lacrimal bone has an opening through which tear ducts enter the nasal cavity. (See Fig. 5-5.)

In addition to the 14 facial bones, three tiny bones called **ossicles** are located in each middle ear. The single U-shaped **hyoid bone** is an in-

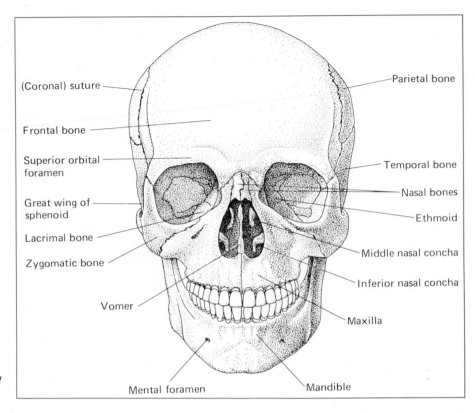

Figure 5-5 **Anterior view of the skull.**

teresting part of the axial skeleton because it does not attach to any other bone. It is discussed along with the other bones of the skull because of its anatomical function. It is located in the neck above the larynx. The hyoid bone serves as the attachment for the muscles of the tongue. (See Fig. 5-6.)

The vertebral column. The vertebral column (Fig. 5-7) is made up of a series of 26 individual bones called *vertebrae.* They are linked together by cartilage and ligaments to allow flexibility, and to provide support of the trunk. The vertebral column also serves as protection for the spinal cord. The column is divided into the *cervical* (neck), *thoracic* (chest), *lumbar* (small of the back), *sacral* (between the hip bones), and *coccygeal* (tail bone) regions.

There are 33 separate vertebrae in the embryo. How does a total of 26 bones result in the adult? The five sacral vertebrae and the four coccygeal vertebrae fuse, each group forming a single bone at the time of birth. The vertebral column is not a straight line when seen from the side. The thoracic and sacral curves appear before birth, but the others develop later. The cervical curve appears when a child begins to sit up and hold its head erect. Another curvature develops in the lumbar region when the child begins to walk. These changes in the curvature of the column result in shifts in the body's center of gravity.

Figure 5-6 **The hyoid bone.**

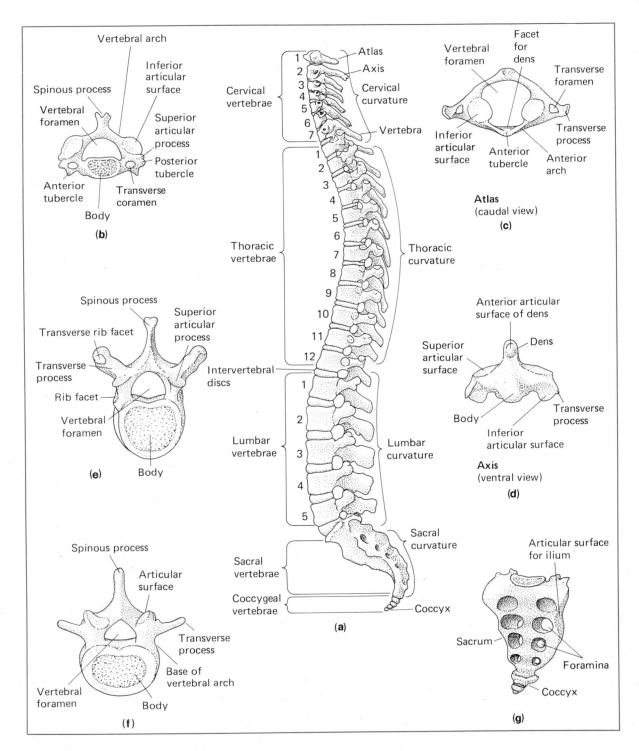

Figure 5-7 (a) Vertebral column. (b) Typical cervical vertebrae. (c) Atlas and (d) axis of the cervical vertebrae. (e) Typical thoracic vertebrae. (f) Typical lumbar vertebrae. (g) Sacrum and coccyx.

A typical vertebra (Fig. 5-7) is composed of a **body** with an arch on the posterior surface. The arch roofs over an opening, the **vertebral foramen,** through which the spinal cord passes. There are several processes which stem from the arch to form joints or attachments. Each vertebra has a **spinous process** for the attachment of ligaments and muscles of the back. Two **transverse** processes project on either side for the attachment of muscles. **Articular processes** form the joints connecting the individual vertebrae. The surfaces of these processes are covered with cartilage.

There are seven **cervical** vertebrae consisting of a large foramen in each transverse process and a small body. However the top two vertebrae are different. The first cervical vertebra, the **atlas,** supports the skull. It has no body and therefore appears as a ring with two transverse processes. The second cervical vertebra, the **axis,** has a protruding upper surface. This process, the *dens*, forms a pivot around which the atlas rotates to allow movement of the head.

The 12 thoracic vertebrae are distinguished by the presence of **rib facets** or **costal pits.** The facets provide attachment of the ribs (Fig. 5-7) with the vertebrae. Most thoracic vertebrae have three facets.

The five lumbar vertebrae have the largest bodies since most of the weight of the body is supported by this region. The *sacrum*, a triangular-shaped bone, contains foramina along the original five sacral bones. These openings are the passageway for nerves and blood vessels. The coccyx bone is considered as the vestigial remains of the four fused *coccygeal* vertebrae.

The sternum and ribs.   Organs in the **thorax** are supported and protected by the *thoracic* vertebrae, the **sternum,** and the **ribs.**

The sternum is a thin flat bone that is divided into three parts (Fig. 5-8): the upper part, or **manubrium;** the long narrow **body;** and the cartilaginous part, the **xiphoid process.** Ligaments form the attachments of the clavicles (collar bones) to each side of the top of the sternum. Seven pairs of **costal cartilages** form the attachment of the ribs to the body of the sternum.

There are 12 pairs of ribs in the human body. These curved bones articulate posteriorly with the thoracic vertebrae and anteriorly with the sternum. The first seven pairs are known as *true ribs* because each has a direct costal cartilage connection with the sternum. The next three pairs are called *false ribs* since their cartilage connections join indirectly with the sternum. The last two pairs of ribs have no connection with the sternum and are called the *floating ribs.*

The Appendicular Skeleton

The **appendicular skeleton** consists of 126 bones and includes the bones of the arms and legs and those of the shoulder and pelvic girdle. The word "appendicular" refers to a structure and its connection to the

axial part of the body. In the human skeleton, the appendages are attached to the axial skeleton by means of a series of bones that form the
shoulder or **pectoral girdle** and the hip or **pelvic girdle** (Fig. 5-9).

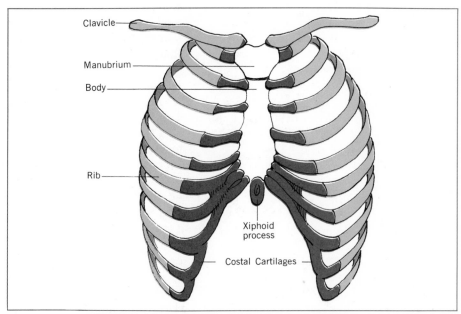

Figure 5-8 The ribs and sternum from the front. Costal cartilages join the true ribs
to the sternum. The false ribs join the seventh true rib. The floating ribs are not
shown.

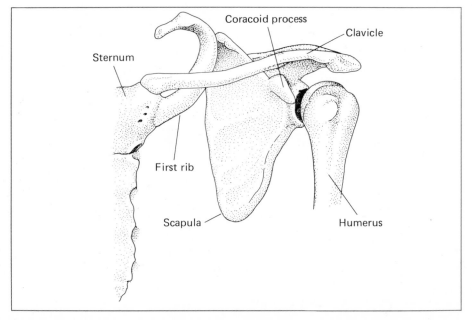

Figure 5-9 **The pectoral
girdle.**

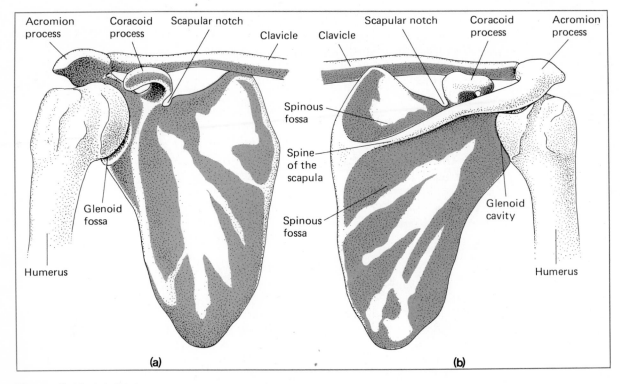

Figure 5-10 (a) Right scapula—anterior view. (b) Right scapula—posterior view.

The shoulder girdle.   The shoulder girdle consists of two pairs of bones, the *scapulae* (shoulder blades), and the **clavicles** (collar bones). These bones provide for the attachment of muscles which firmly bind the arms to the trunk of the body. The muscles also permit free movement of the arms.

The scapula is a flat, triangular-shaped bone. The prominent shelflike ridge, extending obliquely across the posterior surface of the bone, is the **spine** of the scapula. The spine ends in a flattened projection called the **acromion process.** The clavicle articulates with the acromion of the scapula. The other end of the clavicle rests against the top of the sternum. The acromion can be felt as the slight bony projection on the upper surface of the shoulder. Beneath the acromion is a rounded concave fossa called the **glenoid cavity.** The head of the humerus (upper bone of the arm) articulates in this cavity and forms the joint at the shoulder. Underneath the clavicle is the **coracoid process.** This process projects forward and serves as the attachment for several muscles and ligaments. (See Fig. 5-10.)

The arm and forearm.   Each arm consists of three principal bones: the **humerus,** the **ulna,** and the **radius** (Fig. 5-11). These bones form joints which enable movement of the arm. The humerus is the long bone of the upper arm. It extends from the shoulder to the elbow. The smooth rounded head of the humerus articulates with the scapula. On the lateral surface of the humerus is a roughened area. Muscles and ligaments form attachments on this surface to hold the humerus in the socket. In addition, there are several sites along the shaft of the bone where other muscles are attached. The bones of the forearm articulate on smooth surfaces at the lower end of the humerus. On the lower posterior surface of the humerus is a notch or depression called the **olecranon fossa.**

The ulna is the longer bone along the back of the forearm. At the upper end of the ulna is the **olecranon process.** This projection fits into the olecranon fossa of the humerus when the arm is extended. The articulation of these bones forms the joint at the elbow. On the lateral side of the ulna is a shallow notch for the head of the radius. The *head* of the ulna articulates with the radius at the lower ends of the individual bones. The flattened surface at the lower end of the radius enables it to rotate around the ulna.

The wrist and hand. The wrist (carpus) consists of eight small bones which are tightly held together by ligaments. The **carpal bones** are arranged roughly in two rows as to allow flexion of the wrist. On their inner (palmar) surface are attached some of the short muscles that move the thumb and little finger. The palm of the hand consists of five **metacarpal bones.** The metacarpals articulate with the bones of the wrist, and at the rounded ends they connect with the bones of the fingers.

The bones of the fingers contain 14 **phalanges.** Each of the four fingers has three phalanges and the thumb has only two. The fingers can move in a vertical plane. In addition the attachment between each phalanx forms a hinge and enables the parts of the fingers to bend. The thumb is more flexible than the other four fingers because the distal end of its corresponding metacarpal bone is more rounded. It is also attached to many of the muscles in the hand itself. These features permit the thumb to cross the palm of the hand. Occasionally, small sesamoid bones are found within the tendons of the hand.

The pelvic girdle. The pelvic girdle consists of two large hip bones joined to each other. These connect with the sacrum to form the bony pelvis. Each hip bone is composed of three fused bones: the **ilium,** **ischium,** and **pubis.** The ilium is the largest bone and forms the upper portion of the hip bone. The ischium forms the posterior portion, and the anterior portion is formed by the pubis. As each hip bone develops, the three bones fuse into a single bone. Eventually the two sets of bones meet in front and form the **symphysis-pubis joint.** The **sacroiliac joint** forms in the back by the articulation of the hip bones and the sacrum. There are some structural differences between the male and female pelves. (See Fig. 5-12.) In the male, the angle of the two pubic bones

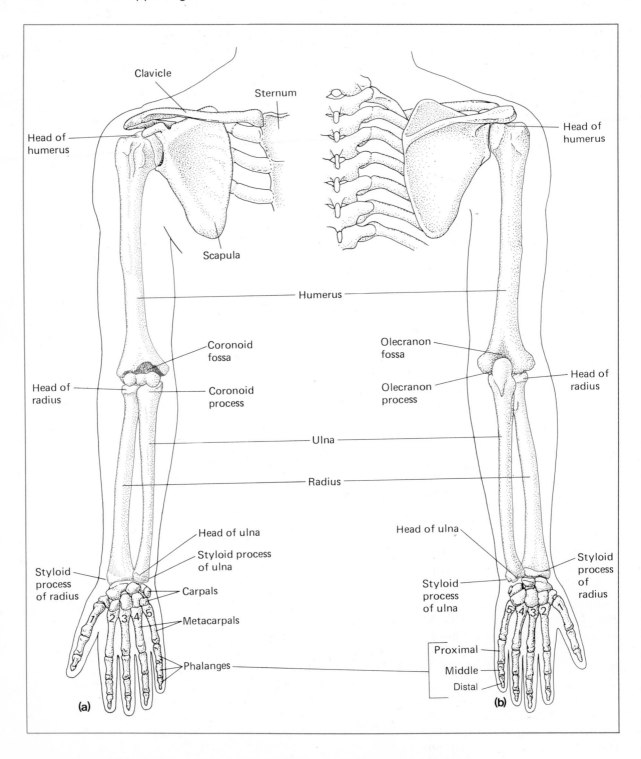

Figure 5-11 **Right shoulder girdle and upper extremity. (a) Anterior view. (b) Posterior view.**

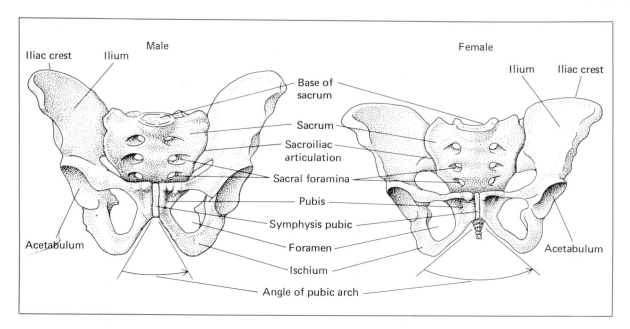

Male

Iliac crest   Ilium

Female

Ilium   Iliac crest

Base of sacrum

Sacrum

Sacroiliac articulation

Sacral foramina

Pubis

Symphysis pubic

Foramen

Ischium

Angle of pubic arch

Acetabulum

Acetabulum

Figure 5-12 **The pelvis. Note the differences between the male and the female pelvis.**

(pubic arch) is more pointed and narrow. In the female, the pubic arch is much wider and rounded. The pelvis of the male is deeper and the cavity is smaller. The female pelvis is shallow and has a larger cavity. The comparison shows that the development of the female pelvis is adapted to the function of childbearing.

The thigh and leg. In the upper leg is the thigh bone or **femur** (Fig. 5-13). The femur is the longest and strongest bone in the human skeleton. The head of the femur is rounded and smooth and articulates with the ilium at an indented surface called the **acetabulum.** (See Fig. 5-12.) The two bones form a *ball-and-socket* joint. Below the head of the femur is the neck. When these are joined to the main shaft a wide angle is formed. The neck region tends to become more porous in older persons and therefore is a fairly common site of fracture.

The lower end of the femur expands into a large flattened area with two bony processes on each side. The larger bone of the lower leg, the **tibia,** joins with both processes to form the knee joint. The *patella* (knee-cap), a flat sesamoid bone, is located just in front of this joint. The patella develops in the tendon of the front thigh muscle, the *quadriceps femoris.* A ligament attaches the patella to the tibia. On the outside of the tibia is a long, slender bone, the **fibula.** The blunt end of the fibula articulates with the tibia *below* the knee joint.

There are similarities in comparing the anatomy of the bones of the lower leg with those of the forearm. The fibula would correspond to the radius of the arm in position and mode of development. Similarly, the tibia is associated with the ulna. The difference, however, is that both the tibia and fibula have nonrotating joints. Although the lower

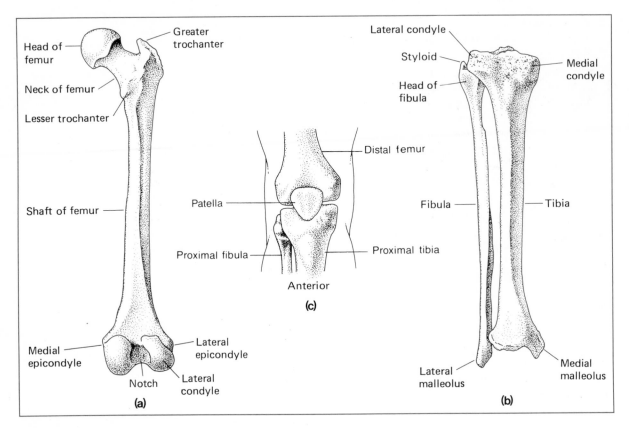

Figure 5-13 **Thigh and leg bones.** (a) Right femur—posterior view. (b) Right tibia and fibula—anterior view. (c) Patella, or kneecap.

leg cannot rotate in the same manner as the forearm, it is more suitable because the leg is an organ for locomotion and weight-bearing. The lower end of the tibia forms the projecting ankle bone on the inside of the leg. At the end of the fibula a lateral projection, the **lateral malleolus,** completes the ankle joint at the side.

The ankle and foot. The ankle and foot (tarsus) are composed of seven **tarsal bones** (Fig. 5-14). The tarsal bones are analogous to the bones of the wrist. Both the tibia and fibula articulate with a broad-topped tarsal bone called the **talus.** As in the wrist, this type of articulation allows the foot to move up and down in a single plane. The largest bone of the tarsals is the heel bone, or **calcaneus.** This bone helps to support the weight of the body and serves as the attachment for muscles of the calf of the leg.

The five **metatarsal bones** are similar to the metacarpal bones of the hand. Each of the metatarsal bones articulates with at least one of the tarsal bones, and, in some cases, with each other. The foot is shaped to form two main arches in its architecture. (See Fig. 5-14.) These curves are formed by the joining of the metatarsals with the tarsal bones. One of the arches is **longitudinal,** and lying perpendicular to it is the **transverse arch.** Together they strengthen the foot and act as a spring to cushion certain movements. Because of any one of a variety of reasons, including poor prenatal nutrition, excessive weight, fatigue, or incorrectly fitted shoes, lowered arches (flat feet) can result. This condition throws unnatural stress and strain on the muscles of the foot, and in walking, may cause fatigue and pain.

The **phalanges** of the toes are similar in the number and in the relative position to the bones of the fingers. Four of the toes each have three phalanges and the great (first) toe has only two. Unlike the thumb, the great toe is nonopposable; that is, it cannot be brought across the sole of the foot.

Figure 5-14 **The bones of the right ankle and foot and the arches of the right foot.**

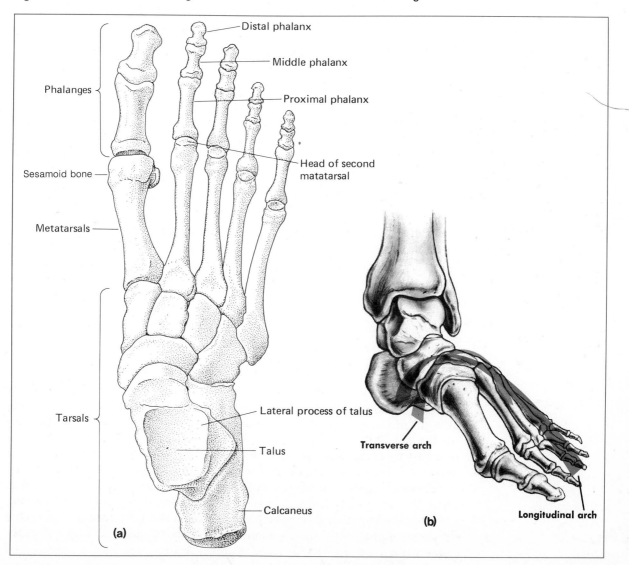

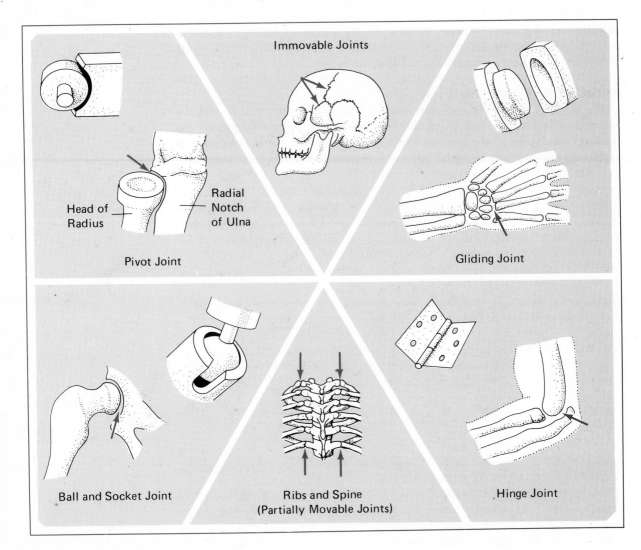

Figure 5-15 **Examples of different types of joints.**

PHYSIOLOGY OF THE SKELETON

Bones are capable of withstanding tremendous forces as well as supporting great weights. Some measurements have been made on the amount of compression the femur can withstand before breaking. The results show different quantitative relationships which are probably due to the variability of the specimens and the methods employed. However, some common observations indicate that a pressure between 15,000 and 19,000 pounds per square inch is needed to break the femur bone. Cylinders cut from the shaft of a bone and subjected to similar tests can withstand a pressure over 27,000 pounds per square inch. Fracture of the bone occurs when the femur is subjected to a twisting motion, or receives a blow from the side with a pressure of a few hundred pounds per square inch. These are examples which illustrate the rigidity and toughness of the skeletal system.

Joints and Movements

The junction of two or more bones is called a *joint* or articulation. The type and range of movement a joint provides is determined by the way in which the bones are connected. The study of joints is known as **arthrology,** and the term *arthritis* refers to an inflammation of a joint. There are three classes of joints that describe the amount of movement between the skeletal bones. The first class provides no movement and is called **synarthroses,** or more simply, **immovable joints.** A *suture* is an example of this type of a joint. Sutures join two serrated edges of different bones such as those found in the skull. **Amphiarthroses** is the second class of joints, which permit only a slight movement. The disc of fibrocartilage between the pubic bones and the intervertebral discs between the bodies of vertebrae are examples of slightly movable joints. Most of the joints in the body belong to the third class, called **diarthroses,** or *freely movable* joints. There are several types of diarthritic joints. (See Fig. 5-15.)

arthro = joint

- **Ball-and-socket** joints allow the freest movement. The surface of the rounded head of one bone moves in a cuplike cavity of the other. The joints of the hip and shoulder are the only two of this type in the body.
- **Hinge joints,** such as the articulation at the elbow or knee, allow free movement in a single plane.
- **Pivot joints** provide a rotary movement in which a bone rotates on a ring or a ring of bone rotates around a central axis. The turning of the skull on the spine is a type of pivot joint.
- **Gliding joints** are found between the carpals of the hand or the tarsals of the foot. The articulating parts of one carpal bone slide from side to side over another carpal without angular or rotary motion.
- **Angular joints** are formed by the articulation of an oval-shaped surface with a concave cavity, such as in the wrist. This type of joint permits movement in two directions.
- **Saddle joints** are similar to the angular joints in their range of movement. However, each of the bones that form the saddle joint has a *concave* and a *convex* articular surface. This joint is found only in the thumb.

Action terms.   There are several terms which describe the movement of joints. It is important to learn the movement at each joint before understanding the action of muscles (Chapter 7).

- **Flexion** is bending or decreasing the angle between the parts, as when the calf bends back toward the thigh.
- **Extension** is the opposite action, stretching out, as when the leg is straightened toward 180° (a straight line).

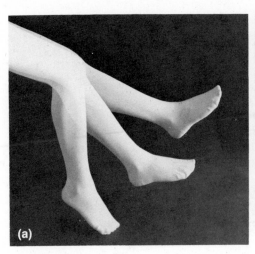
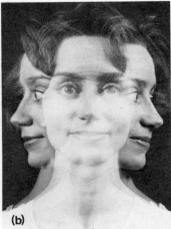

Figure 5-16 **Examples of body movements. (a) Flexion-Extension. (b) Rotation. (c) Abduction-Adduction.**

● ***Rotation*** is turning on an axis, much as the earth turns on its axis, except that in the body complete rotation is impossible because blood vessels, nerves, and other tissues would be torn.

● ***Abduction*** is drawing away from the midline of the body, as the lifting of the arm away from the body.

● ***Adduction,*** the reverse, is moving toward the midline of the body, as when the arm is brought toward the trunk. These last two examples describe the action of the shoulder joint.

Lubrication of joints is most important if the body is to move freely. The movable joints of the skeleton are enclosed by a *fibrous capsule.* The end of each bone is covered by the articular cartilage. The space between the cartilages contains a small amount of lubricating material called the **synovial fluid.** This fluid is secreted by the *endothelial membrane* (synovial membrane) lining the capsule. The fibrous capsule is strengthened by bands of ligaments. Another type of lubricating and cushioning saclike structure is a **bursa.** A bursa can be found between the bones of the elbow, knee, hip, shoulder and ankle joints. Bursae are found in those joints where pressure may be exerted or where the attachment of a tendon rides over a bone.

Bones as levers.   The movements of the muscles and joints provide action in which the bones act as levers. A **simple lever** is a rod that can be moved about a fixed point, or **fulcrum.** A lever acts to exert a force and results in the movement of an object. In the skeletal system, the bones serve as the levers and the joints as the fulcra. Muscle action supplies the force. Simple levers are classified in one of three groups depending on the relation of the effort (E) and the load (R), or resistance, to the fulcrum (F). In Fig. 5-17, these three classes are illustrated by the different parts shown. In these cases, different bones are involved in all three instances, but as the effort is applied in different directions, the type of lever changes.

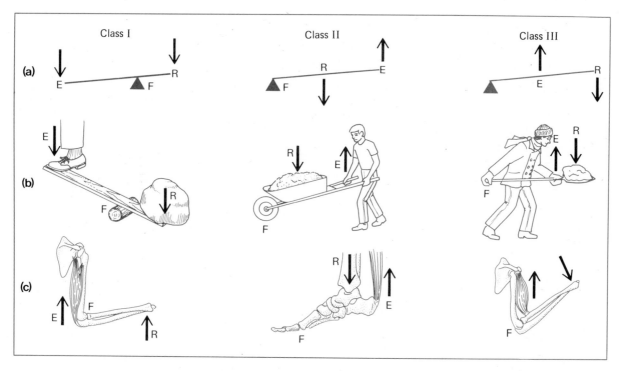

Development and Maintenance of Bone

To a large extent the physiology of bone is involved with the metabolism of calcium and phosphorus mineral salts. In the human body, the process of bone formation is known as **ossification.** The general form the adult skeleton will take is determined long before birth. In the early embryo (sixth week of life), the potential skeletal framework is laid down in two forms: **fibrous membranes** and **flexible hyaline cartilage.** As many as seven months before birth, bone cells begin to replace these materials with bone tissue. The process of ossification continues throughout life but the major activity occurs in the first 25 years. Thereafter, bone tissue is capable of local growth as in the healing of a fracture.

Intramembranous ossification. Of the two forms of ossification, the simpler type involves the bones formed from fibrous membranes. These include the flat cranial bones of the skull that protect the brain, the lower jaw, and parts of the clavicle (collar bone). At various sites in the fibrous membrane, bone-forming cells, called **osteoblasts,** appear. In these *centers of ossification*, the bone-forming cells lay down **spicules** which radiate outward from each center. The spicules develop into fibers, and calcium and phosphorus mineral salts are deposited in the matrix between them. Calcification continues forming an open framework characteristic of spongy bone. The surface of this newly formed

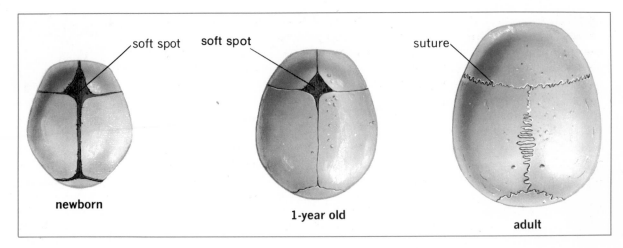

Figure 5-18 **The skull develops from a membrane that forms separate plates of bone. The fontanels are indicated as the *soft spot*.**

bone is covered by a periosteum. A thin layer of compact bone is produced directly under the periosteum forming a hard surfaced covering to the spongy bone underneath. As these bony plates grow toward each other, they will eventually meet forming the "zipperlike" immovable joints that are called ***sutures.*** At birth, the bones of the cranium have not completed their ossification. Membranes cover spaces called ***fontanels,*** left where the bones have not met (Fig. 5-18).

intra = within or inside

osteo = bone

Intracartilaginous ossification. Most of the bones of the body are formed from cartilage which is called ***chondrial tissue.*** Thus, this form of ossification is also known as ***endochondrial ossification*** (Fig. 5-19). The cartilage of the embryo is not strong enough to bear the weight of the human body though it serves as an excellent model for the laying down of the strong bony tissue that characterizes the mature skeleton.

Cartilage consists of cartilage cells called ***chondrocytes.*** These cells are housed in small spaces, the ***lacunae,*** which are surrounded by a large amount of ***gelatinous matrix.*** As the young cells mature, they produce a chemical substance that brings about the calcification of the matrix. The cells are surrounded by this hardened material and are cut off from their source of nutrition. They eventually die, and most of the calcified matrix dissolves.

The membrane surrounding the cartilage is called the ***perichondrium.*** From it, the bone-forming cells, osteoblasts, move rapidly into the area and begin to manufacture their own ground substance, the bone matrix. This deposit is made in long columns called ***trabeculae.*** Once the perichondrium starts producing osteoblasts, it is known as the ***periosteum.*** It is interesting to note that though bone matrix also becomes calcified, the bone cells do not die, as do the cartilage cells. This is because the osteoblasts have long processes that come in contact with the processes of other bone cells and with the capillaries. During the formation of the bone matrix, these provide channels for the passage of nourishment and oxygen to the cells. After the matrix is laid down

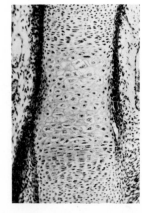

Figure 5-19 Endochondral bone is a result of ossification.

and calcification begins, the processes are drawn back into the cell, leaving the **canaliculi** or small canals. Through these small canals, life-maintaining substances are distributed continually. This gives bone the characteristic of a living tissue in spite of the fact that about 70 percent of bone is composed of mineral salts of an extremely hard consistency. (See Fig. 5-20.)

Ossification commences in the center of the diaphysis and proceeds toward the epiphyses. After this primary center of ossification has its work well under way, a similar ossification center arises in each epiphysis and begins its slower replacement of cartilage. In the X-ray photographs in Fig. 5-21 the left one shows the hand of a child in which the epiphyses appear to be separated from the bone. Actually there is a band of cartilage, the **epiphyseal plate,** connecting them to the shaft. The right one shows the hand of an adult in which the epiphyses have become fused to the diaphysis.

The age at which the epiphyses first appear and then later become firmly attached to the shaft varies somewhat with different individuals as well as with the bones themselves. As a general rule the epiphyses appear earlier in females and fuse to the diaphyses earlier in life than they do in males. Referring to the photograph of the child's hand (Fig. 5-21), we find that the epiphysis on the distal end of the metacarpal

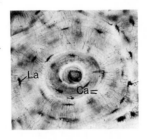

Figure 5-20 A light micrograph of ground bone. The lacunae (**La**) house the osteocytes, and the canaliculi (**Ca**) represent the former location of osteocyte processes.

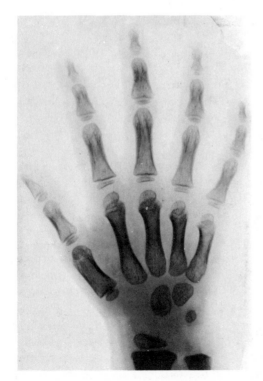 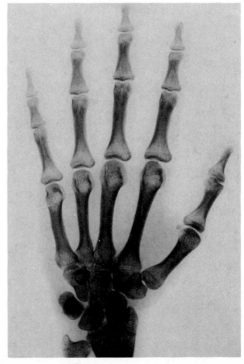

Figure 5-21 (a) A child's hand. Note the absence of solid bone in the wrist. (b) An adult's hand. The cartilage has been replaced by the formation of bone.

bone will appear between the ages of 10 months and 2 years. It becomes solidly attached to the shaft between the ages of 14 and 21 years. In a similar fashion, the epiphysis at the end of the radius makes its appearance between three months and one and one-half years. In girls, this bone becomes attached to the radius at an average age of 17, but in boys this happens at an average age of 19. It is at about this age that bone growth stops; however, bone continues to become stronger and heavier until about 25 years of age.

While replacement of cartilage continues in the interior centers of ossification, osteoblasts between the periosteal membranes are depositing layer after layer of thin bone, forming a collar around the shaft. This type of ossification differs from that of replacing cartilage. The bone matrix is laid down by osteoblasts and is calcified directly.

Because all bone tissue, when first formed, is of the spongy type, these external layers of the long bones and the first bony formations of the skull are only temporary. They must be replaced by compact bone by the processes of destruction and rebuilding. Destruction is probably the work of the **osteoclasts,** the large cells with many nuclei that are thought to dissolve bone tissue. In the interior of the long bones, these cells carve out a large cylindrical space, the **medullary canal,** which fills with marrow. At this stage, newly formed bone is very porous with many spaces between the trabeculae. This new bone must now be converted to more compact bone by building the layers into a more closely woven network with fewer and smaller spaces.

Haversian system.   In the process of bone reformation, the concentric layers, **lamellae,** of bone are laid down on the inside surface of the channels between the trabeculae. As these layers are added, the channel is gradually narrowed to form the **Haversian canals.** The canals contain

Figure 5-22 **The Haversian system.**

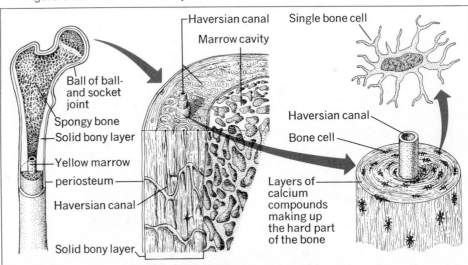

blood vessels and lymphatics. The osteoblasts come to lie in cavities (lacunae) between the lamellae and are then called **osteocytes** (bone-maintaining cells).

The unit of structure of compact bone, then, is the Haversian system (Fig. 5-22), with its lamellae in concentric layers around the *Haversian canal,* the osteocytes lying in the lacunae between the layers of bone, and the tiny canaliculi that channel the tissue fluid between the cells and the blood vessels in the canal. Because of this intricate, interconnecting system of canals there is an extensive blood supply throughout the bone.

Red bone marrow primarily consists of red blood cells, most white blood cells, and the platelets (cells that aid in the clotting of blood). The process of blood cell formation, called **hematopoiesis,** starts in the early embryo in the yolk sac, liver, and spleen. As red bone marrow develops, the formation of most blood cells is carried on by this tissue and continues throughout adult life. (See Chapter 22.)

Tendons and ligaments.   The transmission of muscular pull to bones is achieved through the attachment of tough cords called **tendons.** These are composed of fibrous connective tissue to which muscle fibers are fused, and they are attached to the surface of the periosteum. Through the tendon, the muscle exerts traction on the bone to which it is attached and is thus able to control the movement of the bone.

A ligament is a strong band of connective tissue which connects the bones of a joint. Its function is to support the joint by holding the bones in place.

DISORDERS OF THE SKELETON SYSTEM

The skeleton, like any other part of the body, is subject to various diseases and injuries. Many bone disorders result from deficiencies in mineral supply or necessary vitamins, or from the lack of certain hormones that regulate bone development and reformation.

The general name given to several conditions involving infection of bones is termed **osteomyelitis.** Bone infections may be caused by the invasion of bacteria from some other region of the body, or from the external environment. Certain types of diseases and infections, such as pneumonia or typhoid fever, or inflammation of the teeth may affect the skeletal bones or bone tissue. Injury to a bone due to a hard blow may also serve as a predisposing condition that may be followed by an infection.

Disorder of Joints

One of the more common types of skeletal disorders is **arthritis.** This term is used to include a large number of different conditions in which joints are affected. Arthritis is usually thought of as a disease of the aged. Only certain types, particularly **osteoarthritis,** occur as a result of aging, irritation of the joints, or the wear and abrasion of joints. The

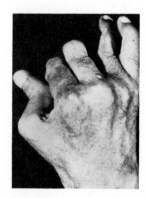

Figure 5-23 **The effects of rheumatoid arthritis.** What deformities do you notice?

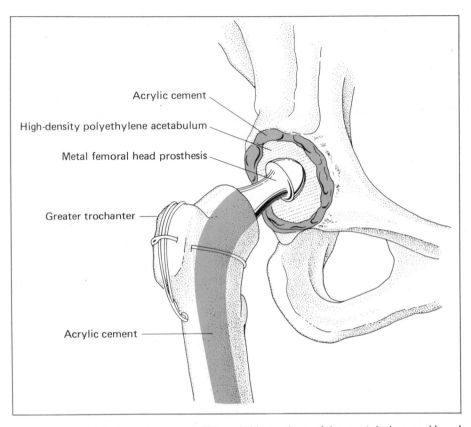

Figure 5-24 Total hip replacement. The arthritic portions of the acetabulum and head of femur are replaced by a prefabricated joint.

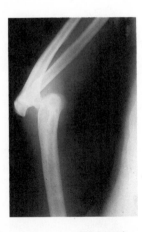

Figure 5-25 Dislocation of the elbow. Note the olecranon process projecting away from the humerus.

most serious forms of arthritis usually develop between the ages of 25 and 50. ***Rheumatoid arthritis*** is a severe form of this disease. It affects about three times as many women as it does men. The joints become swollen and painful due to the inflammation of the *synovial membrane* (Fig. 5-23). The pain causes muscle spasm which may result in a deformity. As the disease progresses, the cartilage separating the bones of the joints is destroyed. Gradually the cartilage is replaced by bands of hard calcium.

Damaged joints may be replaced surgically with artificial joints known as ***prostheses.*** After the diseased portion is removed, the artificial parts are inserted and cemented in place with a special acrylic cement. Considerable success has been achieved with hip and knee replacement. Research is being directed toward substituting other major joints of the shoulder, elbow, wrist and ankle (Fig. 5-24).

Dislocation.   A ***dislocation*** occurs when a bone is forced out of its proper position in a joint. In the case shown in Fig. 5-25, the top of the ulna has been pushed out of position. Note the displaced *olecranon process*, the sharply curved upper part of the ulna. The realignment of the dislocated bones should be performed by a physician.

Sprain.   When a joint is subjected to a sudden, unnatural motion, a sprain may result because of the tearing or straining of the tendons that hold the muscle to the bones or of the ligaments that connect the bones. This is usually accompanied by acute pain and rapid swelling in the region which has been injured.

Vitamin D deficiency.   Vitamin D is essential for normal growth in children and bone replacement in adults. It enables the blood stream to absorb calcium from the intestines. Without vitamin D, calcium and phosphorus cannot be taken into the body. In children, a deficiency of this vitamin D results in a disease called ***rickets.*** The cartilage in the developing bones continues to form rather than being replaced by bone. The bones then remain soft and the weight of the body and muscular action causes the leg bones to become bent and distorted. Other regions of the body may also be deformed. (See the discussion of vitamins in Chapter 14.) In the adult, a vitamin D deficiency causes *demineralization* of the bones, especially of the regions of the hip, legs, and spine. This disease, called ***osteomalacia,*** also results in bowed legs, a flattened pelvis, and spinal deformity because the body weight is too great for the demineralized bones.

Types of Fractures

There are several types of fractures. The names identify the way in which ends of bone are damaged. Most fractures are classified as ***simple*** or ***compound fractures.*** Simple fractures occur when a bone is broken but does not push outward through the skin. In compound fractures, the broken ends of the fractured bone protrude through the skin. The simplest type of fracture is known as a *green-stick* fracture. The broken bone does not separate completely but acts like a sap-filled green stick, the fibers of which separate lengthwise when it is bent. Green-stick fractures are usually found in young children whose bones still contain flexible cartilage. The diagram in Fig. 5-26 shows the various types of fractures of the left femur bone. The *periosteum* and the *endosteum membranes* play important roles in the healing of broken bones. After a fracture has occurred, these membranes immediately begin to produce bone-forming cells. In bones of membranous origin, healing occurs as a direct extension of new growth from each side of the break. However, in long bones there is the formation of a fibrous cartilage model to fill the gap. This model is replaced by bone as in the initial ossification of the long bones. The cartilage tissue and new bone form a ***callus*** or bridge between the broken ends of the bone until healing is complete.

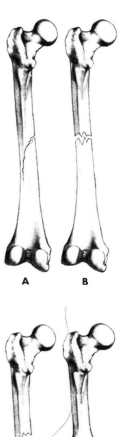

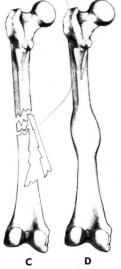

Figure 5-26 Fractures of the left femur bone (posterior view). (a) A greenstick fracture. (b) A closed fracture. (c) A crushed femur bone fragment has penetrated the skin, producing an open fracture. (d) A callus from the healing of a fracture.

SUMMARY

Bone or osseous tissue is made mostly of hard matrix containing calcium and phosphorus mineral salts. The two types of bone are: spongy bone and compact bone. The end of a long bone is called the epiphysis; the medullary canal is in the center of the diaphysis or shaft of the bone. The lining of the medullary canal is the endosteum while the outer covering of bone is called the periosteum. Bones are classified as long, short, flat, and irregular, and their surface marks include projections, depressions, or openings.

The adult human skeleton consists of 206 bones divided into axial and appendicular portions. The appendicular division is subdivided into the pectoral girdle (clavicle—2, scapula—2); the upper arm (humerus—2); the forearm (radius—2, ulna—2); the wrist (carpals—16); the hand (metacarpals—10); the fingers (phalanges—28); the pelvic girdle (ilium—2, ischium—2, pubis—2); the thigh (femur—2); the kneecap (patella—2); the lower leg (tibia—2, fibula—2); the ankle (tarsals—16); the foot (metatarsals—10); the toes (phalanges—28). The human skeleton provides: support, protection, a place for the attachment of muscles, a means for the storage of mineral salts, and for the production of blood cells. Joints provide for flexibility of the body and are classified as: immovable, slightly movable, and freely movable.

The development of bone or ossification occurs in membranes (cranial bones) and cartilage (nearly all others). The former is called intramembranous ossification and the latter, intracartilaginous or endochondreal ossification. The Haversian system is the "unit of structure" within compact bone. The production of blood cells occurs in the red bone marrow (spongy bone) of certain bones. Tendons are composed of fibrous connective tissue and fasten a muscle to the periosteum of a bone.

VOCABULARY REVIEW

Match the statement in the left column with the word(s) in the right column. Place the letter of your choice on your answer paper. *Do not write in this book.*

1. Attaches a bone to a bone.	**a.** bursa
2. Projections on the surface of a bone.	**b.** cartilage
3. Internal columns of bony tissue.	**c.** epiphysis
4. The patella is an example of this type of bone.	**d.** epiphyseal cartilage
5. The type of cell that forms bone.	**e.** fossa
6. A small bony structure found at the ends of a long bone.	**f.** hematopoiesis
7. The forerunner of bone in the majority of the skeleton.	**g.** ligament
	h. membrane bone
8. An immovable type of joint.	**i.** osteoblast
	j. osteoclast

9. A fluid-filled cushion between bones at a joint.

10. The production of blood cells.

k. osteocyte
l. processes
m. sesamoid bone
n. suture
o. tendon
p. trabeculae

TEST YOUR KNOWLEDGE

Group A

On your answer sheet, write the letter of the word that correctly completes the statement.

1. A structure associated with a Haversian canal is a   (a) lamella   (b) bursa   (c) tendon   (d) ligament.
2. Straightening the arm at the elbow is an example of   (a) flexion   (b) distortion   (c) extension   (d) abduction.
3. Bone tissue is an important storage place for   (a) oxygen   (b) white blood cells   (c) chitin   (d) mineral salts.
4. The sacrum is made up of   (a) tarsals   (b) carpals   (c) vertebrae   (d) phalanges.
5. Sutures are found in the   (a) knee   (b) skull   (c) elbow   (d) wrist.
6. The cervical region of the spinal column is found in the   (a) thorax   (b) trunk   (c) neck   (d) pelvis.
7. Movable bones are bound to each other by   (a) ligaments   (b) tendons   (c) epiphyses   (d) joints.
8. Bone-forming cells are known as   (a) spicules   (b) osteocytes   (c) hyaline cartilage   (d) lacunae.
9. The bridge of tissue that forms between the broken ends of a bone is known as   (a) cartilage   (b) callus   (c) periosteum   (d) a prosthesis.
10. The axial skeleton includes the   (a) humerus   (b) femur   (c) tarsus   (d) cranium.

Group B

Write the word or words on your answer sheet that will correctly complete the sentence.

1. The internal skeleton of humans and other vertebrates is called an _____.
2. The intercellular material that is composed of mineral salts and fibers of collagen is called the _____.
3. The human skeleton is divided into two portions, the central structure is the _____ division onto which the _____ division is attached.
4. The junction of two or more bones is called a _____.
5. The membrane that lines the medullary canal is the _____.

CHAPTER 6

MUSCLE TISSUE

Objectives

A. Identify the three types of muscular tissue and describe the general characteristics of each
B. Describe the anatomy of a skeletal muscle
C. Explain the mechanism of muscle fiber contraction and how energy is supplied for it
D. Explain how a nerve impulse stimulates a muscle fiber
E. Describe the properties of a muscle contraction

ANATOMY OF MUSCLE TISSUE

While the jointed skeleton provides a protective supporting framework and permits flexibility, it cannot move by itself. If not for the attached muscles, our bones and joints would be quite useless as far as helping us move. It is muscle tissue that enables humans to maintain an upright position, to hold the head erect, and to assume a variety of positions. Movement enables us to secure food and shelter and to escape danger. The ability to communicate by means of speaking, writing, gesturing, or visual recognition is made possible through muscular activity. Muscles are essential for such life functions as breathing, moving nutrients through the digestive tract, pumping blood through the body, and removing wastes from the body. Finally, part of the energy released by muscular activity is changed into heat, which enables the body to maintain normal temperature. Special properties of muscle tissue make it possible for muscles to respond to electrical impulses and cause movement of some part of the body.

cardio = heart

There are three distinct types of muscle tissue: ***skeletal, visceral,*** and ***cardiac.*** As the name indicates, skeletal muscle is attached to the bones of the body. Its primary role is to move the parts of the skeleton. This type of muscle is also known as ***striated*** or ***voluntary*** muscle tissue. Within these muscles are bundles of fibers. Viewed under a microscope, the fibers contain dark bands or stripes called ***striations.*** Skeletal muscle is capable of being consciously controlled—made to move by intent—to produce and direct movements. Visceral muscle tissue is found in the walls of blood vessels, in the intestines, and in many other internal organs. The spindle-shaped cells of visceral muscle tissue show no stripes or banding and under a microscope appear as **smooth** or **nonstriated** tissue. Visceral muscle cannot be consciously controlled and therefore is referred to as **involuntary muscle.** Cardiac muscle forms the major portion of the walls of the heart. Movement of this muscle tissue is also involuntary. Microscopically, cardiac muscle contains a network of interconnecting fibers with striations similar to those in skeletal muscle. See Fig. 6-1.

Histology of Skeletal Muscle Tissue

Skeletal muscles make up about 35 to 45 percent of the weight of the body. It is the most common single type of tissue. A single muscle also contains other kinds of tissue, such as connective tissue, nerve tissue, and numerous blood vessels.

Each skeletal muscle is covered by a sheath of fibrous connective tissue called the **epimysium.** A cross section of skeletal muscle (Fig. 6-2) shows that the epimysium extends into the muscle tissue and binds the bundles of fibers together. These bundles are called **fasciculi.** Each bundle is covered by a thinner sheath of fibrous connective tissue, the **perimysium.** In a similar manner, the perimysium penetrates the bundle and surrounds the individual muscle fibers. The very thin covering around each muscle fiber is called the **endomysium.** The perimysium and endomysium are interlaced with blood vessels and nerve endings. The blood vessels supply each muscle fiber with the oxygen and the nutrients that are essential for energy-producing reactions. The nerve endings are the receiving stations for nerve impulses that cause muscles to move.

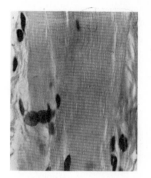

Figure 6-1 The three types of muscle tissue.
top; striated
middle; nonstriated
bottom; cardiac

epi = upon, above

peri = around

endo = in, within

Figure 6-2 A cross section of skeletal muscle shows the various connective-tissue coverings.

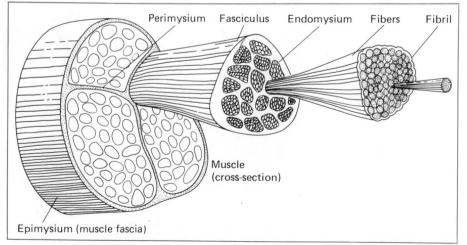

Perimysium   Fasciculus   Endomysium   Fibers   Fibril

Muscle (cross-section)

Epimysium (muscle fascia)

Microscopic View of Skeletal Muscle

Under a light microscope muscle fibers appear as elongated cylindrical cells lying parallel to each other. Each fiber is considered as one cell and can range from 1 to 80 millimeters in length. The cell membrane of a muscle fiber is called the **sarcolemma** and surrounds the **sarcoplasm,** or cytoplasm of the muscle cell. These cells have many nuclei located between the sarcoplasm and the sarcolemma. Numerous mitochondria are also found in this region. Each individual muscle cell contains a number of component bundles of **myofibrils,** illustrated in Fig. 6-3. The myofibrils run lengthwise within the cell and give skeletal muscle its characteristic striations, or alternating pattern of dark and light bands.

Each myofibril is about 1 to 2 microns in diameter. Electron micrographs (Fig. 6-3) show that the myofibrils consist of two kinds of strands (filaments) or **myofilaments,** each of which differs in width and protein

sarco = flesh

myo = muscle

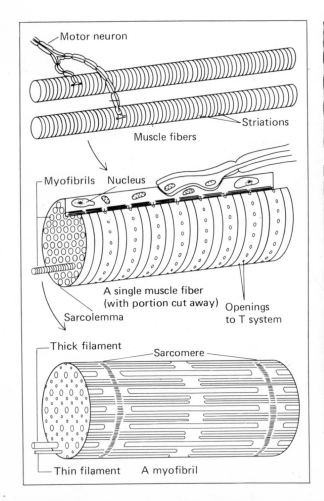

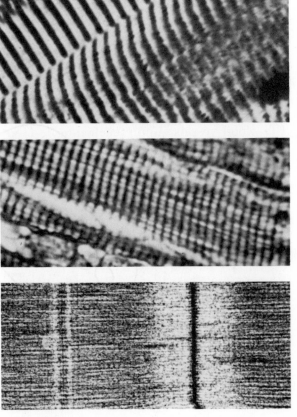

Figure 6-3 Organization of skeletal muscle at various magnifications and in diagram.

material. The thick myofilaments are about 10 millimicrons in diameter and are composed of the protein **myosin.** The thin myofilaments are about 5 millimicrons in diameter and are mostly made up of the protein **actin.** These two protein filaments form a parallel arrangement.

The dark and light areas of a myofibril can be identified by the lettered bands seen in Fig. 6-4. The dark areas are called **A bands** and the light ones are the **I bands.** The A bands (about 1.6 microns long) are the thick filaments containing the protein myosin. A narrow **H band** (about 0.5 micron long), containing only myosin filaments, is seen in the center of the A band. The I bands are the thin filaments (about 1.0 micron long) composed of actin. The I bands are found between and overlap at the ends of the A bands. In the middle of the I band is a dark line known as the **Z line.**

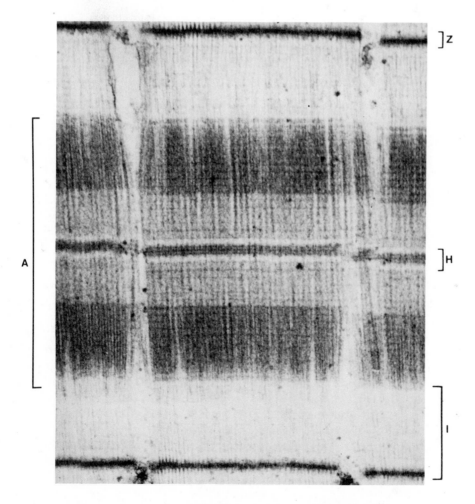

Figure 6-4   Photomicrograph of a sarcomere indicating different bands.

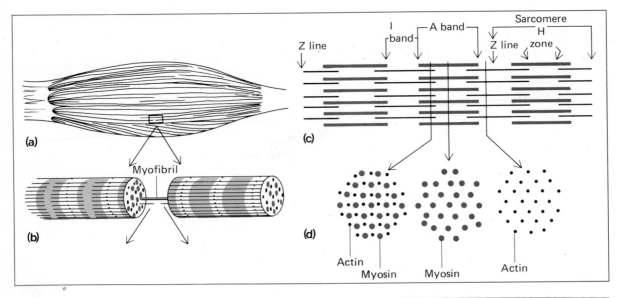

Figure 6-5 (a and b) An illustration of skeletal muscle showing the component parts of a myofibril. (c and d) Cross section of a myofibril through the various bands.

Figure 6-6 Details of a sarcomere showing the thick and thin myofilaments. During contraction, the myosin heads, or cross-bridges, contact the molecules of actin.

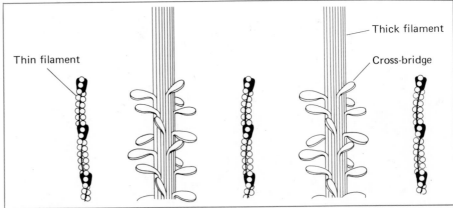

The area between two Z lines defines the structure of a **sarcomere.** The sarcomere is considered to be the functional unit of the contractile system in skeletal muscles. The I bands attach to the Z lines at one end and extend into the adjoining sarcomere. Electron micrographs through a myofibril show that the thick myofilaments of myosin molecules form long rods with knoblike projections. The myosin filaments form **cross-bridges** that connect with the overlapping thin myofilaments. The thin myofilaments of actin molecules have a double-stranded spiral configuration. The muscle cell contains a network of tubules called the **sarcoplasmic reticulum** that is comparable to the **endoplasmic reticulum** of most cells. These tubules surround the myofibrils. Running perpendicular to the sarcoplasmic reticulum are transverse tubules called **T-tubules.** The T-tubules join with others of different myofibrils and eventually open into the surface of the cell through the sarcolemma (Fig. 6-7).

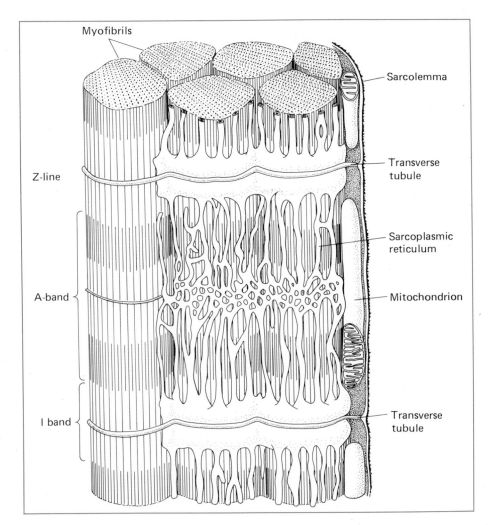

Figure 6-7 **The sarco-plasmic reticulum is a membrane system of the muscle fiber. It is distributed throughout the sarcoplasm, sur-rounding the myofibrils.**

PHYSIOLOGY OF MUSCLE TISSUE

Certain physiological characteristics of living cells are highly specialized in muscle tissue. ***Irritability*** is the ability of living tissue to respond to stimuli. For muscles, it is the ability to contract. When stimulated by impulses, voluntary muscles become shorter and thicker. This special property of ***contractility*** enables muscles to contract and thus produce movement.

Another feature of muscle is its ***extensibility.*** To bend the forearm, for example, the muscle on the *back* of the arm must be extended. This shows that an antagonistic relationship exists between muscle pairs. As one muscle, or group of muscles, contracts, an opposing muscle group responds with the reverse action. Muscles also show ***elasticity*** by returning to their original length when relaxing.

Mechanism of Muscle Contraction

The events that take place during contraction are similar in all muscle cells. According to H. E. Huxley and J. Hanson, the shortening of a myofibril occurs when the thick and thin myofilaments of myosin and actin *slide over one another*. The length of the overlap changes, but the length of the myosin and actin filaments does not change. The length of the A bands remains constant. The Z lines move toward each other as the thin filaments slide toward the center of the sarcomere. To achieve this sliding action, the knoblike projections on the thick myofilament form cross-bridges with the adjacent receptor sites on the thin myofilament. The cross-bridge can oscillate in a backward and forward movement (Fig. 6-8, right). The presence of ATP and calcium ions (Ca^{++}) provides the energy to move the cross-bridge on the myosin filament.

Figure 6-8(a) Contraction of a muscle. Huxley's sliding-filament theory proposes that the A bands, containing the cross-bridges, pull the actin molecules a short distance.

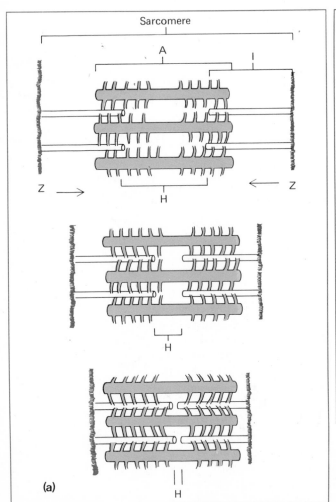

(a)

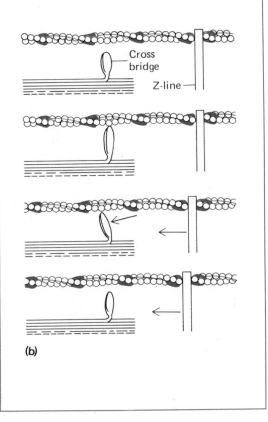

(b)

Figure 6-8(b) Enlarged view. The cross-bridges then swivel to a new position, releasing the molecules. This results in a shortening of the muscle fiber.

In turn, this causes the thin filaments of actin to move inward. When the numerous sarcomeres within the myofibrils shorten, the individual muscle fiber also shortens. As a result, the bundles of muscle fibers pull on the tendons attached to the bones of the body, producing movement.

The myofibril does not remain contracted. The calcium ions, stored in the sarcoplasmic reticulum, are released only when nerve impulses stimulate the fiber. When an impulse arrives on the surface of the sarcolemma (cell membrane), the T-tubules between the myofibrils provide a pathway for the release of the calcium ions. When the nerve impulses stop, calcium is then actively moved back into the sarcoplasmic reticulum and contraction ceases.

The neuromuscular junction.   A nerve cell that stimulates a muscle fiber is called a ***motor neuron.*** The motor neuron extends into a muscle by dividing into many fine terminal branches. Each of the branches connects with a muscle fiber to form a ***neuromuscular junction.*** A motor neuron fiber and the muscle fibers that it stimulates constitute a ***motor unit.*** The terminal end of each neuron fiber flattens to form a disc that fits into a depression on the sarcolemma of the muscle cell. This disc is known as a ***motor end plate.*** Within the neural ending are numerous vesicles, each containing a chemical transmitter called ***acetylcholine (ACh)*** (Fig. 6-9). The arrival of a nerve impulse at the neuromuscular junction causes ACh to be released at the motor end plate. Acetylcholine then diffuses to the receptor sites on the membrane of the muscle cell. The ACh initiates an electrochemical reaction that results in muscular

neuro = nerve

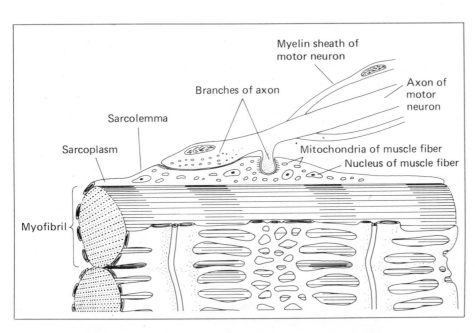

Figure 6-9   A motor end plate and the sarcolemma of a muscle fiber.

ase = enzyme

contraction. A muscle remains in a state of contraction until the production of acetylcholine stops. An enzyme produced in the neuromuscular junction called **acetylcholinesterase** inhibits the action of ACh. This, in turn, allows calcium ions to be reabsorbed by the sarcoplasmic reticulum. Contraction of muscles occurs by means of these chemical reactions. Both processes alternate at rapid speed so that muscular activity can be maintained over a long period of time.

The sliding-filament theory. When a muscle contracts, the *potential energy* of various chemicals is converted into the *kinetic energy* of motion. The energy required for muscular contraction is derived from the high-energy bonds of adenosine triphosphate (ATP). When a muscle cell is stimulated, the electrochemical activity is carried through the cell by the groups of T-tubules known as the **T-system.** The nerve impulses initiate the release of calcium ions (Ca^{++}) by the sarcoplasmic reticulum.

It is known that following the release of the calcium ions, myosin acts as an enzyme (ATPase). In its enzyme role, **ATPase** causes a phosphate group to split from the ATP molecule. The breakdown of the ATP molecule releases a molecule of ADP (adenosine diphosphate), phosphate, and energy.

$$ATP + Ca^{++} \xrightarrow{\quad ATPase \quad} ADP + P + energy$$

Simultaneously, the myosin filaments on the cross-bridge unite with the actin filaments. The supply of energy activates the interaction between the myosin and actin, causing the thin actin filaments to slide over the myosin filaments. The sarcomere shortens and produces the contraction of the muscle cell. When the calcium ions are returned to the sarcoplasmic reticulum, actin and myosin separate and the ADP is resynthesized into ATP.

Sources of Energy for Muscle Contraction

Very little ATP is stored in muscle tissue. Muscle activity of any duration quickly depletes the supply of ATP. If the energy demands of muscle are low, additional ATP can be supplied by cellular respiration. When an adequate supply of oxygen is present, glucose molecules are completely metabolized. This process is called *aerobic respiration* and results in the release of sufficient energy for continuous muscle contraction.

$$Sugar\ (or\ other\ fuel) + O_2 \longrightarrow CO_2 + H_2O + energy$$
$$Energy + ADP + P \longrightarrow ATP\ (for\ muscle\ activity)$$

If muscle activity is strenuous, energy from cellular respiration will be too slow to meet the demand for energy. Fortunately, muscle tissue also contains an energy-rich compound called **creatine phosphate (CP).** This compound is capable of releasing its phosphate group, providing energy to resynthesize ATP from ADP. Most of this intermediate energy is supplied by creatine phosphate within seconds after contraction.

$$ADP + CP \longrightarrow ATP + creatine$$

As the level of CP decreases, additional energy is required. Another source of fuel for prolonged muscular contraction comes from *glycogen*. This complex sugar molecule is stored in skeletal muscle. Glycogen can be converted directly into lactic acid, releasing energy for the resynthesis of creatine phosphate. The process, whereby energy is produced without the presence of oxygen is called *anaerobic respiration*. However, the inadequate supply of oxygen prevents the metabolism of lactic acid. As lactic acid accumulates, it tends to lower the pH of the fluid in muscle tissue. In this condition the muscle loses its ability to contract and *muscle fatigue* will result; an **oxygen debt** is created in the body.

Following exercise, rapid breathing increases the oxygen intake to repay the oxygen debt. There is a short period of time in which the body has to adjust to the new supply of oxygen. This oxygen is used to oxidize one-fifth of the lactic acid into carbon dioxide and water. This releases enough energy to resynthesize the remaining four-fifths of lactic acid back into glycogen, which can be used again. The oxygen-debt mechanism can be summarized as follows:

$$CP \longrightarrow C + P + energy$$
$$ADP + P + energy \longrightarrow ATP$$
$$Glycogen \longrightarrow lactic\ acid + energy$$
$$C + P + energy \longrightarrow CP$$
$$\tfrac{1}{5}\ lactic\ acid + O_2 \longrightarrow CO_2 + H_2O + energy$$
$$\tfrac{4}{5}\ lactic\ acid + energy \longrightarrow Glycogen$$

Properties of Muscle Contractions

The classic method for studying the contraction of a muscle is referred to as **nerve-muscle preparation.** The calf muscle from a frog is removed to demonstrate a simple muscle contraction or **single twitch.** The apparatus used to record muscle activity, shown in Fig. 6-10, is called a **kymograph.** When the muscle is artificially stimulated, it contracts, producing a record of the response.

kyma = wave

A number of different stimuli can be used to cause a muscle to contract. These may include mechanical or physical stimuli, such as striking or pinching the specimen. Chemicals such as ACh, changes in temperature, or electrical stimulation can also be used. The latter is the most common because it has a less damaging effect on the muscle. The duration and strength of the current can be precisely controlled.

The tracing of the muscle twitch, called a **myogram,** is seen in Fig. 6-10 (right). There is a brief period between the application of the stimulus and the beginning of contraction. This is known as the **latent period.** In a frog muscle, this period lasts for about 10 milliseconds (0.01

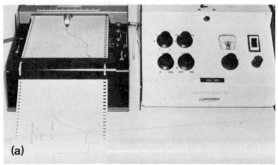

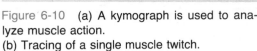

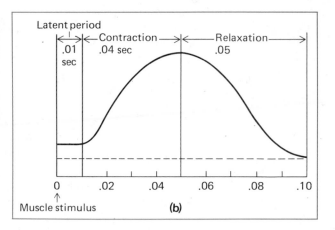

Figure 6-10   (a) A kymograph is used to analyze muscle action.
(b) Tracing of a single muscle twitch.

sec). In warm-blooded mammals, the latent period is shorter, lasting
about 1 millisecond (0.001 sec). The second phase lasts for about 40
milliseconds (0.040 sec). The second phase is called the **contraction
phase** and is indicated by the upward curve. The length of time of the
third phase is about 50 milliseconds (0.050 sec). This phase is called the
relaxation phase and is marked by the downward curve. These three
periods can occur in the absence of oxygen. There follows a short time
interval—the **recovery period**—during which the muscle is supplied
with oxygen. The oxidation of *partially metabolized* products that accumulated during the contraction can be utilized to provide energy for
the next contraction.

Laboratory experiments are usually conducted at room temperature
(20 to 22°C). Using basic laboratory techniques, a graphic record of
muscle action shows that the recovery period is approximately 60 seconds in length.

Temperature is a factor that affects muscle action. The speed of a
contraction increases as the temperature is raised. In the human body,
the effect of temperature is of minor importance because a constant
body temperature is usually maintained. Constant body temperature
is significant in that the chemical processes are carried through more
efficiently. The total time elapsing during a twitch of a muscle in the
human body varies from about 0.01 sec to 0.0075 sec. The duration of
a twitch in the frog muscle is much longer.

A muscle fiber will not contract partially, but once activated with a
stimulus of adequate strength, the fiber will contract to its maximum.
This is called the **all-or-none law** or principle. The weakest stimulus from
a nerve cell that will actually excite a muscle fiber and cause it to contract is called a **threshold** or **liminal stimulus**. Any stimulus below this
sub = below   level will cause no contraction and is referred to as **subthreshold** or
subliminal. However, if a series of subthreshold stimuli are applied
rapidly to a muscle, the muscle may be made to contract. The second
subthreshold stimulus seems to add its effect to the first, and with

repetitive subthreshold stimuli, a threshold can be reached. This additive effect results in a muscle contraction and is called **summation.**

Isotonic vs. isometric muscle contraction.   When a muscle is stimulated, the contraction shortens the muscle and results in moving another structure. A muscle contraction that lifts a load performs mechanical work. This type of contraction is said to be *isotonic* (equal tension). The tension within the muscle and the load remain constant during the movement. On the other hand, if the load is too great to be lifted, the muscle remains contracted, but does not shorten. The strain to lift the load increases the muscle tension. This type of contraction is said to be *isometric* (equal length).

Muscle cells do not multiply but increase in size (diameter) during exercise. This results in larger muscles with greater strength. If the exercise involves lifting weights to a certain height, it is an isotonic exercise. A muscle pushing against an immovable object or against an opposing muscle that prevents movement has an isometric contraction. (See Fig. 6-11.) In both cases, the effort involves two mechanisms:

- A muscle is made up of many motor units; the greater the number of motor units activated, the greater the tension and movement.
- To maintain a sustained contraction, continuous stimulation must be applied.

With rapid and repeated stimulation of a muscle, the individual contractions fuse into a continuous contraction called *tetanus* or *tetanic contraction.* The responses of the muscle to prolonged stimulation finally become weaker and weaker until the ability to contract fails. The result is muscle fatigue (Fig. 6-12). This fatigued state will last until the oxidation and removal of wastes have been completed.

Figure 6-11   (a) The lifting of weights is one example of isotonic exercise.

(b) Exerting the same amount of force against each arm is an example of isometric exercise. Which of these exercises do you prefer?

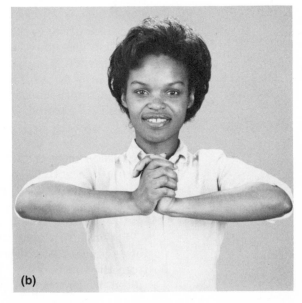

(a)

(b)

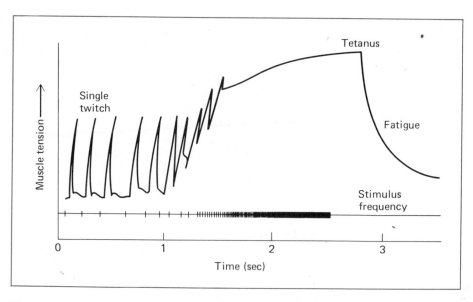

Figure 6-12 Tetanic contraction in skeletal muscle. Contraction is reactivated before relaxation can occur. Prolonged stimulation produces muscle fatigue.

Muscle tone. A normal characteristic of all skeletal muscles is that they remain in a state of partial contraction. At any given time, some motor units are being stimulated while others are relaxed. This results in a tightened or firm muscle but does not cause the muscle to move. Continual contraction or *muscle tone* maintains posture without undue fatigue. This is achieved by the muscles of the neck, back, and legs, which are usually in a state of partial contraction. An unconscious person, however, shows the loss of muscle tone. This is one reason why it is so difficult to carry a person who has fainted. Because of inactivity, a muscle also may lose its tone and become *flaccid.* Often, this is evidenced when a person is confined to a bed for long periods of time. If a muscle is not stimulated for a period of time, as is the case when a bone is fractured or the arm or leg is in a cast, the muscle tends to *atrophy* or become smaller. If normal exercise is resumed, the muscle will recover.

Production of heat during muscle contraction. The heat produced during muscle activity is the principal source of heat in the body. Sixty percent of the energy released during a muscle contraction is generated in the form of heat. *Initial heat* occurs during the actual contraction and relaxation of the muscle. The reaction involves the activities of ATP, which is a nonoxidative process. *Recovery heat* may be produced for as long as 30 minutes following muscular exercise. The heat results from the metabolic activities that restore the supplies of ATP and CP to the precontraction state.

DISORDERS OF THE MUSCLES

Muscles are subject to a number of disorders, with symptoms ranging from paralysis, pain, weakness, and dysfunction to spasms and cramps.

Myasthenia Gravis

Myasthenia gravis is a disease involving weakness and quick fatigue of the voluntary muscles. The eye muscles are most frequently affected, causing drooping eyelids and double vision. There may be difficulty and general weakness in producing such actions as smiling, chewing, speaking, swallowing, and breathing.

This disease is most common in women of child-bearing age, but it can affect anyone. It has recently been discovered that myasthenics develop an abnormal type of immunity in the blood. This causes a degree of nonresponsiveness of muscle receptors to the action of acetylcholine.

With the administration of certain drugs, the level of acetylcholine can be raised, thereby reducing the activity of acetylcholinesterase. This produces a dramatic, though temporary, recovery of muscle strength and may be maintained by the regular use of tablets. Recently, drugs that suppress the abnormal immunity have been found to be even more effective.

Muscular Dystrophy

dys = difficult

Muscular dystrophy includes several types of disorders classified according to the time on onset, symptoms, and cause. This muscular disease is more frequently found in males. Muscular dystrophy is characterized by a progressive wasting away of the skeletal muscles, which shrink to a fraction of their normal size. The onset is usually in childhood, at which time there is a gradual degeneration and atrophy of the muscle cells. The muscle tissue is then replaced by fatty tissue. There is no effective treatment.

Spasms and Cramps

Muscle spasms and muscle cramps are caused by the involuntary contraction of a muscle or group of muscles. Spasms cause weakness and pain and often involve the back and neck muscles. In prolonged cases, the diagnosis usually reveals a pinched spinal nerve associated with vertebral disc disease.

Muscle cramps are tetanic contractions of skeletal or visceral muscle. They may be caused by extreme cold over a prolonged period of time or by severe physical exertion. Cramps may involve a single muscle or a group of muscles. Stretching the cramped muscle usually relieves the pain and results in muscular relaxation.

SUMMARY

The outer covering of a muscle is called the epimysium. The interior of the muscle is divided into muscle bundles and covered by the perimysium. Each muscle cell has a membrane covering called the sarcolemma. Embedded in the sarcoplasm are numerous myofibrils, composed of thick myosin molecules and thin actin molecules. The functional unit of a muscle fiber is the sarcomere. Each sarcomere is divided into bands—the A band, the H band, and the I band. The Z line separates any two sarcomeres. Penetrating the myofibril are the transverse T-tubules.

Physiological characteristics of muscle tissue include irritability, contractility, extensibility, and elasticity. The sliding-filament theory attempts to explain muscle contraction. A neuromuscular junction is the functional connection between motor neuron and muscle fiber.

Acetylcholine initiates an electrochemical reaction resulting in muscle contraction. The primary source of energy for muscle activity is ATP, which is reduced to ADP. A single muscle contraction shows three phases: latent period, contraction period, and relaxation period, followed by a period for recovery. A stimulus not capable of inducing a muscular contraction is a subthreshold or subliminal stimulus. The heat given off during muscle activity is utilized by the body to maintain its temperature.

VOCABULARY REVIEW

Match the statement in the left column with the word(s) in the right column. Place the letter of your choice on your answer paper. *Do not write in this book.*

1. An enzyme produced at the neuromuscular junction.
2. A high-energy bond is split in this material to yield energy.
3. The smallest structural element of a muscle showing cross-striations.
4. The thin protein filament responsible for muscular contraction.
5. Decrease in muscle size due to disuse.
6. The **A** bands contain this protein.
7. A waste product of muscular contraction.
8. The continued contraction of a muscle resulting from continued stimulation.
9. Is surrounded by the sarcolemma.
10. The chemical product of a broken high-energy bond.

a. actin
b. acetylcholine
c. acetylcholinesterase
d. ADP
e. ATP
f. atrophy
g. creatine
h. glycogen
i. lactic acid
j. muscle fiber
k. myofibril
l. myosin
m. striation
n. tetanus

TEST YOUR KNOWLEDGE

Group A

On your answer sheet, write the letter of the word that correctly completes the statement.

1. A neuromuscular junction is the place of contact between a nerve and a   (a) muscle fiber   (b) nerve fiber   (c) bone   (d) sense organ.
2. Voluntary muscle fibers contain the structural protein   (a) pectin   (b) adenosine   (c) actin   (d) creatine.
3. The chemical that transmits a nerve stimulus to a muscle fiber is (a) acetylcholinesterase   (b) cholesterol   (c) acetylcholine (d) adenosine triphosphate.
4. That which covers the entire muscle is called the   (a) perimysium (b) endomysium   (c) epimysium   (d) sarcolemma.
5. A state of prolonged muscle contraction is   (a) fatigue (b) tetanus   (c) tibia   (d) myasthenia gravis.
6. A chemical substance closely involved in muscle contraction is (a) DNA   (b) RNA   (c) ATP   (d) DDT.
7. The functional unit of the contractile system in skeletal muscle is the   (a) sarcomere   (b) sarcolemma   (c) Z line   (d) sarcoplasmic reticulum.
8. The physiological characteristic which allows living tissue to respond to stimuli is   (a) extensibility   (b) irritability   (c) elasticity (d) contractility.
9. The shortest period in the "muscle twitch" is the   (a) latent period (b) contraction period   (c) relaxation period   (d) recovery period.
10. Most of the energy released during muscle activity is used for (a) movement   (b) recovery   (c) replacing ATP   (d) heat.

Group B

Write the word or words on your answer sheet that will correctly complete the sentence.

1. The most common single type of tissue found in the human body is _____ tissue.
2. In a sarcomere, the thick filaments are made of the protein _____. The thin filaments contain the protein _____.
3. H. E. Huxley and J. Hanson proposed the _____ theory to explain how muscle fibers shorten.
4. The immediate source of energy required for muscle contraction is supplied by _____.
5. Following strenuous exercise, an athlete breathes deeply for a period of time to pay back an _____.

THE SKELETAL MUSCLES

Objectives

A. Define the terms *origin* and *insertion*

B. List five types of muscle movement and state an example of each

C. Locate superficial skeletal muscles and relate their structure to the type of movement they produce

D. Explain the function of antagonistic muscle groups

E. Describe the function of the muscles within the major groups of superficial muscles

PRINCIPLES OF SKELETAL MUSCLES IN ACTION

In the previous chapter, several properties of muscle tissue were discussed. A special feature of muscle tissue is its ability to convert potential (chemical) energy into the kinetic energy of contraction. This fundamental property, along with the other properties of muscle tissue, is responsible for the movement of the skeleton. Muscle tissue with its connective tissue coverings, nerve supply, and means of distributing blood and lymph forms the organs of the *muscular system*. Smooth muscle tissue is found in organs of the digestive system, for example the stomach and intestines, as well as in the urinary bladder, an organ of the urinary system. Cardiac muscle tissue is located in the heart, an organ of the circulatory system. Organs containing smooth muscle and the heart will be discussed in other chapters.

For the present, we are concerned with a few of the 600 or more muscles that comprise the skeletal muscular system in the human body. This large number of skeletal muscles makes up approximately 40 percent of the body's weight. Resting muscles use only about 20 percent of the energy consumed by the body. An active muscle needs up to 300 times more energy for short periods.

Types of Attachment

Skeletal muscles produce movement by shortening or pulling on body parts. The contraction of these muscles exerts a force that can be used to perform mechanical work. It is important to remember that skeletal muscles can produce action only by **pulling**, not by pushing. Skeletal muscle may be directly attached to bone by the extension of its connective-tissue covering (*the epimysium*). This tissue appears to fuse directly to the periosteum that covers the bone. Another form of attachment is by means of cords of white fibrous connective tissue called **tendons** (Fig. 7-1). Muscles may be attached to each other or to bone by a broad sheet, or apronlike attachment, of dense connective tissue called an **aponeurosis.** The skeletal muscles cause movement at the places where bones articulate or form joints.

Origin and insertion. When a muscle contracts, it produces movement by changing the position of one bone in relation to another. In most instances, one of the bones remains stationary while the other bone moves. Each muscle, therefore, must have at least two attachments. The proximal attachment is on the immovable bone, called the **origin** of the muscle. The distal attachment, called the **insertion** of the muscle, is on the bone that is moved. When muscles contract, the insertion moves toward the origin. The portion of the muscle that is found between the origin and insertion is referred to as the *body* or *belly* of the muscle. The body of the muscle is usually found covering the immovable bone.

Figure 7-1 Diagram of the connective-tissue fibers (of a tendon) that attach skeletal muscle to bone (left).

Figure 7-2 The biceps and triceps muscles of the arm with their origins and insertions (right).

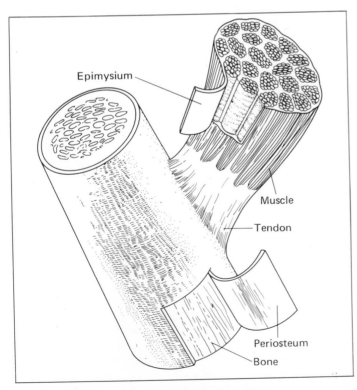

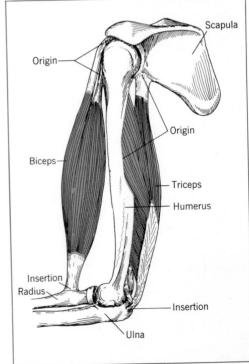

Action of Muscle Groups

The movement produced by the action of specific muscles depends upon the attachments of the origin and insertion to the bones that form the joint. For example, the ***biceps brachii,*** which lies on the front of the upper arm, acts to flex and rotate the forearm. In Fig. 7-2 the biceps and the ***triceps brachii*** (on the back of the upper arm) each have more than one origin. As their names indicate, the triceps has three points of origin and the biceps has two. The action of the biceps and the triceps works together to effect the movement at the elbow. This illustrates an important characteristic: ***Most skeletal muscles work in groups.***

The combined action of various muscles coordinate and control movements of body parts. Seldom do muscles work individually. When certain muscles contract, other muscles in the group relax. The muscles that are primarily responsible for the action are called ***prime movers*** (***or agonists***). The opposing muscles extend or stretch when the prime movers contract. For example, when the biceps contracts, flexing the forearm, another group of muscles extends. Opposite actions are

brach = arm

Figure 7-3 The various types of movements that muscles produce allow the body to perform numerous kinds of activities.

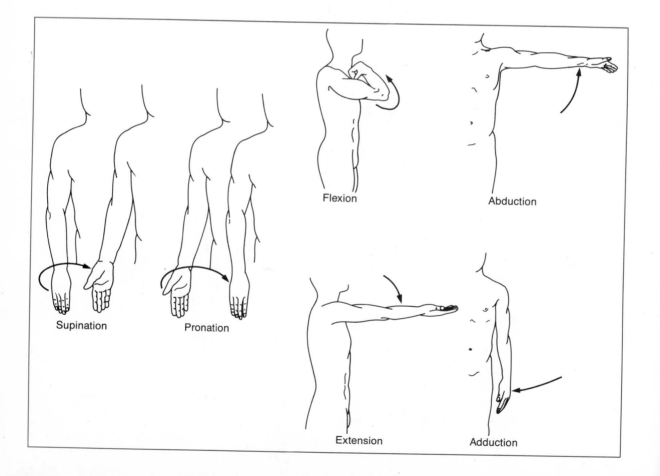

Supination

Pronation

Flexion

Abduction

Extension

Adduction

brought about by the muscles called **antagonists.** This interaction of muscles allows for the smooth and continuous action of movements.

Types of movement.   There are a number of different motions produced by muscle groups.

1. Movement produced by a muscle crossing in front of a joint, bringing the body part forward, is **flexion.** The opposite movement, **extension,** moves the body part to the rear (see Fig. 7-3).
2. Movement of an appendage laterally from the midline (midsagittal plane) is called **abduction;** moving the arm or leg toward the midline is called **adduction.**
3. Movement of an appendage, held completely straight, in a circular motion around the joint is called **circumduction.**
4. Rotating the hand, with the palm downward, is **pronation;** the same motion, with the palm upwards, is **supination.**
5. The turning of the toes of the foot inward is called **inversion;** the opposite, the turning of the foot outward, is **eversion.**

ex = out, away from

circum = around

Naming skeletal muscles.   Many muscles in the human body have English names; others have Latin names. However, the name of a muscle can give a clue to understanding its function, shape, size, or other characteristics. The following can be identified by the names of muscles:

- action: flexors; extensors; abductors; adductors
- direction of fibers: rectus, or parallel to the midline (straight); transverse, or perpendicular to the midline (across); oblique, or diagonal to the midline (inclined)
- location: tibialis anterior or posterior (in front of or behind the lower leg); intercostal (between the ribs); temporalis (near the temporal bone of the skull)
- shape or size: trapezius (trapezoid); deltoid (triangular); minimus (small); maximus (large); longus (long)
- number of attachments: biceps (two origins); triceps (three origins); quadriceps (four origins)
- points of attachment: sacrospinalis (sacrum and spinal vertebrae); sternocleidomastoid (the sternum, the clavicle, and the mastoid process of the temporal bone)

ANATOMY OF THE SUPERFICIAL MUSCLES

Many of the skeletal muscles pass unnoticed because they lie below the surface of the body. These are known as *deep muscles.* The major skeletal muscles are found near the surface of the body and are called **superficial muscles.** These muscles are responsible for the body's contour and are essential in performing physical movements. The focus of the discussion will be on the superficial muscles whose action we can readily observe in carrying out daily activities (Fig. 7-4).

super = above

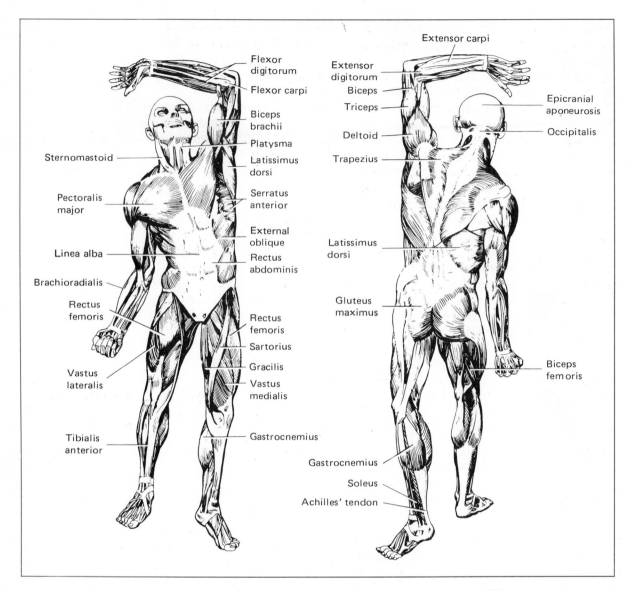

Figure 7-4   Distribution and relation to each other of the superficial muscles of the human body.

Muscles of the Axial Skeleton

The muscles of the face and neck enable you to frown, smile, raise your eyebrows, squint, and perform a variety of conscious and unconscious facial movements. In effect, the facial muscles are sometimes referred to as the "muscles of expression."

Muscles of facial expressions.   The facial muscles differ from other skeletal muscles because of their method of formation and attachment. The ***epicranius*** (or ***occipitofrontalis***) may be considered to be two muscles. The ***occipitalis*** covers the occipital bone of the skull, and the ***frontalis***

covers the front of the cranium. The two muscles are connected by a thin, strong *aponeurosis.* This fibrous tissue extends over the top of the cranium and is appropriately named the **epicranial aponeurosis.** The muscles of the epicranial, which draw the scalp backward and raise the eyebrows, are used when expressing surprise or horror.

The **orbicularis oculi** is the muscle surrounding the eyes. Its fibers have a circular arrangement, forming the muscle layer around the orbit of the eye and the eyelid. A slight contraction of this muscle keeps the eyelid closed as in sleeping. A strong contraction makes the eyelids close tightly, as in squinting in bright sunlight. Similarly, the **orbicularis oris** contains numerous muscle fibers that encircle the mouth. This muscle allows the lips to protrude as in pouting.

A powerful muscle on the side of the face, the **masseter,** is responsible for many of the actions of the jaws. When you clench your teeth, this muscle is involved and can be felt bulging at the angle of the lower jaw. The masseter functions with the **pterygoideus medialis** and the **pterygoideus lateralis** (the pterygoid muscles are not visible from the surface of the body). These two muscles assist in opening and closing the jaws. They also hold the lower jaw in place and permit a rotary motion, which is used in chewing.

Muscles of the neck.   The **platysma** is a muscle of the neck that originates in the covering to the **pectoralis major** and **deltoid** (shoulder) muscles. The platysma is a broad sheet of muscle that extends upward toward the lower jaw. Its action is to draw the outer edges of the lower lip downward, especially when the head is thrown backward.

Figure 7-5 **The superficial and deep muscles of the face and neck (lateral view).**

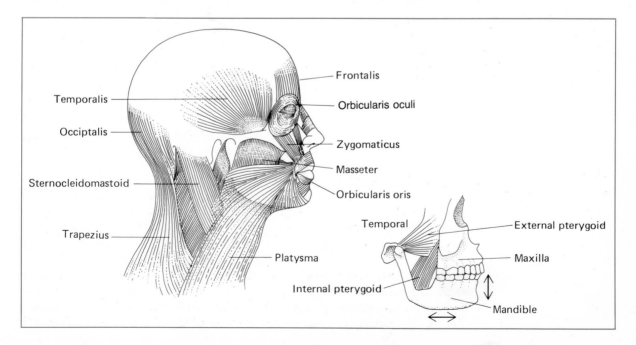

Table 7-1 Muscles of the Face

Muscle	Origin	Insertion	Function
Epicranius (Occipitofrontalis)			
Occipitalis	Occipital bone and mastoid process	Epicranial aponeurosis	Draws the scalp backward
Frontalis	Epicranial aponeurosis	Skin directly above eyebrows	Raises eyebrows and wrinkles forehead
Orbicularis oculi	Nasal portion of frontal bone and frontal process of maxilla	Skin around eye	Closes eyelids
Orbicularis oris	Muscle fibers around mouth	Skin at angle of mouth	Closes lips and aids in speech
Masseter	Maxilla and zygomatic arch	Angle and lateral surface of mandible	Raises lower jaw and clenches teeth when mouth is closed
Pterygoideus:			
Medialis	Maxilla	Ramus of mandible	Closes jaws
Lateralis	Sphenoid	Neck of mandible	Opens jaws and assists in side-to-side movements of jaw

The most conspicuous muscle on the front and side of the neck is the *sternocleidomastoideus*, usually called the **sternomastoid.** The long name given to the muscle simply indicates that it has two points of origin and that its insertion is on the mastoid process behind the ear. It arises from the inner border of the clavicle and the upper end of the sternum. This strong, prominent band of muscle passes obliquely over the side of the neck to a point behind the ear. There are two such muscles, one on each side of the head. When both muscles contract, the head is flexed onto the chest. If one muscle acts, it draws the head toward the shoulder and rotates the chin upward.

semi = half

On the posterior side of the neck are a number of muscles that extend the head. One of these is the **semispinalis capitis.** This muscle arises from the last four cervical and upper five thoracic vertebrae and inserts onto the occipital bone. When both right and left muscles contract, the head is then extended. If one muscle acts alone, the head is rotated toward the contracting muscle's side.

Table 7-2 Muscles of the Neck

Muscle	Origin	Insertion	Function
Platysma	The fascia of the pectoral and deltoid muscles	Mandible and skin around mouth	Extends lower lip and wrinkles skin of neck
Sternocleidomastoideus	Medial third of clavicle and top of sternum	Mastoid process of temporal bone	Draws head toward shoulder on contracting muscle side; when both act, head is flexed onto chest
Semispinalis capitis	Upper five or six thoracic and lower four cervical vertebrae	Occipital bone	When both muscles act, head is extended; when one is acting alone, head is rotated toward side of contracting muscle

Muscles of the abdominal wall. The muscles comprising the front of the abdominal wall serve not only to support the internal organs but also to flex the trunk. Of the superficial muscles involved in these activities, two are of special interest. The ***rectus abdominis*** is a straplike muscle that originates on the pubis and extends up the front of the abdomen. It inserts onto the costal cartilages of the fifth, sixth, and seventh ribs. The ***linea alba,*** or *midline* of the abdomen, is a tough fibrous band formed by the connective tissues of the abdominal muscles. The rectus muscle contracts along with the other abdominal muscles to compress the internal organs, thereby aiding defecation and urination. The rectus muscle also flexes the vertebral column by drawing the thorax downward. It also raises the pelvis, permitting such activities as climbing. Acting as antagonists to the back muscles, the rectus muscle is important in maintaining good posture. If one side of this separated muscle acts alone, it bends and rotates the trunk of the body to that side.

The ***external oblique*** muscles are a pair of muscles in the abdominal region. These muscles help to compress the abdomen and bend and rotate the vertebral column. Each originates on the border of the eight lower ribs. The fibers of each muscle insert on the crest of the ilium. The combined action of the pair of muscles flex the vertebral column by drawing the pelvis toward the sternum. When only one of the pair contracts, it bends and rotates the vertebral column to the same side as the contracting muscle, also bringing the shoulder on the same side forward. (See Fig.7-6.)

Table 7-3   Abdominal Muscles

Muscle	Origin	Insertion	Function
Rectus abdominis	Pubis	Cartilages of fifth, sixth, and seventh ribs	Flexes vertebral column and compresses abdominal organs
External oblique	Lower eight ribs	Anterior half of iliac crest	Compresses abdominal organs and rotates and flexes vertebral column

Muscles of the Upper Extremity

The front upper half of the chest is covered by the **pectoralis major** muscles. This large, fan-shaped pair of muscles is primarily concerned with the downward and forward motion of the arms. The origins of each are the side of the sternum, the medial half of the clavicle, and the cartilages of the true ribs. From these points, the muscle fibers converge to form a strong tendon that inserts just below the head of the humerus. When the muscle contracts, the arm and shoulder is brought forward and the arm is adducted and rotated toward the midline of the body. The most highly developed pectoral muscles can be found in birds, for which they serve in the action of flight. These muscles are called the *breast* of the fowl. For a shot-putter, the pectoralis major provides the thrust to give the initial power to the metal ball.

Muscles of the shoulder and upper arm.   The **trapezius** is another muscle that also greatly affects the action of the shoulder and the upper arm. The trapezius is located on the back of the trunk and extends from the base of the neck toward the top of the shoulder. The contraction of this broad sheet of muscle raises the shoulder and draws the scapula backward and downward. This action results in giving added leverage to the shoulder for the powerful forward thrust required in shot-putting.

The opposite (antagonistic) action on the scapula is brought about by the contraction of the **serratus anterior.** This muscle has its origins on the first nine ribs, and its insertion is on the scapula. The action of the serratus anterior is most obvious when a person is moving a heavy object. The muscle appears as a series of ridges because of the way its origins attach to the ribs. Its fibers pass obliquely from the ribs toward the shoulder and draw the scapula forward.

Two other superficial muscles of the upper extremity are the **deltoid** and the **latissimus dorsi** muscles. The deltoid forms a thick triangular pad of muscle that covers the shoulder joint. The contraction of the major part of the muscle abducts the upper arm from the side of the body, while some of its fibers draw the shoulder backward and outward.

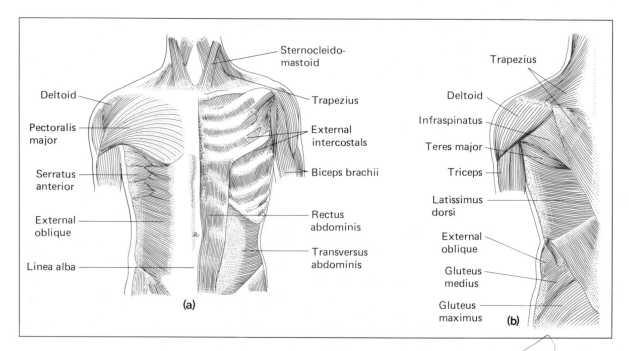

Figure 7-6   (a) Anterior view and (b) posterior view of the muscles of the neck, abdominal area, and arm. Some of the superficial muscles have been removed to show the deep muscles.

This complex action is possible because its origin is partly on the clavicle and partly on the spine of the scapula.

The principal antagonistic muscle of the deltoid is the latissimus dorsi. As its name implies, it is the broadest (*latissimus*) muscle of the back (*dorsi*). Its origin extends from the sixth thoracic vertebra and passes downward to the upper edge of the iliac crest and sacral vertebrae. This broad, flat muscle forms a narrow tendon that inserts along the back part of the upper portion of the humerus. The deltoid muscle raises the arm and draws it forward, whereas the latissimus dorsi pulls the arm downward (adducts) and rotates it backward. Since the raising and lowering of the arm must be accomplished with considerable force at times, both of these muscles can become highly developed. This muscle is primarily used in lifting the body, as in climbing.

Muscles of the forearm.   The muscles that move the forearm produce action at the elbow and at the wrist joint. They also control flexion and extension of the fingers. The muscular action of the biceps and the triceps of the arm have previously been discussed in this chapter.

The muscles of the forearm are responsible for the movements of the wrist and fingers, and some muscles aid in flexing the arm at the elbow. One of the more obvious muscles of this region is the ***brachioradialis.*** This band of muscle runs along the upper part of the forearm on the radial (thumb) side. Its origin is on the lower end of the humerus and extends to the lower end of the radius. Its action is to work with the biceps in flexing the arm.

Table 7-4   Muscles of the Upper Extremity

Muscle	Origin	Insertion	Function
Pectoralis major	Medial half of the clavicle, lateral edge of sternum, and cartilage of upper six ribs	Greater tubercle of humerus	Adducts and rotates upper arm forward
Trapezius	Occipital bone and spines of cervical and all thoracic vertebrae	Lateral end of clavicle and scapula's acromion process	Raises shoulder and draws scapula backward
Serratus anterior	Lateral surface of eight or nine ribs	Lower medial border of scapula	Moves scapula forward and downward
Deltoid	Lateral third of clavicle and spine of scapula	Lateral surface of body of humerus	Abducts arm
Latissimus dorsi	Spines of lower six thoracic, lumbar, and sacral vertebrae plus crest of ilium	Upper anterior surface of humerus	Adducts arm and draws shoulder down and backward

On the front and back surfaces of the forearm are several muscles whose actions are responsible for the motion of the fingers. The tendons from these muscles extend over the wrist and bring about movement at the individual joints of the fingers. These groups of superficial muscles are the *flexor* and *extensor carpi,* some of which have their origins on the lower end of the humerus. The tendons are held down at the wrist by bands of ligaments and prevent an arch from forming between the forearm and the fingers.

Muscles of the Lower Extremity

The muscle groups of the lower limbs present a complex pattern of the actions that result in various movements of the leg. The large muscle at the top of the thigh and covering the hip joint in back is the *gluteus maximus.* The major action of this muscle and two smaller muscles of the same region (the gluteus medius and gluteus minimus) is to extend the thigh and laterally rotate it.

Muscles of the thigh.   The muscles of the thigh itself are responsible for many of the more forceful movements required of the body as a whole.

Table 7-5   Muscles of the Arm and Forearm

Muscle	Origin	Insertion	Function
Biceps brachii	Two heads, one at lateral edge of scapula at the glenoid cavity and the second at the caracoid process	At proximal end of radius	Flexes the forearm and rotates it outward (supinates)
Triceps brachii	One of three, below the glenoid cavity of scapula and two on the posterior side of humerus	At the olecranon process of ulna	Extends the forearm
Brachioradialis	The lateral distal third of the humerus	Lower lateral edge of radius	Flexes the forearm
Flexor carpi muscles	Lower humerus	Various metacarpals	Flexes the hand
Extensor carpi muscles	Lower humerus	Various metacarpals	Extends the hand

On the back of the thigh is a large muscle called the **biceps femoris.** This muscle is one of a group of three posterior thigh muscles, known as the *hamstrings.* The biceps of the leg has two points of origin; one arises on the ischium and the other on the femur. Its principal point of insertion is on the fibula, with a small portion passing to the tibia. Its connection with the fibula can be felt as a heavy tendon on the back of the knee toward the lateral side of the leg. The action of this muscle is twofold: It flexes the leg at the knee and pulls it slightly to the side, and it extends the thigh backward and rotates it.

On the front of the thigh is a very large and powerful muscle, the **quadriceps femoris.** As the name implies, there are four heads of origin, which form four separate muscles. These muscles are the **rectus femoris,** the **vastus lateralis,** the **vastus medialis,** and the **vastus intermedius** (which lies hidden and thus is not a superficial muscle). The three vastus muscles originate on the femur, while the rectus femoris originates on the anterior iliac crest. All four have a common tendon that passes across the patella and inserts on the upper anterior surface of the tibia. These strong muscles extend the lower leg with great force, as needed in kicking a ball or in swimming. The rectus femoris, since it originates above the hip joint, helps to flex the thigh on the pelvis. These muscles help a person stand from a squatting position.

Muscles of the lower leg. In the lower leg are two muscles that form the calf of the leg. The more prominent muscle is the ***gastrocnemius.*** It has its origins on the posterior base of the femur, and its insertion, through the ***tendon of Achilles,*** is on the bone of the heel (*calcaneus*). Closely associated with the gastrocnemius is the ***soleus.*** This muscle originates on the upper posterior surfaces of the tibia and fibula and has a common insertion with that of the gastrocnemius. The action of both of these muscles is to point the foot downward as when standing on your toes. The general form and contraction of these muscles can be noted with each step as the heel is raised from the floor when walking.

The front of the leg is characterized by a complex group of muscles. Some of them are quite obvious in their action, while others are seen as rippling motions under the skin. There is one muscle of special interest because of its ability to flex the foot strongly at the ankle. This muscle is called the ***tibialis anterior.*** It lies just lateral to the tibia, which is its point of origin. The tendon passes over the tibia and inserts on the first metatarsal bone of the foot. When it contracts, the foot flexes up and inverts, allowing the weight of the body to fall on the heel (Fig. 7-7).

Table 7-6   Muscles of the Lower Extremity

Muscle	Origin	Insertion	Function
Gluteus maximus	Crest of ilium, sacrum and coccyx vertebrae	Upper lateral fascia of the femur	Extends femur and rotates it outward
Biceps femoris	Ischium and upper femur	Head of fibula and lateral edge of tibia	Flexes lower leg and extends thigh
Quadriceps femoris	Three vastus heads on femur and rectus on anterior iliac crest	Upper anterior surface of tibia	Extends lower leg (the rectus flexes the thigh)
Gastrocnemius	Lower posterior end of femur	Calcaneus through the tendon of Achilles	Flexes lower leg and extends foot
Soleus	Upper third of fibula and tibia on posterior	Tendon of calcaneus onto heel	Extends foot
Tibialis anterior	Upper lateral half of tibia	First metatarsal	Flexes and inverts foot

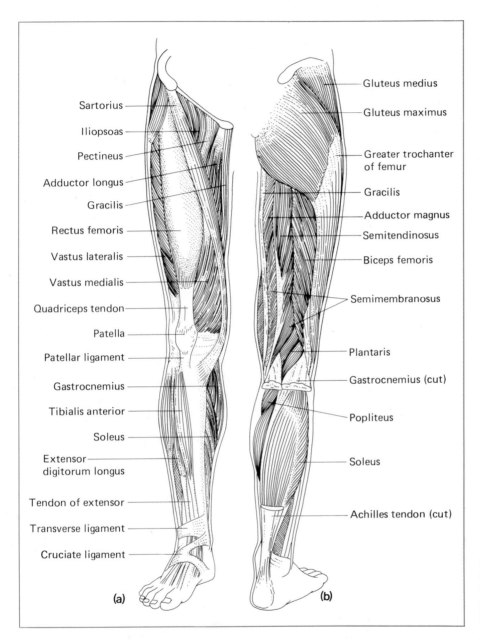

Sartorius

Iliopsoas

Pectineus

Adductor longus

Gracilis

Rectus femoris

Vastus lateralis

Vastus medialis

Quadriceps tendon

Patella

Patellar ligament

Gastrocnemius

Tibialis anterior

Soleus

Extensor digitorum longus

Tendon of extensor

Transverse ligament

Cruciate ligament

Gluteus medius

Gluteus maximus

Greater trochanter of femur

Gracilis

Adductor magnus

Semitendinosus

Biceps femoris

Semimembranosus

Plantaris

Gastrocnemius (cut)

Popliteus

Soleus

Achilles tendon (cut)

(a)   (b)

Figure 7-7 (a) Anterior view and (b) posterior view of the different muscles of the lower extremity.

DISORDERS OF THE MUSCULAR SYSTEM

Muscles are seldom involved with infectious disease. However, since they make up a large portion of the body and are often exposed, they can be more prone to injuries from the external environment. Athletic activities as well as exercise may be the cause of muscular problems.

Pulled Tendons

A sudden, violent contraction of a muscle can cause the fibers of the muscle to tear. More frequently, the musculotendinous connection may partially or completely separate from the bone attachment. This pulled tendon is accompanied by sudden sharp pain and loss of muscular function. Surgical repair may be necessary for this type of injury.

Contusions of Muscles

If a person should fall or receive a severe blow, blood vessels may be broken within a muscle, causing a muscle **contusion** or **muscle bruise.** The injured area becomes discolored and painful. In addition, the muscle is tender when it contracts. In time, the swelling goes down and the discoloration disappears. There is relatively little that can be done externally other than to rest the affected area.

Hernia or Rupture

A *hernia* is usually related to an organ protruding through the wall of the cavity in which it is housed. However, a hernia or rupture can occur within individual muscles. The most typical forms occur when there is a weakness in the abdominal wall or where there are openings through which blood vessels and nerves pass into the legs.

Hernia conditions.   An *umbilical hernia,* common in children, occurs when a portion of the small intestine protrudes through the area of the navel when the abdominal wall is weak. When an infant cries, additional force is exerted in this area and a swelling may be noted under the skin. This condition may correct itself as the child grows older.

An *inguinal hernia* occurs in males. During fetal development, the testes descend through a canal (the inguinal canal) into the scrotum, where they are housed. While this canal normally closes after a child is born, it can remain a weakened area where a portion of the intestine may push through. A hernia of this type is generally corrected by a surgical procedure that draws the muscular wall together.

A *femoral hernia* forms behind the inguinal region, where there is an opening through which the femoral artery, vein, and nerve pass from the abdominal cavity into the leg. A portion of the intestine may be forced through this opening. A femoral hernia occurs more frequently in women than in men. This condition can be caused by a congenital weakness of the abdominal muscles in the area of the groin. A sudden or prolonged increase in the pressure within the abdominal cavity, as occurs in coughing, heavy lifting, or prolonged strain, can also be considered an underlying cause of this kind of hernia.

Summary

Movement occurs when muscles exert their contracting force on the bones of the articulate skeleton. Attachment to a bone can be directly to the periosteum, by a tendon, or by an aponeurosis. The portion of the muscle that acts as an anchor is called the origin. The insertion of the muscle is attached to the movable bone. Muscles work in pairs or groups by opposing or antagonistic actions. Muscles found on the surface of the body are called superficial muscles. A few of the disorders that occur in muscles include pulled tendons, muscular contusions, and various types of hernia conditions.

The types of opposing motion include abduction and adduction, flexion and extension, and pronation and supination. Some of the muscles mentioned in the text are the muscles of expression, muscles of the neck, muscles of the upper extremity, muscles of the arm, muscles of the abdomen, and muscles of the lower extremities.

Vocabulary Review

Match the statement in the left column with the words in the right column. Place the letter of your choice on your answer paper. *Do not write in this book.*

1. The process involved in bending the arm at the elbow.
2. The end of the muscle that is attached to a relatively immovable part of the body.
3. A muscle that draws the scalp backward and raises the eyebrows.
4. The kind of action produced when one muscle of a pair relaxes while the other contracts.
5. The type of attachment made by the end of the biceps brachii that is connected to the radius.
6. The opposing action of pairs of closely associated muscles.
7. Movement of an arm or leg away from the midline of the body.
8. An apronlike attachment of muscle.
9. Muscles that give the body its external contour.
10. The condition that results when a portion of the intestine pushes through the abdominal wall.

a. abduction
b. antagonism
c. aponeurosis
d. extension
e. flexion
f. hernia
g. insertion
h. origin
i. muscular coordination
j. epicranial muscles
k. superficial muscles
l. tendon

Test Your Knowledge

GROUP A

Write the word or words on your answer sheet that will correctly complete the sentence.

1. The sternomastoid muscle   (a) moves the sternum   (b) is concerned with chewing   (c) is easily seen when the head is bent to the side   (d) moves the shoulder.
2. Movement produced by a muscle crossing in front of a joint, bringing the body part forward, is called   (a) flexion   (b) pronation   (c) eversion   (d) adduction.
3. Chewing motions result from the action of the   (a) soleus   (b) masseter   (c) orbicularis oris   (d) orbicularis oculi.
4. Muscles that clench the hand are   (a) in the palm of the hand   (b) on the back of the upper arm   (c) on the front of the forearm   (d) in the wrist.
5. The contracting muscle of a pair   (a) is the prime mover   (b) lacks insertions   (c) lacks tendons   (d) is only found in the arms and legs.
6. A point of attachment of a muscle is the   (a) belly   (b) termination   (c) origin   (d) bursa.
7. The muscle around the eye is the   (a) orbicularis oris (b) masseter   (c) orbicularis oculi   (d) pterygoideus.
8. A muscle that is *not* associated with arm movement is the   (a) pectoralis major   (b) deltoid   (c) trapezius   (d) rectus femoris.
9. A pair of antagonistic muscles is the   (a) masseter and platysma   (b) biceps and triceps   (c) gluteus maximus and gastrocnemius   (d) vastus lateralis and soleus.
10. Choose the one that is *not* a muscle: (a) flexor carpi   (b) external oblique   (c) linea alba   (d) latissimus dorsi.

GROUP B

Write the word or words on your answer sheet that will correctly complete the sentence.

1. The _____ is the place where a skeletal muscle is attached to an immovable bone.
2. The muscles of the shoulder are known as _____ muscles.
3. A muscle must be able to exert its force on a bone at a joint by _____.
4. The part of the muscle found between its attachments is called the _____.
5. When you swing your arm around in a circle, this movement is called _____.

UNIT 3

CHAPTER 8

NERVOUS TISSUE

Objectives

A. Describe the general structures of neurons and neuroglial cells
B. Explain how structure and function are used to classify different kinds of neurons
C. Explain the transmission of a nerve impulse over a neuron
D. Explain how an impulse is transmitted across a synapse

ORGANIZATION OF NERVOUS TISSUE

The ability of the human body to respond quickly and coordinate its activities to changes in both the internal and external environment requires a highly organized *communication system*. There must be a means of relaying this information to centers of recognition that will respond to the stimuli, such as muscles and glands. This function is the responsibility of the *nervous system*.

Although the nervous system is rather complex, it is composed of only two principal kinds of cells. One is a **glial**, or **neuroglial**, **cell**. A glial cell is a nonconducting cell considered to be essential for the protection, structural support, and metabolism of the nerve cell. The second kind of cell is called a **neuron**, which represents the **structural and functional unit of all nervous tissue**. All the functions that are characteristic of the nervous system, such as thinking, activating muscles, and regulating glands, are dependent upon these neurons.

Neuroglial Cells

There are probably 10 times more neuroglial cells in the nervous system than there are neurons. Although an understanding of their functions requires further investigation, much has been learned about these special cells. (See Fig. 8-6)

In the **central nervous system (CNS)**, consisting of the brain and spinal cord, there are a number of different types of neuroglia. **Oligodendroglia** are cells that appear to form a fattylike insulative sheath

glia = glue

neuro = nerve

around the fibers of CNS neurons. Another type, called **microglia,** protect neurons by engulfing invading microorganisms that might cause disease. **Astrocytes** are involved in the active transport of substances from the blood to the brain. They provide a barrier to protect neurons from harmful substances found in the bloodstream. (See Fig. 8-1).

The fibers of most neurons outside the brain and spinal cord are covered by **Schwann cells.** In some instances, Schwann cells enclose the fiber with only a thin layer of fattylike material. More frequently, Schwann cells continue to grow, entwining the neuron fiber with a dense insulative covering. This protective sheath appears as segments along the fiber. Schwann cells are found in the **peripheral nervous system (PNS)** (that portion found outside the CNS). There they play many of the same roles neuroglia play in the central nervous system.

Structure of the Neuron

Nerve cells, or neurons, are structures that vary in size, shape, and manner of branching, yet they have certain basic similarities. Each neuron consists of a **cell body** and two types of cytoplasmic extensions: **dendrites** and **axons.**

Cell body.   The cell body, or **perikaryon,** consists of **neuroplasm** (the cytoplasm of a neuron), which surrounds a well-defined nucleus. Within the nucleus is a single prominent **nucleolus.** The shape of the cell body can vary for different types of neurons. A cell body can be round, as in various sensory neurons, or can have a diamond or pyramidal shape, as do those on the surface of the brain. Still others may be star-shaped, such as those of various motor neurons, which activate muscles. (See Fig. 8-2, which shows neurons of various types and shapes.)

Located within the neuroplasm are the organelles typical of all living human cells. (See Chapter 3.) In addition, there are tiny structures (unique to neurons) called **Nissl bodies.** These granular pieces of endoplasmic reticulum with numerous ribosomes are the site of protein synthesis. Newly formed proteins pass from the cell body into the axon to replace those proteins utilized during metabolic functions. This cytoplasmic activity is called **axoplasmic transport. Neurofibrils** are also found scattered throughout the perikaryon. They appear to congregate at the **axon hillock,** a thick-layered area where the axon meets the cell body. These long, thin filaments, composed of microtubules, appear to facilitate axoplasmic transport and provide support.

Processes.   The cytoplasmic extension of the cell body is called a **nerve fiber** or **process.** The types of processes a neuron has determine in which direction the neuron conducts impulses. A dendrite is typically short with many branches. It receives and conducts impulses *toward* the cell body, and thus is referred to as an **afferent process.** An axon is a single fiber that carries impulses *away from* the cell body and is thus known as an *efferent process.* The axon may have one or more branches, called

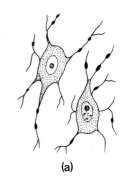

(a)

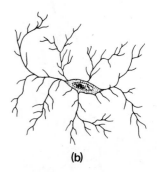

(b)

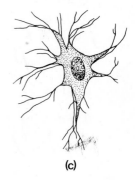

(c)

Figure 8-1 **The kinds of neuroglia found in the nervous system. (a) Oligodendroglia. (b) Microglia. (c) Astrocytes.**

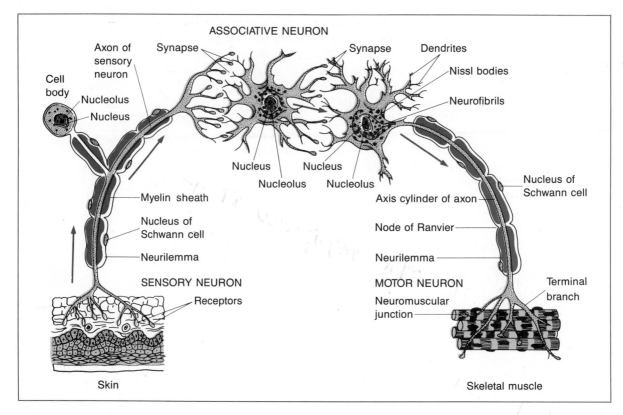

Figure 8-2 Diagram of the pathway of nerve impulses from receptors in the skin, to an associative neuron, to nerve endings in muscle cells.

collaterals, that extend laterally from the main axis. The axon is a long process that arises from the cell body and varies in length depending upon the location of the neuron. Axons and their collaterals terminate into fine branches called *terminal filaments* or *telodendria.* On the ends of these branches are small swellings call *synaptic knobs* or *end bulbs.* Within these knobs are vesicles that contain a transmitter substance. This substance is released when nerve impulses are relayed from one neuron to the next.

The central core of the nerve fiber is called the *axis cylinder.* In many PNS neurons this cylinder is surrounded by a cellular *myelin sheath.* This sheath is formed by the Schwann cells and serves as a protective covering around the axis cylinder. The plasma membrane of a Schwann cell wraps tightly around the axis cylinder, giving it a laminated appearance.

The sheath is sectioned off between Schwann cells, and the nerve fiber appears to have indentations called the *nodes of Ranvier* (Fig. 8-3). The myelin sheath is chiefly composed of a lipid substance, which gives the process a pearly white appearance. Myelinated fibers conduct impulses more rapidly than those without myelin (unmyelinated). In nerve activity on myelinated fibers, impulses jump from one node to another, resulting in an increased rate of conduction. This type of impulse transmission is known as *saltatory conduction* (discussed on page

139). Most of the neurons inside the brain and on the outside portion of the spinal cord are covered with a myelin sheath. These two regions along with PNS neurons make up the **white matter** of the nervous system. Bundles of neurons lacking myelin sheaths have a dull appearance and are darker than myelinated fibers. Cell bodies of neurons and unmyelinated fibers are found covering the surface of the brain and in the central part of the spinal cord. The nonmyelinated areas form the **gray matter** of the central nervous system.

On all PNS neurons the myelin sheath is surrounded by a delicate outer covering, the **neurilemma.** This thin membrane is formed by the Schwann cells and surrounds the myelin. The neurilemma comes in contact with the axis cylinder only at the nodes of Ranvier. The neurilemma is important in the repair of damaged nerve processes. If a PNS neuron is cut, the section of the axis cylinder farthest away from the cell body *degenerates* (Fig. 8-5). The Schwann cells remain, and in time the processes can regenerate themselves by the same pattern of growth. A new axis cylinder forms, guided in its growth by the tunnel-like neurilemma. Within the brain and spinal cord, most neurons lack a neurilemma and thus are unable to regenerate themselves (repair themselves) if injured. Since the cell body cannot reproduce itself (undergo mitosis), any damage to the neuron is always permanent.

Classification of Neurons

The number of processes that extend from the cell body provide a structural basis for classifying neurons.

- A neuron may be unipolar, consisting of a single process that extends from the cell body. This fiber may split, with one portion acting as

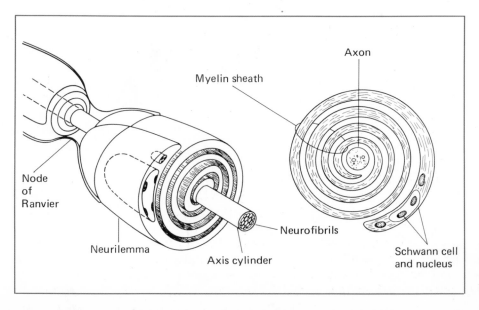

Figure 8-3 Myelinated neuron process showing wrapping of one Schwann cell.

Myelin sheath

Axon

Node of Ranvier

Neurofibrils

Neurilemma

Axis cylinder

Schwann cell and nucleus

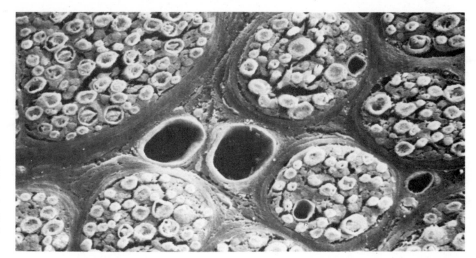

Figure 8-4 **Cross section through a bundle of nerve cell fibers composing a nerve. The insulation surrounding the fibers can be clearly seen.**

a dendrite and the other branch acting as an axon. This type of neuron is associated with the sense organs.

- A *bipolar* neuron has only one dendrite and one axon extending from the cell body. These neurons are found in the retina of the eye, in the inner ear, and in the receptors that respond to smell.
- Most frequently, neurons are *multipolar.* This kind of neuron has dendrites that are like branches of a tree and a single axon, all attached to the cell body. These neurons are found within the brain and spinal cord.

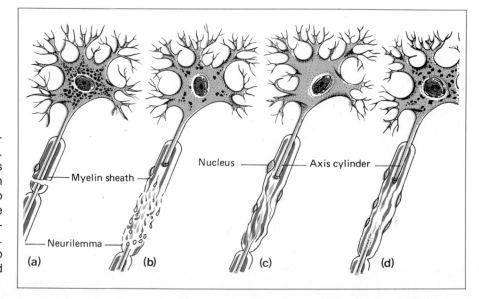

Figure 8-5   Regeneration of a nerve fiber. (a) The nerve fiber is cut. (b) The myelin sheath breaks up into droplets of fat. (c) The myelin sheath and neurilemma begin to form. (d) Neurofibrils begin to sprout from the cut end of the original fiber.

Neurons may also be classified functionally according to direction and destination of the impulse.

- **Sensory** neurons transmit impulses from the ends of special sensory neurons called **receptors** toward the central nervous system. Receptors located in such sense organs as the eye or ear receive stimuli from the external environment. These are known as **exteroceptors.** It is also important for the brain to know when muscles are contracting and the amount of tension they are producing. Such information is transmitted to the central nervous system by receptors located in muscles, tendons, and joints. These receptors are called **proprioceptors.** A third type of receptor, called an **interoceptor,** receives stimuli concerning internal activities such as digestion, circulation, excretion, hunger, thirst, and feelings of sickness or well-being. All of these receptors send impulses toward the brain or spinal cord.
- **Associative neurons,** also known as **connector** or **internuncial** neurons, are found exclusively within the central nervous system. They are activated by impulses arriving by way of the sensory neurons or by other associative neurons within the brain or spinal cord. These neurons provide the link between sensory input and muscular or glandular activity.
- **Motor** neurons transmit impulses from associative neurons within the brain or spinal cord to **effectors,** muscles or glands, which bring about a coordinated response. (See Fig. 8-2 for a study of the three types of neurons.)

PHYSIOLOGY OF THE NEURON

You may have been confronted with an emergency situation, such as a car suddenly pulling away from the curb directly in front of you. The response to move away quickly from the vehicle occurs almost automatically. However, there is a lapse of time between recognizing the danger and the muscular reaction. This time period is referred to as the **reaction time.** How does the brain perceive this sudden problem and cause the muscles to react?

A number of steps are involved in such a response. First, **excitability** (also known as *irritability*) enables neurons to respond to a sudden change in the environment, called a **stimulus.** Thus the visual receptors in the eyes are stimulated by the movement of the car. **Conductivity** enables neurons to encode the stimulus in the form of impulses. Hence, in the second step, the stimulus of the moving car is encoded as impulses that are conducted to the brain. Different parts of the central nervous system sort out the incoming impulses and interpret their meaning in order to provide the proper actions. This process is called **integration.** In the third step, then, the car is perceived as heading straight for you. Finally, instructions (in the form of impulses) from the

Figure 8-6 A −70-millivolt resting potential is established inside the cell membrane when Na⁺ is pumped out and K⁺ is pumped in (black arrows). The membrane is relatively impermeable to Na⁺ but permeable to K⁺ (broken arrows). Most of the negative ions remain inside the membrane.

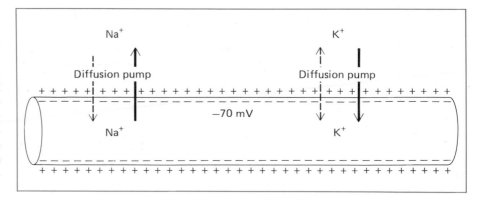

Figure 8-6 A −70-millivolt resting potential is established inside the cell membrane when Na^+ is pumped out and K^+ is pumped in (black arrows). The membrane is relatively impermeable to Na^+ but permeable to K^+ (broken arrows). Most of the negative ions remain inside the membrane.

brain are carried by various motor neurons to selected muscles for the appropriate response. Thus you move out of the way. This is one example of how nervous tissue exhibits the physiological properties of excitability, conductivity, and integration.

Transmission of the Nerve Impulse

The passage of an impulse over a neuron was once thought to resemble the flow of an electric current through metal conductors. However, two very important differences are observed between conduction of a nerve impulse and a current: (1) the speed of an impulse (0.5 to 120 meters per second) is very slow compared to the rate of speed of an electric current (3×10^5 kilometers per second); and (2) once the neuron is stimulated, there is an electrochemical change across its membrane surface that is self-propagating.

The neuron itself supplies energy for the transmission of the impulse. On the other hand, the wire is a passive conductor that is dependent on an outside energy source for the continuous movement of electrons. The electrochemical change that occurs across the membrane surface of the neuron is dependent upon an unequal distribution of ions between the outside and the inside of the plasma membrane. For example, a resting nerve fiber (one not conducting an impulse) can be described as containing an excess of negative charges on the inside and an excess of positive charges on the outside of the membrane. The unequal distribution of charges creates an *electrical gradient* (difference in concentration of charge) across the membrane. In effect, there is an **electric potential difference** between the outside and the inside of the plasma membrane. When electric charges are separated in this way, they have a potential for doing work (expressed in millivolts). The ions that create this potential difference are sodium (Na^+) on the outside and potassium (K^+) on the inside of the nerve membrane. A specific number of negative organic ions on the inside are also involved. The **resting membrane potential** measures about −70 mV (1 millivolt equals 1/1,000 of a volt), and the membrane is said to be **polarized.**

The plasma membrane is selectively permeable. That is, at rest few sodium ions (Na$^+$) diffuse into the cell. Those that do enter are forced out again by the active-transport mechanism called the **sodium pump** (see Fig. 8-6). Potassium ions (K$^+$) pass through the neuron's membrane more easily, but a potassium-pump mechanism keeps the potassium ions inside. The negatively charged organic ions are relatively nondiffusible because of their large size and also remain inside the cell. The sodium-potassium pump continually maintains this unequal distribution of ions between the sides of the membrane, resulting in a **resting potential.**

Action potential.   When an adequate stimulus, one of at least threshold level, is received by a polarized neuron, the plasma membrane becomes permeable to sodium ions (Na$^+$), allowing them to diffuse into the cell. This accounts for the drop in the potential difference on the surface from -70 mV to zero. These electrical changes cause the membrane to become **depolarized.** As the sodium ions continue to flow into, and the potassium ions out of, the cell, there is a reversal of polarity. The depolarization of the membrane causes the adjacent area to reverse its charge distribution, thus creating a chain reaction that continues down the length of the nerve process. The series of changes in electrical polarity *along* the membrane is the **action potential** or **nerve impulse.**

de = remove, decrease

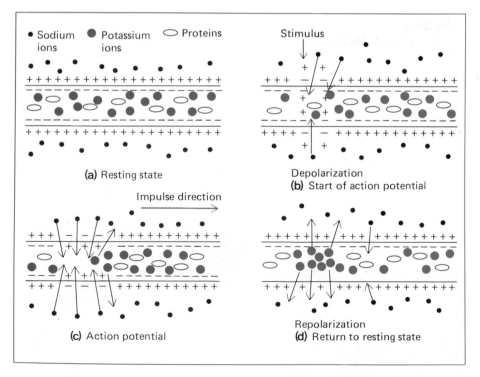

Figure 8-7 **Movement of ions during the generation of an action potential, followed by a series of electrical changes, represents the conduction of an impulse.**

Within 0.001 second, the membrane is restored to its original condition by a rapid outflow of potassium ions through the membrane. This diffusion restores the resting potential, and the membrane is said to be **repolarized.** During recovery the sodium ions that moved in through the plasma membrane must now be pumped out again, while the potassium ions must once again move inward. This is accomplished by the sodium-potassium pump. The energy required for the exchange of electric charges comes from ATP within the cell. The membrane of the nerve fiber is now ready for the next impulse. (See Fig. 8-7.)

During the period of depolarization, no other impulse can be transmitted. This period of time is referred to as the **absolute refractory period,** and its length (about a millisecond) determines the number of impulses that can be transmitted by a given neuron. Following the absolute refractory period, there is a period of time when a second action potential response could be produced. This can occur only if the stimulus strength is greater than threshold level. This period of time, referred to as the **relative refractory period,** can last from one to 10 milliseconds.

Characteristics of nerve impulses.   Under laboratory conditions, nerve fibers (usually axons) may be isolated for stimulation, and an impulse applied to it can be recorded, either visually with an oscilloscope or by the kymographic device. From such studies, nerve action is found to resemble the activity of muscle cells (see Chapter 6). Before a neuron will conduct an impulse, the stimulus must be adequate. When the stimulus is strong enough to activate the neuron, it is called a **threshold** or **liminal** stimulus. Any stimulus weaker than subthreshold will, by itself, produce no action potential. However, if a series of subthreshold stimuli are applied in quick succession to a neuron, their total could reach a threshold and initiate an impulse. This additive process is

supra = above, excess   called **summation.** A **suprathreshold** stimulus, one that is greater than threshold, will produce an impulse that is no different from one produced at the liminal level. This indicates that *the degree and duration of the impulse is independent of the strength of the stimulus.* Therefore, nerve impulse reactions follow the **all-or-none law.** How, then, can stimuli of different intensities be identified? For example, how can one tell the difference between a light touch and a strong slap, or little pain and intense pain? The answer lies in the frequency of impulses sent over the neuron and the number of neurons activated. Although the speed at which the action potential moves along a given neuron is constant, a stronger stimulus will produce a greater number of impulses per unit of time. Also, the greater the intensity of the stimulus, the greater the number of neurons triggered.

Speed of impulses.   Different fibers may have different rates of conduction. These differences are associated with the size, structure, and

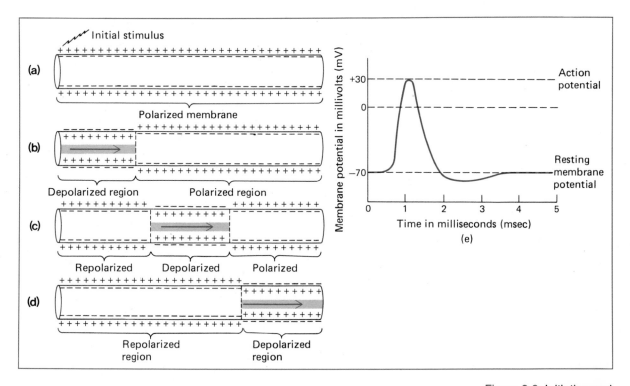

Figure 8-8 **Initiation and transmission of a nerve impulse. (a–d) The colored area represents the region of the membrane that is transmitting the nerve impulse. (e) Record of the potential changes of a nerve impulse.**

physiological makeup of the nerve. Large fibers that are myelinated conduct impulses much faster than thin unmyelinated fibers. As previously learned, Schwann cells entwine the process in segments, forming nodes about 1 millimeter apart. Depolarization of the fiber cannot take place on the axis cylinder below the myelin sheath because of its insulation. Therefore, the depolarization takes place only at the nodes of Ranvier, where the myelin is missing. The impulse appears as a wave that travels along the process from node to node. This form of transmission is *saltatory conduction*. These large myelinated fibers are able to conduct impulses up to 120 meters per second. Such fibers are found in sensory neurons that carry information toward the brain and spinal cord and in large motor neurons, which innervate muscles. Thinner myelinated fibers conduct impulses at the rate of about 10 meters per second. Unmyelinated small fibers transmit impulses at the even slower rate of 0.5 and 2 meters per second.

Synaptic Transmission

Impulses must travel over pathways involving many different neurons. The junction between the axon terminals of one neuron and the dendrites, or cell body, of the next neuron is called a **synapse.** A synapse is similar to the neuromuscular junction found in muscles (see Chapter 6). The axon ends with many terminals, and each one enlarges at its end to form a *synaptic knob*. This knob lies opposite a *receptor* site on

syn = binding, together

the dendrite or cell body of the next neuron. The plasma membrane of the synaptic knob is called the **presynaptic membrane**; the receptor site is part of the **postsynaptic membrane** of the next neuron. The space between the two membranes is called the **synaptic cleft.** The synaptic cleft is about 200 angstroms wide. Fig. 8-9 shows an electrophotomicrograph of a synaptic knob. A close look at the knob reveals mitochondria and the numerous sacs or vesicles that contain molecules of a transmitter substance. When the impulse moves down the axon and reaches the synaptic knob, some of the vesicles move to the inner surface of the presynaptic membrane. Here the vesicles discharge their content of transmitter substance into the synaptic cleft. This chemical diffuses across the gap to large receptor molecules (*macromolecules*) that act as receptor sites on the postsynaptic membrane. The transmitter substance changes the permeability of the postsynaptic membrane. As a result, sodium ions diffuse through and a new action potential is initiated.

Not all transmission of impulses across a synapse results in exciting the postsynaptic neuron. Depending on the identity of the transmitter substance and the way it reacts with the receptor site, the results may be either *excitation* or *inhibition* of the postsynaptic neuron. For example, if a transmitter substance affects permeability of potassium ions through the postsynaptic membrane, additional positive charges will move outward. The interior will then have a greater negative potential.

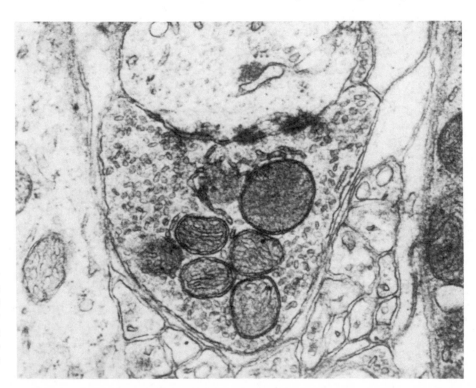

Figure 8-9 Electron micrograph of a synaptic knob filled with synaptic vesicles. This is a motor neuron synapsing with a muscle fiber (magnification approximately 20,000 ×).

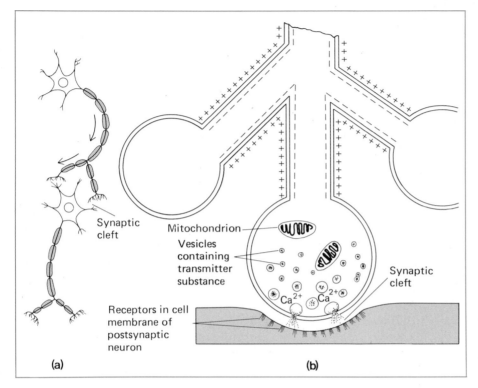

Figure 8-10 Transmission of an impulse between neurons. (a) The wave of depolarization is unable to jump across the synaptic cleft between the two neurons. (b) The release of transmitter substance from vesicles within the synaptic knobs diffuses across the synaptic cleft and may trigger an impulse in the postsynaptic neuron.

This inhibits rather than excites the neuron. Acetylcholine is known to be one type of excitatory transmitter in certain parts of the nervous system. The enzyme *acetylcholinesterase* deactivates acetylcholine, and synaptic transmission of impulses stops. Another excitatory neurotransmitter is **norepinephrine.** This substance may inhibit one type of postsynaptic neuron, such as those found in the brain. Yet, it may excite neurons associated with the smooth muscle of the arterioles, constricting the vessels. A number of organic compounds have been identified in the central nervous system as excitatory or inhibitory transmitter substances, such as *serotonin, gamma-aminobutyric acid (GABA),* and *glutamic acid.*

DISORDERS INVOLVING NEURONS

Multiple Sclerosis

In multiple sclerosis the myelin sheath in neurons that are found in the brain and spinal cord is progressively destroyed. The sheaths harden and form scars or plaques. These plaques interfere with the conduction of impulses over the neuron and cause short-circuiting along the pathways. Multiple sclerosis usually occurs between the ages of 20 and 40. Symptoms include jerky bodily movements (especially of the limbs), double vision, speech difficulty, and partial paralysis of voluntary muscles.

As the disease progresses, the gradual loss of motor control results in the patient's being confined to a wheelchair or bed. The course of the disease is uncertain. The progress of the disease may be interrupted by periods of remission, during which the symptoms disappear. In some cases the disease develops very rapidly, while in others it remains unchanged for years. Multiple sclerosis remains incurable.

Poliomyelitis

Poliomyelitis, also known as **infantile paralysis,** is a disease caused by a virus. It is an acute infection involving destruction of the cell body of motor neurons in the central nervous system. The symptoms of poliomyelitis may vary, but generally they include headache, fever, sore throat, marked listlessness, and stiffness in the back and neck. There may be deep muscle pain and weakness, and in severe cases paralysis and death.

Poliomyelitis may attack people of all ages; however, it is most common in children and occurs most frequently during summer months. In recent years Salk and Sabin immunization has had remarkable success in eradicating the disease in those areas where the vaccine is used.

SUMMARY

The nervous system is composed of two basic kinds of cells: glia, which are nonconducting, supportive cells, and neurons, which are true nerve cells. A neuron is composed of a cell body, as a controlling center, and fibers, which extend outward from the cell body. Those fibers (processes) that carry impulses toward the center of the neuron are called dendrites; the single process that carries impulses away from the cell body is the axon. Processes of peripheral neurons are insulated by a fattylike myelin sheath, which in turn is covered by a neurilemma.

Receptors, located on sensory neurons, initiate impulses when stimulated by some sudden change in the environment. Associative neurons interpret the information and determine the outcome by activating motor neurons, which conduct impulses to effector organs. An impulse or action potential is an electrochemical change on the surface of the plasma membrane of a neuron and is self-propagating. In order for impulses to activate a second neuron, they must cross a synapse, the junction between two neurons.

VOCABULARY REVIEW

Match the statement in the left column with the word(s) in the right column. Place the letter of your choice on your answer paper. *Do not write in this book.*

1. Conduct nerve impulses away from the central nervous system.
2. Nonconductive cells that protect and support neurons.
3. The treelike branched process of a neuron.
4. Neurons that conduct impulses toward the CNS.
5. The other name for the cell body of a neuron.
6. Insulating covering of a neuron.
7. Outer covering of a neuron in the PNS.
8. When impulses jump from node to node on an insulated PNS neuron.
9. The kind of neuron that connects sensory neurons to motor neurons.
10. The time it takes to see danger and react to it.

a. afferent neuron
b. associative
c. axon
d. dendrite
e. efferent neuron
f. glia
g. myelin sheath
h. neurilemma
i. perikaryon
j. reaction time
k. saltatory conduction

TEST YOUR KNOWLEDGE
Group A

On your answer sheet, write the letter of the word that correctly completes the statement.

1. The ability of a receptor to respond to a stimulus is called (a) excitability (b) conductivity (c) integration (d) response.
2. In a polarized neuron the ion found more abundantly outside the membrane than inside is (a) K^+ (b) organic (c) Na^+ (d) H^+.
3. Receptors are (a) centers in the brain that receive nerve impulses (b) specialized nervous tissue sensitive to changes in the environment (c) nerve centers in the spinal cord (d) effectors.
4. Nerve fibers conducting impulses away from the central nervous systems are (a) efferent fibers (b) receptor fibers (c) afferent fibers (d) dendrites.
5. Conduction in neurons depends on (a) movement of electrons as when current passes through a wire (b) an electrochemical process that is self-propagating (c) an entirely chemical process (d) hormones.

6. A stimulus strong enough to activate a neuron is described as (a) subthreshold   (b) absolute refractory period   (c) relative refractory period   (d) threshold.

7. A synapse is   (a) the junction between the axon of one neuron and the dendrite or cell body of the next neuron   (b) the junction between cranial and spinal nerves   (c) the junction between the axons of two different neurons   (d) only found in the peripheral nervous system.

8. The greatest rate (speed) of conduction in neurons within the human body is about   (a) 0.5 meter per second   (b) 5 to 10 meters per second   (c) 100 to 120 meters per second   (d) the speed of light.

9. Synaptic vesicles discharge their transmitter substance through the (a) presynaptic   membrane   (b) postsynaptic   membrane (c) myelin sheath   (d) receptors.

10. Gray matter is made up of neurons that lack a   (a) terminal branch   (b) neurilemma   (c) axon hillock   (d) myelin sheath.

Group B

Write the word or words on your answer sheet that will correctly complete the sentence.

1. _____ transmit impulses from receptors to the CNS.

2. Motor neurons transmit impulses from the brain or spinal cord to _____.

3. Another name for the impulse is the _____.

4. When a neuron is ready to conduct impulses, there are more _____ ions on the outside of the membrane than inside and more _____ or _____ ions on the inside of the membrane than outside.

5. A series of subthreshold stimuli reinforce each other until threshold is reached. This is called _____.

CHAPTER 9

THE NERVOUS SYSTEM

Objectives

A. Name the two major divisions of the nervous system and describe their functions
B. Describe the structure and function of the spinal cord
C. Identify the parts of the human brain and their functions
D. Name the major parts of the peripheral nervous system
E. List the 12 pairs of cranial nerves

ANATOMY OF THE NERVOUS SYSTEM

Humans possess an extremely complex **nervous system**. The functioning of billions of neurons and an even greater number of glial cells that form the brain makes humans unique among all creatures. The brain gives us consciousness, memory, perception, and the ability to evaluate situations and to feel emotions. These abilities would be incomplete, however, without the coordinated and efficient functioning of all of the parts of the body. In order to have this coordination between various parts of the body we need: (1) a method for monitoring activities that are taking place in different parts of the body, (2) a communication network to exchange information between all the parts, and (3) a control center to gather, integrate, store, and recall information and command the necessary actions. In the human body, the nervous system is the monitor, part of the communication network, and the control center. The network of neurons all over the body serves as monitor and communication channels. The brain and spinal cord together act as the control center for all the activities of the body. The nervous system also works closely with the *endocrine system* in maintaining homeostasis. As you learned in Chapter 8, the nervous system controls the body through electrochemical impulses carried over neurons. The endocrine system regulates bodily functions by releasing chemical messengers, called *hormones*, into the bloodstream. These two systems share the responsibility for the data-processing activities of the body.

In both structure and function, the human nervous system is a complex machine. Although all of its parts function as an integral unit, the system is usually divided for study into two portions: the *central nervous system (CNS)* and the *peripheral nervous system (PNS).*

peri = around

The Central Nervous System

The central nervous system (CNS) serves as the control center for the body and consists of the spinal cord and the brain. Stimuli from the internal and external environment activate receptors that relay information to the CNS. As a result of its processing activities, this control center then generates impulses that cause muscles or glands to respond appropriately to the stimuli.

Structure of the spinal cord. The spinal cord is continuous with the brain and emerges from the cranial opening in the base of the skull, the *foramen magnum.* The cord extends downward through the spinal canal and ends at a level near the first and second lumbar vertebra. In the adult, the length of the cord varies from 42 to 45 centimeters (17 to 18 inches). The growth of the spinal cord lags behind the development of the vertebrae so that it does not extend as far in length as the vertebral column. From the end of the spinal cord numerous spinal nerves continue down to the lower lumbar region. Collectively, these resemble a horse's tail and are therefore appropriately named the **cauda equina** (horse tail).

The inability of nerve cell bodies to regenerate implies that any injury to the spinal cord could result in permanent loss of some coordinating function. Hence the bones constituting the vertebral column offer the cord protection. The diameter of the spinal cord (about 13 millimeters) is less than that of the canal in which it is housed. This serves as an advantage in that the spine may be moved freely without injuring the cord. Three membranes, collectively called the **meninges,** also cover and protect the spinal cord. The outermost membrane, the **dura mater,** is a tough protective covering composed of dense, fibrous connective tissue that lines the vertebral canal. The middle membrane, the **arachnoid,** is a thin, transparent tubular sheath that lies immediately under the dura mater. Its name comes from the membrane's spider-web appearance under the microscope. The innermost membrane is known as the **pia mater.** It is a delicate fibrous membrane that adheres to the surface of the spinal cord. The pia mater contains many blood vessels and plays an important part in providing nourishment to cells of the cord.

cerebro = brain

Between the arachnoid membrane and the pia mater is a space called the **subarachnoid space.** This space is filled with **cerebrospinal fluid (CSF),** which protects the spinal cord from mechanical injury by acting as a shock absorber. This fluid also aids in the diagnosis of diseases of the meninges. A sample of CSF can be taken by a spinal puncture in the lower lumbar region and analyzed for signs of bacteria or viruses.

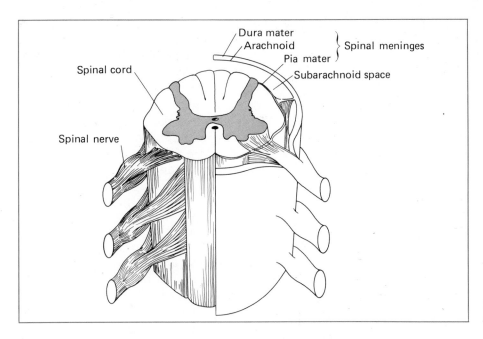

Figure 9-1 **Spinal cord and meninges.**

In this way, such infections as poliomyelitis and meningitis can be detected. Spinal anesthetics are also injected into the subarachnoid space. (See Fig. 9-1.)

A cross section of the spinal cord reveals that it consists of two types of nerve tissue. The central area is darker and appears to have the shape of the letter H. This is gray matter, which consists of nerve cell bodies and unmyelinated dendrites and axons of associative and motor neurons.

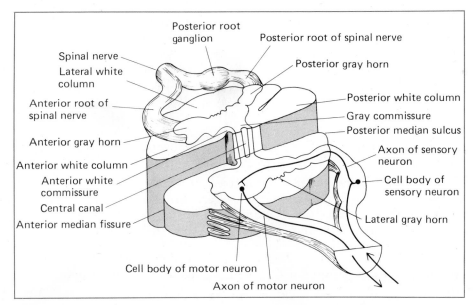

Figure 9-2 Organization of spinal cord as seen in a cross section.

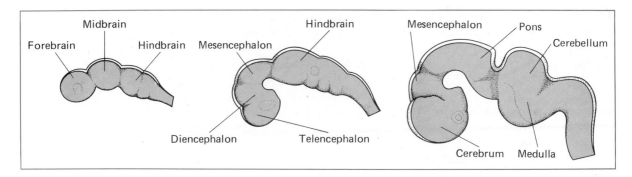

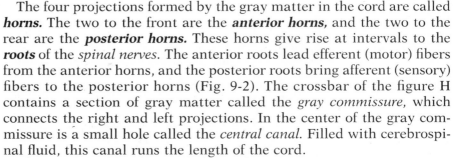

Figure 9-3 Development of the brain in the embryo (above). The adult brain and spinal cord (below).

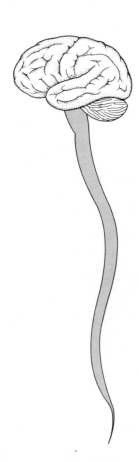

The four projections formed by the gray matter in the cord are called **horns.** The two to the front are the **anterior horns,** and the two to the rear are the **posterior horns.** These horns give rise at intervals to the **roots** of the *spinal nerves.* The anterior roots lead efferent (motor) fibers from the anterior horns, and the posterior roots bring afferent (sensory) fibers to the posterior horns (Fig. 9-2). The crossbar of the figure H contains a section of gray matter called the *gray commissure,* which connects the right and left projections. In the center of the gray commissure is a small hole called the *central canal.* Filled with cerebrospinal fluid, this canal runs the length of the cord.

Surrounding the horns of gray matter and giving an oval shape to the spinal cord is the *white matter.* On each side of the cord, it is organized into three columns: the *anterior white column,* the *posterior white column,* and the *lateral white column.* Within each column are bundles of myelinated fibers, called *fasciculi* or *tracts,* that run the length of the spinal cord. Ascending tracts carry sensory impulses up to the brain; descending tracts carry motor impulses down from the brain. Anterior to the gray commissure is the *anterior white commissure,* which joins the right and left sides of white matter.

Structure of the brain. There is probably no structure in the universe more highly organized than the human brain. Although having only 1.4 kilograms (about 3 pounds) of tissue, the brain is estimated to contain nearly 100 billion cells. It occupies the cranial cavity, which is formed by the eight cranial bones of the skull. This hard, tough covering makes it the best protected organ in the body. Further protection is afforded by the **cranial meninges,** which are continuous with the meninges of the spinal cord. Adhering to the cranial bones and acting as the periosteum is the **dura mater.** Attached to the surface of the brain is the **pia mater,** and in between both is the **arachnoid membrane.** Between the pia mater and arachnoid membrane in the subarachnoid space is the cerebral portion of the cerebrospinal fluid. CSF is very important in the nutrition and protection of the brain.

Development of the brain. During the third and fourth weeks of embryonic development, the human nervous system originates as a flat sheet of cells, the **neural plate.** This plate subsequently folds into an elongated hollow **neural tube.** From the anterior end of the tube, three swellings emerge that will eventually form the brain, or **encephalon.** The smallest of the three is called the **hindbrain,** or **rhombencephalon,** and is directly connected to that portion of the neural tube that will form the spinal cord. The hindbrain is ultimately divided into the **medulla oblongata (myelencephalon)** and the **pons (metencephalon).** Anterior to the hindbrain is a larger swelling called the **midbrain (mesencephalon).** The medulla, pons, and midbrain are collectively called the **brainstem.** The **cerebellum** is posterior to the medulla and is connected to the pons. The third and largest swelling in the neural tube forms the **forebrain (prosencephalon),** which later divides into the **cerebrum (telencephalon)** and the **interbrain (diencephalon)** (Fig. 9-3).

cephalo = head

Deep within the brain are a number of interconnected cavities, called **ventricles,** that are continuous with the central canal of the spinal cord. These spaces are also filled with cerebrospinal fluid. The largest cavities, called the lateral ventricles (ventricles I and II), are located within each half of the cerebrum. They communicate with the third ventricle, located in the center of the diencephalon, by an oval opening called the

Figure 9-4 **The midsagittal surface of the brain.**

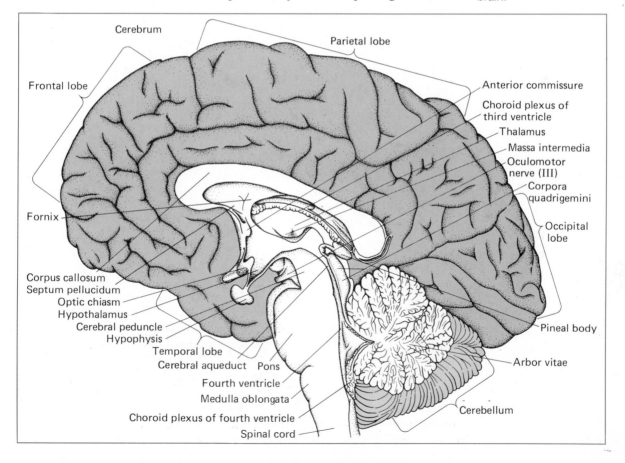

Cerebrum

Parietal lobe

Frontal lobe

Anterior commissure

Choroid plexus of third ventricle

Thalamus

Massa intermedia

Oculomotor nerve (III)

Corpora quadrigemini

Occipital lobe

Fornix

Corpus callosum
Septum pellucidum
Optic chiasm
Hypothalamus
Cerebral peduncle
Hypophysis
Temporal lobe
Cerebral aqueduct   Pons
Fourth ventricle
Medulla oblongata
Choroid plexus of fourth ventricle
Spinal cord

Pineal body

Arbor vitae

Cerebellum

foramen of Monro. Located in front of the cerebellum, the third ventricle is joined to the fourth ventricle by the **cerebral aqueduct** (*aqueduct of Sylvius*). From the fourth ventricle a number of openings communicate with the subarachnoid space of the meninges that surround the brain and cord.

The brainstem.   The **brainstem** is much larger and more complex than the spinal cord. It contains not only the important origins of the cranial nerves, but also the mechanisms for coordinating and integrating all in-coming information. The brainstem includes the **medulla, pons,** and **midbrain.**

pons = bridge

The lowest part of the brainstem, called the **medulla oblongata,** is an enlarged continuation of the spinal cord. It is superior to the foramen magnum, extending upward and slightly forward to the inferior surface of the pons, to which it is attached. The medulla measures about 3 centimeters (1 inch) in length. It contains tracts of myelinated fibers (white matter) that conduct impulses between the spinal cord and brain. Many of these tracts cross over as they pass through the medulla. Dispersed throughout the white matter are centers of gray matter. It is in the medulla that we find the origins of four pairs of cranial nerves: numbers IX, X, XI, and XII.

Just above the medulla, the brainstem enlarges to form the **pons.** The cerebellum connects to the posterior surface of the pons and the midbrain is found above it. The pons is about 2.5 centimeters (less than 1 inch) in length. This part of the brainstem is mostly composed of white matter in the form of large fiber tracts called **peduncles.** The nuclei of cranial nerves numbers V, VI, VII, and VIII are found in the pons.

The shortest part of the brainstem is the **midbrain,** which lies just above the pons. It consists primarily of white matter in the form of great fiber tracts. The midbrain connects the cerebrum and interbrain above with the cerebellum, hindbrain, and cord below. These tracts, found ventrally in the midbrain, are called **cerebral peduncles.** A canal, called the **cerebral aqueduct,** passes through the midbrain and connects the third ventricle with the fourth ventricle. Dorsal to the aqueduct are two pairs of rounded structures called the **corpora quadrigemina.** Cranial nerves III and IV also originate from the ventral surface of the midbrain.

The cerebellum.   The **cerebellum,** the second-largest portion of the brain, is located above the medulla and is connected to the posterior surface of the pons. The internal structure of the cerebellum differs from that of the brainstem in that its gray matter, or cortex, is on its external surface, while its white matter, called the *arbor vitae,* is internal and takes on the shape of tree branches. Deep within the white matter are masses of gray matter, the **cerebellar nuclei.** Numerous furrows called **gyri** are located transversely on the outer surface of the

gyrus = circle

cerebellum. These are ridges that greatly increase the area of gray matter. The cerebellum has two *lateral hemispheres* divided by a central lobe called the **vermis.**

vermis (Lat.) = worm

The diencephalon—interbrain.   Located above the midbrain, the **diencephalon** is almost completely covered by the hemispheres of the cerebrum. It consists principally of two parts, the **thalamus** (the dorsal two thirds) and the **hypothalamus.**

hypo = below, under

The thalamus is a large mass of gray matter that is partially separated into right and left portions, each forming part of a lateral wall of the third ventricle. The thalamus contains many nuclei, which serve as relay stations for incoming sensory impulses.

The hypothalamus is the lower part of the diencephalon and forms the floor and lower lateral walls of the third ventricle. It is partially protected by the sphenoid bone. Extending below the hypothalamus on a stalk (the infundibulum) is the pituitary gland. Hypothalamic neurons send axons to the posterior pituitary, which releases some important hormones.

Running through the core of the brainstem and extending into the basal portion of the diencephalon is the **reticular formation.** It extends from the upper part of the spinal cord, through the medulla, pons, midbrain, and terminates in the hypothalamus. The reticular formation is composed of small masses or centers of gray matter and numerous neurons of various sizes that extend to and from the cortex of the cerebrum, hypothalamus, and various cranial nerves.

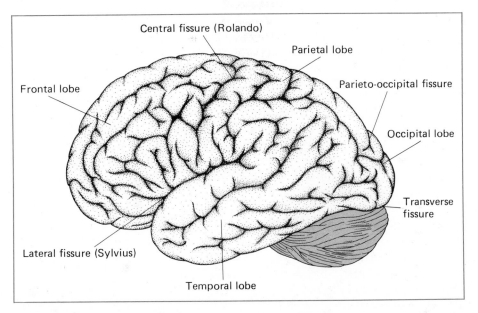

Figure 9-5 **The lobes of the brain are separated by sulci and gyri.**

The cerebrum.   The largest and most prominent part of the human brain is the *cerebrum.* It occupies most of the cranial cavity and accounts for over two thirds of the weight of the entire cranium. It is estimated that more than 15 billion neurons, and an even larger number of glial cells, are found within the cerebrum.

Its large mass of nervous tissue is partially divided by a deep groove, the *longitudinal fissure,* into right and left halves, the *hemispheres.* Covering the outer surface of these hemispheres is a layer of gray matter called the *cerebral cortex.* This covering has the appearance of a walnut, with its many ridges and shallow grooves. Each shallow groove is called a *sulcus,* and the ridge between any two sulci is called a gyrus or, more commonly, a *convolution.* The surface area of the cerebral cortex is greatly increased by these ridges and depressions, and their pattern is apparently unique to each individual.

sulcus = furrow

Each hemisphere is further divided into four lobes. The name of each lobe corresponds to the name of the cranial bone adjacent to it. Thus is derived the *frontal, parietal, temporal,* and *occipital lobes.* These lobes are separated by three fissures that serve as boundary lines: The *central fissure (fissure of Rolando)* separates the frontal lobe from the parietal lobe; the *lateral fissure (fissure of Sylvius)* separates the frontal lobe from the temporal lobe; and the *parieto-occipital fissure* separates the parietal lobe from the occipital lobe. Another prominent fissure called the *transverse fissure* separates the cerebrum from the cerebellum. (See Fig. 9-5.)

The cortex.   The human cortex can be subdivided into several areas. One area protrudes forward on the base of each hemisphere and consists of primitive cortical cells. It becomes swollen at its end to form the *olfactory bulb,* a structure associated with our sense of smell. Another area deep within the cerebrum of each hemisphere contains the *limbic system,* a ring of cortex and associated structures that encircles the ventricles. In the same general area as the limbic system, additional masses of gray matter form the *basal ganglia* or *cerebral nuclei,* which are associated largely with motor activity. The major portion of cortex (about 90 percent) is found on the outer surface of the cerebral hemispheres. This surface, ranging from 2 to 5 millimeters (less than ¼ inch) in thickness, contains 70 percent of all the neurons in the central nervous sytem. Under the microscope, the cerebral cortex is seen as six layers of neuron cell bodies.

The major volume of the cerebrum consists of myelinated fibers (primarily axons), which form the white matter. These myelinated fibers can be classified as: (1) *Association fibers* interconnect different neurons within the same hemisphere, (2) *Commissural fibers* interconnect the neurons of one area in a hemisphere with neurons in a similar area of the opposite hemisphere. Three major groups of commissural fibers are the *corpus callosum,* the *anterior commissure,* and the *posterior commissure.* (3) *Projection fibers* form ascending and descending pathways that interconnect different levels of the central nervous system.

The Peripheral Nervous System

Basically, the peripheral nervous system (PNS) consists of the cranial and spinal nerves and groups of nerve cell bodies called *ganglia.* Within the PNS is a *somatic division* and an *autonomic division.* The somatic division includes *sensory neurons,* which transmit the impulses to the central nervous system for interpretation, and *motor neurons,* which carry impulses from the CNS to activate effectors. The autonomic division acts to maintain the dynamic equilibrium of the body's constant temperature, control smooth muscles, and regulate the heartbeat. Since the structures it innervates cannot normally be consciously controlled, the autonomic nervous system is considered involuntary. The autonomic system is further subdivided into *sympathetic* and *parasympathetic* components.

The somatic division.   *Receptors* are special sensory end organs that make us aware of changes in our external environment. Receptors are located on the peripheral ends of sensory neurons or in the various sense organs, such as the eye and ear. They may be naked, free neuron endings for pain and certain chemical senses. They may also contain a light-sensitive chemical that initiates impulses for vision, or they may be specialized nerve endings sensitive to certain mechanical stimuli. Receptors vary widely in structure according to the sense they serve and also according to their location in the various tissues of the body. Details of the various sense organs will be discussed in subsequent chapters.

A *nerve* consists of numerous individually insulated nerve cell fibers. All the fibers of a nerve may be responsible for conveying sensory impulses toward the central nervous system. A group of these afferent fibers is called a *sensory nerve.* Again, all the fibers of a nerve may carry motor impulses from the CNS to particular muscle fibers or glands. Groups of these efferent fibers form a *motor nerve.* Finally, a nerve may consist of both sensory and motor fibers. To this type has been given the name *mixed nerve.* (Fig. 9-6)

The spinal cord is divided into 31 segments. At each segment, the anterior and posterior roots join a short distance outside of the cord to form a trunk, the *spinal nerve.* The 31 pairs of spinal nerves are mixed nerves, combining fibers from the anterior, or motor, roots and the posterior, or sensory, roots of the spinal cord. Thus spinal nerves are named according to the regions of the spine from which they emerge. The uppermost eight pairs of the spinal nerves are cervical; the next twelve pairs are thoracic. Then come five pairs of lumbar nerves, followed by five pairs of sacral nerves. The last pair of spinal nerves is called the coccygeal nerves. On the posterior root of each nerve is a small swollen area. This area, the *posterior root ganglion,* contains the cell bodies of sensory neurons. Note that there are no anterior root ganglia since the cell bodies of the motor neurons lie within the anterior horn of gray matter within the spinal cord.

A short distance from the cord, just outside the intervertebral foramina, the spinal nerve splits, giving off two branches, the **posterior** and **anterior primary rami.** The posterior primary rami go to the muscles and skin of the posterior surface of the back. The anterior primary rami of all spinal nerves, except those in the thoracic area, run forward to join in a series of complex nerve junctions called **plexuses.** There, they are rearranged and combine to form a network of nerves that supply the muscles and the skin in the arms and legs. These plexuses are named according to the areas they supply: *cervical, brachial, lumbar,* and *sacral.* The posterior primary rami in the thoracic region do not form a plexus but go directly to the muscles between the ribs, called the intercostal muscles, and the skin of the anterior and lateral chest wall. The distribution of the spinal nerves is illustrated in Fig. 9-7.

Nerves arising from the brain are called **cranial nerves.** Each of these nerves is designated by a name or a Roman numeral that corresponds to its point of origin on the brain. With the exception of cranial nerve **X,** or **vagus nerve** (which supplies the heart, lungs, and abdominal organs), all of the cranial nerves function in the head and neck areas. Some are entirely sensory; some are entirely motor. Others (mixed nerves) contain both sensory and motor fibers. In addition, there are some cranial nerves that carry *autonomic fibers,* which innervate involuntary (smooth) muscles, glands, and blood vessels. Table 9-1 lists the cranial nerves and their functions.

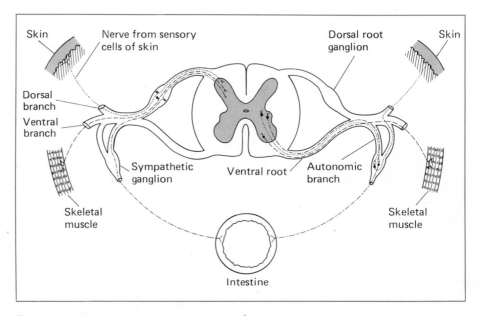

Figure 9-6 Diagram of primary types of sensory and motor neurons of the spinal nerves and their connections with the spinal cord. For convenience, sensory neurons are shown on the left and motor neurons on the right, though both kinds are found on each side of the body.

Table 9-1   The Cranial Nerves

Number (Location)	Name of Nerve	Origin of Sensory Fibers	Effector Innervated by Motor Fibers
I	Olfactory (sensory)	Olfactory mucosa of nose (smell)	None
II	Optic (sensory)	Retina of eye (vision)	None
III	Oculomotor (motor)	Proprioceptors of eyeball muscles (muscle sense)	Muscles which move eyeball (with IV and VI); muscles which change shape of lens; muscles which constrict pupil
IV	Trochlear (motor)	Proprioceptors of eyeball muscles (muscle sense)	Other muscles which move eyeball
V	Trigeminal (sensory and motor)	Teeth and skin of face	Some of muscles used in chewing
VI	Abducens (motor)	Proprioceptors of eyeball muscles (muscle sense)	Other muscles which move eyeball
VII	Facial (motor)	Taste buds of anterior part of tongue	Muscles of the face; submaxillary and sublingual glands
VIII	Auditory (sensory)	Cochlea (hearing) and semicircular canals (senses of movement, balance and rotation)	None
IX	Glossopharyngeal (sensory and motor)	Taste buds of posterior third of tongue, lining of pharynx	Parotid gland; muscles of pharynx used in swallowing
X	Vagus (sensory and motor)	Nerve endings in many of the internal organs—lungs, stomach, aorta, larynx	Parasympathetic fibers to heart, stomach, small intestine, larynx, esophagus
XI	Spinal accessory (motor)	Muscles of shoulder (muscle sense)	Muscles of shoulder
XII	Hypoglossal (motor)	Muscles of tongue (muscle sense)	Muscles of tongue

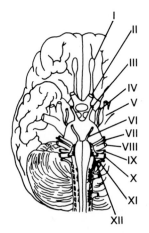

Figure 9-7 Ventral view of human brain showing the 12 pairs of cranial nerves.

The autonomic division (system).   While the somatic division of the nervous system is concerned with the external environment's influence on the human body, the **autonomic nervous system** is responsible for the control of our internal environment: the rate of heart beat, the peristaltic movements of the stomach and intestinal tract, the response the body makes to temperature changes, and the contraction of the urinary bladder. In addition, responses to emergency situations such as

auto = self

a dry mouth, a pounding heart, and sweating palms are under the control of the autonomic nervous system, as is the red blush of embarrassment or anger.

The cell bodies of the efferent nerves of the autonomic nervous system are located in areas ranging from the midbrain to the sacral region of the spinal cord. Unlike the CNS, where it may take only one neuron from the spinal cord to innervate an effector organ, the autonomic system always has at least two neurons involved in the innervation of an effector organ. One neuron runs from the CNS to a ganglion, while the second runs directly from the ganglion to the effector. The first neuron in this autonomic pathway is called a *preganglionic neuron.* Its

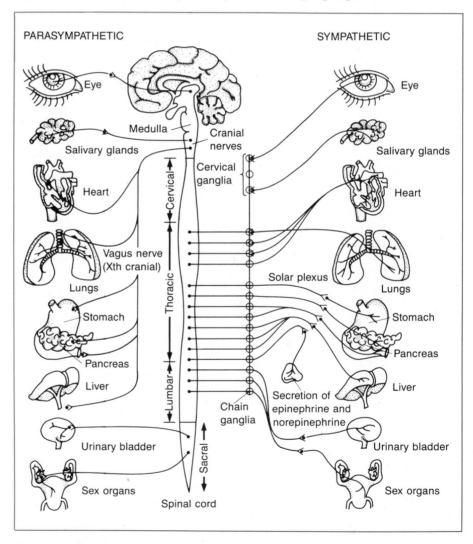

Figure 9-8 Diagram showing the two divisions of the autonomic nervous system and the organs they innervate.

cell body is in the brain or spinal cord and the cell's myelinated axon, the **preganglionic fiber,** passes out of the central nervous system by way of a cranial or spinal nerve. After the axon leaves the nerve, it enters an autonomic ganglion, where it forms a synapse with the dendrite or cell body of the second neuron. The second neuron in the autonomic pathway, called the **postganglionic neuron,** runs entirely outside of the CNS and finally terminates in a visceral effector. The axon of a postganglionic neuron is called the **postganglionic fiber** and is unmyelinated.

On the basis of both physiological and anatomical characteristics, the autonomic nervous system is divided into two parts: the **parasympathetic** and **sympathetic** divisions, also called the **craniosacral** and **thoracolumbar** divisions, respectively.

The parasympathetic division takes its origin from preganglionic fibers arising from cells in the brainstem (midbrain, pons, and medulla) as well as from fibers arising from the second, third, and fourth segments of the sacral spinal cord. The four cranial nerves that carry parasympathetic neurons are the oculomotor (III), facial (VII), glossopharyngeal (IX), and the vagus (X), which has the most extensive distribution of the parasympathetic neurons. As was mentioned above, two neurons are involved in the innervation of visceral organs. In this division, the parasympathetic ganglia are very close to or within the wall of the visceral organ. Thus the preganglionic fibers are very long and the postganglionic neurons have relatively short axons. Because parasympathetic neurons originate in the brain and the sacral region of the cord, this division is sometimes called the **craniosacral outflow.**

Neurons of the sympathetic, or **thoracolumbar,** division originate in the gray matter of the spinal cord in all thoracic segments and in the first two segments of the lumbar region. Preganglionic neurons exit through anterior roots of spinal nerves and run to sympathetic ganglia located near the cord. The sympathetic ganglia are arranged in two long cords more or less segmentally running downward from the neck, thorax, and abdomen to the coccygeal region. However, the ganglionic cord only receives preganglionic fibers from the thoracic and lumbar regions. These short sympathetic neurons synapse in the ganglia with postsynaptic neurons whose long axons continue on to the various visceral organs that they innervate. See Figure 9-8.

THE PHYSIOLOGY OF THE NERVOUS SYSTEM

The major function of the spinal cord is to act as a two-way conducting pathway, carrying sensory impulses from the periphery up to the brain and conducting motor impulses from the brain down to the muscles and glands of the body. A second major function is to act as a center for reflex activity.

Most animals are capable of analyzing sensation and controlling movement. What distinguishes the human brain from that of other living organisms, however, is the variety of more specialized functions it is capable of learning. Probably the most important basis for human learning is language. The ability to speak and understand a language requires a high level of comprehension, one that only the human brain attains. In addition, humans have artistic abilities, which are brain functions seen in no other living organisms.

Conduction Pathways of the Spinal Cord

The tracts or fasciculi in the columns of the spinal cord are individually named according to their origin and destination in the spinal cord. The anterior columns contain descending fibers from the motor areas of the brain (*anterior corticospinal*) and carry voluntary motor impulses to anterior-horn cells. They also contain ascending fibers that carry sensory impulses to the brain (*anterior spinothalamic tracts*).

The lateral columns contain the main descending motor tracts from the motor areas of the brain (*lateral corticospinal*), as well as ascending tracts carrying sensations of heat, cold, and pain to the brain (*lateral spinothalamic tracts*). Other tracts (*spinocerebellar tracts*) are pathways concerned with unconscious muscle sense and positioning of the body. These fiber tracts do not reach the cerebral cortex and are indirectly involved in sensation.

The posterior columns contain only ascending tracts (*fasciculus gracilus and fasciculus cuneatus*). These tracts carry sensations of touch and muscle-to-joint positioning, which keep us consciously aware of the position of the limbs with relationship to the body.

Reflex activity.   The term **reflex** is derived from the word *reflect*, meaning "to turn back." When a receptor is stimulated, as, for example, when the finger touches a hot stove, an impulse is carried to the cell body in the central nervous system. The axon of this sensory neuron usually makes contact with an associative neuron (connector neuron) in the spinal cord. This in turn contacts a motor neuron, which then sends an impulse down its axon to muscles that lift the burned finger off the stove. All of this happens in a split second, even before pain or the danger of the situation is perceived by the brain. Thus, reflexes are involuntary and often unconscious. Furthermore, in order for this action to take place, more than one sensory neuron is stimulated and more than one motor neuron is activated.

The coordination of sensory and motor neurons involved in a reflex act is known as a **reflex arc** (illustrated in Fig. 9-9). Reflexes can be of great protective value. Some examples of protective reflexes are the corneal reflex of the eye, which makes us wink involuntarily when foreign material touches the cornea, or sneezing or coughing, which happens when an irritating substance gets into the nose or trachea. Such reflex actions are effective in guarding the body against injury.

Motor and Sensory Functions of the Brain

The activities of the *brainstem* are mostly below the level of consciousness. This lower portion of the brain is the continuation of the spinal cord and a conduction pathway between the cord and the cerebrum.

The medulla contains many vital centers essential for survival. These include: (1) the **cardiac center,** which controls heart rate; (2) the **vasomotor center,** which controls the constriction or dilation of blood vessels; and (3) the **respiratory center,** which controls the rate and depth of breathing. In addition, there are other centers in the medulla for coughing, sneezing, vomiting, and swallowing.

vaso = vessel

The pons is nearly all white matter in the form of tracts or **peduncles.** The cerebellum attaches to the posterior surface of the pons which forms the major connection of the cerebellum with the brainstem. The **superior peduncles** connect the cerebellum, through the pons, with the parts of CNS by way of the ascending tracts. The **middle peduncle** connects the two hemispheres of the cerebellum. Finally, the **inferior peduncle** connects the cerebellum with the spinal cord by way of descending tracts that pass through the pons.

The midbrain, also mostly white matter, consists of the **cerebral peduncles,** which form the principal pathways between the forebrain and the hindbrain. In addition, visual and auditory reflexes that result in moving the eyeball and turning the head toward auditory stimuli are all controlled by the midbrain.

The cerebellum is constantly receiving impulses concerned with balance, motion, and muscle tone from all parts of the body. When the cerebellum is damaged by disease or tumor, muscular movements become jerky, uncontrolled, and unpredictable. Fine movements also become impossible. Muscles become weak, tremble, and lack tone.

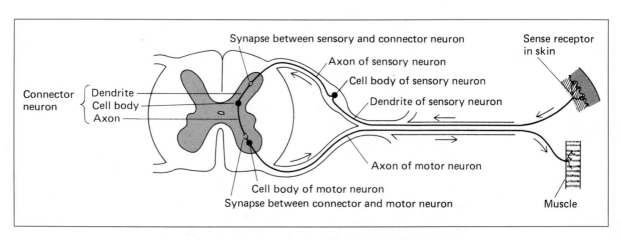

Figure 9-9 Diagram of a reflex arc, showing the pathway of an impulse and indicated by arrows.

Although motor impulses do not originate in the cerebellum, it is responsible for coordinating impulses originating in the highest centers of the brain. As a result, muscular activities are smoothly and efficiently controlled. In addition, the cerebellum aids in maintaining the muscular tone and balance of the body. All of these activities are carried on below the level of consciousness.

The thalamus acts as an important relay center for incoming impulses from all parts of the body and is a sensory-integrating organ. It is believed that the thalamus is the part of the brain involved in our first sensations of touch, temperature, and pleasure or displeasure.

Visceral control.   Despite its small size, the hypothalamus controls many body activities, most of which are related to homeostasis. Thus it is an important center for autonomic activity. For example, it stimulates smooth muscle, regulates the rate of the heartbeat, and controls many glands. In addition, the hypothalamus receives impulses from the viscera and responds appropriately to maintain homeostasis. It has certain secretory functions, such as releasing hormones that are stored in the posterior pituitary. It also regulates the activity of the anterior pituitary by producing chemicals called *regulating factors*. The hypothalamus is associated with emotions of rage and anger. It controls normal body temperature and serves as a thermostat by controlling the diameter of blood vessels. It is in the hypothalamus that one finds the appestat, or *appetite center*. Thus it controls the desire for food or inhibits hunger once food is consumed. The hypothalamus contains the *thirst center* and indirectly controls the water level of the body. The hypothalamus is also concerned with normal *sleeping-waking mechanisms*.

Higher Functions of the Central Nervous System

The neurons of the reticular formation receive and integrate impulses from many sensory pathways as well as from other regions of the brain. The "centers" mentioned previously, for heartbeat rate, respiratory control, vascular activity, swallowing, and vomiting, are all within the reticular formation of the brainstem. The state of consciousness and alertness required for thought processes is believed to be controlled by the reticular formation.

The cerebrum.   The primitive cortex of the cerebrum (the limbic system) is believed to be responsible for many of our emotions, such as the subjective feelings of fear, love, anger, joy, anxiety, hope, and sexual arousal. Another portion of the primitive cortex is the basal ganglia, paired masses of gray matter deep within each hemisphere. Little is known about their function other than that they are believed to be concerned with bodily movement such as the subconscious movement of skeletal muscles that swing the arms in walking. They are also thought to be involved in the planning and programming of movement.

The functional centers of the cerebral cortex are numerous and complex but are usually divided into *motor, sensory,* and *associative areas.* The primary motor areas are located just anterior to the central fissure in the frontal lobe. Not only does stimulation of a specific point of the motor area result in muscular contraction on the opposite side of the body, but also those areas at the top of the frontal lobe initiate response in the most distal parts of the body, such as the toes and feet. As one moves down the side of the motor cortex, the areas in the body involved move upward (the leg, trunk, arm, hand, fingers, neck, and face).

Learning and memory.   The *general sensory areas* of the cerebral cortex are in the parietal lobe, posterior to the central fissure. This sensory area receives impulses from the cutaneous, muscular, and visceral receptors in various parts of the body. Posterior to the general sensory area in the parietal lobe is an area that records our sense of texture, shape, and size of objects. This sense is called the **somesthetic sense.** The primary visual center is located in the occipital lobe. Hearing is recorded and interpreted in the temporal lobe, where is also found the sense of smell. Just above these areas in the lowest portion of the parietal lobe is the center for the sense of taste.

The association areas involve the greater portion of the lateral surfaces of the occipital, parietal, and temporal lobes and the frontal lobe anterior to the motor areas. They are involved in conscious thought, reasoning, judgment, will power, memory, intelligence, and personality.

It is now recognized that no single area of the cerebrum is entirely responsible for total intellectual capacity. However, if an animal's entire cerebral cortex is removed, it will remain conscious, express rage or fear, eat, and respond to loud noises or visual stimuli. Yet it will show no real intelligence; its emotional reactions (shame or rage, for example) will be uninhibited, because lower centers, such as those in the thalamus and hypothalamus, are no longer under the controlling influence of higher brain levels.

Intelligence develops, in part, through storage, in various cortical areas, of impressions received by our special senses (sight, hearing, smell, taste, and touch). These stored impressions are linked by association fibers so that they can, in most cases, be recalled to conscious level at will, or remembered. Memories may exist for a lifetime, which suggests that memory depends on some permanent change in the neurons. This is especially true of past experiences or learnings. Recent memory of newly acquired information is less stable.

Conditioned reflexes.   Earlier in the chapter we considered **unconditioned reflexes.** These reflexes are present at birth, and most of them occur even in the lowest animals. They are not altered by, nor do they depend on, past experience. These include such responses as are required for taking food, for removing waste, and for defense from danger.

Any reflex that is modified as a result of experience or training is known as a **conditioned reflex.** The Russian physiologist Ivan P. Pavlov (1849–1936) developed methods whereby conditioned reflexes could be studied in animals. For example, he found that the food placed in the mouth of a very young puppy that had not been fed previously would bring secretion of saliva. Such a response, of course, is an unconditioned reflex. If, however, a bell was rung every time the animal was offered food, eventually saliva would flow every time it heard the bell, even in the absence of food. The puppy was now conditioned to the sound of the bell; that is, it remembered the association between the bell and food and thus salivated. Conditioned reflexes obviously require the participation of the cerebral cortex of the brain, although some very simple types may involve subcortical structures only.

Another type of conditioning, called **instrumental,** or **operant, conditioning,** is illustrated by the following example. An animal is placed in a cage in which there is a small lever. As it moves about, it accidentally strikes the lever, which then releases a small pellet of food. Eventually it associates striking the lever with the release of food and thereafter does so with a purpose. It *remembers* or has learned the association between operating the lever and receiving food. The lever might then be fixed so that moving it only in a certain direction would release the food. If the animal, by being rewarded, learned to push the lever in the right direction each time, it would have performed on a higher plane than mere conditioned reflex. Many types of human behavior result from operant conditioning.

Sleep.   Humans suspend their activity for a fairly long period of time, usually once every 24 hours. When this sleep-wake cycle occurs within every 24-hour period, it is called a *circadian rhythm*. Why humans must sleep is an unanswered question. Deprived of sleep, humans lose efficiency, cannot maintain concentration, and may become very irritable and fatigued. Following a period of sleep, these symptoms disappear.

When brain waves of sleeping individuals are analyzed experimentally, it is found that there are recurrent periods of heightened activity in the brain. At this time the most characteristic features found during sleep is a period of rapid eye movement (REM), which coincides with rapid irregular low-voltage brain waves. If the subject is awakened during this period of REM, the individual invariably reports that he or she was dreaming. During a night's sleep, REM occurs about every 90 to 110 minutes and lasts for about 10 to 20 minutes. Researchers have found that REM is a period when sleep is most relaxing. During non-REM sleep, the metabolic rate and breathing rate slow down and blood

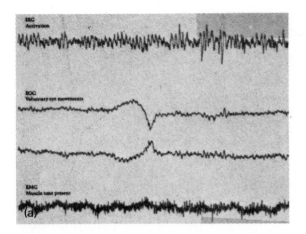

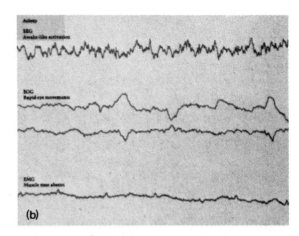

Figure 9-10 An electroencephalograph (EEG) records brain-wave patterns. Electrodes glued to a person's scalp transmit brain waves to a machine that records the awake state (a) and then REM sleep (b).

pressure is lowered. Nervous activity in the brain and spinal cord diminishes.

Functions of the Autonomic Nervous System

The two divisions of the autonomic nervous system, parasympathetic and sympathetic divisions, are largely antagonistic to each other, although they have independent functions as well. If you drive a car with one foot on the accelerator and the other on the brake pedal, you are in a position either to speed up or slow down very quickly. Similar quick changes are achieved by the autonomic nervous system. If you observe the iris of someone's eye, you will note frequent changes in the diameter of its pupil. The nerve control of this reaction can be traced to the autonomic nervous system. The parasympathetic division is responsible for the contraction while the sympathetic is responsible for the dilation of the pupil. In general, parasympathetic impulses increase digestive activities whereas sympathetic impulses inhibit them.

At times of stress the sympathetic division is of great importance in preparing the body for action. It sends accelerator nerves to the heart, vasodilator nerves to blood vessels in the heart and skeletal muscles, and vasoconstrictor nerves to superficial blood vessels. Thus this system speeds the action of the heart, dilates the blood vessels in the muscles (including the coronary vessels of the heart muscle), and constricts the vessels in the skin. This diverts a greater blood supply to muscles and other vital organs. The sympathetic division also dilates the pupils of the eyes for maximum peripheral vision, quiets the gastrointestinal tract, and stimulates the production of epinephrine from the adrenal glands. Epinephrine liberates sugar from the liver and frees fatty acids from adipose tissue for immediate use in energy production.

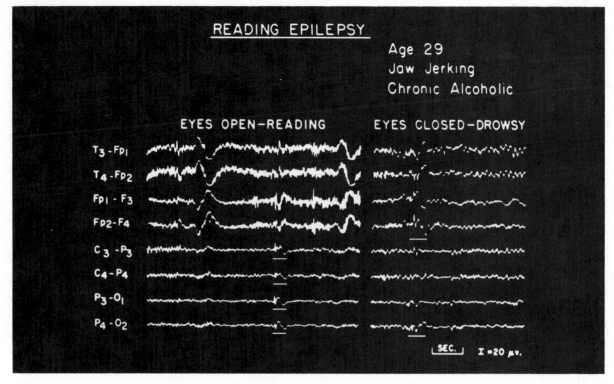

READING EPILEPSY

Age 29
Jaw Jerking
Chronic Alcoholic

EYES OPEN—READING   EYES CLOSED—DROWSY

T_3-Fp_1
T_4-Fp_2
Fp_1-F_3
Fp_2-F_4
C_3-P_3
C_4-P_4
P_3-O_1
P_4-O_2

SEC. $I = 20 \mu v.$

Figure 9-11 The abnormal brain wave pattern of a person with epilepsy.

Autonomic chemicals.   Until recently, the effects of autonomic nerves on the muscles and glands they stimulate have been thought to be due to the nerve impulses themselves. Now, the autonomic nervous system is known to produce various chemical transmitters at synapses as well as at points of contact between nerves and effector organs. The parasympathetic nerve endings throughout the body have been found to liberate acetylcholine. Sympathetic nerve endings have been found to liberate norepinephrine. This has led to the use of the terms **cholinergic** for those fibers that liberate acetylcholine and **adrenergic** for those that liberate norepinephrine.

DISORDERS OF THE NERVOUS SYSTEM

Cerebral Palsy

Cerebral palsy is characterized by *spastic* shaking movements or tremors as well as by loss of motor control. This condition is caused by damage to the motor areas of the brain during fetal development, at birth, or in early infancy. This disorder does not become more severe as time goes on and thus is not progressive. However, once damage to the motor areas occurs, it is irreversible.

Epilepsy

This disorder is due to a sudden burst of irregular electrical activity in the brain, sometimes referred to as a "storm." Diagnosis of this disorder is often achieved by recording brain waves with the electroencephalograph. The most common types of epilepsy are **petit mal, grand mal,** and **psychomotor epilepsy.** If motor areas are involved, there can be severe convulsions associated with the seizure. Various combinations of modern medications have brought complete control of convulsions in many epileptics.

Parkinson's Disease

Parkinson's disease, also known as Parkinson's syndrome, is linked to a disorder in the basal ganglia. It is a progressive disorder associated with a lack of a body chemical called dopamine. It is characterized by unnecessary skeletal movements that interfere with voluntary actions, resulting in shaking or tremors. The face may take on a "masklike" appearance because of the rigidity of facial muscles. Symptoms are often relieved by the administration of the drug L-dopa.

Cerebral Apoplexy

Cerebral apoplexy or stroke is also called cerebral vascular accident (CVA). In this disorder there is a serious disruption of blood circulation to the brain as a result of a blood clot (thrombus) or broken blood vessel (hemorrhage). When the blood supply to the brain is hindered, the brain doesn't receive sufficient oxygen, resulting in the death of irreplaceable neurons. Paralysis and loss of feeling may result, the paralysis appearing on the side of the body opposite to the site of brain damage. The symptoms are most severe at the onset, but as pressure on the brain is relieved, there may be some recovery.

SUMMARY

The brain and spinal cord are protected by the bones of the skull and spinal vertebrae, respectively, the cerebrospinal fluid, and the meninges, which consists of the outer dura mater, the middle arachnoid membrane, and the inner pia mater. The spinal cord serves as a center for reflex activity as well as being a two-way conducting pathway.

The brainstem is found above the foramen magnum and includes the medulla oblongata, pons, and midbrain. The medulla contains vital centers for heart rate, breathing, and vasomotor control. The pons contains the superior, middle, and inferior peduncles, or tracts. The midbrain controls many visual and auditory reflexes. The cerebellum functions to coordinate muscular activity that originates in the cerebrum and maintains muscle tone and physical balance.

Summary Table of the Autonomic System

Organ Innervated	Action of Sympathetic System	Action of Parasympathetic System
Heart	Strengthens and accelerates heart beat	Weakens and slows heart beat
Arteries	Constricts arteries and raises blood pressure	Dilates arteries and lowers blood pressure
Digestive tract	Slows peristalsis, decreases activity	Speeds peristalsis, increases activity
Urinary bladder	Relaxes bladder	Constricts bladder
Muscles in bronchi	Dilates passages, makes for easier breathing	Constricts passages
Muscles of iris	Dilates pupil	Constricts pupil
Muscles attached to hair	Causes erection of hair	Causes hair to lie flat
Sweat glands	Increases secretion	Decreases secretion

VOCABULARY REVIEW

Match the statement in the left column with the word(s) in the right column. Place the letter of your choice on your answer paper. *Do not write in this book.*

1. The large opening in the skull at the base of the brain.
2. The membranes that cover and protect the spinal cord and brain.
3. The H-shaped center of the spinal cord and the outer-surface nervous tissue of the cerebrum.
4. The scientific name for the brain.
5. The lowest part of the brainstem.
6. The spaces within the brain filled with cerebrospinal fluid.
7. That which divides the cerebrum into two hemispheres.
8. The outer surface of nervous tissue on the cerebrum.
9. Special sensory end organs.
10. A term used to describe a nerve made up of both sensory and motor neurons.

a. cortex
b. encephalon
c. foramen magnum
d. gray matter
e. longitudinal fissure
f. medulla oblongata
g. meninges
h. mixed
i. receptors
j. ventricles
k. white matter

TEST YOUR KNOWLEDGE

Group A

On your answer sheet, write the letter of the word(s) that correctly completes the statement.

1. Neural foramina are (a) nerve junctions (b) openings in the vertebral column through which spinal nerves pass (c) nerve cells in the spinal cord (d) coverings for the brain and cord.
2. The dura mater is (a) found covering peripheral neurons (b) a thin, delicate covering of the surface of the spinal cord (c) a tough, protective covering lining the vertebral canal and skull (d) a membrane containing blood vessels.
3. The heart, lungs, and the abdominal areas are served by the (a) olfactory nerve (b) oculomotor nerve (c) vagus nerve (d) commissural fibers.
4. The somatic division of the PNS includes (a) sympathetic and parasympathetic components (b) motor and sensory neurons (c) association fibers (d) all the facial nerves.
5. The spinal nerves are all (a) efferent nerves (b) afferent nerves (c) mixed nerves (d) unmyelinated.
6. The respiratory center in the brain is located in the (a) medulla (b) cerebellum (c) midbrain (d) middle peduncle.
7. A simple reflex arc represents (a) a learned response (b) an unlearned response (c) a partially learned response (d) intelligence.
8. The cerebellum is concerned with (a) balance, coordination of motion, and muscle tone (b) coordination of impulses between the cortex of the brain and the thalamus (c) sleep and emotion (d) conscious sensory activity.
9. The autonomic nervous system is responsible for (a) involuntary control of our internal environment (b) voluntary control of respiration (c) coordination of muscular activity (d) sense perception.

Group B

Write the word or words on your answer sheet that will correctly complete the sentence.

1. The brainstem includes the _____, _____, and _____.
2. Extending into the basal portion of the hypothalamus is the _____.
3. Within the cerebrum, myelinated fibers that join different parts in the same hemisphere are called _____.
4. _____ are located on the peripheral ends of sensory neurons.
5. Groups of afferent fibers form _____ nerves.

SEEING WITH LASERS AND ULTRASOUND

For many years blind people have used guide dogs or canes to help them travel independently. Scientists have now developed new devices that can be used alone or with a guide dog to help make the blind person more aware of his environment. Some devices use ultrasound — with frequencies of sound that are too high to be heard by the human ear. One such device is a box that hangs around the neck, leaving the hands free. This device can be used by people with multiple handicaps because it gives signals both by tones and by vibrations. The box sends out a beam of ultrasound directed in front of the user. If the beam hits an obstacle, some of it is reflected back to the box, where it is converted to tones that can be heard. A low-pitched buzz indicates that an object is between three and six feet away; a high-pitched tone indicates that an object is within three feet. If the person is also deaf, the neck strap will vibrate if the object is between three and six feet away. The whole box will vibrate if the object is closer than three feet. Other sonic devices are fitted into eyeglass frames and can easily detect objects, such as tree branches, that are located at head height. ·

The laser cane uses three beams of infrared light to detect obstacles. One beam is directed straight ahead and detects objects that are five or twelve feet from the cane (the user selects the distance). When the beam encounters an obstacle, it is reflected back to the cane, where electronic circuits either cause a pin to vibrate or emit a tone as a signal to the user. A second beam is directed upward to detect objects at head height at a distance of 2½ feet. This reflected beam causes the loudspeaker in the crook of the cane to emit a beep. The third beam is directed down to detect drop-offs such as curbs or stairs that are deeper than five inches. When the beam is not reflected, the loudspeaker emits a low tone. If for some reason the electronics malfunction, the user is not stranded because the cane can still be used as a traditional long cane.

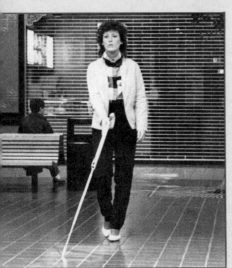

CHAPTER 10

THE EYE AND VISION

Objectives

A. Identify the structures of the eye and describe the function of each
B. Name the two kinds of light-sensitive cells and describe the functions of each
C. Summarize the way in which light waves passing through the eye are converted into images in the brain
D. Define the terms *refraction, convergence*, and *accommodation*
E. Describe visual defects and how they may be corrected

ARCHITECTURE OF THE EYE

Various kinds of sense receptors were discussed in Chapter 8, including: **exteroceptors**, which detect stimuli from our surroundings; **proprioceptors**, which send impulses to the brain from our muscles, tendons, and joints; and **interoceptors**, which are sensitive to changes that occur in our internal organs. The latter two operate primarily below the level of consciousness. We are not aware of them, although they are vital to our well-being. Without them we could not have coordinated movement or maintain internal homeostasis. The receptors activated by changes in the external environment are those with which we are most familiar. Whatever the stimulus, whether mechanical, thermal, electromagnetic, or chemical, conscious interpretation of our environment, or sensation, is made possible through the action of **sensory units**. These units consist of: (1) receptors, (2) sensory pathways through which impulses from the receptors are transmitted to the brain, and (3) brain centers that interpret these impulses. An example of activation due to a change in the environment is the way the ear responds to sound waves reaching it. The sense of hearing involves transforming these mechanical vibrations into nerve impulses which travel to the brain, where they are interpreted as sound. Thus the ear acts much like a ***transducer***, a device that converts one form of energy (the kinetic energy of sound waves) into another (the electrochemical energy of nerve impulses). In a similar manner, the sense organs of

trans = across

169

sight, smell, taste, touch, and equilibrium (associated with the sense of hearing) also act as transducers.

Of all the special sense organs, those that provide us with the greatest knowledge of our environment are the **eyes.** Eyes have highly specialized receptors that convert light energy into nerve impulses. These impulses are transmitted along the **optic nerves** to the brain. The cortex of the occipital lobe of the cerebrum interprets these impulses as **vision,** or **sight.** The eye may be exposed to light and then initiate an impulse along the optic nerve. However, if the impulse does not reach the visual center of the cortex, vision is not possible.

Accessory Organs of the Eye

The accessory organs of the eye serve to house and protect it from external hazards. The eyes are located within bony sockets of the skull called **orbital cavities.** Each orbital cavity is surrounded by seven different skull bones that give it the form of a cone-shaped depression. In front, each orbital is bordered by three bones: the *frontal*, the *superior maxillary*, and the *zygomatic bones.* The bones forming the posterior wall of the orbital cavity are the *sphenoid, lacrimal, ethmoid*, and a small portion of the *palatine.* These skull bones provide a protective and rigid housing for each eye. (Refer to Fig. 5-5.)

Two openings are found in the bones forming the posterior wall of the orbital cavity. The smaller opening, the **optic foramen,** allows both the **optic nerve** and the **ophthalmic artery** to enter the eye. The second opening is a narrow slit called the **superior orbital fissure.** This opening serves as a pathway for veins, other arteries, and those nerves that carry impulses to the muscles of the eye. The orbital cavities also contain:

ophthalmo = eye

- The extrinsic eye muscles, which control movements of the eyes
- The lacrimal apparatus, which is responsible for producing tears
- Fasciae, or tissues, which hold structures in position
- A heavy padding of fat tissue, which cushions and absorbs most of the shock from blows received by the external surface of the eye

Eye muscles. Two groups of muscles operate in the process of seeing. One group, the **intrinsic muscles,** is found within the eye. The second group, the **extrinsic eye muscles,** control certain eye movements; the properties of this group will be discussed first. (See Fig. 10-1.)

Six striated muscles are responsible for the movement of the eye. The insertions of these muscles lie along the midline, or equator, of the eyeball. Their origins *except* for one, that of the **inferior oblique muscle,** are near the apex of the orbital cavity. Two pairs of these muscles are at right angles to each other. Depending on their position, they are called the **superior rectus; inferior rectus; medial,** or **internal, rectus;** and **lateral,** or **external, rectus.** The remaining two muscles of each eye are set at an angle to the other four. Before attaching to the upper part of the eye, the *superior oblique* passes through a small fibrous ring,

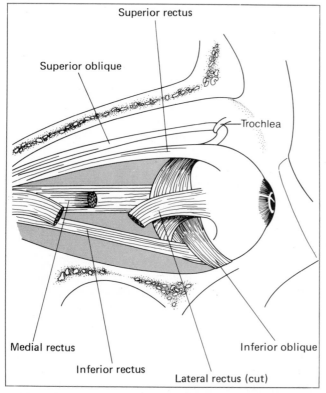

Figure 10-1 Lateral view of right eye showing the extrinsic eye muscles.

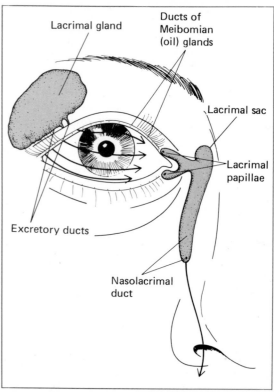

Figure 10-2 The lacrimal apparatus.

called the **trochlea,** located on the upper medial surface of the orbital cavity. The *inferior oblique* arises from that part of the orbital wall formed from the superior maxilla and inserts on the inferior surface of the eyeball between the lateral and superior recti muscles. Table 10-1 summarizes features of the extrinsic eye muscles (page 187).

The lacrimal apparatus.   The production of tears is the function of the **lacrimal gland.** The gland is located along the lateral superior wall of the orbital cavity in a slight depression of the frontal bone. Tears have a germicidal effect and are composed mostly of water and a small amount of salt and mucus. The secretion flows from the gland through several excretory ducts to lubricate the surface of the eye. Some of the tears evaporate from the eye's surface, but those that do not are drained into the nasal cavity by way of two **lacrimal ducts,** located at the inner margin of the eye. These ducts connect to form the **nasolacrimal duct,** which leads to the nasal cavity. When the production of tears increases, because of pain or some emotional state, the lacrimal ducts are unable to carry away the added quantity, and tears flow over the lower lid of the eye (see Fig. 10-2).

The eyebrows, eyelids, and eyelashes.   The *eyebrows* are arched coarse hairs located on the upper circumference of the orbital cavity. They serve to protect the eyes from foreign particles, perspiration, and direct rays of light.

The two **eyelids,** or **palpebrae,** are folds of skin that protect the eye by covering its surface. Sphincter-type muscles, the **orbicularis oculi,** open and close the eyelids. A small muscle called the **levator palpebrae superioris** aids in elevating the upper eyelid. Although the eyelids are under voluntary control, the conscious movement of them usually is not fast enough to prevent injury to the eyes. The eyes are therefore controlled by an automatic reflex action that assists in protecting the eyes. The visual receptors within the eye detect an approaching object (stimulus) and send impulses to the brain. Without conscious thought, the brain responds by sending impulses to the appropriate muscles, which quickly close the eyelids. This reflex action occurs in less than a second.

Attached to the anterior margin of the eyelid are rows of short curved hairs, the **eyelashes.** The eyelashes receive lubricating secretions from the sebaceous glands. If one of these glands becomes infected, a *sty* may develop. Posterior to the lashes are larger oil-producing glands called **tarsal,** or **Meibomian, glands.** Their oily secretions prevent the eyelids from sticking together.

The conjunctiva.   Attached to the inner surface of the eyelids, and continuous over the anterior surface of the eyeball, is a thin transparent mucous membrane, the **conjunctiva.** Various irritants or lack of sleep can produce inflammation of this highly vascular membrane. This condition is known as conjunctivitis.

Components of the Eye

The eyeball (Fig. 10-3) has the shape of a sphere about 24 millimeters (1 inch) in diameter. It is slightly longer than it is wide. This difference is due to the bulge caused by the presence of the **cornea** at the anterior surface of the eyeball.

sclero = hard

The sclera and cornea.   The wall of the eyeball is composed of three layers. The outer layer, called the **sclera,** or **sclerotic coat,** is a white, or opaque, tough, fibrous membrane that helps to maintain the spherical shape of the eye. The six extrinsic eye muscles insert onto this layer. The sclerotic coat continues to the front of the eyeball to form the transparent and colorless **cornea,** that part of the eye through which light waves pass. Thus the cornea is often called *the window of the eye.* The cornea is supplied with many free nerve endings and therefore is extremely sensitive to particles that come in contact with its surface. The cornea and sclera both lack blood vessels. At the junction of the cornea and sclera are venous sinuses, or venous passages, called the **canals of Schlemm.** These canals drain the fluid that accumulates behind the cornea into the venous system of the eyeball.

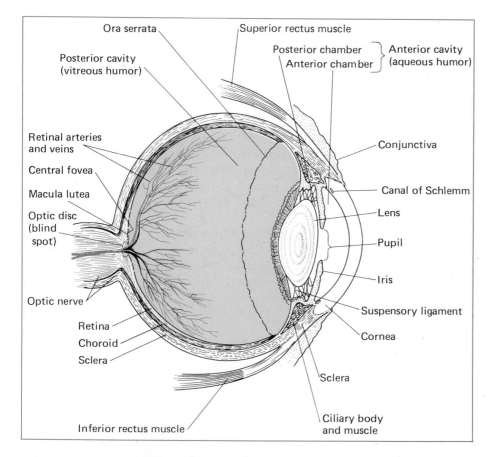

Figure 10-3 **Structure** of the eyeball seen in a sagittal section.

The uvea.   The middle layer covering the eyeball consists of the **choroid coat, ciliary body,** and **iris.** These parts are collectively known as the **uvea.** The choroid supplies blood vessels to the layers of the eye. The choroid coat also contains pigment granules that prevent the reflection of light within the eye. The granules function in the same manner as the black interior of a camera does—they absorb stray light that wou~~ld~~ blur the image. The ciliary body contains one of two smooth _intri~~nsic~~_ _muscles._ The fibers of this muscle support and modify the shape ~~of the~~ lens.

Attached to the anterior portion of the ciliary body is the iris. This highly vascular tissue contains the other smooth intrinsic muscle. This muscle contains two kinds of fibers. A circular arrangement of fibers form the _sphincter layer._ Other muscle fibers _radiate_ from the iris. The hole in the center of the iris is the **pupil.** The pupil is not a structure, but an _opening_ for light to enter the eye. Contraction of the sphincter muscle decreases the opening of the pupil. Contraction of the radial fibers increases the opening of the pupil. Thus the amount of light entering the pupil of the eye is regulated by the muscles of the iris. A decrease in pupil size becomes evident when a bright light is flashed in the eye. If the light is dimmed, the pupil becomes larger.

The iris is the colored portion of the eye, and its coloration depends on the amount and distribution of pigment. If no pigment is present, the iris appears pink, as in an albino. The presence of pigment is found to be in two distinct layers. The layer toward the back of the eye contains large granules of blue pigment. The layer in the front may either totally lack pigment or contain a certain amount of the pigment *melanin.* If this layer lacks melanin, the iris is blue; if melanin is present, the iris could be gray, green, brown, or even black, depending on the amount of melanin present. The eyes of many newborn babies are blue since it takes a number of days for the melanin to form. The amount of melanin found in the iris is controlled by the inheritance of specific genes that determine eye color.

The retina. The *retina* is the innermost layer of the eye. It completely lines the inner surface of the eye up to the ciliary body and contains the cells responsible for converting light into impulses that are transmitted to the central nervous system. These cells are found in three

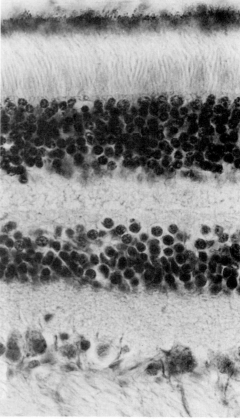

Figure 10-4 Schematic diagram of a enlarged section of the retina and a photograph of this region.

layers in the retina, as shown in Fig. 10-4. Light rays enter from the top of the diagram. The rays travel through the layers of the retina to the pigment layer, where they are absorbed. The layer closest to the choroid contains photoreceptor neurons responsible for converting light waves into impulses. These cells are called **rods** and **cones** because of their microscopic appearance. The axons of the rods and cones synapse with *bipolar neurons* in the second or middle layer of the retina. Finally, the axons of the bipolar cells synapse with neurons in a layer containing multipolar *ganglion cells*. This third layer is found closest to the inner chamber of the eye. Axons of these ganglion cells all pass across the inner surface of the retina and exit in one area at the posterior of the eye. These axons form the *optic nerve* (cranial nerve II), which leads to the brain.

Rods and cones.   These are elongated nerve cells that contain light-sensitive molecules. The rods are more cylindrical than the cones and measure between 40 and 60 microns in length and about 2 microns in width. Each rod is divided into an inner and outer segment. The outer segment contains a photosensitive chemical called **rhodopsin**, known also as **visual purple**. There are many more rods than cones. It is estimated that there are as many as 125 million rods in each retina.

The cones are conical, or flask-shaped, and their pointed end is situated nearer the choroid layer. Like the rods, these cells have an inner and outer segment. There are three kinds of cones each containing a slightly different photochemical substance within its outer segment. These differences cause the cones to be stimulated by light of different colors. Cones are therefore responsible for color vision. F. W. Campbell and W. A. H. Rashton at the University of Cambridge, England, devised the first successful method of distinguishing between the cone pigments and called the three pigments **erythrolabe**, **chlorolabe**, and **cyanolabe**. The cones measure about 25 microns in length and up to 6 microns in width. There are from 6 to 7 million cones in each retina, concentrated primarily in the posterior portion of the eye. (See Fig. 10-4.)

An *ophthalmoscope* is an instrument used to examine the retina. In the center of the retina is a yellow disk, called the **macula lutea**. See Figure 10-5. In the center of the disk is a small depression called the **fovea centralis**. This is the region for acute (sharp) vision as well as color sensitivity since it contains only cones. The area away from the fovea is the **extrafoveal** or peripheral region and contains a greater density of rods. This area is primarily concerned with contrast sensitivity. Slightly medial to the macula is a circular area, white in color, where the optic nerve leaves the eye. This is called the **optic disc**. It contains no rods or cones, so there is no visual reception in the disc. It is referred to as the **blind spot**. The *central retinal artery and vein* with their numerous branches can be seen in the center of the optic disc. These supply the interior of the eye with oxygen and nutrients and carry away waste products.

scope = vision; also, instrument for visual examination

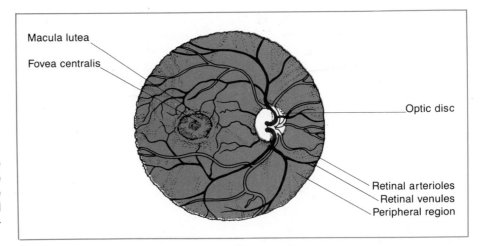

Macula lutea

Fovea centralis

Optic disc

Retinal arterioles
Retinal venules
Peripheral region

Figure 10-5 **View of the posterior surface of the retina. This part of the eye can be examined with an ophthalmoscope.**

Optical structures.   Just behind the pupil and iris is a transparent cellular body, the **crystalline lens.** The lens is disc-shaped, somewhat elastic, and about 9 millimeters in diameter. Its two surfaces are convex, thus forming somewhat of a biconvex lens (although the posterior surface has a greater curvature than the anterior surface). The crystalline lens is held in place by a fibrous *suspensory ligament.* This ligament is attached to the ciliary muscle, which controls the amount of tension exerted on the lens. Located between the cornea and the iris is the **anterior chamber.** This chamber is filled with a transparent watery fluid called **aqueous humor.** This fluid helps to maintain the shape of the cornea and also has a nutritive function. The aqueous humor is produced by the ciliary body and is constantly replenished by the blood vessels in the **posterior chamber,** a small area between the iris and the lens. Aqueous humor is reabsorbed by a venous structure, the **canal of Schlemm**, located along the peripheral border of the anterior chamber.

PHYSIOLOGY OF THE EYE

The retina responds to varying intensities of light, both **achromatic**, or colorless, and **chromatic**, or colored, light. The receptors responsible for detecting black-and-white images are the rods. The cones are responsible for color vision. The previous discussion of the anatomy of the eye helps explain the physiology of vision.

Mechanics of Vision

When light energy stimulates the rods and cones, nerve impulses are produced. These impulses are transmitted to the visual center in the cortex of the brain, where they are interpreted. Understanding the physiology of vision requires learning some of the physical properties of light.

Light is a form of **radiant energy**. The wavelengths of radiant energy needed to stimulate the eye vary from 390 millimicrons (390 millionths of a millimeter) to 760 millimicrons (760 millionths of a millimeter). This range is often written as 3900 to 7000 Ångstrom units. (One ten-millionth of a millimeter equals 1 anstrom unit.) These wavelengths are very short and travel at the tremendous speed of 3×10^5 kilometers per second.

Refraction. Light rays traveling from a less dense medium and entering a more dense medium are **refracted** (bent) toward an imaginary line perpendicular to the surface of the denser medium. (See Fig. 10-6.) The angle between the refracted rays and the perpendicular is called the **angle of refraction**. The more dense the medium, the greater the angle of refraction. However, if the rays strike perpendicularly to the surface, they are not refracted, regardless of the density of the media.

Figure 10-6 The straw appears to bend as it enters the water because of the refraction of light.

When light passes through a transparent object, the rays are first refracted when entering the object and then are refracted again when leaving the object. The angle of refraction depends not only on the density of the object but also on its curvature. If a transparent object, such as a lens, varies in thickness, the light rays will bend toward the thicker portion. A lens that has a surface that is uniformly curved and thicker at the edge than in the middle is a **concave** lens. Parallel rays of light passing through this type of lens will *diverge*, or spread outward. A **convex** lens is uniformly curved and thicker in the middle than at the edge. Parallel light rays that pass through a convex lens will converge and be brought to a single **focal point**, or **focus.** (See Figs. 10-7 and 10-8.)

The distance from the center of a lens to the focal point is called the **focal length** of the lens. If both sides of the lens have a similar curvature, the lens is called *biconcave* or *biconvex*. The first refracting surface encountered by light entering the human eye is the transparent cornea, which accounts for nearly 75 percent of the total refraction. The remaining 25 percent of refracting power is attributed to the crystalline lens. The lens acts to focus images on the retina. As in a camera, the image that falls on the retina is always upside down and reversed. The brain corrects the position and interprets the image.

Convergence. Human beings are able to use two eyes to focus on one object, giving what is known as single binocular vision. This is achieved as light rays reaching both eyes from the same object are bent so that they strike equivalent locations on both retinas. As the object moves closer to the eyes, it is necessary for the eyeballs to turn medially in order for the rays to continue striking equivalent spots on both retinas. If the object is brought very close to the eyes, convergence reaches the point where the eyes appear "crossed."

Accommodation. Of greater importance in achieving a sharp focus on the retina is the ability of the lens to change its shape, or **accommodate.** Many theories have been proposed to explain accommodation. The theory most widely accepted is that contraction of the ciliary muscle re-

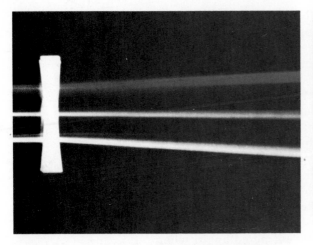

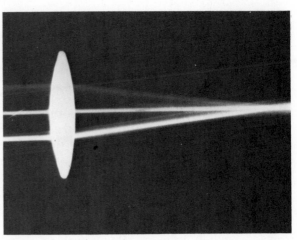

Figure 10-7 Parallel rays of light passing through a concave lens are refracted and diverge. Rays passing through the center of the lens are not refracted.

Figure 10-8 Parallel rays of light passing through a convex lens are refracted and converge. Rays passing through the center of the lens are not refracted.

leases tension on the lens and the lens becomes *more* convex. The action of the ciliary muscle is transmitted to the lens by means of the suspensory ligament. This adjustment is made when objects are closer than 6 meters. When the ciliary muscles are relaxed, the suspensory ligament exerts tension on the lens and *decreases* the lens' curvature. This action produces a sharper image of more distant objects.

The entire process of accommodation is automatic. When the object changes position and the image becomes blurred, sensory impulses are carried to the brain. The brain in turn sends motor impulses to the ciliary muscles to adjust their contraction. As a result, the image is focused clearly again.

The size of the opening of the iris plays an important role in controlling the amount of light that enters the eye. During focusing of the eye, the pupil (opening) changes size and the refractive system of the eye focuses the image. Constriction of the pupil is under the control of the parasympathetic nervous system. Dilation of the pupil is under the influence of the sympathetic nervous system. The pupil also plays a part in the process of accommodation. Contraction of the sphincter muscles constricts the pupil of the eye. The pupil then becomes smaller, thus preventing rays of light from entering the periphery of the lens. This results in a sharp image on the retina, even though the lens has greater curvature.

The Visual Fields

The area that can be viewed without moving the eyes is called the **field of vision.** Visual attention is primarily focused directly in front of the eyes. Yet, we are aware of movement on either side of the eyes.

Peripheral vision.   Peripheral vision gives the eyes a wider view without involving movement of the head. However, on those areas of the retina associated with peripheral vision, a very sharp image is not formed. These areas do not contain many cones, and thus peripheral vision does not include bright color. The field of peripheral vision can be determined by a simple experiment. Hold both hands together and move them about 30 centimeters away from the front of the nose. Looking straight ahead, move the hands slowly apart to the side in the form of an arc toward the back of the head until they disappear from view. The angle that is formed by this movement of the hands is considered to be the field of peripheral vision.

peri = around

Binocular vision.   When one eye is closed, the brain can receive visual impulses only from the open eye. This is termed **monocular vision.** Normally, however, the brain receives impulses from both eyes simultaneously. Although the eyes are set at a slight distance apart, the two separate images fuse into a single image. To understand the phenomenon of binocular vision, we must examine the route taken by the retinal impulses in going from the eye, through the optic nerves, and finally to the visual centers in the brain.

The retina is divided into a nasal, or medial, portion, and a temporal, or lateral, portion. The visual field from a point directly in front of the eye to where the bridge of the nose blocks the view is recorded on the temporal (lateral) half of the retina. The area from straight ahead to the margin of our peripheral view (approximately 95°) is recorded on the nasal (medial) half of the retina.

The optic nerves associated with the temporal half of the retina *do not* cross at the **optic chiasma.** Instead, their fibers go directly to the occipital lobe on the same side. The fibers from the nasal half of each retina *cross* at the optic chiasma and enter the optic tract with the temporal fibers from the opposite eye. Thus, both images of an object on the right of the viewer are received in the left occipital lobe, and therefore create a single visual impression. The terminations of all these fibers in a particular brain area compose the field of interpretation in the visual center.

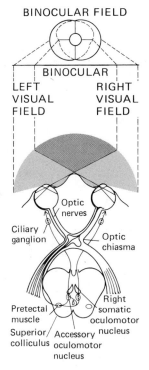

Figure 10-9 **The route of the optic nerves indicates the visual pathways.**

Depth perception.   One of the remarkable facts about the visual process is that an object can be seen in three dimensions—its vertical plane, its horizontal plane, and its vertical, or lateral, distance. To visualize these measurements, the eyes must be equipped for **depth perception,** or **stereoscopic vision.** Since the eyes are set slightly apart, each eye views a single object from a slightly different angle. The images formed on the retina are thus not quite identical. However, the visual centers in the brain interpret these images as one image that has depth and thickness.

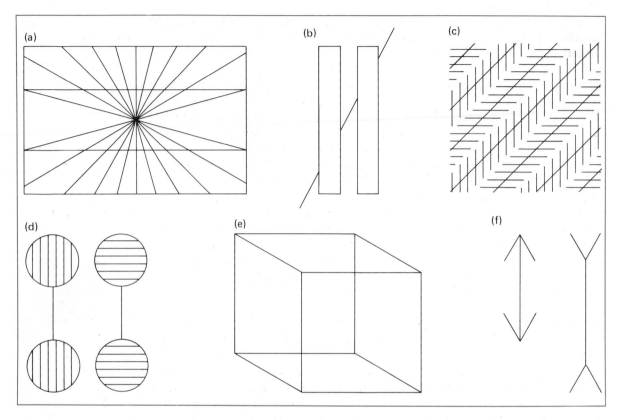

Figure 10-10 Visual illusions. (a) The horizontal lines are parallel. (b) The diagonal line is not staggered. (c) The diagonal lines are parallel. (d) The two pairs of circles are equally bright and the lines within them are equally sharp. (e) The cube seems to change in perspective (not size) when stared at. (f) The vertical lines are equal in length.

Judgment of spatial relationships. Our judgment of distance is a conditioned, or learned, capability and is dependent upon many different clues that are interpreted by the brain. Knowledge of the size of familiar objects is used as a means of judging their distance. For example, we can recognize the difference in height between a telephone pole and a mailbox post. If we see a telephone pole that appears to have the height of the mailbox, we have learned that the telephone pole must be a considerable distance away.

The degree of accommodation and convergence made by the eyes also gives the brain clues as to the relative position of objects. Differences in color between near and distant objects, as well as differences in the shadows found on their surfaces, are other clues used in learning how to judge depth. Such clues are part of our visual memory for judging spatial relationship of objects. However, changing certain reference points can produce **optical illusions,** which result from errors in visual judgment. Figure 10-10 gives some examples of optical illusions.

The Chemistry of Vision

The actual process of converting nerve impulses into sight is still not completely understood, but many of the chemical changes that occur are now known. The rods, which contain the chemical *rhodopsin*, or visual purple, are responsible for vision in dim light. Scientific research on rhodopsin indicates that it is a compound of a protein, **opsin,** combined with **retinene,** a pigment.

Dark and light adaptation.   Vision is reduced when going from bright sunlight to an area of dim light. After a period of time, however, the eyes are able to distinguish objects. This process is known as **dark adaptation.** Rhodopsin is stable in dim light, but when exposed to bright light, it dissociates into opsin and retinene, or *visual yellow.* The speed of this reaction depends upon the brightness of the light and length of time the rhodopsin is exposed. When the light intensity decreases, the reaction is reversed and rhodopsin is re-formed.

Vitamin A from the blood supply to the retina converts retinene back into rhodopsin. A lack of vitamin A will cause a slower return of dark adaptation, referred to as *nightblindness.* Intense light bleaches rhodopsin very rapidly. This explains why bright light may dazzle the eye and cause pain for a short period of time. **Light adaptation** occurs when returning from darkness into bright light.

Color vision.   The cone receptors in the retina contain a light-sensitive pigment. When the eyes are light adapted, they see objects in fine detail and in color.

What kind of light produces a rainbow? When sunlight passes through a triangular glass prism, an arrangement of colors (spectrum) appears on a nearby surface. White light is a combination of all the wavelengths of this spectrum. The same spectral effect is generally produced by refraction. Thus, rays of the sun striking droplets of water still in the air right after a rain are refracted to produce a rainbow. The wavelengths of colors that make up white light range from 3900 Å to 7600 Å. From the spectrum diagram shown in Fig. 10-11, it can be seen that any one color is actually a name for light of a particular wavelength. For example, the label *red* is given to the visual sensation stimulated by wavelengths starting at 7000 Å. The other colors range from orange to yellow, green, blue, and violet, which has a wavelength of about 4000 Å. From this list, one should note that the colors in the red end of the spectrum have the longest visible wavelengths, while those with the shortest wavelengths fall in the violet end of the spectrum.

Radiation with wavelengths shorter than those of the visual spectrum are called **ultraviolet rays.** True ultraviolet rays cannot be seen by the normal human eye. Their presence can be detected when they strike certain chemicals that contain fluorescent qualities. Exposed to such rays, these chemicals appear to glow or emit their own light.

Radiation with wavelengths longer than those within normal visual range are sensed as heat and are called **infrared rays.** Photographic film that is infrared-sensitive will take pictures in total darkness. The image on the film results from differences in the amount of heat given off or reflected by different parts of the photographed object.

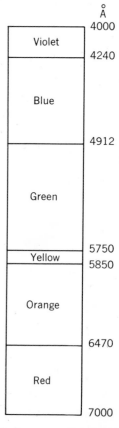

Figure 10-11 **The wavelengths (in Ångstrom units) of visible light and the colors associated with them.**

ultra = beyond

infra = below

White light is a combination of all the wavelengths of the visual spectrum. However, if only three of these wavelengths of light are superimposed upon a light surface, white light is produced. The colors used are red, green, and blue-violet. The actual wavelengths of these three colors have been assigned by an international standard: 7000 Å for red, 5460 Å for green, and 4358 Å for blue-violet. We call these three the **primary colors of vision.** To this list yellow, with a wavelength of 5800 Å, is often added.

All of the other colors represent certain combinations of these primary colors and are thus called **compound colors.** The primary colors of vision should not be confused with the primary colors of pigments. Primary pigment colors are red, blue, and yellow. From these three, nearly all colors can be produced.

The colors we see in objects around us are light waves reflected by the surfaces of those objects. The waves that are reflected stimulate the cones of the retina, and we say the surface is red, blue, green.

Many theories have been proposed to explain color vision. One of the oldest and most widely accepted theories was proposed by Young and Helmholtz early in the nineteenth century. According to their theory, there is a corresponding type of receptor in the retina for each of the three primary colors of light. The three kinds of cone receptors contain different photochemical substances, each chemical sensitive to one of the three wavelengths of the primary colors.

In recent years, through the experiments of MacNichol and other investigators, the three photosensitive pigments predicted by Young and Helmholtz have now been shown to exist. Investigations have been made with microelectrodes placed in different parts of the retina. Recordings show that illumination of the retina decomposes the light-sensitive pigments and changes in electric potential occur. Each pigment, contained in the cones, is affected by specific wavelengths of light.

Visible color is produced by combining the various stimulations of the cones in the proper proportions. In this way three-color information is processed and coded in the retina. Equal stimulation of the three types of cones is perceived as white light. These signals are relayed to the visual cortex where final analysis of color vision takes place.

It is possible to determine the zones of primary-color sensitivity on the surface of the retina. By recording measurements of the field of view for the three primary colors, it is possible to map the color receptors on the retina. Such a map also shows that the peripheral area of the retina is insensitive to any intense light.

As the intensity of white light gradually decreases—at dusk, for example—there is a shift from cone reception to rod reception. This shift from light adaptation to dark adaptation is called the **Purkinje shift.** The shift also occurs when light increases, as at dawn, and a change in the opposite direction returns rod vision to cone or color vision.

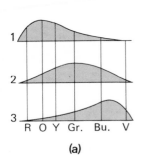

1

2

3

R O Y Gr. Bu. V

(a)

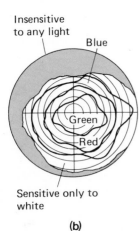

Insensitive to any light

Blue

Green

Red

Sensitive only to white

(b)

Figure 10-12 (a) Diagram of the three primary color sensations (Young-Helmholtz theory): (1) Represents the red, (2) the green, and (3) the blue color sensation. The lettering along the base line indicates the colors of the spectrum. (b) Diagram showing the distribution of color perception in the retina.

THE HUMAN BODY

SKELETAL SYSTEM

BONES

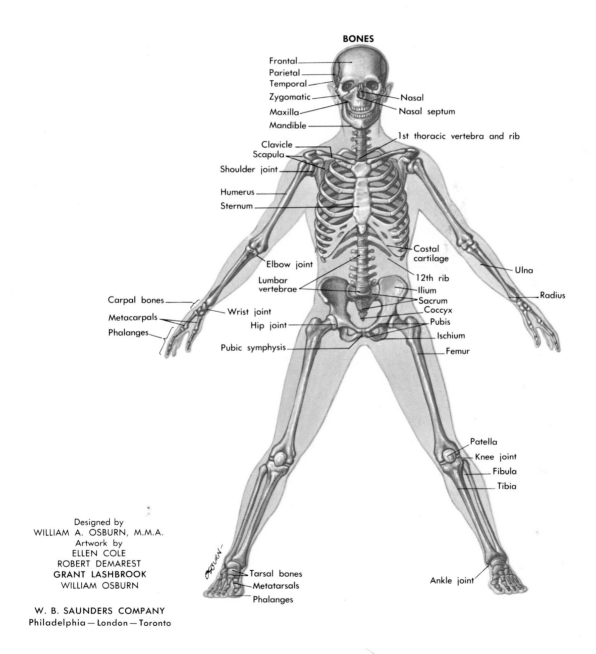

Frontal
Parietal
Temporal
Zygomatic
Maxilla
Mandible
Nasal
Nasal septum
1st thoracic vertebra and rib
Clavicle
Scapula
Shoulder joint
Humerus
Sternum
Costal cartilage
Elbow joint
Ulna
Lumbar vertebrae
12th rib
Ilium
Sacrum
Coccyx
Radius
Carpal bones
Wrist joint
Pubis
Metacarpals
Hip joint
Ischium
Phalanges
Pubic symphysis
Femur
Patella
Knee joint
Fibula
Tibia
Tarsal bones
Metatarsals
Phalanges
Ankle joint

Designed by
WILLIAM A. OSBURN, M.M.A.
Artwork by
ELLEN COLE
ROBERT DEMAREST
GRANT LASHBROOK
WILLIAM OSBURN

W. B. SAUNDERS COMPANY
Philadelphia — London — Toronto

SKELETAL MUSCLES

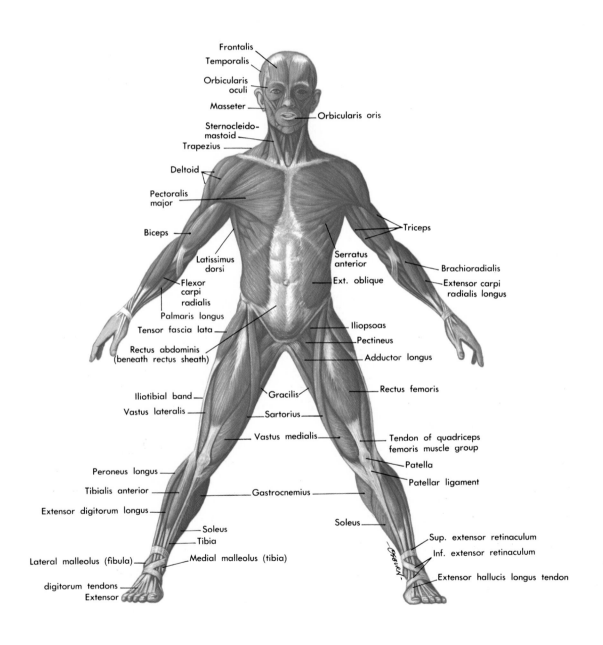

Frontalis

Temporalis

Orbicularis oculi

Masseter

Orbicularis oris

Sternocleido-mastoid

Trapezius

Deltoid

Pectoralis major

Triceps

Biceps

Serratus anterior

Latissimus dorsi

Ext. oblique

Brachioradialis

Flexor carpi radialis

Extensor carpi radialis longus

Palmaris longus

Tensor fascia lata

Iliopsoas

Pectineus

Rectus abdominis (beneath rectus sheath)

Adductor longus

Rectus femoris

Iliotibial band

Gracilis

Vastus lateralis

Sartorius

Vastus medialis

Tendon of quadriceps femoris muscle group

Patella

Peroneus longus

Patellar ligament

Tibialis anterior

Gastrocnemius

Extensor digitorum longus

Soleus

Soleus

Tibia

Sup. extensor retinaculum

Inf. extensor retinaculum

Lateral malleolus (fibula)

Medial malleolus (tibia)

Extensor hallucis longus tendon

digitorum tendons

Extensor

RESPIRATION AND THE HEART

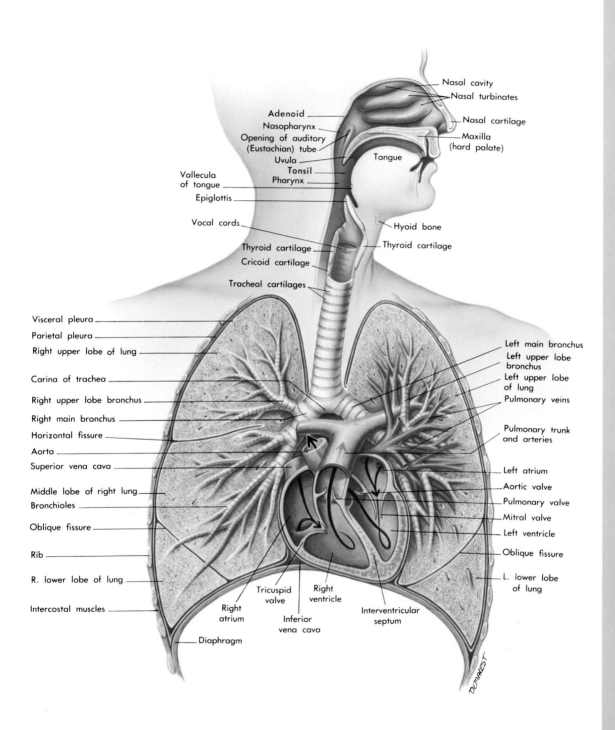

Nasal cavity
Nasal turbinates
Nasal cartilage
Adenoid
Nasopharynx
Maxilla
(hard palate)
Opening of auditory
(Eustachian) tube
Tongue
Uvula
Tonsil
Vallecula
of tongue
Pharynx
Epiglottis
Hyoid bone
Vocal cords
Thyroid cartilage
Thyroid cartilage
Cricoid cartilage
Tracheal cartilages

Visceral pleura
Left main bronchus
Parietal pleura
Left upper lobe
bronchus
Right upper lobe of lung
Left upper lobe
of lung
Carina of trachea
Pulmonary veins
Right upper lobe bronchus
Right main bronchus
Pulmonary trunk
and arteries
Horizontal fissure
Aorta
Left atrium
Superior vena cava
Aortic valve
Pulmonary valve
Middle lobe of right lung
Bronchioles
Mitral valve
Left ventricle
Oblique fissure
Oblique fissure
Rib
L. lower lobe
of lung
R. lower lobe of lung
Tricuspid
valve
Right
ventricle
Intercostal muscles
Right
atrium
Interventricular
septum
Inferior
vena cava
Diaphragm

DEMAREST

BLOOD VASCULAR SYSTEM

VEINS

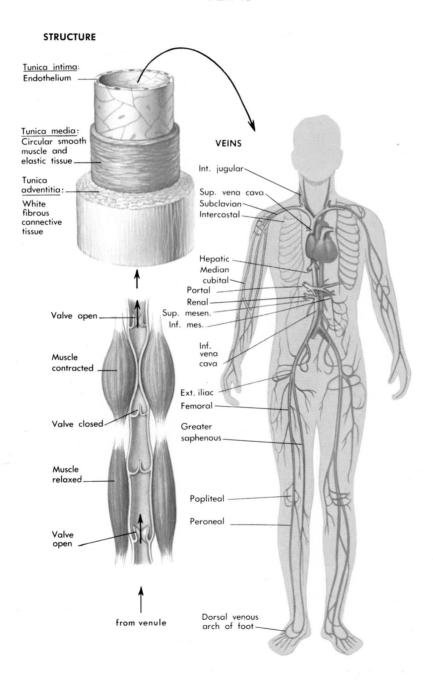

STRUCTURE

Tunica intima:
Endothelium

Tunica media:
Circular smooth
muscle and
elastic tissue

Tunica
adventitia:

White
fibrous
connective
tissue

Valve open

Muscle
contracted

Valve closed

Muscle
relaxed

Valve
open

from venule

VEINS

Int. jugular

Sup. vena cava
Subclavian
Intercostal

Hepatic
Median
cubital
Portal
Renal
Sup. mesen.
Inf. mes.

Inf.
vena
cava

Ext. iliac
Femoral

Greater
saphenous

Popliteal

Peroneal

Dorsal venous
arch of foot

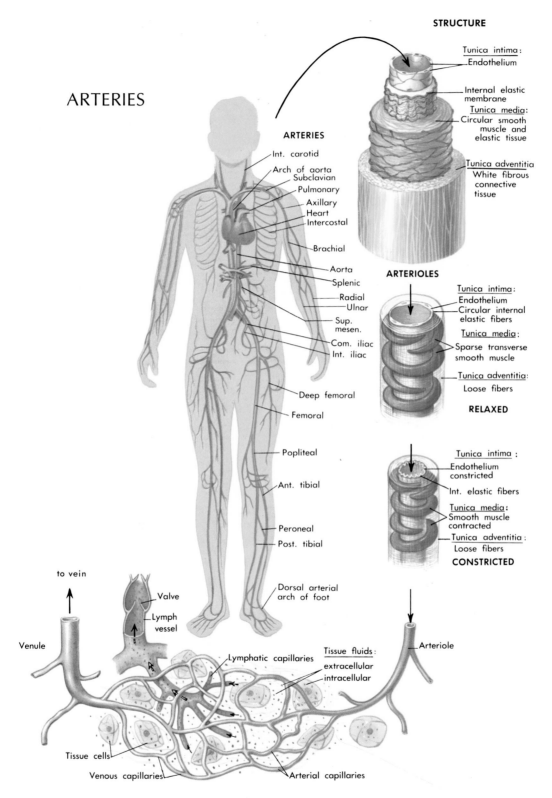

ARTERIES

STRUCTURE

Tunica intima:
Endothelium

Internal elastic membrane

Tunica media:
Circular smooth muscle and elastic tissue

Tunica adventitia
White fibrous connective tissue

ARTERIES

Int. carotid
Arch of aorta
Subclavian
Pulmonary
Axillary
Heart
Intercostal
Brachial
Aorta
Splenic
Radial
Ulnar
Sup. mesen.
Com. iliac
Int. iliac
Deep femoral
Femoral
Popliteal
Ant. tibial
Peroneal
Post. tibial
Dorsal arterial arch of foot

ARTERIOLES

Tunica intima:
Endothelium
Circular internal elastic fibers

Tunica media:
Sparse transverse smooth muscle

Tunica adventitia:
Loose fibers

RELAXED

Tunica intima :
Endothelium constricted

Int. elastic fibers

Tunica media:
Smooth muscle contracted

Tunica adventitia :
Loose fibers

CONSTRICTED

to vein

Valve

Lymph vessel

Venule

Lymphatic capillaries

Tissue fluids:
extracellular
intracellular

Arteriole

Tissue cells

Venous capillaries

Arterial capillaries

A CAPILLARY BED

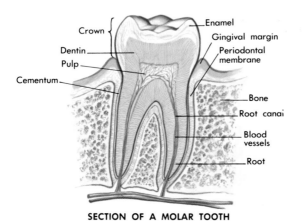

Crown
Enamel
Gingival margin
Dentin
Periodontal membrane
Pulp
Cementum
Bone
Root canal
Blood vessels
Root

SECTION OF A MOLAR TOOTH

DIGESTIVE SYSTEM

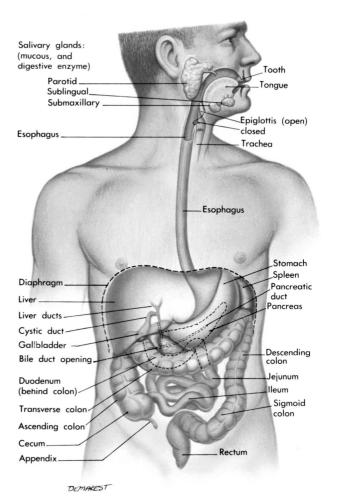

Salivary glands: (mucous, and digestive enzyme)
Parotid
Sublingual
Submaxillary
Esophagus

Tooth
Tongue
Epiglottis (open) closed
Trachea

Esophagus

Diaphragm
Liver
Liver ducts
Cystic duct
Gallbladder
Bile duct opening
Duodenum (behind colon)
Transverse colon
Ascending colon
Cecum
Appendix

Stomach
Spleen
Pancreatic duct
Pancreas
Descending colon
Jejunum
Ileum
Sigmoid colon
Rectum

DEMAREST

BRAIN AND SPINAL NERVES

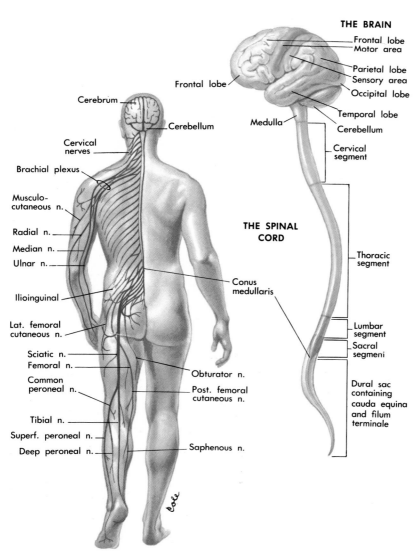

THE BRAIN

Frontal lobe
Motor area
Parietal lobe
Sensory area
Occipital lobe
Temporal lobe
Cerebellum
Cervical segment

Frontal lobe

Medulla

Cerebrum

Cerebellum

Cervical nerves

Brachial plexus

Musculo-cutaneous n.

Radial n.

Median n.

Ulnar n.

Ilioinguinal

Lat. femoral cutaneous n.

Sciatic n.

Femoral n.

Common peroneal n.

Tibial n.

Superf. peroneal n.

Deep peroneal n.

THE SPINAL CORD

Conus medullaris

Obturator n.

Post. femoral cutaneous n.

Saphenous n.

Thoracic segment

Lumbar segment

Sacral segmeni

Dural sac containing cauda equina and filum terminale

THE MAJOR SPINAL NERVES

ORGANS OF SPECIAL SENSE

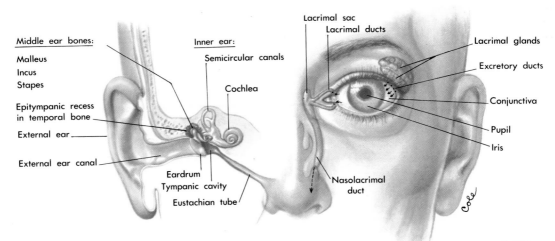

Middle ear bones:

Malleus

Incus

Stapes

Epitympanic recess
in temporal bone

External ear

External ear canal

Eardrum

Tympanic cavity

Eustachian tube

Inner ear:

Semicircular canals

Cochlea

Lacrimal sac

Lacrimal ducts

Lacrimal glands

Excretory ducts

Conjunctiva

Pupil

Iris

Nasolacrimal
duct

THE ORGAN OF HEARING

THE LACRIMAL APPARATUS AND THE EYE

HORIZONTAL SECTION OF THE EYE

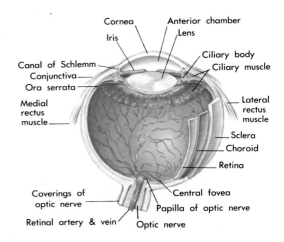

Cornea

Iris

Anterior chamber

Lens

Canal of Schlemm

Conjunctiva

Ora serrata

Medial
rectus
muscle

Ciliary body

Ciliary muscle

Lateral
rectus
muscle

Sclera

Choroid

Retina

Coverings of
optic nerve

Retinal artery & vein

Central fovea

Papilla of optic nerve

Optic nerve

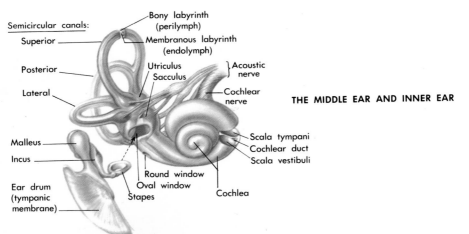

Semicircular canals:

Superior

Posterior

Lateral

Malleus

Incus

Ear drum
(tympanic
membrane)

Bony labyrinth
(perilymph)

Membranous labyrinth
(endolymph)

Utriculus

Sacculus

Acoustic
nerve

Cochlear
nerve

Round window

Oval window

Stapes

Scala tympani

Cochlear duct

Scala vestibuli

Cochlea

THE MIDDLE EAR AND INNER EAR

Afterimages.   Whenever we see a motion picture, we are actually seeing a series of still pictures flashed on a screen in which the objects appear to move. The activity in the photochemical substances within the receptors cannot keep pace with changes in visual stimuli and thus there is a carryover of visual impressions, known as *afterimages.* If the carryover is exactly the same as the original stimulus pattern, it is then a ***positive afterimage.***

By viewing a very bright object (for example, electric light) and then closing the eyes, the same visual image consciously remains even though the external stimulus has stopped. Staring at a colored object and then closing the eyes, one is conscious of the image but it is sometimes in the complementary color. A red image will appear to be green, since green is the complementary color of red. This ***negative afterimage*** is probably due to the fatigue of the originally stimulated receptors.

Characteristics of Color Vision

There are three qualities of color vision: *hue, saturation,* and *brightness.* Hue refers to the sensory response made by the eye to various wavelengths. If the eye is stimulated by light of a wavelength ranging from 6750 Å to 7000 Å, it corresponds to the color red. If light of a wavelength from 5000 Å to 5500 Å enters the eye, it is the color green.

Each wavelength within the visible spectrum represents a different hue. A hue of one single wavelength is said to have "complete saturation." Such a color is seldom encountered in nature, although it can be produced under controlled conditions in a laboratory. The light reflected from the objects around us contains varying amounts of white light. As a pure color is mixed with white light, it becomes less saturated and appears pale, or pastel.

Brilliance or brightness is a quantitative factor determined by the intensity or energy of the light waves reflected by a surface. In terms of pigments, brilliance depends on the amount of black added to the color. The appearance of a colored object is also affected by the brightness of the light falling on it. Objects that seem brilliantly colored in bright light will appear a darker color under dim illumination.

Visual acuity.   Not only are the cones responsible for color vision but also they are involved with the quality of vision. ***Visual acuity*** is the sharpness or degree of detail the eye can see. It is often measured in terms of the smallest distance that can be seen between two vertical black lines on a white background.

Many factors influence visual acuity:

- The brightness or intensity of illumination
- The size of the objects
- The color of the objects
- The retinal area on which the image of the object falls

Certainly if the image falls outside of the fovea centralis, visual acuity is greatly reduced.

Doctors use a simplified visual acuity test when they ask you to read the letters on an eye chart. The **Snellen eye chart** contains a series of letters, each line of the series having a different standard size. The normal eye can read all the letters when 20 feet away. Visual acuity in this case is given as 20/20. If visual acuity is poor, the smallest line read at 20 feet may be a line that the normal eye can read at 100 feet. Visual acuity is then recorded as 20/100. (See Fig. 10-13.)

DISORDERS OF THE EYE

Abnormalities of vision can be classified into two general groups: those caused by defects characteristic of all convex lenses and those caused by structural defects of the eye itself.

Defects in Lens Refraction

In all convex lenses, rays of parallel light that pass through the outer edge of the lens are refracted more than the rays passing through the center of the lens. This is called **spherical aberration.** It results in a series of focal points rather than the single focal point that is required for acute vision. If the aberration is uncorrected, the central visual field may be in focus, while the perimeter of the image is fuzzy or out of focus. The eye automatically corrects for spherical aberration by decreasing the size of the pupil and allowing only those rays of light near the center of the lens to pass through.

A distortion very similar to spherical aberration is **chromatic aberration.** White light passing through the perimeter of a convex lens breaks up into individual wavelengths. The short wavelengths (violet) are refracted more than the long wavelengths (red). In between these extremes, there is a progressive distribution of colors, as in a rainbow. The distortion causes a *halo* effect of colors that appear to surround the image. This defect is corrected by contracting the iris—the size of the pupil becomes smaller and prevents light rays from passing through the edges of the lens.

Structural Defects of the Eye

In an optical instrument such as a camera, a microscope, or a telescope, the quality of the image is largely determined by how perfectly the lenses are ground and polished. In a human eye the quality of the image is affected by the shape of the cornea, the lens, and the eyeball. Any slight irregularity of the cornea or lens or of the eyeball itself may result in abnormalities that may require correction. Nearly 95 million people in the United States wear contact lenses or eyeglasses to correct problems such as astigmatism, myopia, and hypermetropia.

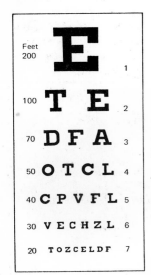

Figure 10-13 Snellen chart. Do you know your visual acuity?

Astigmatism.   The usual cause of astigmatism is a cornea, or more rarely a lens, that has imperfect curvature. As a result, the refracting power of the cornea or lens is not uniform at all points and light fails to come to a single, sharp focal point on the retina. Eyestrain and frequent headaches may result.

The eye specialist may use numerous devices to determine how to correct astigmatism. One such device is an **astigmatic dial**. It consists of radiating lines which all have the same width and darkness. A person with an astigmatism will see some lines more clearly than others. See Fig. 10-14. Astigmatism can usually be corrected by the use of properly prescribed eyeglasses or contact lenses.

Myopia (nearsightedness).   In this condition, parallel rays of light from distant objects come into focus in *front* of the retina instead of on the retina. This will occur if the refractive power of the cornea or lens is too great for the length of the eye or the eye itself is too long for the amount of refraction caused by the cornea or lens. It should be noted that as objects move closer to the eye, accommodation can bring the image into focus on the retina. Thus, a person may be able to read a book without corrective lenses. For distant vision, a *concave lens*, which causes parallel waves of light to diverge, is placed in front of the eye. This causes the focal point to fall on the retina and results in a clear, sharp image of distant objects. (See Fig. 10-15.)

Hypermetropia (farsightedness).   In this condition, light rays from near objects are brought to focus *behind* the retina. The eyeball is either too short for the refractive power of the lens, or the lens has insufficient refractive power. Distant objects on the other hand have a greater chance of being brought to a focus on the retina. The use of *convex lenses* aids the eyes in bringing images of nearby objects to a sharp focus. (See Fig. 10-16.)

Figure 10-14 **Test for astigmatism: Hold book at reading distance. Close one eye. Look at top figure. If lines seem darker than in bottom figure, you may have an astigmatism.**

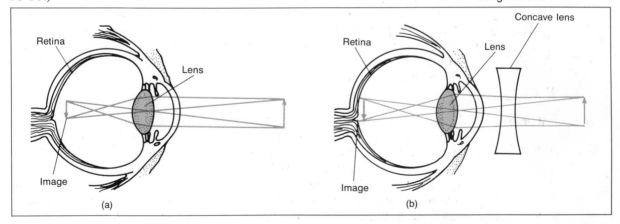

Figure 10-15 (a) Nearsighted (myopic) eye, uncorrected. (b) Nearsighted (myopic) eye, corrected.

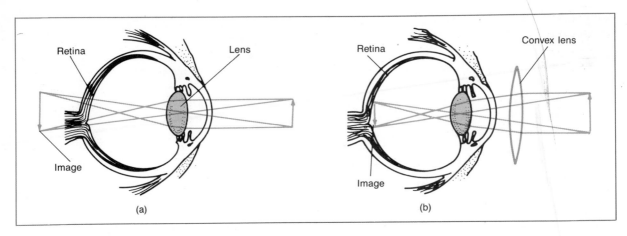

(a)   (b)

Figure 10-16 (a) Far-sighted (hypermetropic) eye, uncorrected. (b) Farsighted (hypermetropic) eye, corrected.

opia = vision

Presbyopia.   Nearly all muscles gradually lose some of their contractive power as they age. This is also true of the ciliary muscles in the eyes. In addition, the crystalline lens loses some of its elasticity with age and is no longer able to assume the maximum convexity necessary for near vision. As a result, there is a decline in the accommodation that can occur in each eye. This condition is called **presbyopia.** From the onset of this condition, eyeglasses with convex lenses are necessary for reading.

Color blindness.   Deficiency in color perception is called **color blindness.** Persons who have this hereditary condition may be classified into two general groups. One group, which is very rare, is totally color-blind. Individuals in this group are called **monochromats** and see objects as if they were viewing a black-and-white television screen. A second group, the **dichromats,** may be red-green or blue-yellow color-blind, the latter seldom occurring. In this second group, the defect may be due either to a problem in the receptors sensitive to red, green, or yellow or to some neurological problem that causes confusion of color perception.

Individuals are called **trichromats** if they are receptive to all three visual primaries. This is normal color vision. There may be a slight deficiency in determining closely related wavelengths of light, but this defect is negligible and certainly causes little hardship.

Cataract.   A prevalent disorder of the human eye in older people (but not exclusively) is the clouding of the lens of the eye, resulting in **cataract.** Well over one million people throughout the world suffer loss of transparency of the lens of the eye every year. The usual remedy, though not a cure, is the removal of the impaired lens and substitution with corrective glasses or contact lenses or the implantation of a substitute plastic lens.

Glaucoma.   If the amount of fluid in the anterior chamber of the eye increases, the eyeball becomes very hard because of the increased pressure. The aqueous fluid does not return to circulation through the canal of Schlemm as quickly as it is formed. As a result, the optic disc is depressed, and there is restriction of the field of vision and possible blindness. Drugs that constrict the pupil, or surgery of the iris, may allow the aqueous humor to drain more easily, thus reducing the pressure.

Conjunctivitis.   Any inflammation of the conjunctiva membrane, which lines the inside of the eyelids and covers the cornea, it called **conjunctivitis,** or pink eye. It can be caused by any irritation from pollen, dust, smoke, wind, pollutants in the air, and chlorine in swimming pools. However, of primary concern is a highly contagious form caused by bacteria or viruses. The condition may become acute or chronic. It is more typical in children and quite contagious. Antibiotics are very effective in its treatment.

itis = inflammation

Table 10-1   Extrinsic Eye Muscles

Muscle	Origin	Insertion	Action	Innervation
Superior rectus	On posterior wall of the orbital cavity, near apex	Superior surface of eyeball	Turns eye upward	Oculomotor nerve (III)
Inferior rectus		Inferior surface of eyeball	Turns eye downward	Oculomotor nerve (III)
Internal, or medial, rectus		Medial surface of eyeball	Turns eye inward	Oculomotor nerve (III)
External or lateral, rectus		Lateral surface of eyeball	Turns eye outward	Abducens nerve (VI)
Superior oblique	Posterior wall of orbital cavity	On top of eye between superior and lateral recti	Rotates eyeball so cornea turns downward and laterally	Trochlear nerve (IV)
Inferior oblique	Superior maxilla near nose	On upper lateral surface of eyeball between superior and lateral recti	Rotates eyeball so cornea rotates upward and laterally	Oculomotor nerve (III)

SUMMARY

Accessory structures associated with the eye include the orbital cavity in the skull, which is surrounded by seven bones; the six extrinsic eye muscles, which move the eyeball; the lacrimal apparatus, which produces tears; the eyebrows; the eyelids; the eyelashes; and the conjunctiva membrane.

The wall of the eyeball is composed of the sclera and cornea, the choroid, the ciliary body, the iris, and the retina.

The fovea centralis in the retina contains only cones and is responsible for detailed color vision. The extrafoveal region, containing primarily rods, is responsible for peripheral vision. The optic disc is a blind spot where no rods or cones are present. Rod receptors are responsible for vision in dim light. Cone receptors react to high-intensity light and provide our perception of color vision.

Refraction is the physical process of bending light rays when they pass from a medium of one density to a medium of a different density. In the human eye, the cornea and lens are refracting surfaces.

Night vision requires the activity of the rods, which contain rhodopsin. Color vision occurs in bright light, in which the cones are activated. Color sensation is produced when various wavelengths of light activate different kinds of cones.

VOCABULARY REVIEW

Match the phrase in the left column with the correct word in the right column. *Do not write in this book.*

1. Chemical responsible for photosensitivity of the rods.
2. Plays an important role in the production of tears.
3. Bony socket in which the eye is located.
4. Middle layer of wall of the eye.
5. Increase in amount of aqueous humor, resulting in increased pressure.
6. Thin transparent membrane that lines eyelids and covers anterior surface of eye.
7. Decline in visual accommodation with age.
8. Visual response by the brain from receiving impulses from both eyes.
9. Distortion of vision due to abnormal curvature of cornea or lens.
10. Near-sightedness.

a. astigmatism
b. binocular vision
c. choroid
d. conjunctiva
e. glaucoma
f. hyperopia
g. iris
h. lacrimal gland
i. monocular vision
j. myopia
k. optic disc
l. orbital cavity
m. presbyopia
n. rhodopsin
o. vitreous body

TEST YOUR KNOWLEDGE
Group A

Select the correct statement by letter.
1. Color vision is due to the activity of   (a) rods   (b) cones   (c) rhodopsin   (d) vitamin A.
2. Peripheral and dim vision is due to the activity of   (a) rhodopsin in the rods   (b) rhodopsin in the cones   (c) opsin in the cones   (d) retinene.
3. The blind spot is due to   (a) the fovea centralis   (b) deficient action of the extrinsic eye muscles   (c) glaucoma   (d) lack of receptors in the optic disc.
4. The anterior chamber of the eye is filled with   (a) vitreous humor   (b) mucus   (c) aqueous humor   (d) tears.
5. When parallel light rays pass through a concave lens, they   (a) converge   (b) diverge   (c) go straight   (d) stop.
6. The optic foramen allows entry to the eye of both the optic nerve and the   (a) ophthalmic artery   (b) maxilla   (c) retina   (d) rods.
7. Night-blindness is due to   (a) a lack of vitamin D   (b) a lack of rentinene   (c) a bacterial infection   (d) a lack of vitamin A.
8. Light rays traveling from a less dense medium to a more dense medium are   (a) converged   (b) refracted   (c) accommodated   (d) optical illusions.
9. The shortest wavelength is that of   (a) infrared   (b) ultraviolet   (c) green   (d) radio waves.
10. When the image focuses behind the retina, the result is   (a) myopia   (b) astigmatism   (c) hyperopia   (d) presbyopia.

Group B

Fill in a word or words that best complete each statement. *Do not write in this book.*
1. The layer of the eye that makes up the white portion as well as the anterior transparent window is called the _sclera_.
2. In the fovea centralis we find primarily photoreceptors called _cones_, while in the extrafoveal region the principal kinds of photoreceptors are _rods_.
3. Light waves bend when they pass at an angle from a transparent medium of one density through a transparent medium of a different density. This process is called _refraction_.
4. _white_ light is a combination of all the wavelengths of a spectrum or rainbow.
5. If the eyeball is too long from front to back, parallel light waves will cross in front of the retina, causing the image to be blurred. This condition is called _nearsightedness (myopia)_

CHAPTER 11

THE EARS: HEARING AND EQUILIBRIUM

Objectives

A. Describe the structure and functions of the parts of the external and middle ears

B. Describe the structures and functions of the parts of the inner ear, including those that help maintain equilibrium

C. Explain the physical properties of sound waves as they affect the human ear

D. Explain how sound waves entering the ear are converted into nerve impulses

E. Describe some of the disorders of the ear and their causes

ANATOMY OF THE EAR

Next to seeing, hearing is probably our most important special sense. Hearing is directly related to the capacity to experience the environment and to learn. Through "sound" we have the means of communication and can learn languages, enjoy music, and share ideas. All of this contributes to our perception of the "real" world. Being able to distinguish between certain sounds helps us to judge the distance and location of objects. Detecting other sounds may protect us by making us aware of danger in certain situations.

The ear is a sensitive mechanical receptor capable of detecting the physical movement of air in the form of vibrations (sound) and converting (transducing) them into neural signals that are sent to the brain for interpretation.

In addition to providing the mechanism for hearing, the ear contains the apparatus for detecting stimuli that makes us aware of our movements and enables us to orient the body for equilibrium, or balance. The structure of the ear is organized into three divisions: The **external ear**, the **middle ear**, and the **inner ear**. Anatomically, the receptors for hearing and equilibrium are located in the same division within the ear, although functionally they serve entirely different purposes. Impulses originating from sound detection and those for movement and equilibrium are transmitted over the same cranial nerve (the VIIIth cranial, or auditory, nerve) but to different destinations in the brain.

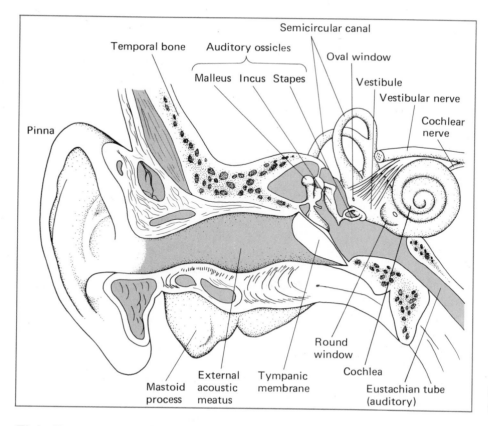

Figure 11-1 **Anatomy of the ear—frontal section.**

The External Ear

The **external**, or **outer, ear** extends beyond the lateral surface of the head. The **pinna** (auricle), or visible part of the ear, has a flexible cartilaginous frame covered by skin. The opening that is seen near the center of the pinna is the **auditory canal** or **external acoustic meatus.** The design of the external ear serves to gather sound waves, while the auditory canal directs them inward to the other parts of the ear. (See Fig 11-1.)

The auditory canal extends about 2.5 centimeters into the temporal bone. The canal is closed at its inner end by the **tympanic membrane,** or **eardrum.** Since the canal forms an S-shape, the eardrum cannot be clearly seen without using special instruments. The passageway of the canal is lined with a thin skin that also covers the tympanic membrane. The outer portion of the canal contains numerous wax-producing **ceruminous** glands and stiff hairs. The secretions of earwax have a sticky consistency and, together with the hairs, prevent foreign objects from entering the ear.

The Middle Ear

The second division of the auditory apparatus is the **middle ear,** or **tympanic cavity.** This small air-filled chamber is located within the spongy portion of the temporal bone near the mastoid process. The bony walls of the tympanic cavity are lined with a mucous membrane similar to that of the nasal cavity. The membrane also covers the inner surface of the eardrum. Between the tympanic membrane and the inner wall of the middle ear are a chain of three small but exceedingly important bones, the **ossicles.**

The ossicles.   The vibrations of sound waves striking the eardrum move along to the ossicles of the middle ear. The bones of the ossicles—the *malleus* (hammer), *incus* (anvil) and *stapes* (stirrup)—are responsible for transmitting the vibrations across the middle ear cavity. The mallet-shaped handle of the hammer, or malleus, attaches to the inner surface of the eardrum so that the tip of the handle ends at the apex of the membrane. The head of the malleus fits into a shallow socket at the base of the incus (anvil). The ligaments between these two bones bind them firmly together. The long process of the incus forms a joint with the head of the stapes (stirrup). This joint moves freely and results in a rocking action of the stapes. The base of the stapes oscillates against the membrane covering an opening, the *oval window,* in the inner wall of the middle ear. At birth the ossicles are completely developed and do not change in size.

The *oval membrane* covering the oval window has a very much smaller surface area (3.2 square millimeters) than the tympanic membrane (64 square millimeters). The sound energy is concentrated on the smaller surface area (increases pressure) and produces a greater amplification of vibrations. Two small striated muscles are responsible for controlling the extent of movement of the ossicles. The *tensor tympani* muscle has its insertion near the head on the handle of the malleus, and its origin is on the floor of the tympanic cavity. When this muscle contracts, the handle of the malleus is pulled inward and produces tension of the eardrum. The second muscle is the *stapedius* muscle. It arises near the roof of the cavity and inserts on the posterior surface of the neck of the stapes. It acts in opposition to the tensor tympani muscle.

The Eustachian (auditory) tube.   Equalization of air pressure must be maintained between the tympanic membrane of the middle ear and the outer air. If the tension on either side of the eardrum is unequal, the membrane will not vibrate normally and hearing will be impaired. The air pressure between the middle ear and the outer air is equalized through the *Eustachian,* or *auditory tube.* This tube leads from the middle ear to the nasopharynx. It is about 40 millimeters in length, and its narrowest part is about 3 millimeters in diameter. Its oval pharyngeal opening has inner edges that normally touch, thus effectively closing the tube. In swallowing or yawning, muscles cause the edges to open and air is allowed to enter or leave the middle ear, depending upon external air pressure. If the entrance to the Eustachian

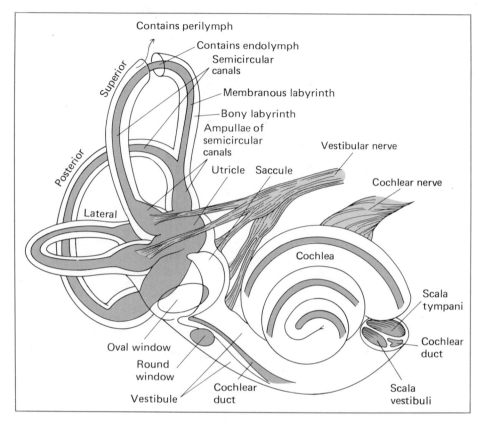

Contains perilymph

Contains endolymph

Semicircular canals

Membranous labyrinth

Bony labyrinth

Ampullae of semicircular canals

Vestibular nerve

Cochlear nerve

Utricle   Saccule

Cochlea

Scala tympani

Cochlear duct

Oval window

Round window

Vestibule   Cochlear duct

Scala vestibuli

Superior

Posterior

Lateral

Figure 11-2 **Details of the inner ear showing the membranous labyrinth.**

tube is inflamed or congested as a result of infection, these edges may not open. Inequalities of pressure are not adjusted for, and the eardrum cannot move freely. Slight temporary deafness and discomfort result.

The Inner Ear

The inner ear is the most important part of the auditory apparatus. Within it are found receptors initiating the nerve impulses that are interpreted by the brain as sound. In addition, the inner ear contains specialized receptors concerned with equilibrium, or balance. The inner ear is divided into a series of three fluid-filled tubes or canals, collectively called the **bony labyrinth.** The three parts of the labyrinth are the **vestibule,** the **cochlea,** and the **semicircular canals** (Fig. 11-2).

Within the bony labyrinth is a tubular membrane that follows the same shape as the bony canal and is called the **membranous labyrinth.** The membrane does not adhere to the bony wall but is separated from it by a fluid called the **perilymph.** The tubular membranous labyrinth is filled with another fluid, the **endolymph.**

The vestibule.   The vestibule is connected to the middle ear by the oval window. It acts as an entrance to the other two divisions of the inner ear—the semicircular canals and the cochlea.

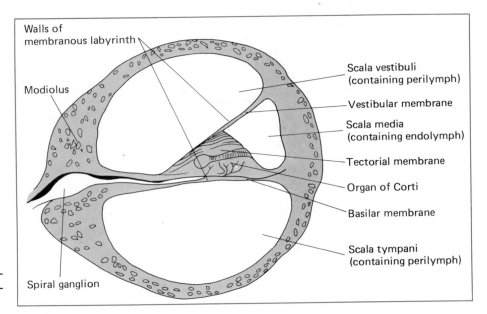

Walls of membranous labyrinth

Modiolus

Spiral ganglion

Scala vestibuli (containing perilymph)

Vestibular membrane

Scala media (containing endolymph)

Tectorial membrane

Organ of Corti

Basilar membrane

Scala tympani (containing perilymph)

Figure 11-3 **Cross section through the cochlea.**

Found within the vestibule are two membranous sacs filled with endolymph—the **utricle** and **saccule.** The saccule communicates with the cochlear duct. The utricle and saccule each contain an area of hair cells from which receptor cells arise. Some of the cells contain neurons with short hairs that extend into a gelatinous material. Other types of cell provide support. The hair cells serve as end organs for fibers of the vestibular nerve, a branch of the auditory nerve (cranial nerve VIII). Entangled in the brushlike hairs are **statoconia** (otoliths), which are small, solid particles of calcium carbonate.

The cochlea.   The cochlea resembles a snail's shell with its spiral canal coiled about two and a half turns around a central pillar, the **modiolus.** Within the modiolus is a branch of the auditory nerve that is called the **cochlear nerve.** This branch is concerned with conducting sensory impulses to the brain for the sense of hearing. Projecting from the central pillar, like the threads of a screw, is a small shelf of bone called the **spiral lamina.** This shelf is continuous within the spiral canal and partially divides it in two. Attached to the outer border of the spiral lamina is a thin membrane, the **basilar membrane.** This membrane extends across to the cochleal wall and separates the canal into two main passages. The two passages then join with one another at the apex of the cochlea. The upper passage is the **scala vestibuli,** and the passage below the lamina that is attached to the basilar membrane is the **scala tympani** (see Figure 11-3). The oval window connects with the scala vestibuli, and the **round window** is the membrane covering the opening of the scala tympani.

From the upper edge of the bony spiral lamina across to the outer edge of the canal above the basilar membrane extends the extremely thin **vestibular membrane,** or the **membrane of Reissner.** This mem-

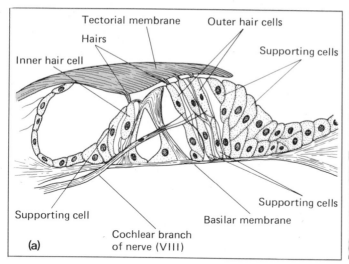

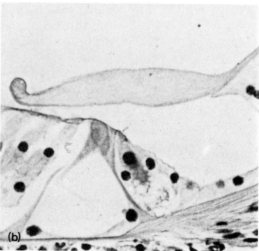

Figure 11-4 (a) Enlargement of the spiral organ or organ of Corti. (b) Light photomicrograph of the organ of Corti in section.

brane divides the scala vestibuli unequally, forming a smaller, triangular canal called the **scala media.** The scala media is filled with endolymph. Because of its extreme thinness, the vestibular membrane does not obstruct the movement of sound waves from the scala vestibuli into the scala media. Thus, as far as sound transmission is concerned, the two passages can be considered a single canal.

The organ of Corti.   The sense receptors for sound are found within the **organ of Corti.** This organ is located on the upper surface of the basilar membrane in the scala media and is just lateral to the bony spiral lamina. The cells of the organ of Corti include a series of modified columnar epithelial cells, out of which small hairlike endings protrude. These hair cells lie in two rows running the entire length of the canal. Covering these cells is a jellylike canopy that is attached to the lamina and that arches over the hair-cell endings. This membrane is called the **tectorial membrane.** On the lower surface of the tectorial membrane is a gel coating in which the hair-cell endings are embedded, as seen in Fig. 11-4. The upward movement of the basilar membrane moves the hair cells inward, bending those endings that are embedded in the tectorial membrane. Then the basilar membrane moves downward, and the hair cells move outward, once again bending the hairs. The movements of the basilar membrane stimulate the hair cells of the organ of Corti and initiate nerve impulses in the fibers of the auditory nerve.

The semicircular canals.   Three looped tubes, the semicircular canals, are embedded in the temporal bone anterior and superior to the vestibule. They are about 4 millimeters in diameter and vary in length from 12 to 22 millimeters. The semicircular canals are connected to the vestibule at five points. As seen in Fig. 11-2, two of the canals join to form a common canal that connects with the vestibule. The most obvious structural characteristic of the semicircular canals is the fact that the three loops are situated at right angles (90°) to each other in three different planes. Two of these structures, the anterior canal and posterior canal, are vertical and at right angles to each other. They are also situated at 45° angles to the midline of the head. The third canal is horizontal and at a right angle to the two vertical canals.

Within the semicircular canals is a portion of the membranous labyrinth. This tubular membrane is filled with endolymph and is held in place by connective tissue and the surrounding perilymph. The diameter of the **semicircular ducts,** as these tubular membranes are called, is about one-fourth the diameter of the bony canals that surround them. At their junction with the vestibule, the semicircular ducts join the utricle at the five openings mentioned above.

One end of each canal connects with the vestibule and causes a bulging called the **ampulla.** In the ampulla, the semicircular ducts contain a crest-shaped group of hair cells called the **crista ampullaris.** These cells can detect changes in acceleration and deceleration. Covering their hairs is a gelatinous substance called the **cupula,** into which the hairs protrude (Fig. 11-5). The branch of the vestibular nerve that innervates these receptor cells leads to the medulla of the brain. Here it forms synapses with other neurons, some of which carry impulses to the motor nerves that control eye movement. Other neurons carry impulses directly to the cerebellum.

Figure 11-5 Semicircular ducts and microscopic view show the position of the crista for equilibrium or balance of the body.

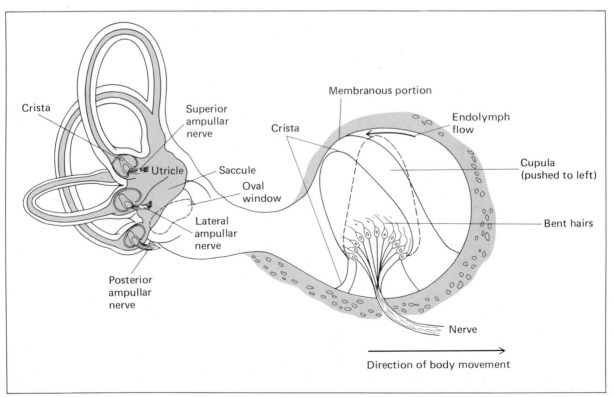

PHYSIOLOGY OF HEARING AND EQUILIBRIUM

When an object vibrates in air, it produces regions in which the air molecules are squeezed together (**compressions**) alternating with regions in which the air molecules are farther apart (**rarefactions**). Thus, sound waves are alternate regions of compressed and rarefied air. Sound waves move outward in all directions like the waves produced by a rock thrown into a quiet pool of water. Sound waves may be reflected in the same way light waves are reflected when they come in contact with a nonabsorbing surface. If such a surface reflects the sound waves directly back to the source, the sound is heard at the source as an echo.

The Nature of Sound

The speed at which sound waves travel through different substances varies according to the elasticity and density of the substance. In air at sea level with a temperature of 25°C, sound travels about 346 meters per second, or about 775 miles per hour. In all cases, the speed of sound is much slower than the speed of light. Thus smoke is seen from a starter's gun at a track meet before the gun shot is heard and the distant flash of lightning is seen before the thunder is heard. Table 11-1 lists the speed of sound in various substances in meters per second at 25°C.

Characteristics of Sound

The physical properties of sound waves as they affect the human ear are called: **loudness, pitch,** and **quality.**

Loudness.   When recording the compressions and rarefactions of sound waves into a line graph by using an oscilloscope, certain characteristics of sound can be illustrated. The line moves upward during the compression phase and downward during the rarefaction phase (Fig. 11-6). The distance the line moves up or down is called the **amplitude** and provides us with an illustration of *loudness*, or *intensity*, of sound. The greater the amplitude, the louder the sound.

Table 11-1   Speed of Sound

Medium	Velocity (m/s)
Air (0° C)	331.7
Air (25° C)	346
Water (distilled)	1487
Water (sea)	1531
Brick	3650
Glass	4540
Marble	3810
Iron	5200

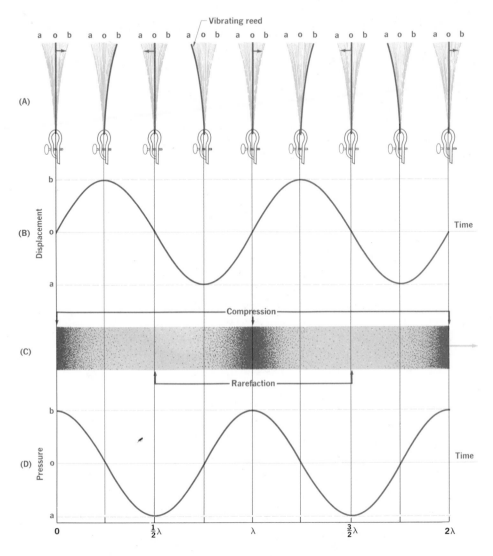

Figure 11-6 Vibrating reed in (a) produces displacement variations with time in (b). Compression and rarefaction of the sound wave are shown in (c). Pressure variations of the sound wave with time are seen in (d). Displacement and pressure curves are 90° out of phase.

In determining the loudness of a sound, a comparative scale is used that has a starting point at the threshold of barely audible sound. The units in the scale are called **bels,** named after the great experimentalist Alexander Graham Bell (1847–1922). The unit *decibel* (¹⁄₁₀ of a bel) is usually used since the decibel seems to be the smallest change in sound intensity the human ear can distinguish. Table 11-2 gives decibel levels of various sounds.

Table 11-2   Intensity Levels of Sound (in thousands of hertz)

Type of Sound	Intensity (decibels)
Threshold of Hearing	0
Ordinary breathing	10
Whispering	10–20
Very soft music	30
Average home	40–50
Conversation	60–70
Heavy street traffic	70–80
Thunder	110
Threshold of pain	120
Jet engine	170

Pitch.   The time interval separating a point on a compression curve of a sound wave from the same point on the next compression curve is called the *cycle* of the sound. *Pitch*, or *frequency*, is the number of sound waves produced by a vibrating object per second. Formerly expressed as cycles per second, or *cps*, frequency is designated in **hertz (Hz).** (One hertz equals one cycle per second.) When striking the keys of a piano, moving to the right on the keyboard, each succeeding string vibrates at a faster rate than those before, and thus each note is higher in pitch. Just as the eye has upper and lower limits in perceiving light waves, so the human ear has two thresholds of pitch. The human ears are unable to respond to sound with a frequency less than 16 Hz—our low threshold—nor can most people detect sounds with frequencies above 15,000 to 20,000 Hz—our high threshold. The human ear has its greatest sensitivity between 2000 and 4000 Hz.

Quality.   If several different musical instruments all play a note of the same pitch, it is quite easy to distinguish one type of instrument from another. The quality of tone gives instruments their characteristic sound. In describing sound, one may say that a cello is "rich," a trumpet is "harsh," a celesta has a "tinkling" sound, and a bass drum "booms." All of these terms are attempts to describe the *quality*, or *timbre*, of the instrument. Graphically, quality or timbre is represented by the smoothness of the sound wave, as seen in Fig. 11-7. If the line produced is a simple sine wave, or curve, the tone is classified as a **pure tone.** Pure tones can be artificially produced, but they are never found in a musical instrument or in the human voice. If the sound wave produces a curve that contains many smaller deviations, the tone is composed of a number of sound waves. These are called **overtones,** or **harmonics.** If these overtones are produced in a consistent pattern, the sound is pleasing. If they are spasmodic and irregular, the sound is perceived as noise. The form of the sound wave determines the quality of the sound.

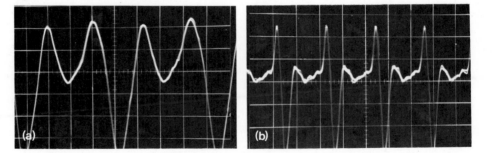

Figure 11-7 Oscillograms of (a) the sound of a French horn and (b) the sound of a trumpet.

Hearing

What happens after sound waves set the tympanic membrane and the chain of middle ear bones in motion? The stirrup transmits the vibrations via the oval window to the fluid in the vestibular canal. This motion initiates vibration in the elastic basilar membrane in the form of a traveling wave that moves rapidly from the base to the apex of the cochlea. Maximum displacement of the basilar membrane takes place in that part of the cochlea where the wave stops traveling. The lower the frequency of the sound waves, the longer the distance the wave travels, and the closer the maximum displacement is to the apex. The higher the frequency, the shorter the distance the wave travels, and the closer the maximum displacement is to the base of the cochlea (to the oval window). Thus there is for each region along the cochlea a *best frequency*—that is, a frequency that makes the membrane vibrate and stimulate hair cells to generate nerve impulses.

Place theory. The mechanism just mentioned above briefly describes the *place theory* of audition, or hearing. This theory is based on the assumption that our discrimination of sound frequency and quality is based primarily upon the *place* where maximum movement of the basilar membrane occurs. Impulses seem to be initiated when the receptors of the organ of Corti are stimulated at that part of the basilar membrane corresponding to a particular pitch. It is estimated that by this mechanism we can distinguish over 10,000 different frequencies or tones. To accomplish this remarkable feat, we have an estimated 15,000 to 30,000 hair cells, each contributing a fiber to the cochlear nerve. Impulses from this nerve travel to the auditory cortex in the temporal lobe of the cerebrum.

Equilibrium

Two different groups of organs are involved in maintaining proper balance. The two fluid-filled membranous sacs (the utricle and saccule) found in the vestibule represent one group; the second group is represented by the semicircular canals.

Static equilibrium.   When the head is tilted, gravity causes the otoliths found in the utricle and saccule to come in contact with the hair cells in each of the macula. These send out impulses that normally enable us to determine the position of the head with respect to gravity. Thus the utricle and saccule make us aware of **static equilibrium**—that is, the orientation of the body relative to the ground. The nerve impulses from the hair cells are carried through a branch of the eighth cranial nerve, called the **vestibular nerve.** These impulses are conveyed to the medulla and then to the cerebellum, where efferent responses result in changing the body position by muscular contraction.

Dynamic equilibrium.   When there is a change in horizontal acceleration, as when we move forward or rotate rapidly, the endolymph within the semicircular ducts press against the gelatinous cupula, which in turn causes the hairs to bend in a direction opposite to the direction of motion. When a constant speed is reached, the fluid moves at the same rate as the body and the cupula swings back into the upright position. When the body is suddenly stopped, deceleration produces displacement of the endolymph in the direction the body was moving, and the cupula swings in a direction opposite to that during acceleration. This movement of endolymph stimulates the hair cells, which produce sensory impulses that make us aware of **dynamic equilibrium**—that is, the orientation of our body in relation to movement along the horizontal plane.

Associated with the function of the semicircular canals and the vestibular apparatus in maintaining equilibrium are the kinesthetic and visual senses, which also provide clues to spatial and structural position. The resulting response to these sensations involves coordinated muscular activity under the control of the cerebellum, which keeps the body in a vertical position.

DISORDERS OF THE EAR

The function of the external ear and the middle ear are to collect, conduct, and transform sound waves into mechanical vibrations. The inner ear then converts these mechanical vibrations into impulses that are transmitted to the brain. When there is some difficulty in the external or middle ear, a **conductive** hearing loss occurs. When the problem lies in the inner ear, **nerve,** or **sensory,** *hearing impairment* is the result.

Hearing Impairment (Partial or Total Deafness)

Conductive impairment may result from a blockage of the external ear canal—for example, from compacted earwax, from a perforated eardrum, from middle-ear infection, or from disease involving the ossicles. This type of hearing loss is usually correctable, either by surgically reconstructing the ear or by using a hearing aid.

Sensory hearing impairment may result from disturbances in inner-ear circulation or fluid pressure, or from problems in nerve conduction. The most common form of sensory impairment results from aging, since nerve endings become inefficient in receiving sound as time goes by. Although not correctable, this type of impairment rarely leads to total deafness.

Otosclerosis.   A common cause of partial deafness is the hereditary disease otosclerosis. In this disorder, the ossicles lose their ability to move, and thus sound waves are prevented from reaching the inner ear. It is usually progressive, beginning in early adulthood and increasing in severity as time goes by.

Local treatment to the ear itself or medication will not improve hearing in people with otosclerosis. However, certain surgical procedures, which include replacing the stapes (stirrup) with a wire prosthesis, may restore hearing.

Tinnitus (Head Noise)

The occurrence of **head noise,** or **tinnitus,** is very common and annoying. It may come and go, or be constant. It may be rather mild, or severe, in intensity. It may vary from a low pitch to a highly penetrating ringing sensation. Tinnitus may be caused by cerumen (wax) accumulating in the auditory meatus, by perforations of the tympanic membrane, by fluid in the middle ear, by any change in pressure of the fluid in the inner ear, or by inflammation to the organ of Corti. Tinnitus must always be thought of as a symptom, not a disease. If the causative agent is removed or corrected, the symptoms usually clear up.

Dizziness

Dizziness is a symptom, not a disease. Dizziness can be caused by many different diseases and can vary in severity from a mild unsteadiness to a whirling sensation known as *vertigo.* It may be due to changes in circulation to inner-ear structures or to their central connections. *Ménière's Disease* is a common cause of repeated attacks of dizziness. It is usually due to increased pressure of the inner-ear fluids. The attacks of dizziness may occur suddenly and without warning. Accompanying the vertigo may be nausea and vomiting. After the initial attack, succeeding attacks may be less severe and less frequent, and the individual may be free of symptoms for years at a time. Treatment of Ménière's disease may involve medicines or surgery and is aimed at improving the inner-ear circulation and controlling the fluid pressure changes of the inner ear.

SUMMARY

The ear is divided into three primary parts: the external, the middle, and the inner ear. The external, or outer, ear consists of the pinna, which leads inward through the external auditory meatus to the tympanic membrane (eardrum). The middle ear is an air-filled, mucus-lined cavity containing a lever system of three ossicles—the malleus, incus, and stapes. The Eustachian tube leads from the middle ear to the nasopharynx. It functions to equalize the air pressure on both sides of the eardrum by allowing air to enter or leave the middle ear. The inner ear consists of a bony labyrinth in which is suspended a membranous labyrinth. Surrounded by perilymph, the membranous labyrinth contains endolymph. The inner ear is divided into three parts: the vestibule, the cochlea, and the semicircular canals.

Sound waves are alternate regions of compressed and rarefied air. The speed at which sound waves travel through substances varies with their individual densities. The shape of the wave graphically illustrates characteristics of sound as they affect the human ear—such characteristics as loudness, pitch, and quality. Loudness is represented by the amplitude or height of the wave; pitch, by the number of waves per second; and quality, by the wave's characteristic shape and the distribution of its overtones.

The place theory is based on the assumption that certain areas of the basilar membrane vibrate in sympathy with sound waves of a given pitch. The higher frequencies affect the bottom of the cochlea, while the lower frequencies are registered at the apex of the cochlea.

Vocabulary Review

Match the statement in the left column with the word(s) in the right column. Place the letter of your choice on your answer paper. *Do not write in this book.*

1. Another name for the middle ear.
2. The auditory tube or canal.
3. Conducts sensory impulses to the brain for the sense of hearing.
4. The positional sense involving forward motion or rotation.
5. Another name for earwax.
6. Units used in the scale of loudness of sound.
7. Structure associated with the inner ear.
8. "Ear stones" of calcium carbonate.
9. Alternate compression and rarefaction of air molecules.
10. A unit expressing the number of cycles per second.

a. auditory nerve
b. bony labyrinth
c. cerumen
d. decibels
e. dynamic equilibrium
f. Eustachian tube
g. hertz
h. otoliths
i. sound wave
j. static equilibrium
k. tympanic cavity

TEST YOUR KNOWLEDGE
Group A

On your answer sheet, write the letter of the word(s) that correctly complete(s) the statement.

1. The Eustachian tube   (a) drains fluid from the inner ear into the nasopharynx   (b) keeps up pressure in the inner ear   (c) keeps the pressure equal on both sides of the eardrum   (d) is associated with balance.
2. The semicircular canals are associated with   (a) maintaining equilibrium   (b) hearing   (c) draining fluid from the ear   (d) equalizing pressure.
3. The tiny bone (ossicle) that rocks in the oval window is the (a) incus   (b) malleus   (c) anvil   (d) stapes.
4. The basilar membrane is within   (a) the semicircular canals (b) the vestibule   (c) the cochlea   (d) the auditory meatus.
5. The human ear has its greatest sensitivity between   (a) 15,000 to 20,000 Hz   (b) 2000 to 4000 Hz   (c) 100 to 300 Hz   (d) 0 to 16 Hz.
6. Total deafness, without hope of recovery, may result from   (a) otosclerosis   (b) destruction of the auditory nerve   (c) destruction of the eardrum   (d) accumulation of earwax.
7. The function of balance in the ear depends on the   (a) vestibular nerve   (b) cochlear nerve   (c) trigeminal nerve   (d) facial nerve.
8. The semicircular canals are at an angle to each other of   (a) 180° (b) 45°   (c) 60°   (d) 90°.
9. The speed of sound at sea level is about   (a) 345 meters per second   (b) 500 meters per second   (c) 775 meters per second (d) 186,000 miles per second.
10. The unit hertz is associated with   (a) loudness   (b) pitch (c) quality   (d) timbre.

Group B

Write the word or words on your answer sheet that will correctly complete the sentence.

1. In addition to converting sound into nerve impulses, which are interpreted by the brain as hearing, the ear is also an important organ of _____ sense.
2. The partition that divides the outer ear from the middle ear is the _____.
3. Endolymph is found within the _____.
4. The sense receptors for sound are found within the _____.
5. Sound waves are alternate regions of _____ and _____ air.

UNIT 4

CHAPTER 12

FOOD AND NUTRITION

Objectives

A. Identify the macronutrients required for metabolic activity and describe their roles
B. Describe the functions of minerals and name their food sources
C. Describe the importance of vitamins in terms of their best food sources, functions, deficiency disorders, and risks of megadoses
D. Describe the structure and functions of enzymes and factors affecting their activities

BASIC ASPECTS OF NUTRIENTS

Food is fundamental to the survival of living organisms. Food provides substances for tissue growth, maintenance, and repair. The processes whereby food is obtained, ingested, digested, absorbed, and metabolized are known as *nutrition.* Digestion first breaks food down into substances called *nutrients,* which can be easily absorbed and metabolized by body cells and tissues. Nutrients are used to build tissues and to provide energy. Some nutrients also serve to regulate metabolic processes.

Macronutrients

To date nutrition scientists have identified some 40 to 45 substances as nutrients essential for life. Nutrients can be divided into two general categories. Those needed by the body in relatively large quantities are called *macronutrients* (carbohydrates, lipids, proteins, and water); those needed by the body in relatively small quantities are called *micronutrients* (mineral elements and vitamins). All nutrients except for mineral elements and water are organic substances. Organic substances contain the element carbon. Mineral elements and water are called inorganic substances because they lack carbon.

macro = large
micro = very small

Carbohydrates. Carbohydrates are major constituents of grains, legumes, and potatoes and are the world's least expensive sources of calories. In some cultures carbohydrates may supply 80 percent or more of the total calories in a daily diet. In the United States, carbohydrates supply an average of only 40 to 50 percent of the total calories. A calorie is a unit of heat energy. It is the amount of heat needed to raise the temperature of one gram of water through one Celsius (centigrade) degree. One calorie is a very small unit, so the unit usually used is the *kilocalorie*, which is 1,000 times larger. It is called a *Calorie*, written with a capital *C*. This is the unit usually used in connection with foods.

Carbohydrates are made up of carbon, hydrogen, and oxygen. The proportion of hydrogen to oxygen atoms in carbohydrates is the same as that found in water; two parts hydrogen to one part oxygen.

The building blocks of carbohydrates are simple sugars. Some simple sugars, or **monosaccharides** (one-unit sugars), are *glucose*, *fructose*, and *galactose*. These three monosaccharides have the same chemical formula, $C_6H_{12}O_6$, but each has a different structure. Two monosaccharides can join together chemically, with the loss of a molecule of water, forming a **disaccharide** (a two-unit sugar). See Figs. 2-9, 2-11, and 2-12 for diagrams illustrating these processes. Some examples of the formation of disaccharides from monosaccharides are:

glucose + fructose $\longrightarrow$ sucrose (table sugar)
glucose + galactose $\longrightarrow$ lactose (milk sugar)
glucose + glucose $\longrightarrow$ maltose (malt sugar)

A molecule consisting of many units of simple sugars linked together is called a **polysaccharide**. Polysaccharide molecules may be straight or branched chains, and can be extremely complex in structure. For example, starch is a relatively long-chain polysaccharide made by plants. It consists of over 1,000 glucose units linked together in a branched chain. When starch is broken down, it yields intermediary polysaccharides called **dextrins**. Dextrins in turn can be broken down into maltose and finally into glucose. *Glycogen* is a starch produced by animals. The glycogen molecule is a highly branched chain of glucose units. In the human body, glycogen is stored in the liver and in muscle cells.

Cellulose is another polysaccharide made by plants. It is found in cell walls. Cellulose is composed of glucose units linked in an unbranched chain. The way in which they are linked makes cellulose resistant to human digestive enzymes. Cellulose is broken down in the digestive tract of ruminant animals, such as cows and goats, by the action of bacteria. Cellulose, along with other indigestible material, constitutes *dietary fiber*.

Starch is stored by plants as an energy source and is found in seeds such as wheat, rice, oats, corn, peas, and beans; in tubers (underground stems) such as potatoes; and in roots such as yams, sweet potatoes, and cassava. Sucrose is found principally in sugar cane, sugar beets, and maple syrup. Lactose is found in milk, and fructose in fruits. In some instances sugar may change into starch. This happens to sweet corn after it is picked.

The chief function of carbohydrates in the body is to supply energy. They are an essential source of energy for the central nervous system. Approximately one fifth of the energy required by our basal metabolism is used for brain function. Recent studies with laboratory animals showed that after eating a meal rich in carbohydrates, rats showed an increase in the rate at which their brains synthesized **serotonin.** Serotonin is one of several compounds called **neurotransmitters.** These substances are known to be present in the neurons of mammals and when released transmit signals across nerve synapses to other neurons within the brain or spinal cord. Researchers postulate that an increase in the secretion of serotonin after rats eat a high-carbohydrate meal modifies their brain functions, such as learning. However, the high serotonin levels may not contribute to the learning ability of a rat. In addition, it is not known if the evidence from laboratory investigations is applicable to humans.

When they are in short supply, amino acids can be manufactured from carbohydrates. Carbohydrates also are said to have a *protein-sparing* effect. That is, when carbohydrates are plentiful in the diet, they are used as a source of energy instead of proteins, thereby making proteins available for building and repairing body tissue. It is much more efficient to use carbohydrates for energy than to use proteins.

Carbohydrates and proteins can combine to form compounds called **glycoproteins,** which have important functions in the body. Glycoproteins can serve as synovial fluid in joints and as components of nails, bone, cartilage, and skin.

Lipids (fats). *Lipids* and *fats* are usually considered as foods of the affluent people and cultures of the world. Economically deprived people are less likely to consume foods with a high-fat content, such as meat and dairy products. More commonly, such people consume low-fat foods, such as grains and vegetables. This contrast in diets exists between the United States and the Orient, for example. Most of the population of the United States gets 40 to 45 percent of its calories from fats; that of the Orient gets only 10 percent from fats.

Foods contain many different kinds of fats. The building blocks of fats are fatty acids. They are attached to a substance called **glycerol** [$C_3H_5(OH)_3$], an organic alcohol. Simple fats are named according to the number of fatty acid molecules attached to the molecule of glycerol: **monoglycerides** contain one fatty acid; **diglycerides,** two fatty acids; and **triglycerides,** three fatty acids. About 98 percent of the fats in foods are triglycerides. This is also the most common form of fat in the body. Monoglycerides and diglycerides are used as food additives to stabilize and emulsify fats. These fats are frequently listed on food labels. They are also formed in the body as the by-products of triglycerides during the digestive process.

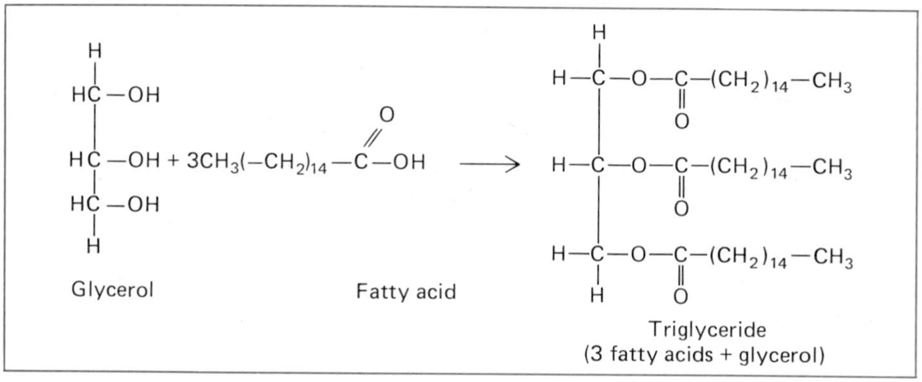

Figure 12-1 **A triglyceride is formed when glycerol combines with three fatty acids.**

Fatty acids are made up of carbon, hydrogen, and oxygen. There are 12 common fatty acids. The amount of carbon and hydrogen found in the molecules of each differs from that found in the others. A fatty acid with a small number of carbon atoms (4 to 6) is called a short-chain fatty acid; one with many carbon atoms (more than 12) is called a long-chain fatty acid; and one with an intermediate number of carbon atoms (8 to 12) is called a medium-chain fatty acid. The length of the carbon chain can affect its digestion in the body. For example, infants are able to digest short- or medium-chain fatty acids more readily than long-chain fatty acids.

The terms **saturated** and **unsaturated** are used to describe the amount of hydrogen found in fatty acids. When a fatty acid has all available carbon atoms joined to hydrogen atoms, it is called a *saturated* fatty acid. When the carbon atoms in a fatty acid are capable of holding more hydrogen atoms but instead form double or triple bonds with other carbon atoms, the fatty acid is considered *unsaturated*. If unsaturated fatty acid can hold more hydrogen at only one place on its carbon chain, it is called **monounsaturated.** If it can hold more hydrogen at more than one place on its carbon chain, it is termed **polyunsaturated.** A fat that has a high proportion of saturated fatty acids is called

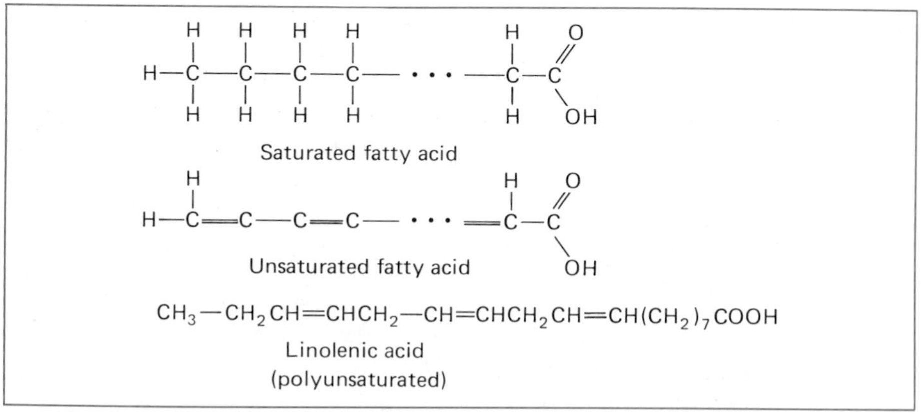

Figure 12-2 **Molecular structures of saturated and unsaturated fatty acids.**

a *saturated fat;* one with a high proportion of polyunsaturated fatty acids is called a *polyunsaturated fat.*

A saturated fat is usually solid at room temperature and most often comes from animals. Such a fat contains a large proportion of saturated fatty acids, such as myristic or palmitic acid. Some saturated fats are found in beef, lamb, pork (as lard), milk (as milk, fat, or butter), and coconut and palm oils. Derived from plants, the last two form liquids because they contain fewer carbon atoms. These liquid fats are still considered to be saturated. The most common monounsaturated fatty acid is oleic acid. Oils high in monounsaturated fatty acids are olive and peanut oils.

Polyunsaturated fats are usually liquid at room temperature and usually come from plants. They are high in polyunsaturated fatty acids (often referred to as *PUFA*), such as linoleic and linolenic acids (both from plants) and arachidonic acid (from animals). Some examples of polyunsaturated fats are safflower, soybean, corn, and cottonseed oils. Animal foods that are relatively rich in polyunsaturated fats are poultry and fish.

Evidence from research indicates that saturated fats may cause the formation of cholesterol and raise the level of cholesterol in the blood. Polyunsaturated fats, on the other hand, appear to have the opposite effect and lower blood cholesterol. High cholesterol levels in the blood are believed to be a factor in the development of coronary heart disease.

Two other fatlike substances that are obtained from food and that can also be synthesized in the body are *phospholipids* and *sterols.* Phospholipids contain phosphate joined with one or two fatty acids and often a nitrogen-containing substance. They are one of the fatty substances found in the blood. *Lecithin* is the best known example of a phospholipid. It is synthesized by both plants and animals and is found in foods such as egg yolk and soybeans. Often lecithin is added to foods (for example, salad dressings) as an *emulsifier.*

Cholesterol is the best known of the sterols. It is synthesized in the body and is found in all animal foods. Cholesterol is an essential constituent of cell membranes. It is utilized by the body in the formation of vitamin D, steroid hormones, and bile salts, which are needed for the digestion and absorption of fats. Cholesterol does not need to be obtained from food since it can be synthesized in the body.

Fats are concentrated sources of energy that make it possible for living organisms to conveniently and efficiently store energy. A gram of pure fat provides nine calories, as compared to the four calories per gram provided by carbohydrate or protein. This means that fats are two-and-a-quarter times higher in caloric content than carbohydrates or proteins. For people who need concentrated sources of energy, the high caloric content of fats is an advantage. For those who need to reduce the caloric content in their diets, however, high-fat foods have an adverse effect. When the body takes in a surplus of carbohydrates, proteins, or fats, it converts the surplus to fat for storage. This stored fat can provide a reserve supply of energy in emergency situations, such as during periods of illness or starvation. Fat also helps cushion body organs by acting as a shock absorber.

Fats tend to slow down the digestion process because they are digested more slowly than other nutrients. Therefore a meal high in fat has a high *satiety* value—that is, it makes one feel satisfied longer than does a meal containing little or no fat. Fats obtained from food contain other essential nutrients, primarily the **fat-soluble vitamins** A, D, E, and K. Fat also contains an important fatty acid—*linoleic acid*—which the body does not synthesize and must obtain from food. Within the cell, fats work with other nutrients to carry out vital metabolic functions. Fats are involved in the formation of cell membranes, blood vessels, and tissues; in the synthesis and regulation of hormones and hormonelike substances (prostaglandins); and in the transmission of nerve impulses and in memory storage.

Proteins. Proteins have two general functions in the body. They are the basic structural material of cells, and, as enzymes, they regulate the chemical processes in the body. In addition to carbon, hydrogen, and oxygen, all proteins contain nitrogen. Most also contain sulfur, many contain phosphorus, and some have various other elements. Protein molecules are large and complex. The building blocks of proteins are amino acids.

An amino acid is composed of an organic acid group ($-COOH$), an amino group ($-NH_2$), and an organic radical, or group (Fig. 12-3). The organic group can vary in size and shape and distinguishes one amino acid from another. For instance, the organic group on some amino acids contains sulfur; on others a ring-type structure is attached. Amino acids are joined in long, unbranched chains. The attraction and repulsion between charges of units on the chain cause the chain to bend into twisted and tangled shapes. The sequence of amino acids in the chain determines the characteristics of the protein. Some proteins, such as those in hair, skin, nails, and bones, are long and stringlike. Others are soft and globular, such as those in blood. Various agents such as heat and alcohol can break or unwind the protein chains. When this happens, the protein is said to be denatured. It cannot be returned to its natural state. The solidification of the albumen in an egg when it is subjected to heat is an example of denaturation. Some proteins are more easily digested once denatured.

When two amino acids join together, the amino group of one combines with the carboxyl group of the other, forming a **peptide bond** ($CO - NH$) and releasing one molecule of water (H_2O). During the process of digestion, proteins are broken down by the splitting of these peptide bonds. Eventually all the peptide bonds are broken and the amino acids are released.

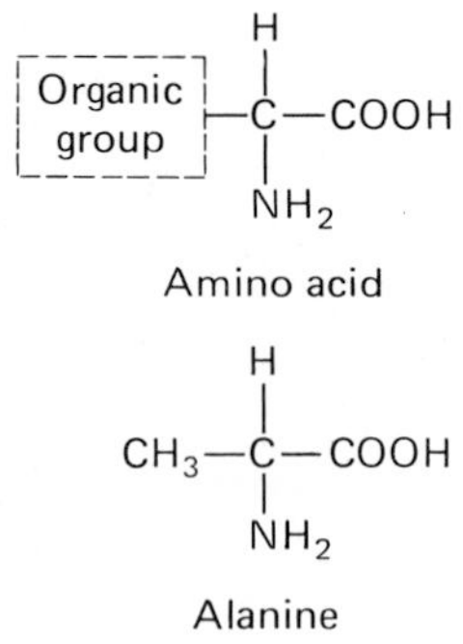

H
|
[Organic group]—C—COOH
|
NH$_2$

Amino acid

H
|
CH$_3$—C—COOH
|
NH$_2$

Alanine

Figure 12-3 Molecular structure of amino acid and a type of amino acid—alanine.

A compound with two or more amino acids is called a **peptide.** When two amino acids combine, they form a **dipeptide;** three amino acids form a **tripeptide;** and when many amino acids link together, they form a **polypeptide.** When a protein is made up of only amino acids, it is called a *simple protein* (although it can be a very complex compound). If amino acids are united with other compounds, such as fats (lipoproteins), carbohydrates (glycoproteins), or phosphates (nucleoproteins), they are called **conjugated proteins.**

Building and repairing body tissue are primary functions of proteins since they form the architectural basis of every cell. The most apparent need for protein is during growth of an organism, when the total amount of protoplasm increases. Such periods occur in infancy, childhood, adolescence, and pregnancy. During lactation, mothers need extra protein to produce milk. Infants and children need the most protein in proportion to size. The lack of protein in early childhood can lead to the deficiency disease *kwashiorkor,* and eventually death. The lack of both protein and calories can result in another possibly fatal deficiency disease called *marasmus.*

In addition to supplying raw material for the formation of new tissues and for the maintenance of existing structures, proteins are also needed to replace tissue that is lost, either through stress, burns, hemorrhage, serious illnesses, or natural daily destruction. The human body is not a static structure. Cells are constantly being worn down and rebuilt. The rate at which this happens varies in different parts of the body. For example, the tissue that lines the intestine forms new cells every one to three days, whereas red blood cells regenerate every 120 days. Liver cells have a high renewal rate, but muscle cells renew themselves much more slowly. The renewal rate of brain cells is negligible. This degradation of body tissues provides amino acids that can be reused by the body for rebuilding protein.

Another function of protein is to maintain water balance in the body. Proteins in the blood (plasma proteins) are too large to pass through cell walls. Thus they exert an osmotic pressure that draws intracellular fluid into the blood. This is important because during the process of oxidation of glucose in the cell, excess water is formed, which if not removed could damage the cell. If plasma protein levels are low, fluid may accumulate in the tissues and **edema** (swelling) results. Starving adults and children are often seen to have this retention of extra water in their tissues.

The pH of blood and most body tissues is slightly alkaline, with a pH range of from 7.35 to 7.45. Proteins function to maintain this pH level by acting as acids and bases. This *amphoteric* property of proteins is due to the amino groups, which act as bases, and the carboxyl groups, which act as acids. Consequently, proteins can unite with either acids or bases to reduce excess acidity or alkalinity in body fluids and thereby maintain the pH of the blood.

Proteins are used for many purposes in the body. They serve in the formation of enzymes, hormones, and antibodies. When the protein intake is low, fewer antibodies are produced and the body is more vulnerable to attack by disease organisms. One amino acid, ***tryptophan,*** is a *precursor* (a form that precedes) of the B vitamin niacin.

Any extra protein available in the body after it is distributed for essential functions can be used as a source of energy. The number of calories provided by proteins is similar to that of carbohydrates—4 calories per gram. If the body can produce enough energy from other sources, proteins provide amino acids for metabolic activities. Amino acids are not stored. On the other hand, if energy is not provided by carbohydrate and fat sources, then protein is metabolized for energy that is needed to keep body processes functioning. When the body fat stores are depleted and the diet lacks sufficient energy reserves, then the proteins will be drawn from body tissue.

The amino acid composition of proteins ingested determines the body's ability to synthesize new proteins. Of the amino acids found in proteins, eight are called ***essential amino acids.*** These must be obtained from food since they cannot be formed by the body. Cells can synthesize most of the amino acids from the temporary pool of amino acids left after proteins are metabolized or from foods that contain the necessary nitrogenous ingredient. These ***nonessential amino acids*** are synthesized from foods. These amino acids can be recycled or formed by the body. See Table 12-1 for the complete list of both essential and nonessential amino acids.

The key factor in determining a biological value of a food protein is its amino acid composition. Those foods having all of the essential amino acids in sufficient amounts for body needs are called ***complete proteins.*** Complete proteins are those obtained from animal foods such

Table 12-1
Amino Acids

Essential	Nonessential
Isoleucine	Alanine
Leucine	Arginine
Lysine	Aspartic acid
Methionine	Cysteine
Phenylalanine	Glycine
Threonine	Glutamic acid
Tryptophan	Hydroxylysine
Valine	Hydroxyproline
Histidine (more in infants)	Proline
	Serine
	Crystine
	Tyrosine

as eggs, milk, meat, fish, and poultry. Foods in which one or more of the essential amino acids are present in amounts insufficient for the needs of the body are called **less complete proteins** (formerly referred to as *incomplete proteins*). These are the proteins obtained from plants, legumes (beans, peas, lentils), nuts, grains, and vegetables.

When eating foods to furnish protein needs, it is important to keep some critical factors in mind. Most significant is that essential amino acids must be provided from food sources. In addition, all of the essential amino acids must be present at the same time, or within a very short time, for the body to synthesize proteins. If one or more essential amino acids are present in only limited amounts, the ability of the body to utilize other amino acids in synthesizing body protein is limited accordingly. This concept, called the **law of the minimum,** illustrates that an adequate quantity of all the essential amino acids is required to synthesize and maintain the total amount of proteins necessary.

Since amino acids are not stored anywhere in the body, as are fats and carbohydrates, the amino acid reserves are depleted in just a few hours. However, the amino acid pool available to tissues comes not only from dietary (*exogenous*) sources, but also from the breakdown of normal tissue (*endogenous*) sources. During body metabolism, amino acids from these two sources are indistinguishable. Thus the body itself forms a reserve pool of protein.

Water.   Water, an inorganic compound, is classified as a nutrient and thus is an essential component of every cell. However, it is not a source of energy like fats and carbohydrates, nor does it build or repair body tissue as do proteins. Instead, it functions primarily as a solvent and transport vehicle for substances in the body and as a medium in which all cellular reactions take place. People can live without food for a period of time—until depletion of stored carbohydrate and fat and of a large portion of body protein occurs. People cannot survive without water for more than a few days. Water comprises between 55 and 65 percent of the body weight. The loss of just 10 percent of body water results in serious consequences. Survival is not possible with a loss of 20 to 22 percent of body water.

Water is obtained from three sources: (1) liquids; (2) foods, which contain varying amounts of water; and (3) from the metabolism of carbohydrates, fats, and proteins. Water is lost through three mechanisms: (1) evaporation of water from the skin, (2) expiration of water vapor from the lungs, and (3) elimination of water in urine and feces. Usually the intake of water and the loss of water is finely balanced, averaging between 2 to 2.5 liters per day. If the loss of water exceeds water intake, as in hot weather, during strenuous exercise, or as a result of vomiting or diarrhea, nerve impulses are transmitted to a thirst center in the brain, which causes an individual to compensate for the water loss. If the intake of water is insufficient to replace the loss, a thirst stimulus will trigger hormone action and cause the kidneys to conserve water. Under these circumstances the kidneys produce a more concentrated form of urine. If a person consumes more water than needed, the kidneys will excrete a greater quantity of diluted urine.

Water serves as the liquid medium for the transport of all substances in the body such as hormones, nutrients, antibodies, and other cellular chemicals. The physical and chemical properties of water makes it possible for both dissolved and suspended substances to be transported from one part of the cell to another. The strong bonds between the atoms in water molecules provide an environment of chemical stability for many chemical reactions to occur without involving the water itself. Water is indispensable for digestion since the hydrolysis of foods requires that water molecules enter into the digestive processes.

Water also serves to regulate body temperature. Since water absorbs and releases heat very slowly, it helps to maintain a constant body temperature. Evaporation of the water in perspiration cools the skin and helps to prevent a rise in body temperature. In addition, water helps to maintain the proper concentration of substances in the cell through its ability to move freely across cell membranes.

Micronutrients

Mineral elements are the inorganic substances that are essential components of both plant and animal tissues. Like vitamins, mineral elements are needed in the body in small amounts and contribute to the regulation of body processes.

Mineral elements. Normally we do not eat minerals in their pure or chemical form but rather as components of other nutrient substances. They may be either of organic or inorganic origin, such as calcium carbonate, a mineral found in water. Scientists divide these essential elements into two general categories, depending on the amount needed by the body. **Macroelements** are those that are present in relatively large amounts in the body. Thus, large amounts of these are needed in the diet (more than 100 milligrams). **Microelements,** or trace elements, are those needed in very minute amounts. (See Table 12-2.)

Different minerals often work together or in combination with vitamins, fats, proteins, or carbohydrates. Some minerals are antagonistic to one another; an excess of one can nullify the effects of another. In addition, excess amounts of essential minerals often can be toxic. Thus it is important that diets contain minerals in properly balanced amounts. The physiological functions of minerals can be grouped into two categories: *building* and *regulating*. Building functions affect the growth and maintenance of the skeleton, teeth, blood, and all soft tissues. Regulating functions affect the following: heartbeat, blood clotting, nerve responses, osmotic pressure of body fluids, acid-base balance, and the transport of gases to and from the lungs.

Minerals also function as integral parts of enzymes and hormones. They are involved in such processes as the transmission of nerve impulses, the contraction of muscle, and the production of energy within cells.

Table 12-2
Function of Mineral Elements

Mineral Element	Function	Excellent Food Source
MACROELEMENTS		
Sodium	1. Exerts osmotic pressure in fluids to help maintain normal water balance 2. Maintains the pH of the body through the kidney 3. With potassium, responsible for transmission of nerve impulses	Enough naturally present in foods and water
Potassium	1. Functions on inside of cells exerting osmotic pressure to help maintain normal water balance 2. Helps kidney control pH in body fluids 3. With sodium, maintains transmission of nerve impulses	Almost all foods, both plant and animal
Chlorine	1. As part of hydrochloric acid (HCl) maintains pH of the stomach 2. Combines with sodium and potassium to maintain osmotic pressure and water balance. 3. Important in *chloride shift,* necessary in maintaining pH of blood	Naturally occurring in foods
Calcium	1. In combination with phosphorus forms mineral substance that is used in formation of bones and teeth 2. Essential for blood clotting 3. Needed for nerve and muscle function 4. Activates a number of enzymes	Milk and milk products; sardines; canned salmon eaten with bones; dark-green leafy vegetables; citrus fruits; dried beans and peas
Phosphorus	1. Works with calcium in mineralization of bone 2. Important buffer in blood in form of *phosphates* 3. Key component in forming nucleic acids, phospholipids, adenosine mono-, di-, and triphosphates, niacin-containing coenzymes, and enzymes involved in many body processes	Red meat; poultry; fish; eggs; dried beans and peas; milk and milk products; whole-grain cereals
Magnesium	1. Involves a large number of enzyme functions 2. Involved in protein synthesis 3. Plays a role in nerve and muscle function 4. Promotes activity of thyroid hormone	Leafy, green vegetables (eaten raw); nuts (especially almonds and cashews); soybeans; seeds; whole grains
Sulfur	1. Involved in many oxidation-reduction reactions in the body	Beef; wheat germ; dried beans and peas; pea-

Mineral Element	Function	Excellent Food Source
	2. Detoxifies many substances formed in body	nuts; clams

MICROELEMENTS OR TRACE ELEMENTS

Mineral Element	Function	Excellent Food Source
Iron	1. Structural component of hemoglobin 2. Component of myoglobin in muscles 3. Important part of enzymes needed in oxidation-reduction reactions in cells	Liver; kidneys; red meats; egg yolk; green, leafy vegetables; dried fruits, beans, and peas; potatoes; enriched and whole-grain cereals
Copper	1. Needed for enzymes involved in the storage and release of iron 2. Key role in connective-tissue metabolism 3. Part of enzyme involved in glucose metabolism 4. Needed in formation of phospholipids in nerves	Oysters; nuts; cocoa powder; beef and pork liver; kidneys; dried beans; corn-oil margarine
Iodine	1. Part of thyroid hormone, which regulates basal metabolism	Seafood; saltwater fish; seaweed; iodized salt; sea salt
Zinc	1. A cofactor with many enzymes 2. Cofactor with hormone *insulin* 3. Component of hair, nails, connective tissue, and blood 4. Plays a role in the sense of taste	Meat; liver; eggs; poultry; seafood; legumes; nuts; whole grains; milk
Manganese	1. Synthesis of complex carbohydrates 2. Utilization of glucose 3. Synthesis of fats and cholesterol 4. Development of pancreas 5. Contraction of muscles	Nuts; whole grains; vegetables; fruits; tea; instant coffee; cocoa powder
Fluorine	Not essential, but plays a role in prevention of dental caries	Fish; most meats; fluoridated water; foods cooked in fluoridated water
Cobalt	Part of vitamin B_{12}	Almost all foods
Molybdenum	Component of some essential enzymes	Legumes; cereal grains; liver; kidney; some dark-green vegetables
Selenium	Antioxidant; part of enzymes preventing breakdown of fats and other body chemicals; interacts with vitamin E	Seafood; whole-grain cereals; meat; egg yolk; chicken; milk

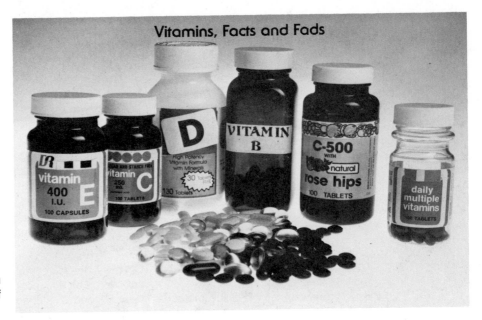

Vitamins, Facts and Fads

Figure 12-4 **Have you ever purchased any of these products?**

Vitamins.   *Vitamin* is a term that is well known. The extensive publicity given vitamins in advertisements and in news reports has conveyed to the public the importance and beneficial effects of these substances. For too many people, however, proper nutrition means only having a sufficient supply of vitamins. This general attitude toward vitamins does not lead to good nutrition. People should understand what vitamins actually are and how they function in the body.

Like the other macronutrients, vitamins are complex organic compounds needed for growth, maintenance, and regulation of metabolic processes. However, unlike the macronutrients, vitamins are needed in minute amounts. They are not involved in the digestive process but are directly absorbed into the small intestine. As organic compounds, vitamins are found only in living things. The original sources and syntheses of most vitamins are plants. Some vitamins are also synthesized by microorganisms. Consequently, animals require plants as a source of vitamins. Humans, in turn, get vitamins from both plants and animals. Many vitamins function as agents that are necessary to activate enzymes in order for specific metabolic processes to occur in the body. However, unlike enzymes, vitamins are utilized for nutritive functions and must be resupplied from dietary sources.

Vitamins are divided into two groups: the *fat-soluble* vitamins (A, D, E, and K) and the *water-soluble vitamins* (C, niacin, thiamin, riboflavin, folic acid, pantothenic acid, pyridoxine, B_{12}, and biotin). All of the water-soluble vitamins, except vitamin C, are classified as B vitamins. Fat-soluble vitamins are stored in body fat. Thus, taking them daily is less crucial than taking the water-soluble vitamins, which are not stored in the body. Water-soluble vitamins are usually excreted if the body receives excess amounts. These vitamins are less likely to have toxic effects when ingested in large amounts. However, some toxic reactions have been observed when very large amounts are taken over a period of time. Those vitamins that are stored in body fat, especially vitamins A and D, can reach toxic levels in the body, causing serious side effects and possibly death. A reaction to an excess of vitamins is termed **hypervitaminosis.** Most nutrition scientists agree that ingesting vitamins in amounts exceeding the body's needs is dangerous. However, others have advocated use of super-size doses of vitamins (megadoses) to prevent or cure certain conditions. The effects of such vitamin intake remain a controversial issue.

Many of the vitamins were discovered as a result of researchers' trying to find the cause for certain diseases, such as scurvy, beriberi, and pellagra. These diseases were found to be caused by diets deficient in certain vitamins. As a result of research and the overall better quality of diet in the United States, these deficiency diseases rarely occur. More detailed information about vitamins and their functions has been accumulated. To simplify and quickly comprehend the information on vitamins, see Table 12-3 on pages 220–221.

CHEMICAL ASPECTS OF ENZYMES

All chemical reactions in the body require enzymes. An **enzyme** is a protein that can shorten the time required for a reaction from weeks or months to seconds. Generally, each enzyme is specific for one or one kind of reaction. The substance acted upon by an enzyme is known as the **substrate**. The substance or substances resulting from enzyme action are the **products** of the reaction. Enzymes themselves are not changed in the reactions they catalyze.

Composition of Enzymes

Enzymes may require the presence of additional compounds before they can function. These are known as *prosthetic groups, coenzymes,* or *cofactors,* depending on the type of molecule and how it is bonded to the enzyme. Vitamins and minerals often act in this capacity. Vitamins (coenzymes) and minerals (cofactors) help biochemical reactions by keeping the enzyme in its active form or by binding the enzyme to the substrate, a vital step before the reactions can proceed. As cofactors and coenzymes, minerals and vitamins participate indirectly in the reaction. Like enzymes, they do not become part of the substrate or products and are not changed by the reaction. Among the minerals that act as cofactors are calcium, magnesium, manganese, copper, and zinc. Vitamins that act as coenzymes include primarily the water-soluble vitamins such as niacin and riboflavin.

Table 12-3 VITAMINS

	Best Sources	Main Roles	Deficiency Diseases	Risk of Megadoses
Water Soluble				
Thiamin (B₁)	Pork (especially ham); liver; oysters; whole-grain and enriched cereals, pasta, and bread	Helps release energy from carbohydrates; aids in the synthesis of an important nervous system chemical	*Beriberi*	None known
Riboflavin (B₂)	Liver; milk; meat; dark-green vegetables; eggs; whole-grain and enriched cereals, pasta, and bread; mushrooms; dried beans and peas	Helps release energy from carbohydrates, proteins, and fats; aids in the maintenance of mucous membranes	Skin disorders	None known
Niacin (B₃, nicotinamide, nicotinic acid)	Liver; poultry; meat; tuna; eggs; whole-grain and enriched cereals, pasta, and bread; nuts; dried peas and beans	Participates with thiamin and riboflavin in facilitating energy production in cells	*Pellagra*	Duodenal ulcer; abnormal liver function; elevated blood sugar; excessive uric acid in blood, possibly leading to gout
B₆ (includes pyridoxine, pyridoxal, and pyridoxamine)	Whole-grain (but not enriched) cereals and bread; liver; avocado; spinach; green beans; bananas; fish; poultry; meats; nuts; potatoes; green leafy vegetables	Aids in the absorption and metabolism of proteins; helps the body use fats; assists in the formation of red blood cells	Deficiency rare	Dependency on high dosage leading to deficiency symptoms such as skin disorders and kidney stones when resuming normal intake
B₁₂ (cobalamin)	Mainly in animal foods; kidneys; meat; fish; eggs; milk; oysters; nutritional yeast	Aids in formation of red blood cells; assists in the building of genetic material; helps the functioning of the nervous system	*Pernicious anemia*	None known
Folacin (folic acid)	Liver; kidneys; dark-green leafy vegetables; wheat germ; dried beans and peas	Acts with B₁₂ in synthesizing genetic material; aids in formation of hemoglobin in red blood cells	*Megaloblastic anemia*	None identified
Pantothenic acid	In all plants and animals, especially liver; kidneys; whole-grain cereal and bread; nuts; eggs; dark-green vegetables	Helps in the metabolism of carbohydrates, proteins, and fats; aids in the formation of hormones and nerve-regulating substances	Not known	Increased need for thiamin, possibly causing thiamin deficiency symptoms

Vitamin	Sources	Functions	Deficiency	Excess/Toxicity
Biotin	Egg yolk; liver; kidneys; dark-green vegetables; green beans	Aids in the formation of fatty acids; helps release energy from carbohydrates	Not known	None known
C (ascorbic acid)	Citrus fruits; tomatoes; strawberries; melon; green peppers; potatoes; dark-green vegetables	Aids in the formation of collagen; helps maintain capillaries, bones, and teeth; helps protect other vitamins from oxidation; may block formation of cancer-causing nitrosamines	Scurvy	Dependency on high doses, possibly precipitating symptoms of scurvy when withdrawn; kidney and bladder stones; diarrhea; urinary-tract irritation; increased tendency for blood to clot
Fat Soluble				
A	Liver; eggs; cheese; butter; fortified margarine and milk; yellow, orange, dark-green vegetables and fruits	Assists in the formation and maintenance of healthy skin, hair, and mucous membranes; aids in the ability to see in dim light (night vision); needed for proper bone growth, tooth development and reproduction	Night blindness	Blurred vision; loss of appetite; headaches; skin rashes; nausea; diarrhea; hair loss; menstrual irregularities; extreme fatigue; joint pain; liver damage; insomnia; abnormal bone growth
D	Fortified milk; egg yolk; liver; tuna; salmon; cod liver oil; made in skin in sunlight	Aids in formation and maintenance of bones and teeth; assists in the absorption and use of calcium and phosphorus	In children, *rickets*; in adults, *osteomalacia*	In infants, calcium deposits in kidneys and excessive calcium in blood; in adults, calcium deposits throughout the body; deafness; nausea; loss of appetite; kidney stones; fragile bones; high blood pressure; high blood cholesterol; increased lead absorption
E (alpha tocopherol)	Vegetable oils; margarine; wheat germ; whole-grain cereals and bread; liver; dried beans; green leafy vegetables	Aids in formation of red blood cells, muscles, and other tissues; protects vitamin A and essential fatty acids from oxidation	Not seen in humans	None definitely known; reports of headache, blurred vision, extreme fatigue, muscle weakness
K	Green leafy vegetables; cabbage; cauliflower; peas; potatoes; liver; cereals	Aids in the synthesis of substances needed for the blood to clot; helps maintain normal bone metabolism	Hemorrhage, especially in newborn infants	Jaundice in babies

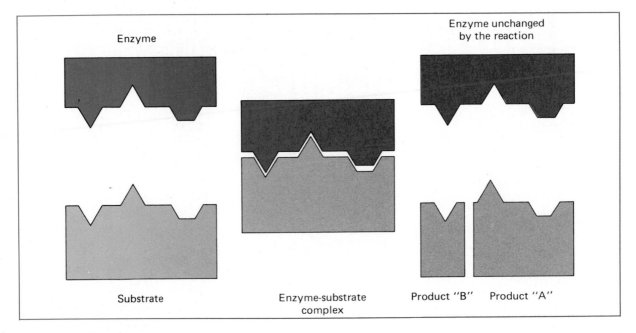

Figure 12-5 **When the enzyme-substrate complex is formed, the substrate breaks into different products but the enzyme remains unchanged.**

Enzyme activity. How does an enzyme increase the rate of a chemical reaction? Enzymes act upon specific substrate molecules and convert them into products different from the substrate. These products also serve as substrates for other enzymes. This process is illustrated in the following diagram:

$$\text{substrate A} \xrightarrow{\text{enzyme } a} \text{substrate B} \xrightarrow{\text{enzyme } b}$$

$$\text{substrate C} \xrightarrow{\text{enzyme } c} \text{end products}$$

Each enzyme in the system is dependent on the enzymes preceding it for its substrate. A lack of one enzyme in the system could prevent the action of all enzymes following it and cause an accumulation of its normal substrate. For example, if enzyme b is absent, enzyme c has no substrate and the normal end product is not formed. In addition, substrate C would accumulate. Thus the action of the entire system depends upon the proper functioning of each component enzyme. The energy required for each chemical reaction to take place is called the **energy of activation.** Enzymes are named in various ways. The major way in which they are named is to add the suffix *-ase* to the substrate with which they react, or to add the suffix to the type of reaction they catalyze. For example, **proteases** react with protein substrates; **oxidases** are enzymes that oxidize compounds.

Specificity of enzymes.   Every chemical reaction in a cell or within the body involves the action of a specific enzyme on a particular substrate molecule. The specificity of enzyme action can be explained by using a lock-and-key hypothesis. The key (enzyme) that is shaped to fit the lock (substrate) will turn in the lock to open it. In the same way, the right combination of a substrate-enzyme complex is activated by the enzyme and the substrate is altered in the reaction.

Enzymes appear to have two types of sites that concern their activity. The site that holds the enzyme and substrate together is known as the **binding site.** The site on the enzyme molecule that is directly responsible for the chemical change of the substrate is the **active site**.

Some enzymes are in an inactive form when first synthesized. The more recent, internationally approved name for such substances is **preenzymes,** although the terms *zymogens* and *proenzymes* are still widely used. A preenzyme is generally activated by splitting a portion of its molecule, which apparently blocks the enzyme activity. This may be a way of inhibiting enzyme activity from occurring at the wrong time or place. For example the digestive enzymes will not be activated until they enter the digestive system. This prevents other cells from being destroyed.

Factors affecting enzyme activity.   Experiments have also shown that the rate of an enzyme reaction is directly proportional to the concentration of the enzyme. Thus, the greater the concentration of enzyme, the faster the reaction takes place. The explanation is that the larger the ratio of enzyme molecules to substrate molecules, the greater are the chances of enzyme molecules colliding with substrate molecules to cause a reaction. If the enzyme concentration remains constant, the rate of the chemical reaction is directly proportional to the substrate concentration. Gradually, the rate of the reaction levels off to a **maximum value.** The maximum value represents the point at which the enzyme activity reaches an *optimum* working level with the substrate molecules. The enzyme is said to be *saturated* with substrate.

It has long been known that the rate of most chemical reactions increases with temperature. The same is true of most enzyme reactions. However, the protein nature of enzymes upon reaching maximum temperature is such that the rate of the reaction decreases. The constant rise in temperature destroys the enzyme and is referred to as **denaturating.** Therefore, it is found that each enzyme has an **optimum temperature** at which it operates best. Most enzymes function at an optimum temperature of 45°C, which is a few degrees above body temperature (37°C). Below 10°C, enzymes are inactive and above 60°C they cannot become activated. The effectiveness of enzymes in increasing the rates of chemical reactions also varies with the pH level in the environment. Most enzymes show maximum activity at a pH of from 6 to 8, which is the normal level of pH in the body.

The term *activator* is generally reserved for specific ions that are required by some enzymes for their activity. Iron, copper, manganese, cobalt, and zinc ions are the most common. Ordinarily, only one kind of ion functions with a specific enzyme. Any substance that interferes with the activity of an enzyme is called an *inhibitor*. Thus, a substance that denatures proteins will automatically become an inhibitor. Interference with or conversion of strategic components of enzymes will inhibit enzyme activity. The effect of drugs on the metabolism of the body is probably the result of direct or indirect inhibition of enzymes.

DIET AND HEALTH

In 1983 the per capita consumption of soft drinks, refined sugar, and fats in the United States was about 151 liters of soft drinks, 32.2 kg of sugar, and 62.7 kg of fat. The figures for 1984 show a downward trend for the sugar and fats (30.6 and 59.6 kg, respectively). This may show an increasing awareness on the part of the public of the importance of diet to health. The U.S. Senate Select Committee on Nutrition and Human Needs was seriously concerned with the nutritional patterns of the average American diet. After reviewing scientific data and volumes of testimony from dieticians, nutritionists, research scientists, and health officials, the committee issued the 1977 report of *Dietary Goals for the United States*. These goals were designed to provide nutrition knowledge with which Americans could begin to plan their diets to minimize health risks.

Dietary Goals

In 1980, the U.S. Department of Agriculture and the Department of Health and Human Services issued a publication called *Nutrition and Your Health—Dietary Guidelines for Americans*. It offers practical advice on ways to choose a nutritious and varied diet. A revised 1985 edition is also available. The guidelines apply to people who are generally in good health. People who have certain diseases or conditions that interfere with normal nutrition need special diets. Suggestions for a diet to maintain good health include the following:

- Eat a variety of foods, including fruits and vegetables; whole-grain and enriched breads, cereals, and grain products; milk, cheese, and yogurt; meats, poultry, fish, eggs; legumes (dry peas and beans). Normally there is no need for excessive consumption of any type of nutrient, nor is there normally a need for using supplementary vitamins or minerals if the diet includes a good variety of foods. Women of childbearing age or who are pregnant or nursing may need supplements. Infants and the elderly may also have special requirements.
- Maintain a reasonable weight by improving eating habits; eat slowly, take smaller portions, avoid "seconds."

- Avoid too much fat, saturated fat, and cholesterol by choosing lean meat, fish, poultry, dry beans, and peas as protein sources; use low-fat milk and milk products; use eggs and organ meats (such as liver) in moderation; limit intake of butter, cream, hydrogenated margarine, shortenings, coconut oil, and foods made from such products.
- Eat foods with adequate starch and fiber by substituting starches for fat and sugar; select foods which are good sources of fiber and starch, such as whole-grain breads and cereals, fruits and vegetables, beans, peas, and nuts.
- Avoid too much sugar by using less of all sugars, including white sugar, brown sugar, raw sugar, honey, and syrup; eat less of foods containing these sugars, such as candy, soft drinks, ice cream, cake, cookies; select fresh fruits or fruits canned without sugar or light syrup rather than heavy syrup.
- Avoid too much sodium by learning to enjoy the unsalted flavors of foods; cook with only small amounts of added salt; add little or no salt to food at the table; limit intake of salty foods, such as potato chips, pretzels, salted nuts and popcorn, condiments, cheese, pickled foods, cured meats.

Research has shown that a person's daily diet may be able to influence the level of risk of developing diseases such as heart disease, stroke, high blood pressure, cancer, and diabetes.

Resource for a balanced diet.   Nutrition scientists maintain that a diet can be evaluated for the presence of certain *key nutrients* that are reliable indicators that other nutrients are contained. The summary table on this page provides a list of the nutrients, their sources, and their functions in the body. The table also serves as a review of chapter material.

Summary Table

SUBSTANCE	ESSENTIAL FOR	SOURCE
A. Inorganic compound		
Water	Composition of protoplasm, tissue fluid, and blood; dissolving substances	All foods (released during oxidation)
Mineral elements		
Sodium	Blood and other body tissues	A variety of foods, especially processed foods
Calcium	Deposition in bones and teeth, heart and nerve action, clotting of blood	Milk, whole-grain cereals, vegetables, meats

SUBSTANCE	ESSENTIAL FOR	SOURCE
Phosphorous	Deposition in bones and teeth; formation of ATP, nucleic acids	Milk, whole-grain cereals, vegetables, meats
Magnesium	Muscle and nerve action	Vegetables
Potassium	Blood and cell activities, growth	Vegetables
Iron	Formation of red blood corpuscles	Leafy vegetables, liver, meats, raisins, prunes
Iodine	Secretion by thyroid gland	Seafoods, water, iodized salt
B. Complex organic substances		
Vitamins	Regulation of body processes, prevention of deficiency diseases	Various foods, especially milk, butter, lean meats, fruits, leafy vegetables; also made synthetically
C. Organic nutrients		
Carbohydrates	Energy (stored as fat or glycogen), bulk in diet	Cereals, bread, pastries, tapioca, fruits, vegetables
Fats	Energy (stored as fat or glycogen)	Butter, cream, lard, oils, cheese, oleomargarine, nuts, meats
Proteins	Growth, maintenance, and repair of protoplasm	Lean meats, eggs, milk, wheat, beans, peas, cheese

VOCABULARY REVIEW

Match the statement in the left column with the word(s) in the right column. *Do not write in this book.*

1. Contains the NH_2 group.
2. A unit of heat content.
3. Cholesterol belongs to this group of macronutrients.
4. Any substance that affects the rate of a chemical reaction.
5. The material affected by an enzyme.

a. catalysts
b. end product
c. enzyme
d. substrate
e. calorie
f. carbohydrate
g. protein
h. lipids
i. amino acid
j. nutrient

TEST YOUR KNOWLEDGE

Group A

Write the letter of the word(s) that correctly completes the statement. *Do not write in this book.*

1. The term *macronutrient* applies to   (a) vitamins   (b) mineral salts   (c) carbohydrates   (d) carbonates.
2. One of the monosaccharides is   (a) sucrose   (b) glucose (c) maltose   (d) lactose.
3. Phosphorus is needed for (a) maintaining the proper pH of the stomach   (b) muscle function   (c) forming nucleic acids (d) controlling the pH of body fluids.
4. Amino acids are derived from   (a) starches   (b) fats (c) proteins   (d) glycogen.
5. Minerals acting as cofactors may (a) become part of the substrate (b) help bind the enzyme to the substrate (c) be changed by the reaction (d) become part of the product.
6. A good source of iron is (a) pork liver (b) milk (c) seafood (d) corn-oil margarine.
7. An enzyme is a   (a) protein compound   (b) carbohydrate (c) fatty acid   (d) mineral element.
8. The optimum operating temperature of most enzymes in the body is about   (a) 0°C   (b) 50°C   (c) 37°C   (d) 98.6°C.
9. One fat-soluble vitamin is   (a) vitamin A   (b) vitamin B (c) vitamin C   (d) folacin.
10. The vitamin associated with clotting of blood is   (a) vitamin A (b) vitamin C   (c) vitamin K   (d) vitamin B_1.

Group B

Answer the following. Briefly explain your answer. *Do not write in this book.*

1. (a) Define an enzyme. (b) Explain the action of enzymes. (c) What are the factors in the environment that affect the activity of an enzyme? Give an example for each factor.
2. (a) How may vitamins be classified? (b) Give two examples of vitamins from each of the groups. (c) Give the best food sources of each vitamin selected. (d) What deficiency is associated with each of these vitamins?
3. Few food advertisements are usually devoted to healthful products. Write a television commercial or a newspaper or magazine advertisement to sell a healthful food, such as your favorite fresh fruit or vegetable. Design any packaging to make the product appealing. Identify the macronutrients and micronutrients in the product.

CHAPTER 13

THE MOUTH, PHARYNX, AND ESOPHAGUS

Objectives

A. Identify the parts of the mouth, pharynx, and esophagus
B. Label the parts of a typical tooth and describe the functions of different types of teeth
C. Explain the development of the deciduous and permanent teeth
D. List the sequence of events in swallowing
E. Describe the location of the salivary glands and the functions of their secretions

GROSS ANATOMY OF THE DIGESTIVE TRACT

The organs of the digestive system perform the vital function of preparing food for cells to absorb and metabolize. Most food when first eaten cannot reach the cells because it is not in a form that can pass through the intestinal wall and into the bloodstream. Food must be reduced from a complex structure to a simple structure. The process of altering the physical state and chemical composition of food so that it can be absorbed and utilized by cells of the body is known as **digestion.** Digestion is accomplished mainly by the action of acid and enzymes secreted into the digestive tract or of enzymes that are contained in cells lining the digestive tract.

The role of the enzymes is crucial in digestion because the action of digestion is primarily one of hydrolysis—that is, of breaking food down through the chemical interaction of water. As protein catalysts, the digestive enzymes accelerate this digestive reaction. When digestion is complete, the resulting small molecules can pass through the cells of the intestinal tract, entering the blood and lymph. This process is called **absorption.**

Foods are finally utilized by the cells by means of a complex process called **metabolism.** Metabolism consists of two major processes, catabolism and anabolism. Both require a series of enzyme-catalyzed chemical reactions known as *metabolic pathways.*

Organization of the Gastrointestinal Tract

The main organs of the digestive system form a tube called the *digestive tract, gastrointestinal tract,* or *alimentary canal.* The tube, a long, muscular structure lined with mucous membrane, extends from the lips to the anus. There are several accessory organs located among, or alongside, the main digestive organs. Each part of the digestive tract and its accessory organs has its own special anatomical characteristics and performs its own particular functions. Each part makes an essential contribution to the process of making food available to the cells of the body.

Arrangement of the layers.   The gastrointestinal tract is essentially a tube consisting of four layers of tissues: a *mucous membrane lining,* a layer of *connective tissue,* a *muscular layer,* and a *fibroserous layer.*

The mucous membrane, or *mucosa,* is the innermost layer of the digestive tube. This membrane is composed of epithelium that varies in structure depending on the function of the part of the tube it lines. For example, it is designed for protection in the mouth, and therefore its structure is stratified squamous epithelium (see Chapter 4). However, the epithelium in the stomach and small intestine is simple columnar epithelium because these structures are primarily designed for digestion and absorption. The mucosa also contains many glands, blood vessels, and lymph vessels.

Just underneath the mucosa is a layer of loose connective tissue that contains networks of blood vessels. This submucous layer, or *submucosa,* serves to connect the mucous membrane with the main muscular layer.

The muscular layer, called the *muscularis mucosa,* is the third main tissue layer in the wall of the digestive tube. It is composed of two very substantial sheets of smooth muscle—an inner, circular sheet and an outer, longitudinal sheet. It functions both to mix the contents of the digestive tube with digestive juices and to move food and digestive products through the gastrointestinal tract.

The fibroserous layer consists of fibrous tissue in that part of the tract above the diaphragm and serous tissue in that part below the diaphragm. The serous membrane, or *serosa,* is known as the visceral peritoneum. Although the same four tissue layers form the various organs associated with the alimentary canal, the structure of three of them varies in different organs.

The Oral Cavity

Food enters the alimentary canal through the mouth. The roof of the mouth is composed of the hard and soft palates, while the tongue serves as part of the floor of the mouth cavity. The lining of the mouth contains nerve endings that respond to pressure, pain, heat, and cold. Thus the mouth has a degree of sensitivity that is not apparent elsewhere in the digestive system.

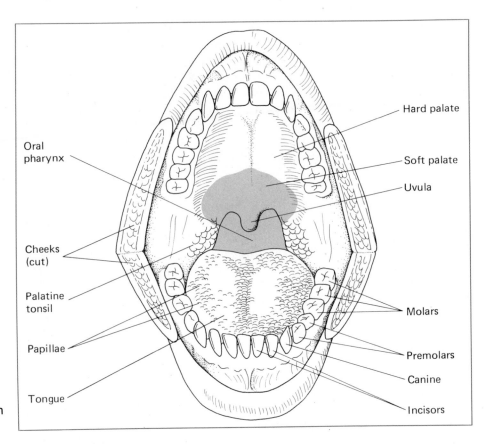

Figure 13-1 **The mouth and permanent teeth.**

The tongue.   The tongue plays a primary role in speech. In addition, it is needed for swallowing and chewing. It also contains nerve endings that are sensitive to sweet, sour, salty, and bitter food and thus serves the sense of taste.

The tongue, a highly muscular organ, consists mainly of muscle bundles arranged so that they lie in several different planes. This arrangement allows the tongue to move in many different directions.

The epithelial surface of the tongue contains nerve endings for receiving sensations of pressure, heat, and cold and for the detection of those chemical substances that give flavor to food. The tongue is covered by small projections, the **papillae,** which give parts of it a velvet-like appearance. Some of the papillae are very small and inconspicuous, but toward the back of the tongue they become large and are raised above the surface. The sense organs of taste, the **taste buds,** are located in the small depressions in the epithelium of the tongue and are more heavily concentrated in the areas surrounding the papillae. A substance must be in solution before its flavor can be recognized. The solution passes through the opening of the taste bud and stimulates the taste cells, giving rise to impulses that are carried to the brain through nerve fibers at the base of the taste bud.

The teeth.   The primary function of the teeth is to divide food into smaller bits. This function, called **chewing,** or **mastication,** increases the surface area of the morsel of food. As a result, the digestive juices come in contact with a greater surface area of food than they could if the food were swallowed without being chewed. Humans, like all mammals, have two sets of teeth. The first set, called the **deciduous teeth,** appear within a few months after birth and are temporary. These are later replaced by a second set, the **permanent teeth,** that remain throughout life. Since the deciduous teeth appear early in life, their formation begins some months before birth. The same is true of some of the permanent teeth: The buds from which they develop are formed in the tissues of the jaws before birth. At about age six, the deciduous teeth begin to be lost and are replaced by the permanent teeth.

When a tooth makes its appearance through the gum, it is said to *erupt*. In the case of a deciduous tooth, this process may cause some pain. It is also usually accompanied by an increase in the amount of saliva formed because of the irritation to the gums and the resulting stimulation of the nerve endings. The eruption of the permanent, or adult, set, with the possible exception of the teeth in back of the mouth, is usually not accompanied by similar discomfort because there is not the same amount of irritation present. These second teeth do not simply push the earlier teeth out of their sockets, as many people believe. Instead, as they grow within the jaw, a group of cells, the **odontoclasts,** form in front of the tip of the tooth and dissolve the base of the first tooth. In this respect, the odontoclasts behave much as the osteoclasts do in forming permanent bone tissues. Finally, the first tooth is held in place by only the tissues of the gum, and final separation from these tissues may occur spontaneously or the tooth may need a little coaxing.

Arrangement of the teeth.   There are different shapes of teeth that are analogous to the function they perform. On each jaw, in front, there are four teeth with edges adapted for biting—the *incisors*. On each side of

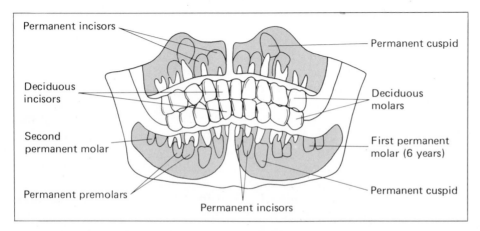

Figure 13-2 **The teeth of a young child. The jaws are exposed to show the roots of the deciduous teeth and the developing adult teeth.**

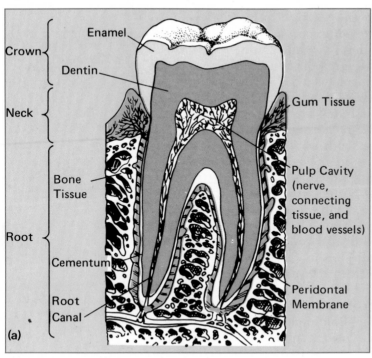

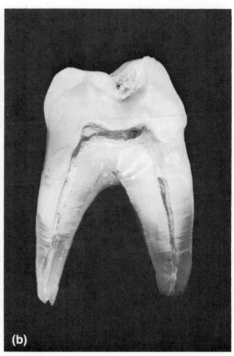

Figure 13-3 (a) Structure of a typical molar (longitudinal section). (b) Photomicrograph of a section of a human tooth.

these is a *canine* tooth that is pointed for tearing food. In back of each canine are two *premolar* teeth. These are sometimes referred to as the *biscuspid* teeth because of the presence of two points, or *cusps*. Since the cusps of the upper and lower premolar teeth mesh in chewing, they act to cut or shear food. Behind the premolars there are three *molar* teeth on each side of the jaw. These teeth are characterized by relatively flat surfaces that permit the food to be ground between them. The third molars, or *wisdom teeth*, are the last to erupt. Normally, the total number of permanent teeth is 32, as compared with the 20 in the deciduous set. Occasionally these numbers vary because of the failure of some of the teeth to erupt.

Formation of teeth.   During the formation of a tooth (Fig. 13-3), two separate groups of tissues add support to its structure. The enamel is formed from specialized types of epithelial cells. The enamel covers only the crown of the tooth and extends a short distance below the gum line. At the same time, **dentin** is formed by the *odontoblasts*. Both of these regions are formed in a manner that is similar to the formation of bone. The teeth, like the bones, are composed of a protein matrix that is made of calcium and phosphorus salts. The enamel is laid down in the form of the microscopic hexagonal pillars that are held together by a cementing substance, *cementum*. This binding material can be eroded away by acids produced in the mouth during the initial stage of tooth decay. Bacteria that get into the mouth produce enzymes and acids that split proteins. This destroys the protein matrix of the enamel and can result in the formation of cavities.

Dentin is the part of the tooth that lies below the enamel. It contains a large number of very fine canals, the dentinal tubules, which are similar to the small bony tubes in which the osteocytes lie (Fig. 13-3). Within each of these tubules is a cellular process from an odontoblast, the body of which lies against the dentin wall of the pulp cavity.

Below the surface of the gum, the tooth is surrounded by the periosteum of the alveolar bone, which fixes the tooth to the connective tissue (periodontal membrane) lining the socket. The fibers of this membrane are firmly attached by the cementum—at one end to the tooth socket in the jaw bone and at the other end to the root. The slinglike arrangement prevents the tooth from being pushed inward by the considerable force exerted on the tooth in chewing. However, in those cases in which the teeth do not oppose each other at the correct angle (malocclusion), there is a lever action of one tooth against another that pushes the teeth out of their normal position. The abnormal pressure may eventually result in disease of the periodontal membrane. After a tooth has been removed from the jaw, bone fills in the cavity that is created.

The pulp cavity contains the blood vessels and nerves essential for the nourishment of the teeth. If bacteria invade these soft tissues, infection can spread rapidly downward through the root canal to the bone. An abscess then forms at the base of the root.

Figure 13-4(a) Location of the three types of salivary glands. (b) Photomicrograph of a compound gland and the secreting cells of the salivary glands.

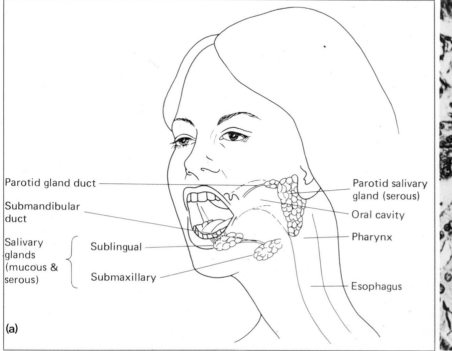

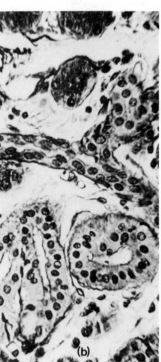

Parotid gland duct

Submandibular duct

Salivary glands (mucous & serous)

Sublingual

Submaxillary

Parotid salivary gland (serous)

Oral cavity

Pharynx

Esophagus

(a)

(b)

Salivary glands.   There are three pairs of salivary glands that secrete their fluids into the mouth: *parotid, submandibular,* and *sublingual.* These accessory organs of the digestive tract manufacture saliva.

In Fig. 13-4a, the parotid gland can be seen lying just in front of and slightly below the level of the opening of the ear. This is the largest of the glands and the one that usually becomes enlarged during an attack of mumps. Below it and near the angle of the lower jaw is the submandibular gland, and under the tongue is the sublingual gland.

The saliva-forming cells are at the ends of small branches leading off from the main duct. They form a compound type of gland. When the gland has been inactive for a time, the cells become filled with granules that disappear after a period of activity (Fig. 13-4b). This indicates that the materials in saliva, other than water, are formed as solid substances within the cells and then are converted into a fluid that can pass through the cell membrane, and in turn by ducts into the mouth.

The secretion of saliva is controlled by the autonomic nervous system. A simple reflex causes the secretion of saliva when food is placed in the mouth, as the food is being thoroughly chewed. The particles of food stimulate nerve endings in the mouth. When the food is in solution, it also stimulates the taste buds. Through fibers of the autonomic nervous system, there is a dilation of the blood vessels passing to the glands, and saliva is produced. The daily flow of saliva in a normal individual amounts to between 1000 and 2000 milliliters. Much of this saliva is produced during periods when there is no food in the mouth.

The Pharynx

The oral cavity opens into the pharynx, a region shared by both the digestive and respiratory systems. The pharynx has three anatomical divisions: the *nasopharynx,* located behind the nose and palate; the *oropharynx,* located behind the mouth from the soft palate above to the level of the hyoid bone in the neck below; and the *laryngopharynx,* which extends from the hyoid bone to its termination in the esophagus.

A small muscular flap of tissue, the *uvula,* hangs from the soft palate and prevents food from entering the nasal cavity during swallowing by closing off the posterior nares. The *tonsils* are masses of lymph tissue on either side of the pharyngeal opening. The pharyngeal tonsils are located in the nasopharynx on its posterior wall opposite the posterior nares. Although the cavity of the nasopharynx differs from the oral and laryngeal division in that it does not collapse, it may become obstructed. If the pharyngeal tonsils enlarge (a condition called *adenoids*), they fill the space behind the posterior nares and make it difficult or impossible for air to travel from the nose into the throat.

Two pairs of tonsils are found in the oropharynx: the *faucial,* or *palatine, tonsil,* located between the pillars of the fauces; and the *lingual tonsil,* located on the base of the tongue. The palatine tonsils are the ones most commonly removed by a tonsillectomy.

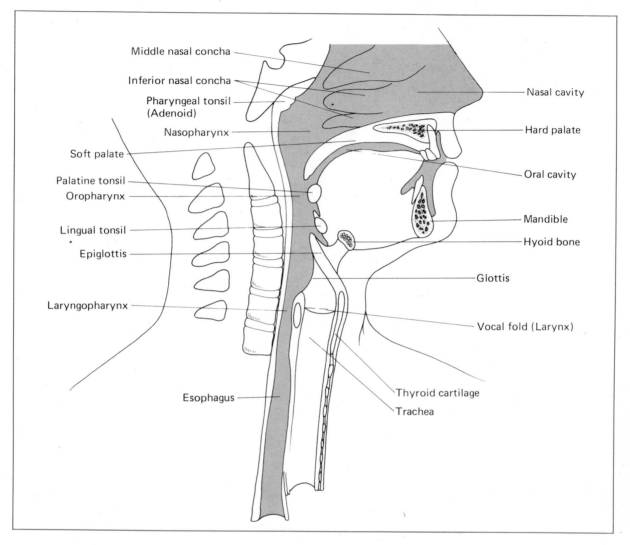

Middle nasal concha

Inferior nasal concha

Pharyngeal tonsil
(Adenoid)

Nasopharynx

Soft palate

Palatine tonsil

Oropharynx

Lingual tonsil

Epiglottis

Laryngopharynx

Esophagus

Nasal cavity

Hard palate

Oral cavity

Mandible

Hyoid bone

Glottis

Vocal fold (Larynx)

Thyroid cartilage

Trachea

Figure 13-5 Sagittal section of the head and neck region showing the divisions of the pharynx and position of the esophagus.

The Esophagus

The esophagus, a collapsible tube about 25 centimeters (10 inches) long, extends from the pharynx to the stomach, piercing the diaphragm in its descent from the thoracic to the abdominal cavity. It lies behind the trachea and heart.

The esophagus is covered by a tough protective tissue. Beneath this are two layers of muscle fibers, the outer of which runs longitudinally (lengthwise) and the inner in a circular direction. It is through the alternating contraction and relaxation of these two layers that the peristaltic wave is set up. Internally, the esophagus is lined by a mucous membrane that is held loosely to the circular muscles by connective tissue. The mucus secretion of the lining serves as a lubricant.

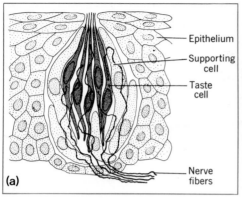

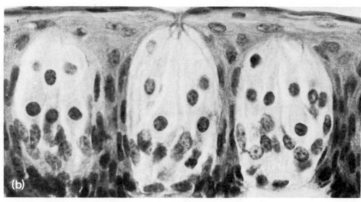

Figure 13-6 (a) A diagram of a taste bud found on the tongue. (b) Photomicrographs of taste buds sensitive to the different flavors.

Epithelium
Supporting cell
Taste cell
Nerve fibers

(a)

(b)

DIGESTIVE FUNCTIONS

Not all surfaces of the tongue are sensitive to all flavors to the same degree. Taste buds are sensitive to sweet, sour, bitter, and salty flavors. Taste buds especially sensitive to sweet flavors are located near the tip of the tongue; those for sour flavors lie along the sides. The receptors for bitter flavors are on the surface of the tongue toward the back. Salty flavors are detected by taste buds at the front and sides of the tongue. If these various groups of taste buds are stimulated individually by a very weak electric current, the stimulus will produce in each group the same sensation as a chemical substance of the flavor would.

Many complex flavors are the result of smelling foods rather than tasting them. Odors are carried to the *olfactory* nerve endings located in the upper part of the nasal cavity, and people are apt to confuse the odor of the substance with its flavor simply because the material is present in the mouth at the time. One can seldom get the real flavor of a food with a bad head cold. With this condition, increased mucus secretions cover the olfactory nerve endings, and one cannot detect the odors of the food.

Motor Functions

Four pairs of muscles are involved in the process of **mastication.** Because of the manner and place of their origins and insertions, the movements of the jaw are due to the structure of the joint between the **mandible** (jaw bone) and the **temporal bone,** which these muscles move. The temporalis raises the mandible, closes the mouth, and clenches the teeth. The **masseter** (the most powerful jaw muscle) also closes the mouth and clenches the teeth. The two **pterygoids,** medial and lateral, open the mouth and move the jaw from side to side.

Mechanics of chewing.   The **lips, cheeks,** and **tongue** also have their roles in the process of *mastication.* When the teeth come together, the food is squeezed out from between them into the mouth cavity. The lips and cheeks are then drawn inward as the jaw is lowered for the next biting motion, and the tongue spreads outward toward the sides. These motions tend to push the food back between the teeth again so that it can be rechewed. The potential force of the muscles responsible for the motion of the jaws is much greater than that actually required to chew the food. It is not the force of the teeth on the food that is important, but rather the grinding motion that serves to break the food down.

An important result of proper mastication of food is the breakdown of the cellulose fibers of plant materials. This allows the digestive juices to act on parts of the food that would otherwise be inaccessible because of their enclosure by cellulose walls. Persons who have lost their teeth or who are equipped with poorly fitting dentures sometimes suffer from nutritional disorders because they cannot chew food sufficiently. On the other hand, excessive chewing of food, as some people recommend, has no special merit.

Secretory Functions

The process of mastication is helped by the saliva, which is formed in and secreted by the salivary glands. This fluid plays several roles. It softens and lubricates the food mass, which can then be more easily chewed and swallowed. It also dissolves some of the food so that it can be tasted. Saliva partly digests the starch in foods. In addition, saliva washes the teeth, helps neutralize mouth acids, and keeps the inside of the mouth flexible and moist. This last function of the saliva is especially important for speech.

Salivary secretion. Saliva is a mixture of different chemical compounds, the composition of which changes as the activity of the gland increases or decreases. Saliva has an average pH value of 6.7, which means that it is slightly acid. Normally, the saliva contains about 99 percent water. In the saliva are dissolved minute traces of several different salts of sodium, potassium, and calcium, as well as organic materials like the salivary enzyme amylase.

Mucins also form important constituents of saliva. They are the major proteins in saliva that lubricate the food particles so that they can be swallowed.

Salivary digestion. Food remains in the mouth for such a relatively short time that little digestion can occur there. Nevertheless, the action of the salivary enzyme amylase should not be underrated. Salivary amylase can decompose starch into simpler products, such as dextrins, and break off some of the disaccharide maltose. Although little of this activity occurs in the mouth, digestion of food by salivary amylase may continue in the stomach. The ball of food (bolus) does not break up

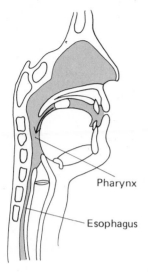

Pharynx

Esophagus

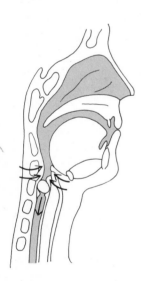

Figure 13-7 **Swallowing.** The bolus is projected down into the upper part of the esophagus. Peristaltic waves push the bolus down the lower end of the esophagus into the stomach.

immediately upon reaching the stomach, so digestion by the action of salivary amylase may continue for as long as half an hour. Eventually, the acid in the gastric juice will stop this digestion; but by then, as much as 75 percent of the starch of potatoes and bread may have been broken down. The digestive process in the small intestine completes the breakdown of carbohydrates through the action of its numerous enzymes. This will be discussed in detail in Chapter 15.

Control of Digestive Functions

The act of swallowing, called **deglutition,** is a complex one involving many of the muscles that form the walls of the mouth and pharynx. It is usually started as a voluntary process, but shortly becomes involuntary as the food comes under control of the smooth muscles of the esophagus. In swallowing, the tip of the tongue arches slightly and starts to push the food toward the back of the mouth. Then a wave of muscular contraction passes over the tongue, which forces the food against the hard palate. At the same time, the uvula and soft palate close the opening to the nasopharynx, preventing food from going into that region. Once the bolus reaches the pharynx, involuntary control takes over the process. As the food moves backward and downward, the soft palate is drawn slightly down, just as the larynx (voicebox) is drawn upward toward the epiglottis (the covering of the opening of the windpipe). As added insurance against the entry of food into the respiratory system, the nervous impulses responsible for breathing stop so that it is almost impossible to breathe and swallow at the same time. Action of the involuntary muscles sends the bolus down the esophagus to the stomach.

Peristalsis. The series of involuntary muscular contractions that move food along the esophagus is known as **peristalsis** (Fig. 13-7). These contractions appear behind the bolus and push it forward. At the same time, the muscles of the walls of the esophagus in the region of the food, and those muscles preceding it, relax, thereby relieving pressure on the bolus so that it can go forward. Solid food of an average consistency requires an average of 9 seconds to pass along the 10 inches of esophagus. When the peristaltic wave reaches the stomach, it causes the muscle fibers that guard the cardiac opening of the stomach to relax. This opening allows for the passage of food from the esophagus into the stomach.

When fluids are drunk, they pass quickly through the slightly dilated esophagus and outrun the peristaltic wave. They then collect at the cardiac sphincter and are able to enter only after the wave of muscular contraction has reached that point. If, however, there is a rapid succession of swallowing movements, as in drinking a glass of water, the opening may remain relaxed. The fluid then passes directly from the esophagus into the stomach.

DISORDERS OF THE UPPER DIGESTIVE TRACT

One pathological defect of the mouth, with which a person can be born (congenital), is cleft lip or cleft palate, or both. In the normal embryonic development of the face, tissue masses that form the upper lip and palate must grow together and fuse. If they fail to fuse, a cleft results. Surgical correction of a cleft lip is important for cosmetic reasons and for clear speech. Surgical correction of a cleft palate may be necessary for normal speech and swallowing.

Some people are born with a segment of the esophagus missing. This is called *esophageal atresia*. Most often, either the upper or lower end of the remaining esophagus is connected to the trachea or windpipe. The danger, then, is that saliva, food, or gastric juices can enter the lung and cause a severe chemical pneumonitis.

SUMMARY

The mouth is formed by the cheeks, hard and soft palates, tongue, and muscles. The tongue is an important organ for mastication, taste, and speech. The three pairs of salivary glands aid the tongue by lubricating the food. The salivary enzyme amylase digests starch.

The teeth are the primary structures for mastication. The deciduous, or baby, teeth are replaced by permanent teeth in the adult.

From the mouth the food is forced into the pharynx, which has three divisions. The pharynx serves as an organ for speech as well as a passageway for the digestive and respiratory tracts. The three pairs of tonsils are found in the pharynx and under the tongue.

The esophagus is a collapsible muscle tube that extends from the pharynx to the stomach. It is located posterior to the windpipe and heart. By means of peristalsis, food is pushed into the stomach for further chemical and mechanical digestion.

VOCABULARY REVIEW

Match the phrase in the left column with the correct word(s) in the right column. *Do not write in this book.*

1. The hardest substance in the body.
2. Contains the nerves and blood vessels of the tooth.
3. The part of the tooth containing the processes of the odontoblasts.
4. The material that attaches the periodontal membrane to the tooth.

   a. cementum
   b. peristalsis
   c. dentin
   d. enamel
   e. incisor
   f. larynx
   g. mastication

5. A region common to both the digestive and respiratory systems.

6. A structure in the neck that moves upward when you swallow.

7. A process resulting from the contraction of smooth muscle fibers.

h. pharynx
i. pulp cavity

TEST YOUR KNOWLEDGE

Group A

Select the correct statement by letter.

1. The parotid glands produce an enzyme that digests   (a) proteins   (b) sugars   (c) starches   (d) fats.
2. Incisor teeth are best fitted for   (a) tearing   (b) biting   (c) grinding   (d) chewing.
3. The mouth opens into the   (a) uvula   (b) soft palate   (c) pharynx   (d) larynx.
4. A basic flavor to which the taste buds are *not* sensitive is   (a) sweet   (b) sour   (c) salty   (d) peppery.
5. A full set of permanent teeth consists of   (a) 20 teeth   (b) 24 teeth   (c) 28 teeth   (d) 32 teeth.
6. Dentin is formed from   (a) odontoclasts   (b) odontoblasts   (c) ameloblasts   (d) orthodontists.
7. A term that does *not* describe a salivary gland is   (a) parotid   (b) sublingual   (c) submandibular   (d) subcutaneous.
8. The covering for the opening of the windpipe is called the   (a) glottis   (b) epiglottis   (c) trachea   (d) larynx.
9. The series of muscular contractions that moves food along the digestive tract is called   (a) perimysium   (b) pericardium   (c) periosteum   (d) peristalsis.
10. In the structure of a tooth, the enamel covers   (a) the whole tooth   (b) only the crown   (c) only the root   (d) only the dentin.

Group B

1. Discuss the statement, "Digestion is both a chemical and mechanical process," from the standpoint of the activities occurring in the mouth.
2. How is the secretion of saliva controlled?
3. Briefly describe the process of deglutition.
4. Make a carefully drawn and fully labeled sketch of a human tooth.
5. Why will proper care of the teeth of a young child help to ensure better teeth in the adult? What factors are involved in this care?

CHAPTER 14

THE STOMACH

Objectives

A. Describe the gross anatomy of the stomach
B. Describe gastric motility and how it is regulated
C. Explain the mechanism by which gastric secretion is regulated
D. Identify the components and functions of gastric juice
E. List the sequence in which digestion takes place in the stomach

GENERAL STRUCTURE OF THE STOMACH

The stomach is located in the upper left quadrant of the abdominal cavity. It lies under the diaphragm and extends slightly to the left of the midline of the body. The diaphragm is the large muscle that separates the thoracic cavity from the abdominal cavity. The middle portion of the stomach is loosely held by the broad, membranous **greater** and **lesser omenta.**

The position of the stomach changes frequently, except where it is firmly fixed at its beginning and its end. For example, the stomach is pushed downward with each inspiration and moves upward with each expiration. When the stomach is distended, as when a large meal is eaten, its size increases, thus interfering with the descent of the diaphragm during inspiration. This results in the uncomfortable feeling of **dyspnea** (difficult breathing), which often accompanies overeating. The stomach is described as a distensible sac; thus it can vary in shape.

dys = difficult

Anatomy of the Stomach

The **fundus,** the **body,** and the **pylorus** are the three divisions of the stomach. The stomach has two openings, one from the esophagus—the **cardiac opening**—and one to the small intestine—the lower **pyloric opening.** The right side of the stomach has a concave shape, known as the **lesser curvature.** Its left side has a convex shape, called the **greater curvature.** The main portion of the stomach is the body, which tapers

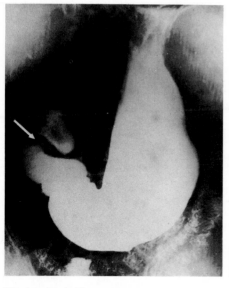

Figure 14-1 X-ray photograph of a normal stomach. The fundus is at the top right, below it is the body, and to the lower left of this is the pyloric portion. The pyloric orifice is marked by the white arrow.

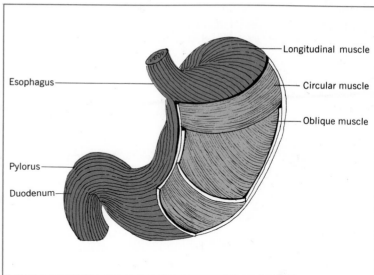

Figure 14-2 The muscular coats of the stomach.

peri = around

off into the narrow portion of the pylorus. The lowermost portion of the body is frequently referred to as the **antrum.** The fundus is the rounded portion of the stomach that appears as the bulge above the cardiac opening.

Sphincter muscles.   The cardiac and pyloric openings are formed by sphincter muscles. These muscles consist of a ring of circular muscle fibers. When relaxed, these fibers form an opening; when contracted, they close the opening. Although the **cardiac sphincter** is not a well-defined anatomical structure, this muscle forms the opening between the esophagus and the stomach. The cardiac opening is sometimes referred to as the **cardia.**

The **pyloric sphincter** guards the opening from the pyloric portion of the stomach into the beginning of the small intestine. This sphincter muscle is usually open and not much of an obstacle to the passage of material.

The layers of the stomach.   The outside layer of the stomach is a fibroserous coat called the **peritoneum.** The peritoneum that lines the entire abdominal cavity is the **parietal** peritoneum. The **visceral** peritoneum covers some organs in the abdominal cavity. Hanging from the greater curvature of the stomach is the large apronlike fold of peritoneum, the greater omentum. The greater omentum contains varying amounts of fat, deposited within or upon it.

The mucous membrane lining the stomach's innermost layer is thrown into folds called **rugae.** This folding of the mucosa is due to the structure of the submucosa, which is a loose, areolar, vascular layer. These folds gradually smooth out and disappear as the stomach becomes distended from food.

The muscular layers of the stomach wall are composed of smooth muscle and include an inner, *oblique* layer; a *middle, circular* layer (the major part of the muscular wall); and an *outer, longitudinal* layer. The ringlike pyloric sphincter is formed by a thickening of the circular muscle fibers.

The gastric glands.   The mucous membrane lining the stomach is covered with a single layer of columnar epithelial cells. This layer also extends down into numerous little wells or openings called **gastric pits.** At the bottom of each pit are long tubular glands called **gastric glands.** These glands produce **gastric juice.** Gastric juice is delivered to the pits, which in turn, conduct it to the surface of the mucosa. Gastric juice contains hydrochloric acid (HCl), digestive enzymes, and mucus. The hydrochloric acid is produced by *parietal cells,* which are located alongside the opening of a gastric gland. Parietal cells are also thought to produce an unidentified substance called **intrinsic factor.** Intrinsic factor is necessary for vitamin B_{12} absorption, and vitamin B_{12} is essential for the normal development of red blood cells.

The enzymes in gastric juice are produced by **zymogenic,** or **chief, cells** located in the base or body of the gastric glands. Mucus is produced by **mucous cells** that are located in the neck of each gland. The mucus serves to protect the inner mucous membrane lining of the stomach.

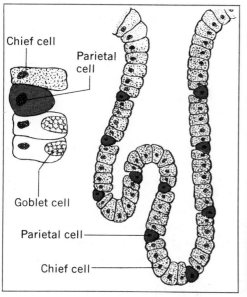

Figure 14-3 (left) A gastric gland showing the three principal types of cells.

Figure 14-4 (right) Crystals of gastric protease, the active protein-digesting enzyme of the gastric system.

PHYSIOLOGY OF THE STOMACH

The quantity of gastric juice produced during an average meal is about 0.5 to 0.7 liters. The secretion of gastric juice is partly under nervous control and partly under the control of chemical regulators belonging to the group of secretions known as **hormones.**

Gastric Secretions

The sight, odor, taste, or even the thought of food will start the activity of the gastric glands. This is called the **cephalic phase** of gastric juice secretion. In this phase the gastric glands are subject to the same type of conditioned reflex that the salivary glands are in the presence of similar stimuli. The gastric juice produced in this phase is about one quarter of the normal flow. When food is introduced through experiment directly into the stomach without the influence of sensory stimuli, the secretion of gastric juice is only three fourths of the normal amount. The question then arises as to what type of mechanism continues the secretion of gastric juice in sufficient quantities to complete the digestion of the meal. Nervous stimulation caused by the presence of food in the stomach is not the answer because when all gastric nerves are severed, secretion is hardly affected.

Control of gastric secretion. Experiments with animals were used to answer the question of how digestion continues after the cephalic phase has stopped. While an animal was under anesthesia, the cardiac end of the stomach was tied off and salt solution was introduced into the cavity of the stomach through a tube placed in the pyloric portion. After an hour the solution was removed and tested for the presence of hydrochloric acid and pepsin (known also as gastric protease). None of these substances was found. However, if small bits of the lining of the pyloric portion were ground up and an extract of this injected into the stomach's blood supply, pepsin and hydrochloric acid both appeared in the salt solution. This indicated that some chemical substance present in the extract of the pyloric lining stimulated the gastric glands to secrete gastric juice. Other scientific research has confirmed these results.

When solid food enters the stomach, cells in the pyloric region are stimulated to produce the hormone **gastrin.** This hormone is absorbed by the bloodstream, which carries it to the cells of the mucosal layer of the lining of the stomach. The gastrin cells stimulate the glands to produce gastric juice. This phase of secretion is called the **gastric phase.** Gastrin belongs to that group of substances known as *secretagogues*—hormones that stimulate secretion in glands.

The **intestinal phase** of gastric juice secretion is less clearly understood than the two previously discussed. However, chemical control, as well as a neural reflex, mechanism is believed to operate. The intestinal phase begins with the presence of stimuli in the lumen of the intestinal tract. These stimuli include *distention, acidity, osmolarity,* and the *end products* of carbohydrate, fat, and protein digestion.

The stimuli evoke neural reflexes and activate the secretion of the hormones **secretin, cholecystokinin (CCK),** and **gastroinhibitory peptide (GIP),** which causes a decrease in the secretion of gastric juice and, in turn, the emptying of gastric contents.

Motor Functions

The motility of the stomach involves three major processes:

- the storing of food until it can be accommodated in the small intestine
- the mixing of food with gastric secretions until a semifluid mixture called **chyme** is formed
- the emptying of food from the stomach into the small intestine at a rate suitable for proper digestion and absorption by the small intestine

Gastric motility.   As food enters the stomach, it forms concentric circles in the body of the stomach. Food most recently ingested lies closest to the opening from the esophagus; food previously ingested lies nearest to the wall of the stomach. Normally the body of the stomach has relatively little muscle tone. Consequently it can bulge progressively outward, storing greater and greater quantities of food. The stomach's capacity for food is thought to be about 1 liter. The pressure in the stomach remains low until this limit is reached.

As previously discussed, the gastric juice is secreted by glands that line the inner mucosal wall of the entire body of the stomach. This secretion comes immediately into contact with the stored food lying against the mucosal surface of the stomach. When the stomach is filled, **mixing waves** move along the stomach wall approximately once every 20 seconds. In general, the waves become more intense as they approach the lower end of the stomach called the *antrum.* The mixing waves tend to move the gastric secretions and the outermost layer of food gradually toward the antrum. Upon entering the antrum, the waves become stronger and the food and gastric secretions become progressively mixed until they form a liquid, chyme. The degree of liquidity of chyme depends on the volume of food ingested and the amount of gastric juice secreted, as well as on the degree of digestion that has occurred.

Basically, the emptying of the stomach is promoted by peristaltic waves in the antrum. These antral peristaltic waves characteristically occur almost exactly three times per minute. With each peristaltic wave, several millimeters of chyme are forced into the duodenum. The rate of peristaltic movement of the pylorus is regulated by stimuli from the stomach itself and from the duodenum. The gastric stimuli that determine the rate of gastric emptying are primarily the degree of distention of the stomach and the volume of its contents. In general, the rate of food emptying from the stomach is almost proportional to the *square root* of the volume of food remaining in the stomach at any given time.

Regulating gastric motility.   Peristaltic movement can be opposed or inhibited by resistance of the pylorus. Such resistance is due to the **enterogastric reflex** from the duodenum. Feedback by certain hormones from the duodenum also inhibits gastric emptying. In the enterogastric reflex, strong nervous signals are frequently sent from the duodenum back to the stomach, especially when the stomach is emptying food into the duodenum. These signals probably play the most important role in determining the rate of stomach emptying. The enterogastric reflex is influenced by a variety of factors: the degree of distention of the duodenum, the acidity of the duodenal chyme, the osmolality of the chyme, and primarily the products of protein and fat digestion present in the duodenum.

When fatty foods are present in the chyme that enters the duodenum, the rate of stomach emptying decreases. This is because the complete digestion of fats requires a longer time. The decrease in rate of stomach emptying allows for the slow digestion of fats in food that has already moved into the small intestine. The precise mechanism by which fats cause this decrease in movement is not completely known. Presumably it is the effect of some feedback mechanism elicited by the presence of fats in the duodenum. The hormones considered to be involved in this activity are **secretin, cholecystokinin,** and **gastroinhibitory peptide.**

gastro = stomach

Researchers disagree on the exact time it takes for food to pass through the stomach. In general, liquids will pass through the stomach more rapidly than solids. It has been estimated that an average of 3 hours is required for a meal consisting of carbohydrates, fats, and proteins to leave the stomach.

Very little food is absorbed into the bloodstream through the walls of the stomach. At this point the food has not been sufficiently digested and reduced in size to pass readily through the membranes. Some materials, such as water, glucose, a few salts, alcohol, and substances with low acidity (aspirin), can be absorbed through the stomach wall. However, their absorption depends upon the volume of food in the stomach.

Alterations in motility.   Vomiting is a means by which the upper gastrointestinal tract rids itself of its contents when the digestive tract becomes excessively irritated, overdistended, or occasionally overexcited. The stimuli that cause the vomiting reflex can be found in any part of the gastrointestinal tract. Most often the stimuli are distention or irritation of the stomach or duodenum. Other stimuli that frequently cause vomiting are rapid changes in the position of the body, such as those produced by the motion of a car, ship, plane, or swing. In these examples, vomiting is due to stimulation of the semicircular canals of the inner ear.

The *vomiting center* is located in the *medulla*. Once the vomiting center has been stimulated, the act of vomiting involves taking a deep breath, raising the hyoid bone for the larynx to pull the esophagus open, closing the glottis, and lifting the soft palate to close the posterior nares. A sudden and powerful contraction of the diaphragm occurs along with contraction of the abdominal muscles. This action squeezes the stomach between the two sets of muscles, and the expulsion of the gastric contents moves upward through the esophagus.

Mechanical and Chemical Digestion

Mechanical digestion in the stomach consists of churning and peristalsis. Churning is the forward and backward movement of the gastric contents, which mixes the food with the gastric juice to form chyme. Peristaltic waves occur in the stomach about three times per minute and sweep the contents of the stomach toward the closed pyloric sphincter. At intervals, strong peristaltic waves press the chyme past the sphincter into the duodenum.

Gastric juice is the most acid fluid of the body because it contains hydrochloric acid (HCl), which in concentrated form can destroy tissue. The concentration of HCl present in the gastric juice (about 0.6 percent) does not affect the normal, living stomach, however. Following death, the mucous lining of the stomach is rapidly destroyed, so there is evidently some protective agent present in the living organ that prevents destruction of the tissue. It has been suggested that protection is afforded by the salts of sodium and potassium, especially when they are combined with chlorides and bicarbonates. The range of acidity of gastric juice is from a pH of 1.0 to 3.5.

Activity of gastric juice.   HCl functions in several different ways during gastric digestion. First, its presence is necessary for the formation of **gastric protease** from the proenzyme, **pepsinogen.** Second, gastric protease is unable to digest proteins unless the gastric contents remain acid. A third activity of HCl is to destroy bacteria that might have entered with the food. This is not always an effective process, since some live bacteria do enter the small intestine and multiply. Finally, HCl activates other enzymes secreted by the gastric mucosa.

ase = enzyme

The active enzyme of the stomach, *gastric protease,* (*pepsin*), digests only proteins. Purified crystals of this enzyme are shown in Figure 14-4. Gastric protease is able to break down large molecules of proteins, including casein in milk, into simpler molecules. These molecules, however, are still too large to enter the bloodstream. These nonabsorbable intermediate products (proteoses and peptones) are peptide fragments that must be further digested by the small intestine.

GASTRIC DISORDERS

itis = inflammation

One of the most common disorders of the stomach is **gastritis.** Gastritis means inflammation of the *gastric mucosa.* This usually can result from the presence of irritant foods on the gastric mucosa, from excessive abrasion of the stomach mucosa by the stomach's own gastric secretions, or occasionally from the presence of bacteria or viruses.

Inflammation of the Mucosa

One of the most frequent causes of gastritis is irritation of the mucosa by alcohol. The inflamed mucosa in gastritis often causes a painful burning sensation under the mediastinal area of the diaphragm. Reflexes initiated in the stomach mucosa cause the salivary glands to secrete intensely. The frequent swallowing of foamy saliva allows air to accumulate in the stomach. As a result, the person usually belches profusely and simultaneously experiences a burning sensation in the throat.

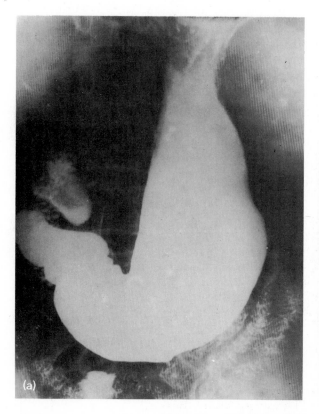 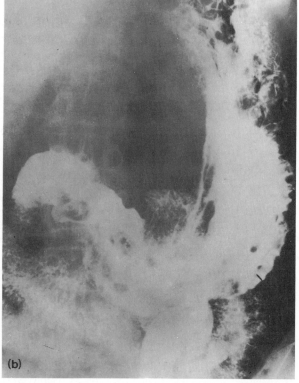

Figure 14-5 (a) Photo showing a normal stomach. (b) Cancerous stomach showing destruction of the lining, muscle layers, and blood vessels.

Cancer of the Stomach

Stomach cancer begins as an open sore. As it spreads, it involves more and more of the stomach wall. The lining, muscle layers, and blood vessels are gradually invaded or destroyed. Slight indigestion and stomach distress after eating can be symptoms of early stomach cancer. Unfortunately, people often waste valuable time trying various self-prescribed treatments. Fragments of cancer then spread through the circulatory system to other parts of the body. This spreading is called *metastasis.* If the cancer is discovered before this happens, part or most of the stomach can be removed.

SUMMARY

The stomach is an elongated pouch that lies in the left upper quandrant of the abdominal cavity. The size of the stomach varies in different individuals and varies according to whether it is distended or not.

The stomach is divided into three regions: the fundus, the portion near the esophageal opening; the body, the central portion; and the pylorus, which is the constricted lower portion. The opening of the esophagus into the stomach is called the cardiac opening, or cardia. The opening of the pylorus into the duodenum is guarded by the pyloric sphincter.

The glands in the stomach contain three different types of epithelial cells: the neck mucous cells of the gastric mucosa, which secrete mucus; the parietal cells, which secrete hydrochloric acid and possibly the intrinsic factor; and the chief cells (zymogenic cells), which secrete the enzymes of gastric juice.

The stomach functions as a food reservoir. In addition, it secretes gastric juice, which converts proteins into intermediate products. The contractions of the stomach break down food into small particles, mix them well with gastric juice, and move the contents on to the duodenum. The stomach also secretes the antianemic intrinsic factor and carries on a limited amount of absorption. Lastly, the stomach secretes the hormone gastrin.

VOCABULARY REVIEW

Match the statement in the left column with the correct word(s) in the right column. *Do not write in this book.*

1. The lower end of the esophagus.
2. Its opening is controlled by the consistency of chyme.
3. The lining of the abdominal cavity.
4. A hormone that slows stomach emptying.
5. Folds of mucosa that line the stomach.
6. Secretion of the parietal cells.

a. cardiac orifice
b. duodenum
c. secretin
d. hydrochloric acid
e. parietal cell
f. gastric protease
g. peritoneum

7. The rounded upper part of the stomach.
8. An enzyme that breaks down proteins.

h.   pyloric sphincter
i.   fundus
j.   rugae

TEST YOUR KNOWLEDGE

Group A

Write the letter of the word(s) that correctly complete(s) the statement. *Do not write in this book.*

1. The esophagus opens into the stomach at the   (a) cardiac orifice   (b) pyloric orifice   (c) duodenum   (d) pharynx.
2. The enzyme gastric protease digests (a) starch   (b) sugar   (c) protein   (d) fat
3. Adult human gastric juice contains   (a) ptyalin   (b) rennin   (c) hydrochloric acid   (d) casein.
4. Cells in the pyloric region of the stomach secrete the hormone   (a) pepsin   (b) amylase   (c) gastrin   (d) lipase.
5. Of the following, the one that is not a general region of the stomach is the   (a) fundus   (b) body   (c) pylorus   (d) gastric mucosa.
6. Resistance to peristaltic movement in the stomach can be caused by the   (a) enterogastric reflex   (b) vomiting center in the medulla   (c) zymogenic cells   (d) intrinsic factor.
7. Gastric protease can break down molecules of   (a) lipids   (b) glucose   (c) maltose   (d) casein.
8. The neural reflex that inhibits the rate at which the stomach empties is called   (a) vomiting   (b) enterogastric   (c) pyloric   (d) gastric.
9. The pH of gastric juice is about   (a) 4   (b) 1   (c) 7   (d) 10.
10. The lining of the stomach contains   (a) parotid glands   (b) parietal glands   (c) peptil glands   (d) gastric glands.

Group B

Answer the following. Briefly explain your answers. *Do not write in this book.*

1. (a) Briefly describe the structure of the stomach, taking into consideration the nature of its walls and the functions of its various regions.   (b) How does the stomach move, and what are the results of this activity?
2. Briefly describe the structures responsible for the production of gastric juice and how its secretion is controlled.
3. Outline what happens to a piece of beef (protein and fat), a slice of bread (carbohydrates, proteins, fat, and mineral salts), and a glass of milk during gastric digestion.
4. Briefly describe the entry of food into the stomach and the factors that control its passage into the small intestine.

CHAPTER 15

THE INTESTINES

Objectives

A. Describe the gross anatomy of the peritoneum, small intestine, liver, gallbladder, pancreas, and large intestine

B. Describe the functions of the peritoneum, small intestine, liver, gallbladder, pancreas, and large intestine

C. Explain the mechanism and factors controlling motility in the small and large intestines

D. List the composition and function of each of the intestinal secretions

E. Describe the mechanisms that regulate enzyme secretion in the digestive system

F. Describe the composition and functions of bile and pancreatic juice

GROSS ANATOMY OF THE INTESTINES AND ACCESSORY ORGANS

The greater part of digestion and absorption occurs in the intestines, primarily in the small intestine. The process is dependent on both endocrine and exocrine secretions and the controlled movement of ingested food materials through the tract so that the digestion and absorption can occur. The accessory organs that secrete digestive juices into the small intestine are the liver, with its biliary apparatus, and the pancreas. In addition, the wall of the small intestine contains glands that secrete enzymes. Collectively, these secretions complete the digestive process so that foods may be absorbed into the bloodstream. As part of the digestive system, the large intestine primarily absorbs water and serves as an organ of elimination. Ingested food materials that cannot be put into an absorbable form become waste material (feces) that is ultimately eliminated from the body. The process of altering the chemical and physical composition of food is complex and requires the function of each part of the digestive system.

Small Intestine

The small intestine is a tube measuring approximately 2.5 centimeters (1 inch) in diameter and 6 meters (20 feet) in length. Its coiled loops fill most of the abdominal cavity. The small intestine consists of three divisions: the **duodenum**, the **jejunum**, and the **ileum.** The duodenum

ileo = ileum

is the uppermost division and is the part into which the pyloric end of the stomach opens. It is about 25 centimeters (10 inches) long and occupies a fixed position in the abdominal cavity. The duodenum becomes the jejunum at the point where the digestive tube turns abruptly forward and downward. The jejunal portion continues for approximately the next 2.5 meters (8 feet), where it becomes the ileum without any clear line of demarcation between the two. The ileum is about 3.5 meters (12 feet) long.

Villi. *Villi* are important modifications of the mucosal layer of the small intestine. Millions of these fingerlike projections, each about 1 millimeter in height, give the intestinal mucosa a velvety appearance. Each villus contains an *arteriole*, a *venule*, and a *lymph vessel (lacteal)*. Microscopic examination of epithelial cells on the surface of the villi shows that they contain 1700 ultrafine **microvilli.** These tiny projections give a brushlike appearance to the surface area. Intestinal digestive enzymes, previously believed to be produced in the **crypts of Lieberkühn** between villi, are now found to be derived from the disintegration of the epithelial on the tips of the villi. The presence of villi and microvilli increases the surface area of the small intestine about 600 times, making it ideally suited for the absorption of digested food. Mucus-secreting **goblet cells** are found on villi and in the crypts.

ole = small

The peritoneum.   The membrane covering most of the organs of the digestive tract and holding them loosely in position in the abdominal cavity is the **peritoneum.** The peritoneum is a large continuous sheet of *serous membrane.* That part of the peritoneum lining the walls of the abdominal cavity is called the **parietal layer.** The part that forms the outer serous coat of the abdominal organs is referred to as the **visceral layer.** In several places the peritoneum forms extensions that bind the abdominal organs together.

The mesentery is a fan-shaped fold of parietal peritoneum that projects from the posterior abdominal wall near the lumbar region. The mesentery allows free movement of each coil of the intestine and helps to prevent the long tube from getting entangled. A similar but less extensive fold of peritoneum called the **transverse mesocolon** attaches the transverse colon to the posterior abdominal wall. The **greater omentum,** a continuation of the serous membrane of the greater curvature of the stomach and the first section of the duodenum, joins with the transverse mesocolon. Spotty deposits of fat accumulate in the omentum and give it the appearance of a lacy apron hanging down loosely over the intestines. In cases of localized abdominal inflammation, such as appendicitis, the greater omentum envelops the inflamed area, walling it off from the rest of the abdomen. The **lesser omentum** extends from the lesser curvature of the stomach and attaches to the liver and to the first part of the duodenum. The *falciform ligament* is a sickle-shaped fold of the peritoneum extending from the liver to the anterior abdominal wall.

meso = middle

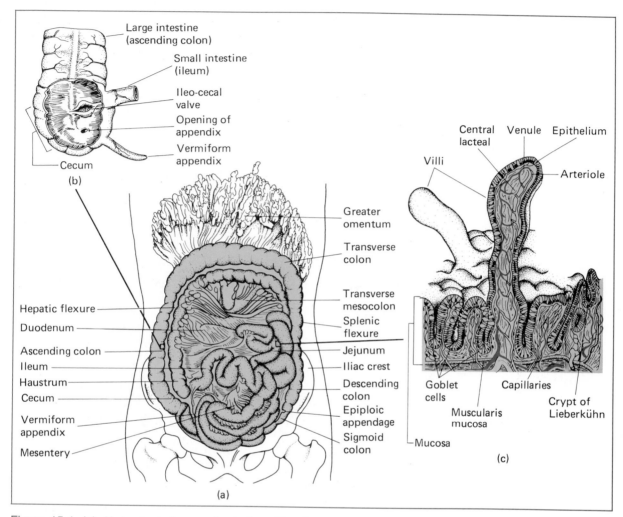

Figure 15-1 (a) Abdominal viscera, front view. The flexures of the colon and the loops of the small intestine are seen. (b) Structural view of the cecum. (c) Cross section of a villus in the small intestine.

Large Intestine

The lower part of the gastrointestinal tract is known as the **large intestine.** Its diameter is somewhat larger than that of the small intestine, averaging 6 centimeters (2½ inches) but varying and decreasing toward the lower end of the canal. Its length, however, is much less—1.5 to 1.8 meters (5 to 6 feet). The first 5 to 8 centimeters (2 to 3 inches) of the large intestine is named the **cecum.** It is a blind pouch located in the lower right quadrant of the abdomen.

Attached to the cecum is the **vermiform appendix,** a tube that resembles a large worm in size and shape. This blind tube averages 8 to 10

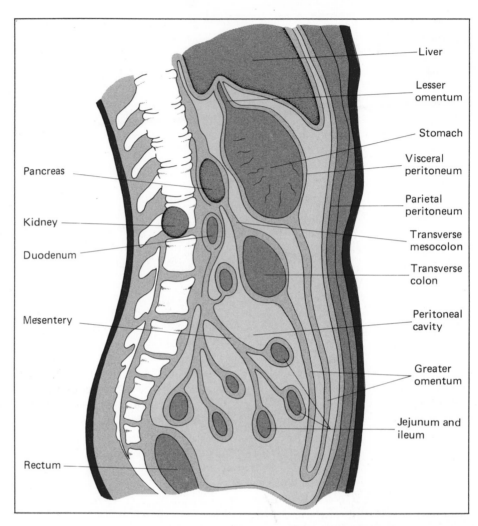

Figure 15-2 **A longitudinal section through the abdominal cavity. Some of the organs, such as the pancreas and the kidneys, lie behind the parietal peritoneum.**

centimeters (3 to 4 inches) in length and extends from the lower portion of the cecum (Fig. 15-1b). The structure of the walls of the appendix is similar to that of the rest of the intestine. The lymphoid tissue in its mucous lining may become inflamed—known as **appendicitis.**

Divisions of the colon.   That part of the large intestine that extends from the cecum to the anus—the **colon**—is divided into the following portions: the **ascending, transverse, descending,** and **sigmoid colons** (see Fig. 15-1a).

The *ascending colon* lies in a vertical position, on the right side of the abdomen. It extends from the cecum up to the lower border of the liver. The *ileum* joins the large intestine at the junction of the cecum and ascending colon, the place of attachment resembling the letter T (Fig. 15-1b). The **ileocecal valve** is the opening that permits material to pass from the ileum into the large intestine. The valve, however, prevents the contents of the colon from flowing back to the ileum.

The *transverse colon* passes horizontally across the abdomen, below the liver, stomach, and spleen. This part of the colon is situated above the small intestine. The transverse colon extends from the **hepatic flexure** to the **splenic flexure.** These two points are where the colon bends, forming 90° angles.

hepat = liver

The *descending colon* also lies in a vertical position but on the left side of the abdomen. It extends from a point below the stomach to the level of the iliac crest on the hip bone, at which point it becomes the *sigmoid colon.* This portion of the large intestine is S-shaped. The lower curve of the S bends to the left to form the **rectum.**

The rectum is 18 to 20 centimeters (7 to 8 inches) long. The last few centimeters of the rectum is called the anal canal. The mucous lining of this canal is arranged in numerous vertical folds known as **anal columns,** each of which contains an artery and a vein. **Hemorrhoids** (or **piles**) are enlargements of the rectal veins in the anal canal. The opening of the canal to the exterior is guarded by two sphincter muscles. The **internal anal sphincter,** composed of smooth muscle, is involuntary. The **external anal sphincter** is composed of striated muscle and therefore is under voluntary control. Young children learn to control the external anal sphincter as they mature. The opening from the anal canal to the exterior is the **anus.**

Arrangement of layers.   The inner, mucous membrane, or mucosa, of the large intestine differs from the inner lining of the small intestine in that it contains no villi and the epithelial cells contain large numbers of goblet cells. Also, the glands of this mucosa secrete mucus but no digestive enzymes.

The external, muscular layer differs from that in other parts of the gastrointestinal tract by the organization of its longitudinal muscle fibers. These fibers are concentrated into three flat bands called **teniae coli.** These bands are not as long as the large intestine. Consequently, the large intestine is gathered (shirred) into *sacculations*, or pouchlike structures known as **haustra.**

The visceral peritoneum that covers the large intestine forms little pouches containing fat. These pouches, called **epiploic appendages,** hang from the large intestine, except in the regions of the cecum, appendix, and rectum.

The Liver

The liver is the largest gland in the body and weighs 1.1 to 1.6 kilograms (about 3 lbs) in the adult. Its major part lies in the upper right side of the abdominal cavity under the dome of the diaphragm.

There are four anatomical lobes in the liver. Two of these lobes can be seen from the front surface of the liver; the other two, smaller lobes are visible on the liver's posterior surface. The left lobe projects downward over part of the stomach. Located under the large right lobe is the gallbladder, the lower margin of which projects slightly below the edge of the lobe. The liver has the consistency of a soft solid, is reddish

(a)

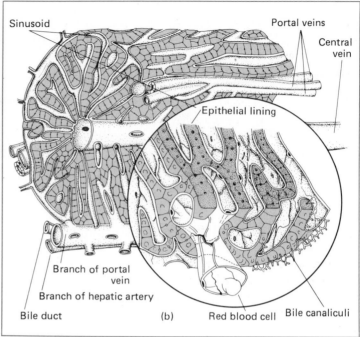

Figure 15-3 (a) Photo of a longitudinal section of the anterior lobes of the liver. In front, the liver consists of two lobes. Not seen in the photo are the two lobes that comprise the posterior of the liver. (b) A diagram of a liver lobule. The enlarged view shows the arrangement of hepatic cells and bile canaliculi.

brown in color, and can be easily broken or cut. The liver is almost completely covered with visceral peritoneum.

The basic functional unit of the liver is the cylindrical **liver lobule,** several millimeters in length and 0.8 to 2 millimeters in diameter. The human liver contains 50,000 to 100,000 individual lobules.

Microscopic examination of the liver.   The liver lobule is constructed around a central vein that empties into the hepatic veins, which empty into the inferior vena cava. The lobule itself is composed of many **hepatic cellular plates** that radiate laterally from the central vein, like the spokes of a wheel. Each hepatic plate is usually two cells thick. Between the adjacent cells lie small **bile canaliculi,** which empty into **bile ducts.** These ducts originate in the partitions (septa) between the liver lobules.

Between the partitions of the lobules are small **portal venules,** which receive blood from the portal veins. From the portal venules blood enters the flat branches of the **hepatic sinusoids,** which lie between the hepatic plates, and continues to flow into the central vein. In this manner, the hepatic cells are continuously exposed to portal venous blood that contains the products of digestion. **Hepatic arterioles** are also present in the interlobular partitions. These arterioles carry oxygen to the liver lobules and to the interlobular partitions. Many of these small arterioles empty directly into the hepatic sinusoids.

The hepatic sinusoids are lined by two major types of cells—**endothelial cells** and **Kupffer cells,** which *phagocytize,* or *engulf,* bacteria or other foreign matter in the blood.

phago = eat

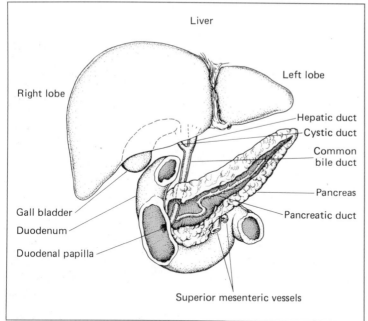

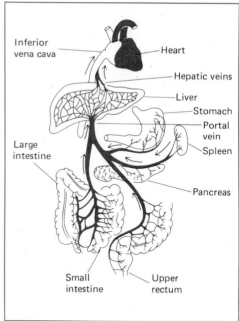

Figure 15-4 The relationship of the duodenum to the pancreas and liver.

Figure 15-5 Diagram of the general distribution of the portal vein of the digestive organs and the liver.

The portal circulation consists of the veins that drain blood from the intestines (except the lower part of the rectum), pancreas, spleen, and gallbladder. Through these veins the blood is transported and distributed to the hepatic sinusoids. The blood is then collected by the hepatic veins, which empty into the inferior vena cava. In the portal circulation, unlike other parts of the circulatory system, the blood passes through *two sets* of minute vessels instead of one continuous branching system. The capillaries in the walls of the intestines, pancreas, spleen, and gallbladder are one set of minute vessels. The hepatic sinusoids in the liver are the other set.

The gallbladder and bile ducts.   The **gallbladder** is a pear-shaped structure lying on the undersurface of the liver. The gallbladder is about 7 to 10 centimeters long and has a volume of from 30 to 50 milliliters. The walls of the gallbladder have three layers: the **outer serosal,** the **middle fibromuscular,** and the **inner mucosal.** The serous coat partially covers the gallbladder. The fibromuscular layer forms a thin but strong framework of dense fibrous connective tissue combined with smooth muscle tissue. The mucosa is loosely connected to the fibromuscular layer and is arranged into minute folds called **rugae.**

The **biliary ducts,** or passageways, consist of two hepatic ducts, right and left, which originate from the right and left lobes of the liver. These ducts join to form the **common hepatic duct.** The common hepatic duct enters the *lesser omentum,* where it joins with the **cystic duct** of the gallbladder. These connecting ducts form the **common bile duct,** which opens into the duodenum at the **duodenal papilla.**

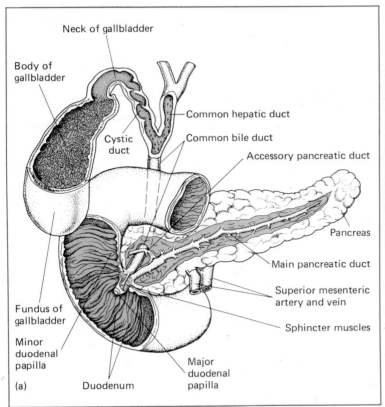

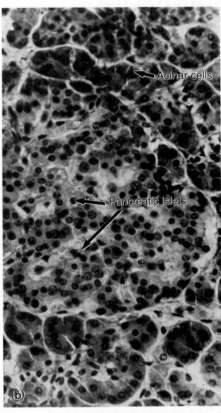

Figure 15-6 (a) The gallbladder and divisions. (b) Photomicrograph of a cross section of the pancreas. The pancreatic islets lie among the enzyme-secreting cells.

exo = out
endo = within

The Pancreas

The pancreas is both an **exocrine** and **endocrine** gland. It lies across the posterior abdominal wall, from the duodenum to the spleen. The pancreas is composed of lobules in which the **exocrine secretory units,** called **acini,** resemble grapes hanging on stems. Each *acinus* consists of a single row of epithelial cells arranged around the opening of a duct. These **acinar cells** are concerned with the production of **pancreatic juice,** the exocrine secretion of this important gland.

Pathway for secretions. The pancreatic juice contains the enzymes for digestion. The secretions of the acinar cells flow into ducts, which join to form the **pancreatic duct.** The pancreatic juice continues to flow through the pancreatic duct and combines with the bile in the **common bile duct,** the channel through which both secretions enter the duodenum.

The glandular tissue of the pancreas contains cells that produce the hormone **insulin.** This hormone plays an important role in carbohydrate *metabolism.* The hormone-secreting cells form the **pancreatic islets.** Within the islets are three different types of cells. The majority of cells are small and store the insulin (as granules). These cells are called the **beta cells.** The other cells are larger and secrete glucagon, which, like insulin, affects carbohydrate metabolism. These are called **alpha cells.** In addition there are **delta cells** that produce **somastatin,** which inhibits the release of both insulin and glucagon.

DIGESTIVE AND ABSORPTIVE FUNCTIONS

The small intestines perform several important functions.

1. They complete the digestion of foods.
2. They absorb the end products of digestion into blood and lymph.
3. They secrete hormones that help control the production of pancreatic juice, bile, and intestinal juice.
4. They control the amount of fluid and electrolytes lost from the body.

Intestinal Motility

In the small intestine the partially digested food is subjected to three different types of intestinal movement.

- Peristaltic contractions (about 11 per minute) move food along the tube at a fairly constant rate of 7.6 centimeters per minute.
- Segmental action breaks up and mixes the food as it moves in the small intestine. Segmentation is accomplished by the contractions of circular muscle that constrict the tube every few seconds. This churning action increases the surface area of the food, exposing more of it to digestive juices.
- Pendular (oscillating) movement involves separate sections of the intestine. The muscular layers contract in such a way that the contents appear to be flung along the loops. Pendular movement does not push food along the intestine but rather mixes it thoroughly with the secretions.

There is another type of motion in the intestine, which is produced by the swaying of villi. These tiny fingerlike projections stir the fluids and the chyme into a more thorough mixture.

The action of smooth muscle is slower than that of striated muscle, and thus smooth muscles are capable of remaining contracted for longer periods of time. They can also contract to a greater degree without suffering permanent injury. These characteristics make smooth muscle tissue more suitable for the intestines. The moving force in the digestive system is a wave of contraction caused by these muscles. The intensity of the contraction is evidenced by the fact that contracting intestinal walls turn pale because the contraction squeezes blood momentarily out of its tissues. Another characteristic of smooth muscles is their ability to contract rhythmically as a result of the steady inflow of nervous impulses. Cardiac muscle can also contract in a rhythmic manner, but it does this without nervous stimulation. Skeletal muscles may contract in this way, but within a short period of time they become fatigued. Smooth muscle cells are more sensitive to mechanical and

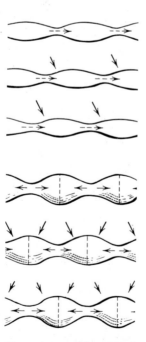

Figure 15-7 Movement in the small intestine: (above) peristalsis; (below) segmentation.

chemical stimuli than either skeletal or cardiac muscle cells. Thus the mere presence of food in the digestive system activates their contraction and continued stimulation maintains movement. These intestinal movements mix the chyme with the digestive juices and aid in the absorption of the end products of digestion.

The autonomic nervous system also controls the movement of the smooth muscles of the digestive tract. **Parasympathetic** stimulation *increases* the motility of the small intestine, while **sympathetic** stimulation *decreases* the contraction of smooth muscle in the gastrointestinal tract.

Activity of Intestinal Secretions

The greatest amount of digestion occurs in the small intestine for the following reasons:

1. The food entering the intestine has already undergone preliminary breakdown due to mechanical and chemical activities of the mouth and stomach.
2. The small intestine receives secretions that provide all of the enzymes necessary for the complete digestion of each major food group.
3. The food remains in the small intestine for a relatively long period of time, thus giving these enzymes ample opportunity to complete their tasks.

Three secretions are emptied into the small intestine: *pancreatic juice, intestinal juice,* and *bile.* The rates of production and flow of these se-

Table 15-1
Gastrointestinal Hormones

Hormone	Where Produced	Agents Stimulating Production	Action
Gastrin	Gastric mucosa of antrum	Distention Peptide fragments	Stimulates gastric motility. Stimulates HCl secretion.
Secretin	Duodenal mucosa	Acid in duodenum	Inhibits gastric motility. Inhibits HCl secretion. Stimulates secretion of bicarbonate-rich pancreatic juice.
Cholecystokinin (CCK)	Duodenal mucosa	Amino acids or fatty acids in the duodenum	Inhibits gastric motility. Inhibits HCl secretion. Stimulates pancreatic secretion. Stimulates gallbladder contraction.
Gastro-inhibitory Peptide (GIP)	Duodenal mucosa	Fatty acids or monosaccharides in the duodenum	Inhibits gastric motility.

cretions are affected by nerve impulses and by hormones produced in the mucosa when acid chyme enters the duodenum. Table 15-1 summarizes the functions of the gastrointestinal hormones.

Chemical digestion in the small intestine is of fundamental importance in the breakdown of all three major food groups. As previously discussed, some digestion occurs in the mouth and in the stomach. The task is completed by enzymes secreted by the pancreas and intestines and bile secreted by the liver. Together, these substances are capable of initiating and carrying out the entire digestive process.

As a result of the action of salivary amylase, carbohydrates enter the duodenum partly in the form of *dextrins* and *maltose*. Undigested starches, glycogen, sucrose, and lactose may also enter. **Pancreatic amylase** aids in the hydrolysis of starches, dextrins, and glycogen, converting them to maltose. Intestinal enzymes then complete the digestion of carbohydrates forming simple sugars: **Maltase** splits maltose to form two molecules of glucose; *sucrase* splits sucrose to form one molecule of glucose and one molecule of fructose; *lactase* splits lactose to form one molecule of glucose and one molecule of galactose.

ase = enzyme

Fats are attacked by **pancreatic lipase** and split into glycerol and fatty acids. Although lipase can and does attack large fat globules, its action is much more effective when fats are first emulsified by the action of bile salts.

Proteins, and protein fragments produced by the action of gastric protease, are hydrolyzed by a series of enzymes in the small intestine. These protein-splitting enzymes break proteins down into amino acids. Generally, the pancreatic proteases initiate the digestion of large protein molecules and protein fragments by splitting them into small fragments. Although a few amino acids are split off by the pancreatic enzymes, the intestinal proteases do the bulk of the work and complete the protein digestion by breaking down proteins into their constituent amino acids.

Because the pancreatic proteases will attack body protein, they must be secreted in an inactive form so as to prevent their digesting the pancreas itself. The inactive forms, **trypsinogen** and **chymotrypsinogen,** become activated only after they enter the small intestine. **Enterokinase,** an enzyme in the intestinal juice, triggers the activation process by converting trypsinogen into its active form, called **trypsin.** Trypsin, in turn, changes chymotrypsinogen to active **chymotrypsin,** another protein-splitting enzyme whose action produces smaller polypeptides.

Intestinal proteases called **aminopeptidase** and **dipeptidase** do not attack body protein and can be produced in their active form. The action of these enzymes is to complete the digestive processes initiated by the pancreatic enzymes. The end products of digested proteins (amino acids) can then be easily absorbed. Review the process of digestion in the Summary Table at the end of this chapter.

Absorption of End Products

Absorption involves the transfer of the end products of digestion through the epithelial lining of the digestive tract and into the blood or lymph vessels. The end products of digestion are transported across the epithelial cells by several mechanisms. A significant amount of material can be moved by *simple diffusion* as long as a concentration gradient exists between the intestinal lumen and the cytoplasm of the intestinal cells. Through *active transport* certain nutrients, such as sugars and amino acids, are moved into the epithelial cells against a concentration gradient. Specialized mechanisms are also present for the absorption of certain vitamins and minerals. Water absorption occurs by the process of *osmosis.* The major transport mechanisms are discussed in Chapter 3.

The greatest amount of absorption takes place in the small intestine and for several important reasons:

1. Food there is in an absorbable form since most of the enzymes required for digestion are in the small intestine.
2. The circular folds, or villi, and the microvilli provide a greater surface area for absorption. In addition, the villi are in constant motion because of the activity of smooth muscle fibers in the mucosa. This helps to stir the intestinal contents and promotes maximal contact between the absorptive surface and the nutrients.
3. The small intestine contains an abundance of blood and lymph vessels.
4. Food remains in the small intestine for a long period of time.

The simple sugars *glucose, galactose,* and *fructose* enter the intestinal epithelial cells by the mechanism of active transport processes. After passing through the epithelial cells, these sugars enter the blood capillaries of the villi and are transported via the portal vein to the liver for processing and storage.

The amino acids are also absorbed by means of active transport mechanisms. The specific mechanism involved in absorbing amino acids is determined by the chemical structure of the amino acids. However, amino acids with similar structures can use the same carrier. Like the simple sugars, amino acids enter the blood capillaries of the villi, from which they are transported via the portal vein to the liver for processing.

Products of fat digestion—glycerol and fatty acids—are absorbed by simple diffusion. However, since fatty acids are not water-soluble, a special mechanism is needed to "ferry" them through the liquid chyme to the absorptive surface of the intestine. *Bile salts* are of primary importance in this process. They combine with the fatty acids to form spherical, water-soluble structures known as **micelles.** When a micelle reaches the epithelium, it is believed that the fatty acids dissolve in the lipid cell membrane. The fatty acids are then able to enter the cytoplasm by simple diffusion. The bile salts thus become free to pick up another load of fatty acids, and the process is repeated.

Within the cytoplasm of intestinal epithelial cells, the fatty acids combine with glycerol to form **triglycerides.** The triglycerides become covered with a protein coat to form water-soluble **chylomicrons,** which diffuse into the **central lacteals** of the villi. They are then transported through the lymph vascular system and into the bloodstream. The fat-soluble vitamins (A, D, E, K) are absorbed along with the products of fat digestion.

Activities of the Large Intestine

The main functions of the large intestine are absorption of water, secretion of mucus, and elimination of the wastes of digestion. Water is absorbed by osmosis. As the nutrients and mineral salts are absorbed, the intestinal fluid becomes **hypotonic,** and thus water is able to be absorbed by the osmotic process. (Refer to Chapter 3.)

The material that enters the cecum is for the most part lacking in usable nutrients. The nutrients have all been digested and their products absorbed during their passage through the small intestine. There are still, however, some substances that the body can use and that the large intestine will retrieve before the waste materials are eliminated. As the contents of the small intestine enter the cecum, they are in a very liquid state. Because water accounts for a large percentage of the total fluids of the body, its loss would represent a serious depletion of fluid reserves. One of the principal functions of the large intestine is to recover this water so that the body's reserves are not depleted.

The large intestine's capacity to absorb water is very great. Some people who suffer from constipation have the idea that to make the evacuation of the bowel's content easier, they should drink large quantities of water. Even though a person may drink as much as 3 liters (about 3.2 qts.) of water per day above normal intake, there is no indication that by so doing the person changes the nature of the feces or that the constipation can be relieved. The additional water is absorbed through the walls of the large intestine, and the amount of urine formed and excreted is proportionally increased.

The large intestine is filled with bacteria, which act on the undigested residues of food. They cause **fermentation** of carbohydrates and **putrefaction** of proteins. Some of the split proteins are eliminated with the feces and others are absorbed. Those that are absorbed are combined with other substances in the liver and then excreted by way of the urine.

Bacteria that inhabit the large intestine synthesize vitamin K, which is then absorbed, after **emulsification** by bile, in the small intestine. Some components of the vitamin B group are produced by the intestinal bacteria, and these are also absorbed and used by the body. Individuals who take antibiotic drugs over a long period of time may develop a vitamin B deficiency if a sufficient number of these helpful bacteria are destroyed by the antibiotics.

Motility in the colon. There is considerable question as to just how the colon moves its contents along its tube. No frequent, rhythmic peristalsis or other movement has been observed there, except in the transverse colon, where there are slow, weak accordionlike movements of the *haustra* rather than actual peristalsis. At certain intervals—two or three times in 24 hours—mass peristalsis occurs in the large intestine. This massive movement often sweeps through the entire length of the colon, pushing the feces before it. These waves are sometimes preceded by peristalsis in the small intestine and by the taking of food into the stomach. The sigmoid colon, which serves as a storehouse for the feces, becomes filled so that during a wave of mass peristalsis the feces may enter the rectum. The lower end of the rectum is equipped with stiff longitudinal folds of tissue that lie just beneath the mucosal lining. These delay further progress of the contents into the anal canal.

After the absorption of water from the contents of the large intestine, bacterial action changes the consistency of the intestinal contents from a liquid to the semisolid state of the **feces.** Some of the contents of the feces are bacterial. Other substances are bile pigments, the products of bacterial action in the intestine, many mineral inorganic salts, mucus, and such indigestible components of food as cellulose. **Cellulose** is the fibrous part of plant food humans are unable to digest. It forms a small percentage of the feces and contributes to its bulk. This bulk stimulates the lining of the intestines and induces peristalsis. Fruits and vegetables provide this type of fiber.

When the anal canal is full, nervous stimuli are set up which result in defecation, or the act of excreting intestinal wastes through the anus. This process is voluntarily controlled by the action of an external sphincter muscle that surrounds the anus. In an infant, the ability to control the act of defecation is lacking, but with patient training it can be established.

Alterations in motility. Constipation is a condition in which normal, regular defecation is delayed. The feces are retained within the bowel for a longer-than-normal period of time. This permits continued bacterial activity, with production of large amounts of gas. During its longer retention in the intestine, the feces lose more water, forming a very solid mass, which is difficult to evacuate. Ignoring the desire to defecate at the habitual time can often be the beginning of constipation problems. However, if a person eats a well-balanced diet and exercises moderately there are few physiological reasons for constipation. Sometimes a person becomes "bowel-conscious" and worries about his or her inability to have bowel movements at regular intervals. Seeking aid from laxatives can eventually reduce the muscle tone of the bowel so that a dependency on laxatives can develop. As a result, habitual use of laxatives may replace the normal functions of the large intestine. For people whose activities and diet are limited by such conditions as long-term illness, the use of laxatives may be required. On the other hand, healthy individuals seldom have a need for them. Regularity of bowel movements varies from two or three a day to one every other day.

The Liver and Bile Secretion

Of all the organs in the human body, the liver is the largest and the busiest. It performs a variety of functions, many essential to life.

Bile is considered to be both a *secretion* and an *excretion*. It is continuously formed by the *hepatic cells*. The average daily volume of bile ranges from 600 to 800 milliliters. Bile appears as a golden yellow fluid and contains *water, bile salts, bilirubin, cholesterol,* and various *inorganic salts.* Of these constituents, only the bile salts that are formed from cholesterol have important functions to perform in the gastrointestinal tract.

Bile salts and pigments.   The bile salts are **sodium glycocholate** and **sodium taurocholate.** Both salts are formed from **cholesterol** and are the constituents of bile concerned with the digestive process. Just as soap and water are used to wash grease off hands, so the digestive system uses these salts as "soap" to detach fat molecules from one another. Droplets of fat are surrounded with a layer of bile salt molecules so that they do not fuse together again. By separating fats and oils into very small droplets of this type, bile salts form an **emulsion.** The breaking of the large droplets of oil into tiny droplets greatly increases the surface area of fats and oil so that they can be exposed more completely to the action of the digestive enzymes. The bile salts also form water-soluble complexes with cholesterol and fatty acids, many of which would not readily dissolve in the intestinal fluid. A third function of these salts is to activate the enzyme lipase produced by the pancreas, and to some extent the lipase produced by the stomach. They also help neutralize the acidic gastric chyme. After their use in the digestive process, about 90 percent of the bile salts are reabsorbed by the portal circulation and returned to the liver for reuse.

Bilirubin, a bile pigment, is excreted by the liver. Bilirubin is red. When oxidized, this compound becomes **biliverdin,** which is green. Bilirubin, however, is the chief pigment of human bile. Bilirubin is the major end product produced by the decomposition of **hemoglobin.** When worn-out red blood cells are removed from the circulating blood, their hemoglobin is broken down and bilirubin is formed. This usually occurs in various organs and tissues, such as the *spleen, bone marrow,* and *lymph nodes.* Bilirubin is first bound to a plasma protein, then travels in the bloodstream to the liver, where it is made water-soluble. This process is called **conjugation. Conjugated bilirubin** is excreted into the bile. Iron is extracted from the hemoglobin and is liberated into the bloodstream for reuse by the body.

hem = blood

Control of bile secretion.   Between meals, bile is stored in the gallbladder. There the bile becomes very concentrated because of the absorption of water, sodium chloride, and other electrolytes.

When it reaches the small intestine, bilirubin undergoes further chemical changes and is converted into **urobilinogen,** a yellowish substance that accounts for the yellow color of the feces (intestinal wastes).

Some of the urobilinogen is absorbed through the walls of the large intestine and is returned to the liver for reuse.

In the forming of bile the liver has both a secretory and excretory activity. A secretion may be defined as a product of cells that is used by the body in its performance of a specific function. Bile aids in the digestion of fats and can thus be considered a secretion. An excretion is a substance that the body eliminates—a waste material. The liver plays an excretory role because it gets rid of worn-out hemoglobin and other waste products. These wastes could cause damage to the body if they accumulated.

Metabolic functions of the liver.   The liver plays a vital role in the metabolism of the three major *macronutrients:* carbohydrates, proteins, and lipids. These functions are only briefly dealt with in this chapter. The details will be considered later in the text.

The liver helps to maintain a normal concentration of glucose in the blood. For example, after meals, when the glucose level of the blood tends to rise, the liver converts glucose to glycogen and stores it until needed. This process is called *glycogenesis.* When the blood glucose level falls, the glycogen is changed back to glucose and released into the blood. This process is known as *glycogenolysis.* Other simple sugars (galactose and fructose) that enter the liver are converted into glucose, which may be stored or used immediately as required by the body. If the blood glucose level continues to fall, the liver has the ability to form glucose from *noncarbohydrates,* such as amino acids and the glycerol portion of fats. This activity is referred to as *gluconeogenesis.*

The blood glucose level is controlled by several hormones. Insulin serves to lower the blood glucose level by increasing the use of glucose in the tissues, by promoting the storage of glucose as glycogen in the liver and muscles, by promoting the formation of fat from glucose, and by decreasing gluconeogenesis from amino acids. An important basic effect of insulin is its ability to increase the transport of glucose, amino acids, and fatty acids through most of the cell membranes of the body.

Secretory functions.   The effects of insulin are counteracted by several hormones. *Epinephrine* stimulates the breakdown of glycogen in the liver and produces an increased blood glucose level. An increased blood glucose level is called **hyperglycemia. Adrenocortical hormones**, particularly the **glucocorticoids,** stimulate gluconeogenesis. Finally, thyroid hormone also stimulates gluconeogenesis.

emia = blood

Glucagon increases the blood glucose level temporarily when stimulated by a falling blood glucose level, a condition called **hypoglycemia.** Glucagon increases the blood glucose level by stimulating rapid glycogenolysis in the liver, which releases glucose into the blood.

The liver plays its most vital role in the metabolism of protein. For example, except for some of the **gamma globulins,** essentially all of the plasma proteins are formed by the hepatic cells. The liver forms important chemical substances, such as **phosphocreatine,** from amino acids. It synthesizes certain amino acids, referred to as the **nonessential amino acids.** The liver also deaminizes amino acids so they can be used for energy or can be converted to glucose or fat. **Deamination** is the enzymatic removal of the amino group (NH_2) with the formation of ammonia and a keto acid. The ammonia is converted to urea and eliminated in the urine. The keto acids are used for energy or converted to other substances.

Although the metabolism of fats probably can take place in most cells of the body, certain activities occur much more rapidly in the liver. For example, fatty acids are an important source of energy, but they must first be broken down to small molecules that can enter the Kreb's cycle and be oxidized. It is believed that about 60 percent of all the preliminary breakdown of fatty acids in the body occurs in the liver. Other specific functions of the liver are the formation of large quantities of cholesterol and phospholipids, the formation of *lipoproteins,* and the synthesis of fat from glucose and amino acids.

The liver is capable of forming vitamin A from precursor substances found in certain vegetable foods. It also stores large quantities of vitamin A, vitamin D, and vitamin B_{12}. Iron is stored in the liver in the form of ferritin. Various hormones (for example, gonadal and adrenocortical) are inactivated in the liver and in the urine. The liver also is concerned with the detoxification of harmful compounds, such as alcohol and certain drugs. These substances can then be eliminated in the bile and urine.

Another function of the liver is the formation of blood clotting factors. The phagocytic Kupffer cells, which are located on the endothelial cells in the liver, are part of the reticuloendothelial system, which is concerned with protecting the body against invasion by foreign material.

The function of the gallbladder is to store and concentrate bile. When food enters the duodenum, the gallbladder contracts, discharging the concentrated bile into the duodenum. The signal for gallbladder contraction is the intestinal hormone **cholecystokinin** (CCK). The stimulus for this hormone's release is the presence of fatty acids and amino acids in the duodenum. It is from its ability to cause contraction of the gallbladder that cholecystokinin received its name: *chole* ("bile") *kystis* ("bladder"), *kinin* ("to move").

chole = bile

Pancreatic Functions

The exocrine function of the pancreas involves the production and secretion of pancreatic juice, in which digestive enzymes are found. Pancreatic juice contains the enzymes that act to *hydrolyze* fat to fatty acids and glycerol, *split* carbohydrates to disaccharides, and *break down* proteins to polypeptides. (Review the activity of intestinal enzymes discussed earlier in this chapter.)

Pancreatic juice is alkaline, with an average pH of 7.5. However, the pH level increases to 8 as the rate of secretion increases. Due to its alkalinity, pancreatic juice stops the activity of acidic gastric juice, and digestion of chyme in the intestine proceeds under entirely different chemical conditions than those found in the stomach. As alkalinity increases, the acidity of the chyme becomes neutralized. Although the pH of pancreatic juice stops the work of gastric juice (pepsin), it also functions to adjust the chemical environment for the action of intestinal enzymes in digestion of foods. The control of pancreatic secretion is similar to the cephalic phase of gastric secretion. Stimuli such as the sight, smell, or presence of food in the mouth can initiate enzyme secretion of the pancreas. However, its major control is due to the hormone known as **secretin.** This hormone is secreted by the intestinal mucosa and travels through the bloodstream to the pancreas. When the intestinal content is acidic, secretin acts as a *chemical messenger* causing the pancreas to secrete enzymes in response. As a result, the alkalinity of the pancreatic juices neutralizes the acid content in the small intestine.

PATHOLOGICAL PROBLEMS

Jaundice is a condition in which the skin, mucous membranes, and other tissues have a greenish yellow appearance. It is caused by excess bilirubin accumulation in the blood.

Liver Dysfunctions

dys = difficult

Bilirubin can accumulate in three ways:

1. The liver normally has a great reserve capacity for conjugating bilirubin. However, an abnormally rapid breakdown of red blood cells (a hemolytic process) can result in overproduction of bilirubin. This causes **hemolytic jaundice.**
2. In **hepatitis** (inflammation of the liver) the hepatocytes malfunction and reduce the ability of the liver to conjugate a normal amount of bilirubin or to excrete bilirubin.
3. If the bile ducts become blocked (as by gallstones), bile is backed up and circulates in the blood. Whatever the obstruction, this type of jaundice is called **obstructive jaundice.**

Gastric Disorder—Gastroenteritis

gastro = stomach

The common **gastroenteritis** (upset stomach) may be caused by bacteria or viruses that may enter the body by way of the mouth and intestinal tract. The term *food poisoning* is sometimes applied to severe cases of acute gastrointestinal upset.

The most common food-borne gastroenteritis is caused by **staphylococcal toxin.** In this instance, it is not the microorganism itself that causes the disease but the product of bacterial metabolism produced during growth of the staphylococcus microorganism. Improper storage of food, especially poultry and cream sauces, permits the growth of toxin-producing staphylococcal strains. The potent toxin causes vomiting and diarrhea one to eight hours after eating. The diarrhea is severe but usually lasts for 24 hours only.

Summary Table
Chemical Digestion

Glands	Secretions	Acts on	Products Formed
Salivary	Salivary amylase	Starch and glycogen Dextrins	Dextrins Maltose
Gastric	Hydrochloric acid Gastric protease Gastric lipase (little importance)	Activates pepsinogen Protein Emulsified fat (butter, cream, etc.)	Gastric protease (pepsin) Proteoses and peptones Fatty acids and glycerol (protein fragments)
Pancreas (acinar cell)	Pancreatic amylase Pancreatic lipase Pancreatic proteases (trypsin and chymotrypsin)	Starch, glycogen, and dextrins Bile and emulsified fats Protein, intact or fragmented (proteoses and peptones)	Maltose Fatty acids and glycerol Protein fragments and amino acids
Intestinal (brush border cells of villi)	Enterokinase Proteases Maltase Sucrase	Activates trypsinogen Protein fragments Maltose Sucrose	Pancreatic protease (trypsin) Amino acids Glucose Glucose and fructose
Microvilli	Lactase	Lactose	Glucose and galactose
Liver	Bile salts (no enzymes)	Large fat globules	Emulsified fats (tiny droplets)

VOCABULARY REVIEW

Match the statement in the left column with the correct word(s) in the right column. *Do not write in this book.*

1. A portion of the large intestine forming a blind pouch.
2. A storable form of carbohydrate.
3. A function of the liver connected with protein conversion.
4. Absorption of fats occurs in this structure.
5. Concentrates and stores bile.
6. Duct that opens into the duodenum.
7. Important in the control of the blood glucose level.
8. The small fingerlike structures primarily responsible for absorption in the small intestine.
9. The longest part of the small intestine.
10. The conversion of glucose to glycogen.

a. cecum
b. common bile duct
c. cystic duct
d. deamination
e. gallbladder
f. glycogen
g. ileum
h. pancreatic islets
i. jejunum
j. lacteal
k. villi
l. glycogenesis
m. glycogenolysis

TEST YOUR KNOWLEDGE

Group A

Write the word(s) on your answer sheet that will correctly complete the statement. *Do not write in this book.*

1. The liver receives blood through the (a) portal vein (b) portal arteries (c) hepatic veins (d) renal arteries.
2. The stomach connects directly with the (a) jejunum (b) ileum (c) duodenum (d) peritoneum.
3. The chemical messenger that causes the pancreas to secrete its digestive enzymes is (a) pepsin (b) secretin (c) glucagon (d) cholesterol.
4. Bile passes into the gallbladder through the (a) hepatic duct (b) cystic duct (c) common bile duct (d) pancreatic duct.
5. Fats are emulsified by (a) bilirubin (b) biliverdin (c) cholesterol (d) bile salts.
6. The intestinal enzymes that digest proteins are (a) amylases (b) lipases (c) proteases (d) peptidases.
7. The digested nutrient that commonly enters the blood through the lacteal is (a) starch (b) sugar (c) protein (d) fat.
8. Peristalsis does not occur in the (a) ileum (b) duodenum (c) jejunum (d) liver.
9. The term that is *not* closely related to the others in the following group is (a) villus (b) vermiform appendix (c) lacteal (d) lymph vessel.
10. Narrow muscle bands in the large intestine produce pouches called (a) rugae (b) feces (c) haustra (d) lobules.

Group B

Answer the following. Briefly explain your answer. *Do not write in this book.*

1. (a) What are the functions of the liver? (b) Briefly describe how each of these is performed.
2. Describe how each of the products formed by the digestion of each of the classes of macronutrients enters the circulatory system.
3. (a) Describe the structure of the small intestine, including the peritoneum. (b) What is the function of each of the parts mentioned?
4. What is the relation of the portal circulation to the digestive system?

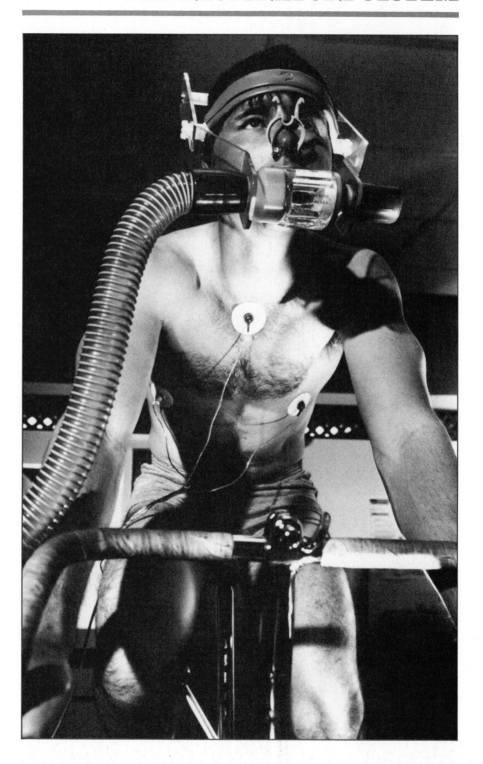

CHAPTER 16

THE RESPIRATORY ORGANS

Objectives

A. Identify the structures of the upper and lower respiratory tracts

B. Describe the functions of each structure of the upper and lower respiratory tracts

C. Explain how sound is produced by the vocal cords and how loudness, pitch, and quality are controlled

D. Explain how gas exchange occurs in the lungs

STRUCTURE OF THE RESPIRATORY TRACT

The organs of the respiratory system serve to distribute air from the external environment throughout the lungs and to exchange gases between the lungs and the body's internal environment. Oxygen must be supplied to the body's cells for metabolic activities and carbon dioxide must be removed. Most of the billions of cells in the body lie too distant from the body's surface to exchange gases directly with the outside air. The cells' air intake must first pass into the bloodstream. The blood then circulates throughout the body so that the cells can exchange gases. The events just described involve the functioning of both the *respiratory* system and the *circulatory* system. These two systems are essential for gaseous exchange in body cells.

For purposes of study, the respiratory system may be divided into upper and lower tracts, or divisions. The organs of the upper respiratory tract are located outside of the thorax, or chest cavity, whereas those in the lower tract are located almost entirely within it.

The upper respiratory tract is composed of the nose, pharynx, and larynx. The lower respiratory tract consists of the trachea, bronchi, and lungs. Functionally, the respiratory system also includes a number of other structures such as the mouth, ribcage, and diaphragm. Together these structures constitute a lifeline supplying oxygen and removing waste gases.

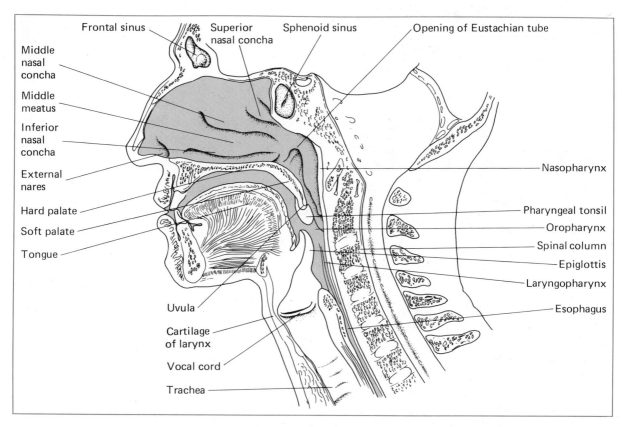

Figure 16-1   Sagittal section of nasal cavity and related air passages.

The Upper Respiratory Tract

This part of the respiratory system acts as an air *distributor*. The organs of the upper respiratory tract consist of the nose, pharynx, and larynx. Together these organs help to provide the air supply for the human body.

The nasal cavity.   Air enters the body through the external openings of the nose and passes into the narrow ***nasal cavities*** (Fig. 16-1). The ***nasal conchae,*** or ***turbinate, bones*** greatly increase the surface area of these cavities. The inner wall of each nasal cavity is folded into three ridges. The conchae cause turbulence in the flowing air, forcing it to move in many different directions before leaving the nasal cavity. The conchae, as well as the entire lining of the nose, are covered by a layer of *mucous membrane.* The turbulence aids the mucous membrane of the conchae to cleanse dust and foreign particles from the air.

The scroll-like conchae of the nasal cavity allow the entering air to be warmed by the blood and moistened and cleansed of small dirt particles by the sticky mucus secreted by the mucous membrane. Small hairs are located at the entrance of the nose and aid in preventing larger bits of dirt from entering. By the time the air reaches the lungs, it is saturated with water vapor, cleansed of large particles, and warmed to body temperature.

Internally, the nasal cavity is divided into two smaller cavities by a wall called the **nasal septum.** The two openings that conduct air into these two cavities are called the **external nares,** or **nostrils.** A thin sheet of cartilage, which is firmly attached to the nasal bones, forms the bridge of the nose. The conchae form the lateral, or side, walls. In the covering of the uppermost of the three turbinate bones, the **superior concha,** are the nerve endings for the olfactory sense (the sense of smell). The cells that received olfactory stimuli are elongated and supported by larger cells. Their free ends have hairlike projections against which the air brushes as it flows along the nasal passage.

para = beside, near

Four pairs of sinuses drain into the nose. These openings, the **paranasal sinuses,** are in the *frontal, maxillary, ethmoid,* and *sphenoid bones.* The paranasal chambers act as resonance chambers in voice production.

The pharynx. From the nasal cavity the air passes into the **pharynx.** This is the region in the back of the mouth that serves as a passageway for both food and air. The pharynx is the site of extensive growth in the early embryo. From its walls a fingerlike projection develops that later divides into two parts to form the lungs and their passageways. In addition, masses of lymphatic tissue develop in this region, which later become the **tonsils.**

The lower portion of the pharynx ends at the **glottis.** This is the opening into the larynx. The glottis is a narrow slit, with its long axis lying in a front-to-back direction. In males this opening extends to about 23 millimeters; in females it reaches about 17 to 18 millimeters in length. Around the rim of the glottis are the **vocal cords,** whose vibrations produce sounds. The **epiglottis** is a flap of cartilage tissue that fits over the glottis during swallowing. (See chapter 13.)

epi = upon, above

The larynx. The glottis opens into a roughly triangular chamber, the **larynx,** or **voice box.** The apex of the triangle points forward and is quite conspicuous in the male. It is commonly called the *Adam's apple.* The walls of the larynx are composed of plates of cartilage derived from embryonic structures called the **pharyngeal arches.** The largest of these cartilages are the **thyroid** and **cricoid** cartilages. The thyroid cartilage is composed of two separate plates that form the **laryngeal prominence** (Adam's apple). Its movement is visible and can be felt during the act of swallowing. Skeletal muscles that are attached to the cartilage help to move the larynx, thus preventing food from entering the trachea.

The cricoid cartilage is shaped like a signet ring with the thickened area directed toward the back of the trachea. It lies below the thyroid cartilage and connects with the upper part of the trachea by means of membranes.

There are seven other cartilages connected with the larynx. Two of these are the **arytenoid cartilages,** which are attached to the vocal cords. Another is the attachment of the epiglottis to the front wall of the thyroid.

Attached to the inner walls of the larynx, at one end by the arytenoid cartilages and at the other by membranes, are the **vocal cords.** The space between the two vocal cords is the opening of the glottis. When the cords are in a relaxed position, as they are during quiet breathing, the glottis is somewhat triangular in shape, with its apex pointing forward. The shape of the glottis can be changed, however, and these changes result in the different qualities of the voice.

The Lower Respiratory Tract

Below the larynx the respiratory system branches into smaller components. The trachea divides into smaller passageways until it reaches the site for air exchange.

The trachea.   The trachea is a tube about 109 millimeters long and 18 to 25 millimeters wide. It is a round tube that is slightly flattened along the rear surface. Its walls are composed of alternate bands of membrane and cartilage, the former supporting and holding the cartilage in place. The bands of cartilage may be either horseshoe-shaped, almost completely encircling the trachea, or less prominently shaped, passing only part of the way around the tube. In either case, their free ends are held together by a tough membrane that contains scattered bands of smooth muscle tissue.

The **esophagus** lies just behind the trachea, aligned with the openings in the cartilage bands. It is possible for the esophagus to swell as a bolus of food passes through it. The enlarged portion of the esophagus projects into the tracheal cavity, causing pressure on the trachea. Swallowing a mass of food that is too large to be easily accommodated by the esophagus can produce discomfort and a momentary feeling of suffocation. The bands of cartilage are firm yet somewhat elastic. The rings that partially encircle the trachea maintain an open passageway for entering air. The lining of the trachea is composed of a *ciliated epithelium* in which there are goblet cells that secrete mucus. The action of the cilia sweeps small particles of debris from the lungs upward to the mouth or nasal cavity so that they can be eliminated.

The bronchi.   The trachea divides at its lower end into two tubes called **primary bronchi.** The right *bronchus* is slightly larger and more vertical than the left bronchus, which explains why objects frequently lodge in the right bronchus. The structure of the bronchi resembles that of the trachea, although the bronchi are smaller. The cartilaginous rings of the bronchi give way to cartilage plates when they enter and branch within the lungs. The bronchi are lined with **ciliated mucosa,** as is the trachea.

broncho = bronchus, trachea

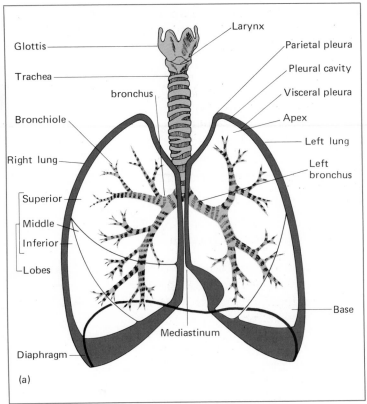

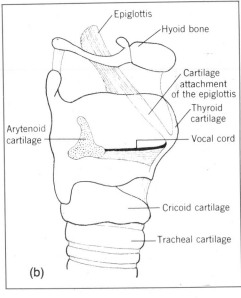

Figure 16-2 (a) The lungs and associated parts. (b) Diagram of the side of the larynx showing the thyroid cartilage.

Each primary bronchus enters the lung on its respective side and immediately divides into smaller branches called **secondary bronchi.** The secondary bronchi continue to branch, forming small **bronchioles.** The trachea and the two primary bronchi resemble a tree trunk with branches, and thus are referred to as the *bronchial tree.* The bronchioles subdivide into smaller and smaller tubes, eventually terminating in microscopic branches that divide into **alveolar ducts.** These ducts terminate in several **alveolar sacs,** the walls of which consist of numerous **alveoli.** The structure of an alveolar duct and its clusters of alveolar sacs resemble a stem surrounded by a bunch of grapes. Some 300 million alveoli are estimated to be present in both lungs.

By the time the secondary bronchi have branched into bronchioles, the cartilaginous rings have become irregular and have disappeared in the smaller bronchioles. The alveolar ducts, the alveolar sacs, and the alveoli consist of a single layer of simple squamous epithelial tissue. The entire tissue layer is very thin and delicate and thus suitable for the exchange of gases in the lungs.

The lungs—external structure. The lungs are cone-shaped organs that fill the **pleural** portions of the thoracic cavity. (See Fig. 16-2.) The **apex** of the lung (upper border) projects slightly above the *collar bones* (clavicles). The lungs are enclosed on three sides by the thoracic cavity. The external surface of each lung is roughly indented to allow space for the heart and other structures. The surface area between the lungs is called the **mediastinal surface.** The indentation is greater on the left than on the right because of the position of the heart.

The lungs—internal structure.   The primary bronchi and the pulmonary and bronchial vessels are bound together to form the root of the lung, or **hilus,** through which they enter and leave the lung. The **base** (lower surface) of the lung is concave and rests upon the *diaphragm.* The *left* lung is divided into the **superior lobe** and the **inferior lobe.** The *right* lung divides into the **superior, middle,** and **inferior** lobes. The interior of each lung contains the innumerable branches that make up the *bronchial tree.* The smallest branches, the *bronchioles,* terminate in the most important structures of the lung, the *alveoli.*

Each lung is covered by a membrane known as the **pleura.** The pleural membrane is composed of two layers—the **visceral pleura** completely covers the lungs, and the **parietal pleura** forms the lining of the thoracic cavity. The two membranes are continuous and together form the divisions of the thoracic cavity. The internal area between the lungs is called the **mediastinum.** Between the pleural membranes is a small (negligible) space called the **pleural cavity.** This potential space is filled with a thin film of serous fluid. The fluid prevents the pleurae from rubbing against one another as the chest cavity expands and air enters. Although small, the pleural cavity is important in respiratory functions.

PHYSIOLOGY OF THE RESPIRATORY ORGANS

The role of the respiratory apparatus is to move air into and out of the lungs. The cells in the body produce carbon dioxide in the course of performing their metabolic activities. In turn, the carbon dioxide must be removed from the body and a continuous supply of oxygen must be provided. The function of respiratory organs is to maintain a constant air supply through their passageways.

Functions of the Upper Tract

The upper respiratory tract serves as the passageway for air going to and coming from the lungs. The nasal passages filter the air of impurities, warm and moisten the air, and remove dust and particles from the membrane lining of the respiratory tract.

The organ of smell.   The nose serves as the organ of smell. The **olfactory receptors** are located in the *nasal mucosa.* The nose also aids in **phonation,** or the production of the proper sounds for speech.

Although the olfactory sense in humans is not as developed as that in many other living things, the olfactory receptors are still able to distinguish hundreds of odors, even those present in small concentrations. The olfactory receptors in the upper nasal cavity can easily be-

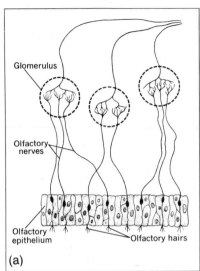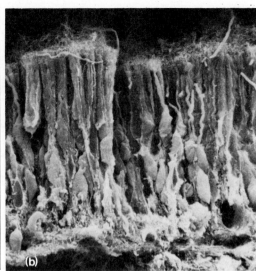

Figure 16-3 (a) Olfactory cells and supporting structures. The nerve fiber from each olfactory cell joins with others to form the olfactory nerve. (b) Photomicrograph of olfactory cells. [Photo from Richard G. Kessel and Randy H. Kardon, *Tissues and Organs: A Text-Atlas of Scanning Electron Microscopy* (W. H. Freeman and Company, © 1979).]

Glomerulus

Olfactory nerves

Olfactory epithelium

Olfactory hairs

(a)

(b)

come insensitive to certain odors. This fact explains why odors that are very noticeable initially are not sensed after a short period of time.

An inflammation of the lining of the nasal passages is characterized by an acute congestion of the mucous membrane and increased secretion of mucus. The swelling of the mucous membrane and the accumulated secretions result in difficulty in breathing through the nose.

The pharynx. The pharynx serves as the *muscular* passageway for the respiratory and digestive tracts. Both air and food must pass through the pharynx before reaching the larynx and esophagus, respectively. The pharynx also plays an important part in phonation. For example, different vowel sounds can be formed only by regulating the shape of the pharynx.

The larynx—voice box. The larynx functions as a vital portion of the airway to the lungs. It protects the airway against the entrance of solids or liquids during swallowing. When food is swallowed, muscles squeeze the laryngeal opening and close the airway. The major function of the larynx is that of voice production. This accounts for its popular name— the voice box.

Formation of sound. As air is expired (exhaled) through the glottis, sound is produced when the vocal cords are made to vibrate by the passing air. The vibration of the vocal cords produces the sound. The **paranasal sinuses** contribute to the sound of the voice by acting as sounding boards or resonating chambers. Thus the size and shape of the *nose, mouth, pharynx,* and *bony sinuses* help to determine the quality of the voice. The muscles of the *pharynx,* the *face,* the *tongue,* and the *lips* are used to modify different sounds into words. The speech center of the *cerebral cortex* of the brain integrates the activity of all the muscles concerned so that intelligible sounds or words may be spoken.

Speech is a uniquely human accomplishment. It requires higher intelligence and the proper vocal apparatus. The almost unlimited variety of sounds that can be produced is indicated by the fact that no two languages make use of exactly the same set of sounds. These differences can be recognized when comparing the speech patterns of various languages. In addition, it is easy to identify regional differences in speech, or accents, within the United States. Another point of interest relates to the early babbling of infants in their attempt to make the sounds of a variety of languages. Later, they eventually learn the sounds they hear repeated by others in their environment.

Properties of sound.   Three characteristics of sound are important in speech. **Loudness** (volume, or amplitude) is the result of changing the force with which air is expelled from the lungs. At the same time, the vocal cords are held in the proper position to produce sound. One need only try to speak during inhalation to realize that the sounds produced are quite indistinct. Thus clear speech is possible only when a column of air is expelled. If the force of exhalation is very strong, the sounds produced will be much louder than when the vocal cords are not subjected to the more vigorous passage of air.

A second property of sound is its **pitch.** The pitch of the voice changes as the vocal cords vary in length. The increase or decrease in length results from the tension exerted on the cords by the *arytenoid cartilages.* The vocal cords, like a violin string, produce a low pitch when relatively relaxed and a high pitch when stretched.

A third characteristic of the voice is its **timbre,** or **quality.** This is a complex aspect of speech in which harmonics (overtones) are formed as a result of vibrations occurring within the cavities of the nose, throat, and thorax. The various cavities of the body give a quality to the human voice that is quite separate from that produced by the vocal cords. If the larynx, chest cavities, and sinuses are large, the quality of the voice is increased. In trained singers, the harmonics that develop in these cavities give particular quality to the voice, which distinguishes professional performers from untrained singers. It is only by intensive training that the full value of the resonating chambers of the body can be attained.

The sounds of the human voice are the result of the vibration of the vocal cords. Depending upon the tension exerted on them, the cords produce sounds different in pitch and volume. Variation in pitch is limited by the size of the vocal cords. In the male the length of the cords is between 2 and 2.54 centimeters, whereas those of a female average about 1.78 centimeters in length. Since this fundamental difference exists, the pitch of the male voice is usually lower than that of

Table 16-1
Ranges of the Human Voice

Classification	Vibrations per Second
Bass	80–250
Baritone	100–350
Tenor	125–425
Contralto	150–500
Mezzo-soprano	200–650
Soprano	250–750, or higher

a female. It is possible, however, for both sexes to alter the length of the cords by voluntarily controlling the movement of the arytenoid cartilages. This contraction and relaxation of the vocal cords results in the range of a person's voice. Based on the rate of vibration of the cords, the ranges of voices may be classified as shown in Table 16-1.

The individual speech sounds are the result of the action of the *tongue*, *teeth*, and *lips* and the partial closing of the *glottis*. In making the sounds of the various vowels, the tongue and the lips play the major roles. The back of the tongue may be elevated in forming some sounds, while the front of the tongue may be used in making others. Thus in making the sound of *ee* (as in cheese), the front part of the tongue is slightly elevated and the lips drawn back. To make the sound *u* (as in rule), the lips are pursed and the back of the tongue raised.

In the formation of consonants, the *lips*, *teeth*, and *tongue* are important. As you say the letter *p* or *b*, notice that the sound is made by the sudden opening of both lips. The difference between the *p* and the *b* sound depends on the action of the vocal cords. The *b* is voiced; that is, it has tone produced by vibration of the vocal cords. The *p* is voiceless; that is, it has no cord-produced tone.

Various combinations of sounds result from different positions of the organs in the mouth region or from the regulation of the rate at which air is expelled. In some instances a nasal tone is produced by the passage of air through the nose.

Functions of the Lower Tract

The lower passageways for air transport act as support structures. Because they have cartilage in their walls, these channels do not collapse when air passes into the lungs.

The trachea. The trachea provides part of the open passageway through which entering air can reach the lungs. Obstruction of this

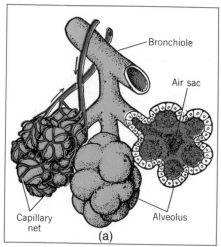

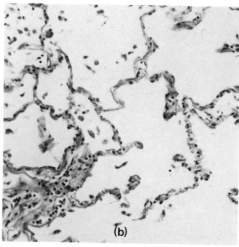

Figure 16-4   (a) Diagram of the air sacs showing capillaries covering the alveoli. An air sac is shown in cross section. (b) Photomicrograph in cross section of an alveolus.

airway, for even a few minutes, can cause death from **asphyxiation** (suffocation). In emergency situations a hole may be made in the trachea to restore the airway. This procedure is called a **tracheotomy**.

tomy = cutting

Air exchange in the lungs.   The tubes composing the bronchial tree perform the same function as the trachea—that of distributing air to the lung's interior. The *alveoli*, enveloped by networks of capillaries, serve for the exchange of oxygen and carbon dioxide between the air in the alveoli and the blood. Certain diseases may block the passage of air through the bronchioles or alveoli and interrupt the exchange of gases between the air and blood.

The lungs perform two important functions: they distribute the entering air and they exchange gases with the blood. The distribution of air to the alveoli is the function of the tubes of the bronchial tree. The exchange of gases between air and blood is the combined function of the alveoli and the blood capillaries that surround them. These two structures, the alveoli belonging to the respiratory system and the capillaries belonging to the circulatory system, together serve as highly efficient gas exchangers.

The respiratory membrane composing the very thin-walled alveoli comes in contact with the equally thin membrane of the pulmonary capillaries. These membranes allow the extremely rapid **diffusion** of gases between the alveolar air and the pulmonary capillary blood. (See Fig. 16-4.)

It has been estimated that if the lungs' 300 million alveoli could be opened up flat, they would form a surface area about the size of a tennis court. It is not surprising, therefore, that large amounts of oxygen can be quickly transferred into the bloodstream, while large amounts of carbon dioxide can rapidly diffuse from the blood into the air.

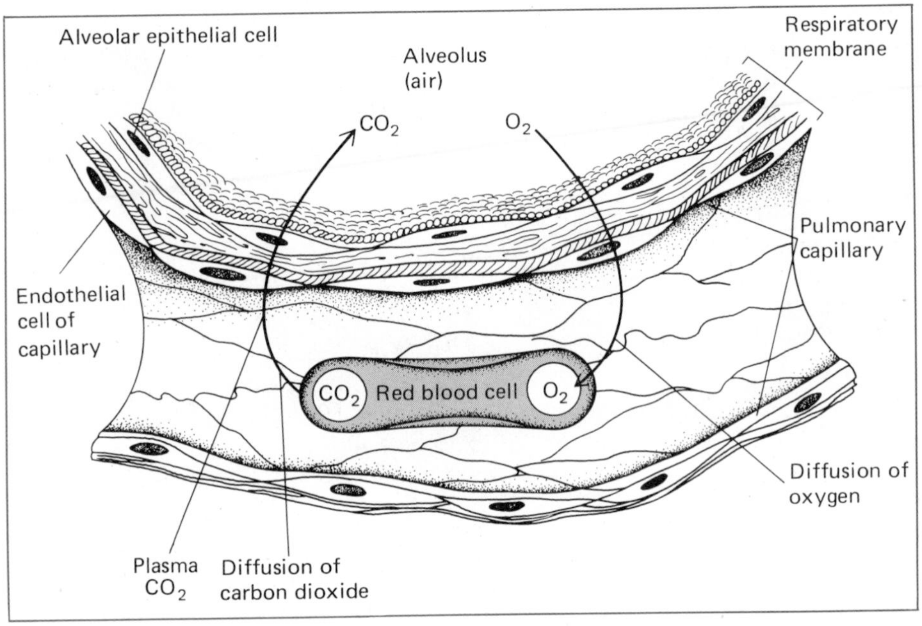

Figure 16-5 Exchange of oxygen and carbon dioxide between the blood and the pulmonary capillary.

RESPIRATORY PROBLEMS

The respiratory tract is vulnerable to a variety of infections. Infections of the respiratory tract, which includes the nasal passages, are usually caused by *viruses.*

Upper Respiratory Disorders

itis = inflammation

The **common cold,** the most familiar of all respiratory infectious diseases, is an upper respiratory tract disorder. Colds are usually caused by a virus or a group of viruses. No other mammal has been found to be more susceptible to the common cold than are humans. It is a uniquely human disease and a very contagious one.

Other upper respiratory tract infections, called **URI,** are **nasopharyngitis** and **laryngitis.** These are frequently caused by *streptococci* and must be treated promptly because of the possibility of rheumatic fever.

Lower Respiratory Disorders

Bronchitis is an inflammation of the bronchi and may be due to an *acute* or a *chronic* infection. Acute bronchitis may result from a *bacterial* or *viral* infection. Chronic bronchitis is usually due to a chronic irritation, particularly cigarette smoking and air pollution.

pneumo = air, lungs

Pneumonia is an inflammation of the lung. The inflammation, as inflammation of any part of the body, results in redness due to dilation of the blood vessels and edema, or swelling, due to the influx of cellular defense substances. In addition, formation of an exudate (tissue fluid and cellular debris) fills the alveolar spaces of the infected part of the lung, thus interfering with air exchange. Pneumonia may be caused by bacteria, viruses, or fungi.

When a pneumonic infection involves an entire lobe of the lung, it is called **lobar pneumonia.** When the inflammation involves the bronchi, it is referred to as **bronchopneumonia.**

SUMMARY

The respiratory system is primarily responsible for the exchange of gases between blood and entering air. The nose serves as a passageway for incoming and outgoing air. The pharynx primarily serves as both the respiratory and digestive passageway for air, food, and liquids. The larynx contains the vocal cords, which vibrate to produce sounds. Speech is a function that involves the production of sound by the vocal cords as well as resonance by the paranasal sinuses and muscular movements of the pharynx, larynx, and tongue.

The major function of the trachea is to provide an open passageway for air going to and coming from the lungs. The trachea divides into two branches—the primary bronchi—and they in turn divide into smaller and smaller branches, which finally terminate in the alveoli. In the lungs, alveolar air and blood come in contact for the rapid exchange of gases.

VOCABULARY REVIEW

Match the statement in the left column with the correct word(s) in the right column. *Do not write in this book.*

1. Folds in a bone that increase the surface area of the nasal passages.
2. Passes downward from the larynx anteriorly to the esophagus.
3. Membranes that respond to currents of air and produce sound.
4. The internal area between the lungs.
5. Divides the nasal cavity into two halves.
6. The trachea divides into these two major branches of the lungs.
7. The tube in the lung that ends in an air sac.
8. The tension exerted on the vocal cords produces this difference.
9. A thin-walled structure through the walls of which the blood absorbs oxygen.

a. alveolus
b. bronchiole
c. bronchi
d. vocal cords
e. glottis
f. pitch of sound
g. larynx
h. mediastinum
i. nasal conchae
j. pharynx
k. septum
l. trachea

TEST YOUR KNOWLEDGE

Group A

Write the letter of the word(s) that correctly completes the statement. *Do not write in this book.*

1. The sense of smell depends on sense organs lying   (a) below the septa   (b) in the antra   (c) on the tongue   (d) above the superior conchae.
2. Human sounds are produced by vibration of the   (a) epiglottis   (b) vocal cords   (c) arytenoid cartilages   (d) cricoid cartilage.
3. The so-called Adam's apple is properly known as the   (a) pharynx   (b) trachea   (c) larynx   (d) bronchus.
4. The characteristic of the human voice that depends on the rate of vibrations is   (a) volume   (b) amplitude   (c) pitch   (d) timbre.
5. Rings of cartilage stiffen the walls of the   (a) esophagus   (b) nasal cavity   (c) trachea   (d) diaphragm.
6. Windpipe is a common name for the   (a) larynx   (b) pharynx   (c) trachea   (d) bronchus.
7. Of the following, the one that is not within the chest cavity is the   (a) heart   (b) lungs   (c) esophagus   (d) larynx.
8. The membrane that covers the lungs is called the   (a) mesentery   (b) pleura   (c) mediastinum   (d) pericardium.
9. Oxygen passes into the blood of the capillaries chiefly through the walls of the   (a) bronchioles   (b) bronchi   (c) alveoli   (d) pharyngeal arches.
10. Of the following terms, the one that is least closely related to the others by structure is the   (a) larynx   (b) trachea   (c) glottis   (d) air sac.

Group B

Answer the following. Briefly explain your answer. *Do not write in this book.*

1. Trace the path taken by a molecule of oxygen from the time it enters the respiratory system until it is absorbed by the blood. Tell over what structures it passes and the effect each has on it.
2. Distinguish between the members of the following pairs of terms: (a) air sac/alveolus   (b) bronchus/bronchiole   (c) glottis/epiglottis   (d) pharynx/larynx.
3. Briefly describe the structure of the larynx and the relationship of the vocal cords to it.
4. A foreign body that obstructs (closes) the glottis may cause death. Why?
5. Why are foreign bodies that have been aspirated (that have entered the airway instead of the esophagus) more frequently found in the right bronchus than the left?

CHAPTER 17

THE MECHANICS OF BREATHING

Objectives

A. Describe the structures of the thoracic cavity

B. Explain the properties of gases and how they relate to the process of respiration

C. Describe the changes that occur in pulmonary pressure during breathing

D. Describe the most common methods of artificial respiration

E. Identify respiratory air volumes during normal and forceful breathing efforts

F. Describe the chemical and neural factors that regulate breathing

THE DESIGN FOR GAS EXCHANGE

The respiratory system carries out a combined set of processes that include breathing, gas exchange in the lungs and tissues, transport of gases by the blood, and oxygen utilization in the cells. **Breathing** is the mechanical process of taking air into the lungs and expelling carbon dioxide from the lungs. **Respiration** is a broad term that is used to refer to the exchange of gases between a living organism and its environment. In a single-celled organism, such as an ameba, respiration is easily accomplished because the exchange of oxygen and carbon dioxide takes place by simple diffusion through its cell membrane. (See Chapter 3.) However, in more complex organisms, such as humans, specialized structures are necessary in order for the body cells to exchange gases with the external environment. Consequently, a system of organs is responsible for the transfer of gases to and from the cells of the body. (Review Chapter 16.)

In respiration two phases are identified: **external respiration,** which involves the exchange of gases between the bloodstream and the air, and **internal respiration,** which is concerned with the exchange of gases that takes place between the circulating blood and the various tissue cells as they utilize oxygen and produce carbon dioxide. This chapter explains the anatomical relationships of external respiration.

The Thoracic Cavity

The **thoracic cavity,** or **chest cavity,** is divided into a right and left pleural cavity. The space between the two is called the **mediastinum.** A wall of fibrous tissue covers the mediastinum, completely separating it from the right and left pleural cavities. Thus the only organs in the thoracic cavity that are not located in the mediastinum are the lungs. The organs that are in the mediastinum and between the lungs include the heart (enclosed in the pericardial sac), the trachea, the right and left bronchi, the esophagus, the thymus, the various blood and lymph vessels, and several important nerves.

The pleura.   As discussed in Chapter 16, each lung is covered by a transparent serous membrane called the **pleura.** This membrane, like the peritoneum, consists of two layers—a **visceral layer** and a **parietal layer.** The pleura, however, forms a closed sac around each lung. The visceral pleura adheres to the surface of the lung and dips into the fissures between its lobes. The parietal pleura lines the inner surface of the chest wall, covering the upper surface of the diaphragm and the other structures in the mediastinum.

Although the space between the visceral and the parietal pleurae is called the pleural cavity, it must be emphasized that this is purely an implied space. Normally, the pleural layers are held in close contact with one another by the cohesive effect of a thin film of serous fluid called **pleural fluid.** This fluid also serves to lubricate the surfaces so that they can move smoothly over one another. To demonstrate the effect of the pleural fluid, place a thin film of water between two glass slides; note that the wet surfaces cling together and the slides cannot be easily separated. However, the water allows for the slippery side-to-side movement between these two surfaces. Pleurisy is an inflammation of the pleura. Pain accompanies every breath and is due to friction between the surfaces of the pleura.

Figure 17-1 Medial area of the thoracic cavity.   Figure 17-2 Thoracic cavity and pleural spaces.

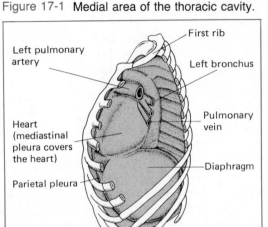

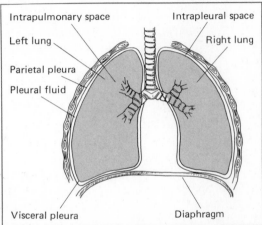

Properties of Gases

Breathing, or **pulmonary ventilation,** is the act of taking *fresh* air into and expelling *stale* air from the lungs. External respiration requires two structures: a large moist surface where blood and air can come in close contact, and a bellows such as that provided by the lungs and the thoracic cage, which has muscles and nerves that move the thorax.

pulmo = lungs

The gas laws.   The *principle of Amedeo Avogadro (1776–1856)* states that *equal volumes of gases at the same temperature and pressure contain the same number of molecules*. Avogadro determined that at **standard temperature and pressure in dry air (STPD)** —that is, 0°C and 760 millimeters of mercury—the molecular weight in grams of any gas will occupy a volume of 22.4 liters. Avogadro's observation is the basis for determining the volume occupied by the composition of gases in a mixture. Provided that temperature and volume remain constant, the pressure exerted by a gas produces a gradient for diffusion depending on the number of molecules of gas. In the body, diffusion of a particular gas is therefore most rapid when its diffusion gradients are high due to adequate ventilation.

Boyle's law states that when a gas is subjected to pressure, its volume varies inversely with the pressure, temperature remaining constant. If the pressure on 100 milliliters of gas is doubled, the volume is reduced by one half. If the pressure is tripled, the volume is reduced to one third. This relation of volume to pressure can be stated mathematically as $V/V' = P/P'$, where V is the original volume of the gas and V' is the volume of the gas at the increased, or decreased, pressure, and where P is the original pressure and P' is the new pressure of the gas. The law is named for the English physicist Robert Boyle (1627–1691), who discovered and formulated this relationship. Any decrease of volume of a gas must therefore increase the pressure of a gas and thus increase the rate of its diffusion or movement.

Charles' law [named after Jacques Charles (1746–1823)] states that the volume of a gas is directly proportional to its absolute temperature. Absolute zero (−273°C) is the temperature at which all molecular motion stops. Therefore, for each rise of 1°C, a gas will expand by 1/273 of its original volume because of the increase of activity of its total number of molecules. This relationship between volume and temperature can be expressed as $V/V' = T/T'$. The capital letter T is used to show that the measurements are based on the absolute, or Kelvin, scale (0°C = 273K). At normal body temperature, 37°C (310K), 1 gram-mole of gas has a volume of 25.4 liters. Thus, the rate of diffusion is more-or-less constant with the temperature of the body.

Charles' law, Boyle's law, and Avogadro's law are all combined in the **ideal gas law,** which can be expressed as $PV = nRT$, where P, V, and T are, respectively, the pressure, volume, and temperature (K) of the gas, and where n is the number of gram moles (a gram mole is the molecular weight in grams) of the gas and R is a constant. An example of the operation of this law as applied to nonliving systems is found in the fact that the air pressure inside an automobile tire rises as the tire becomes hot at high speeds.

Dalton's law of partial pressures states that each gas in a mixture exerts its own pressure in proportion to the volume it occupies in the total volume of the gas mixture. Thus, at sea level the volume of oxygen present in air of average composition is approximately 21 percent. Hence oxygen will exert only 21 percent of the total pressure of this air. At an altitude of 19,312 meters (12 miles), the percentage of oxygen is reduced to about 18 percent of the total volume of the air and will therefore exert a pressure that is 2 to 3 percent below that at sea level. This means that the molecules of oxygen are so widely separated that they are unable to exert adequate pressure on the cell membranes and therefore cannot enter the blood in sufficient numbers. The relation can be stated as follows: $PV = V(p_1 + p_2 + p_3 \ldots)$, where V stands for the volume of the mixture, P for total pressure of the mixture, and p_1, p_2, $p_3 \ldots$ for the pressure of individual gases. In other words, the total pressure of the mixture of gases in a given volume equals the sum of all of the partial pressures. The atmospheric pressure thus represents the sum of all of the partial pressures of gases composing the air around the earth. (See Table 17-1.)

Table 17-1
Composition of the Earth's Atmosphere

Gas	Percentage in Dry Air
Nitrogen	78.09
Oxygen	20.94
Carbon dioxide	0.03
Rare gases (argon, neon, helium, xenon)	0.94

Diffusion of Gases

Henry's law plays an important part in the entrance of gases into the blood. The law states that a gas is dissolved by a liquid in direct proportion to its partial pressure, provided that it does not react chemically with the liquid. Henry's law is important to the understanding of the physical processes in breathing, since all gases that pass through the cell membranes in the lungs must first be dissolved in the cell fluids and in the thin film of fluid that covers the inner surface of the lungs.

A mathematical statement of Henry's law is $P/V = K$. In this equation, P can be the pressure of a gas in the lungs; V, its concentration in the plasma; and K, a constant, the *coefficient of solubility*. The value of this coefficient depends on the units used for expressing P and V. The solubility coefficients for important respiratory gases at body temperatures are as follows:

Oxygen	.024
Carbon dioxide	0.57
Nitrogen	0.012

The *gas laws* are related to movement of gases across the alveolar-capillary and capillary-tissue membranes.

Composition of air. Humans breathe an atmosphere that normally contains the proportions of gases shown in Table 17-1. Air with this composition is inhaled and mixed with the air already in the lungs and respiratory passageways. In stating values of gases in the body, the term *partial pressure* is conventionally used. Partial pressure is expressed in millimeters Hg (mercury) and refers to the portion (or part) of the total pressure of a mixture of gases contributed by one of the gases present in the mixture (Dalton's law). The pressure of a gas may be calculated by multiplying the percentage of the gas in the mixture times the total pressure of the gases in the mixture. For oxygen and carbon dioxide, the symbols p_{O_2} and p_{CO_2} are used to designate partial pressure of oxygen and carbon dioxide.

The composition of the air in various parts of the respiratory system and in the blood and tissues is listed in Table 17-2. Composition of air in the alveoli is controlled by the ventilation of the lungs and the diffusion of gases into and out of the bloodstream. The levels in the bloodstream are determined by lung values and tissue activity. Table 17-2 indicates that the p_{O_2} decreases as air goes from the atmosphere to the alveoli, to the blood, and finally to tissues. That is, diffusion of oxygen will always proceed *down* a diffusion gradient, as it does going from the atmosphere, to the blood, and to the tissues. Conversely, p_{CO_2} is highest in the tissues and its pressure decreases as air goes from tissues, to blood, to lung, and to atmosphere. Nitrogen concentration is nearly constant in all parts of the body, indicating that it is not involved in metabolic reactions.

Table 17-2
Partial Pressures in mm Hg of Gases in Various Areas of the Body

Gas	Atmosphere	Trachea	Alveoli	Blood Arterial	Blood Venous	Tissues
Nitrogen	596	564	573	573	573	573
Oxygen	158	149	100	95	40	40
Carbon dioxide	0.3	0.3	40	40	46	46
Water vapor	5.7(av.)	47	47	47	47	47
Total	760	760	760	755	706	706

THE RESPIRATORY CYCLE

The basic principle underlying the movement of any gas is that it travels from an area of higher pressure to an area of lower pressure. This principle applies not only to ventilation, or the flow of air into and out of the lungs, but also to the diffusion of oxygen and carbon dioxide through the alveolar and capillary membranes of the lungs. The respiratory muscles of the thoracic cage and the elasticity of the lungs (compliance) make possible the necessary changes in the pressure gradient so that air first flows into and then is expelled from the respiratory passages. The phases of the respiratory cycle involve the following types of pressures: **atmospheric, intrapulmonic,** and **intrapleural.**

intra = within, inside

The pressure exerted against all parts of the body by the surrounding air is called *atmospheric pressure.* It averages 760 mm Hg at sea level. Any pressure that falls below atmospheric pressure is called **subatmospheric,** or **negative, pressure** and represents a partial vacuum.

Changes in Pulmonary Pressure

The pressure of air within the bronchial tubes and the alveoli is called **intrapulmonic pressure** (or intra-alveolar pressure). In breathing, this pressure fluctuates below and above atmospheric pressure as air moves into and out of the lungs.

The pressure that exists between the two layers of pleura (the lung cavity) is called the **intrapleural pressure.** During normal breathing, the intrapleural pressure is always *subatmospheric*, or *negative*. This is due to the fact that the lungs have a tendency to collapse because of their elasticity. This elastic recoil, or *compliance*, constantly expands the intrapleural space, thus lowering the pressure in the lung cavity below atmospheric pressure, which exerts a pull on the thoracic walls. The intrapleural pressure undergoes changes during each respiratory cycle, yet retains a negative pressure. It becomes less negative when one coughs or has difficulty in defecating.

Inspiration.   *Inspiration* is the act of taking air into the lungs. Changes in the size of the thoracic cavity and changes in intrapleural and intra-pulmonic pressures result in inspiration.

Contraction of the diaphragm alone, or of the diaphragm and the external intercostal muscles, produces quiet inspiration. The diaphragm and the external intercostal muscles contract when impulses from the central nervous system (CNS) communicate with them by way of the phrenic and intercostal nerves. As the diaphragm contracts, its dome moves downward, enlarging the size of the thoracic cavity from top to bottom. The contraction of the external intercostals raises the ribs and, at the same time, rotates them slightly, thereby pushing the sternum forward. The thoracic cavity thus is enlarged from side to side and from front to back.

As the thoracic cavity enlarges, the parietal pleura tends to pull away from the visceral pleura, thus lowering (or making more negative) the intrapleural pressure. However, pleural fluid causes the strong cohesion between the two pleural layers, and so normally their separation does not occur. Instead, the visceral pleura, and the lung that it covers, expands with the enlarging thoracic cavity. As the lung increases in size, the intrapulmonic pressure immediately falls below atmospheric pressure and air will flow into the lungs through the pharynx, trachea, and bronchi until the intrapulmonic pressure is equal to that of the atmosphere (Boyle's law).

Expiration.   *Expiration* is the act of expelling air from the lungs. Quiet expiration is entirely a passive process that is a result of pressure changes opposite to those of inspiration.

In expiration, the diaphragm and the external intercostal muscles relax because of a cessation of impulses from the CNS. As the muscles return to their relaxed position, the lungs recoil and the thoracic cavity decreases in size.

As the thoracic cavity and the lungs decrease in size, there is a corresponding increase in the intrapleural pressure (that is, it becomes less negative) and in the intrapulmonic pressure, which quickly rises above atmospheric pressure. This compression of the lung creates a pressure gradient between the alveoli and the external atmosphere, and air flows out of the lungs into the air until the intrapulmonic pressure equals atmospheric pressure.

Hypoxia

Hypoxia results whenever oxygen is not delivered to the tissues or when there is an insufficient supply of oxygen to the tissues. The physical and chemical causes for failure in breathing include drowning, electrocution, the use of certain drugs that affect the respiratory muscles, poisonous gases such as carbon monoxide, or a mechanical obstruction of the respiratory passages.

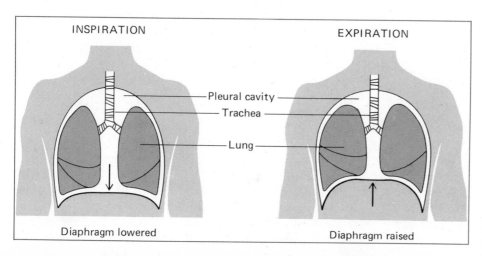

Figure 17-3 Movement of diaphragm during inspiration and expiration.

A complete lack of oxygen in the blood for more than four minutes may result in permanent injury or death. When artificial respiration is required, it should be started immediately and be continued for as long as necessary. Occasionally, victims have been saved from death by continued artificial respiration over a 30-hour period. Thus, it is evident that once started, the process should continue as long as there is a chance to revive the victim.

Artificial respiration.   The best method for artificial respiration is based on the introduction of air, or a mixture of gases, directly into the lungs. This is accomplished by mechanical devices or by breathing directly into the victim's mouth.

One mechanical device that has been in use for some years is the inspirator. This consists of a mask that fits over the mouth and nose of the victim. Through the mask, compressed air is pumped into the lungs and then withdrawn. Although the theory of this method is good, its successful operation depends on the skill of the operator. In the hands of a poorly trained person, it is highly dangerous because of the possibility of exerting too much pressure, either positive or negative, on the thin-walled air sacs.

Modern resuscitators are made to work on the same principle as the inspirator except that they use oxygen and are equipped with a complex series of valves that automatically regulate the pressure of the oxygen entering or leaving the lungs. These machines develop a positive pressure of about 8 millimeters of mercury to force air into the lungs and then withdraw it at about the same pressure. The resuscitator is equipped with one or more tanks of oxygen under a pressure of about

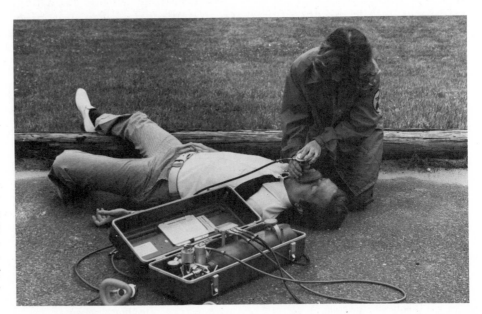

Figure 17-4 **A** demonstration of a resuscitator in use. Resuscitators are valuable in cases of carbon monoxide poisoning, drowning, electrical shock, and other conditions of suffocation.

100 atmospheres. It is able to apply resuscitation for periods of 20 minutes to 1 hour depending on the number and size of tanks available.

It is possible, and at times desirable, to equip a resuscitator with a tank of carbon dioxide so that this gas may be administered with the oxygen. A small amount of CO_2 (5 percent) is considered suitable because of the stimulating effect it has on the respiratory center in the medulla oblongata. Carbon dioxide mixed with oxygen is especially valuable in reviving victims of carbon monoxide poisoning. The proper percentage combination of O_2 and CO_2 has a stimulating effect on the respiratory center.

A preferred technique, which the American Red Cross advocates, is mouth-to-mouth resuscitation, which is believed to be the best method of artificial respiration. As Figure 17-6 shows, this method consists of breathing into the mouth of the victim, which is checked beforehand to ensure that his or her respiratory tract is free of obstructions. This method is a safe one from the standpoint of both the victim and the operator because at no time can the pressure within the lungs of either reach a dangerous level. It also has the advantage of administering a

Figure 17-6 Mouth-to-mouth resuscitation. (a) Tilt position of head back to clear airway. (b) Pinch victim's nostrils together. (c) Blow air into victim's lungs. (d) While taking another breath, allow victim to exhale.

Figure 17-5 The combination of artificial respiration and circulation is called cardiopulmonary resuscitation (CPR) and is administered only when both breathing and heart beat has stopped. CPR should only be administered by trained individuals or medical personnel.

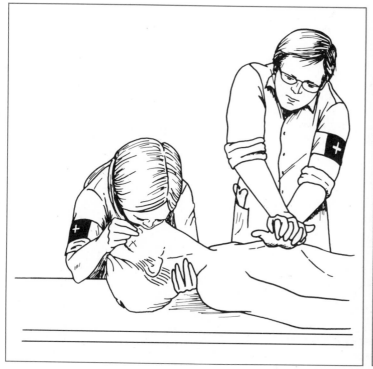

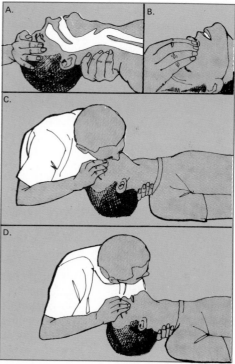

small amount of carbon dioxide (4 percent), which may have a stimulating effect on the medulla. Research has shown that about 1 liter of air is exchanged with each breath by this method. This is considerably more than is possible with the older manual methods.

Changes in Pulmonary Volume

The volume of air that moves in and out of the respiratory passages during inspiration and expiration is measured with an instrument called a **spirometer.** The record produced by a person breathing in and out of a spirometer is called a **spirogram.** From this record information concerning pulmonary volume (lung capacity) can be determined.

meter = measure

The volume of air moved in or out of the respiratory passage during normal, quiet breathing is called the **tidal volume.** The *average tidal volume* during inspiration is 500 ml., and the same amount moves out on expiration. The volume of air that can be inhaled by the deepest possible inspiration in excess of the tidal is the **inspiratory reserve volume.** The average inspiratory reserve is 3000 ml. The volume of air that can be exhaled by the deepest possible expiration in excess of the tidal is the **expiratory reserve volume.** This volume of air is about 1200 ml. The sum of the tidal, the inspiratory reserve, and the expiratory reserve air is called the **vital capacity.** The vital capacity can also be defined as the volume of air that can be exhaled by the deepest possible expiration after the deepest possible inspiration.

Volumes that cannot be determined directly from the spirogram are those involving air that stays in the lungs following forced expiration. **Residual air** is the amount of air remaining in the lungs after the deepest possible expiration, and it cannot be removed by voluntary effort. The average volume of residual air is 1200 ml, and it is eliminated only by collapse of the lungs. (See Table 17-3.)

Minimal air is that air remaining in the lungs after the lungs have collapsed. Minimal air may be important in medicolegal practice to determine if any signs of respiration existed in infants declared dead immediately after their delivery.

Table 17-3
Lung Volumes in Healthy Persons Between 20 and 30 Years of Age

Volume	Male (in ml)	Female (in ml)
Total lung capacity	6000	4200
Vital capacity	4800	3400
Inspiratory reserve	3100	2100
Tidal volume	500	500
Expiratory reserve	1200	800
Residual volume	1200	1000

The Control of Breathing

The regulation of breathing is controlled by two different factors. One of these is **neural** (nervous) control, while the other is **chemical** in nature. Both of these may operate simultaneously, but each is independent of the other. The neural control of the respiratory movements is the result of stimuli affecting the **respiratory center** in the medulla of the brain, while the chemical control depends on a change in the pH of the blood. A decrease in pH stimulates the neurons in the respiratory center to increase breathing.

The respiratory center.   Neural control of breathing was first demonstrated by Galen in the 1st century A.D. He showed that a center in the lower brain was responsible for this activity. More modern experiments carried out under very rigorously controlled conditions have shown that Galen was essentially correct in his conclusions.

Located near the upper part of the medulla is a region with cells that constitutes the respiratory center (Fig. 17-7). Its exact location has been well established by experiments using electrical and thermal stimuli. A slight stimulus of any nature applied to the center causes the rate of respiration to increase or decrease. It is thought that the center is actually composed of two distinct regions, one that controls inspiration and one that controls expiration.

Two nerve pathways are concerned in control of the breathing process. One follows the *phrenic nerves* to the diaphragm and the *intercostal nerves* to the intercostal muscles. This pathway consists largely of motor fibers, although there are sensory fibers that carry impulses to the respiratory center. The other pathway transmits afferent impulses from the lungs along the vagus nerves, which transmit impulses also from the skin, nose, larynx, and abdominal organs. Thus, many different stimuli from various regions of the body help to regulate the breathing process.

Stimuli that arise in the surface membranes of the body can change the breathing rhythm in several familiar ways. The prick of a pin or sudden drenching with cold water makes a person gasp. Irritation of the nasal membranes or the larynx results in a sneeze or a cough. Likewise, a thought, sight, or sound may bring about a change in the respiratory rate. It is also possible to alter voluntarily the rate of breathing by holding the breath. All of these stimuli arise either in body surfaces or in the cortex of the cerebrum.

Impulses that normally regulate the rhythm of breathing travel along branches of the vagus to the lungs. As the lungs expand, nerve endings in their tissues are stimulated and set up impulses that pass to the respiratory center. Here these impulses initiate a series of impulses that eventually pass along the phrenic nerves and the intercostal nerves, causing the diaphragm and external intercostals to relax. When complete exhalation has occurred, the process is reversed and inspiratory impulses travel to the diaphragm and external intercostals, causing

Figure 17-7 **The nerve pathways involved in the control of respiration.**

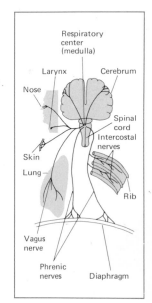

them to contract. The action of these muscles during breathing is another example of reciprocal innervation, such as that illustrated by the biceps and triceps.

Control mechanism of carbon dioxide.   Respiration is chemically controlled through the *carbon dioxide* in the blood. When the blood passes through an active tissue, it picks up carbon dioxide and other products of the oxidations occurring in the tissues. The increase in carbon dioxide stimulates the respiratory center, which is richly supplied with blood vessels.

This chemical regulating mechanism is responsible for the fundamental rhythm of the respiratory center and maintains breathing both day and night without any conscious thought on an individual's part. It is this mechanism that comes to the rescue of upset mothers whose children, in fits of temper, hold their breath until they become "blue in the face." Parents who remain calm know that children will breathe in a few seconds, as an involuntary response due to this chemical control.

RESPIRATORY DISORDERS

There are a wide variety of conditions that can reduce the effectiveness of the breathing mechanism. These conditions may be caused by paralysis of the respiratory muscles, collapse of the lung, or obstruction of an airway.

Damage to Respiratory Organs

Paralysis of the respiratory muscle interferes with breathing; the degree of interference depends upon the muscles involved. Paralysis of the respiratory muscles may be caused by damage to the respiratory center in the medulla, spinal cord injury, or poliomyelitis, which is an acute viral disease. Generally the virus afflicts the motor neurons of the spinal cord, which weakens or paralyzes the respiratory muscles. If ventilation becomes impossible, a mechanical respirator is used.

If air enters the pleural cavity—a condition termed **pneumothorax**—the lung collapses and can no longer function. A lung is sometimes collapsed purposely for the treatment of tuberculosis. The collapsed lung rests, which apparently is helpful in combating the infection. Once the opening into the pleural cavity is closed, the air contained therein is slowly absorbed and the lung expands and becomes active once again.

Allergy Problems

Asthma is most often an allergic disorder. Typically, foreign materials in the air, such as plant pollens, stimulate a reaction in sensitive individuals. As a result, there is edema of the walls of the airways, secretion of mucus, and spasms of the smooth muscles. The airways are narrowed by the swelling and spasms, and the secretion of mucus further blocks the movement of air. As a result, breathing becomes very difficult.

SUMMARY

Respiration refers to the exchange of gases between an organism and the environment. External respiration involves gas exchange between blood and air, while internal respiration involves the exchange between blood and tissue cells. The thoracic cavity has three divisions: two pleural portions, containing the lungs, and the mediastinum, containing the heart, esophagus, trachea, great blood vessels, and lymph vessels. An increase in the size of the thoracic cavity leads to inspiration; a decrease, to expiration.

The gas laws form the basis for study of the respiratory cycle. During inspiration, contraction of the inspiratory muscles enlarges the thoracic cavity, decreasing intrapleural and intrapulmonic pressures, and causing air to flow into the lungs. In expiration, which is essentially a passive process, the inspiratory muscles relax, the elastic tissue of the lungs recoils, and the thoracic cavity decreases in size. These increase the intrapleural and intrapulmonic pressures, causing air to flow out of the lungs.

VOCABULARY REVIEW

Match the statement in the left column with the correct word(s) in the right column. *Do not write in this book.*

1. Refers to the maximum total amount of air contained by the lungs.
2. The volume of air remaining in the lungs following maximum respiratory effort.
3. A mechanical device used to restore breathing when the body's respiratory apparatus fails.
4. A device for measuring lung capacity.
5. The volume of air entering and leaving the body with each breath.
6. Stimulation of this by a change in pH of the blood results in a change in the rate of breathing.
7. Stimuli passing along these afferent nerves control the rate of breathing.
8. The intercostal muscles and diaphragm are stimulated by impulses that enter them through these nerves.
9. The gas law concerned with the effect of temperature on the volume of a gas.
10. The gas law stating the relation between the partial pressure and solubility.

a. resuscitator
b. Boyle's law
c. pulmonary disease
d. Charles' law
e. Dalton's law
f. expiratory reserve volume
g. Henry's law
h. inspiratory reserve volume
i. phrenic and intercostal nerves
j. residual volume
k. respiratory center
l. spirometer
m. tidal volume
n. vagus nerves
o. total lung capacity

TEST YOUR KNOWLEDGE

Group A

Write the word or words that will correctly complete the statement. *Do not write in this book.*

1. The gas in the atmosphere that is most closely concerned with respiration is   (a) nitrogen   (b) water vapor   (c) oxygen   (d) hydrogen.
2. The brain center for the control of breathing is in the   (a) cerebrum   (b) cerebellum   (c) medulla   (d) pons.
3. The gas in the blood that exerts chemical control over the rate of breathing is   (a) oxygen   (b) nitrogen   (c) carbon dioxide   (d) hydrogen.
4. Boyle's law pertains to the   (a) number of molecules in equal volumes of gases   (b) effect of temperature on gas volume   (c) effect of pressure on gas volume   (d) effect of pressure on the solubility of gases.
5. The volume of air taken in with maximal inhalation is called the   (a) inspiratory reserve volume   (b) tidal volume   (c) residual volume   (d) expiratory reserve volume.
6. The gas law concerning partial pressures is attributed to   (a) Avogadro   (b) Charles   (c) Dalton   (d) Henry.
7. The pleural cavity is located between the   (a) visceral and parietal pleurae   (b) ribs and mediastinum   (c) ribs and diaphragm   (d) heart and esophagus.
8. The physical process by which gases enter and leave the blood is   (a) dispersion   (b) absorption   (c) diffusion   (d) osmosis.
9. The pressure within the bronchial tree and the alveoli is called   (a) intrapulmonic   (b) intrapleural   (c) intrathoracic   (d) atmospheric.
10. The pressure within the pleural cavity is   (a) atmospheric   (b) subatmospheric   (c) intra-alveolar   (d) intrapulmonic.

Group B

Answer the following. Briefly explain your answer. *Do not write in this book.*

1. Explain the meaning of each of the following.   (a) tidal air   (b) inspiratory reserve volume   (c) expiratory reserve volume   (d) residual air   (e) minimal air
2. How do changes in the pH of the blood affect the rate of breathing?
3. Distinguish between external and internal respiration.
4. Outline the procedure for carrying out resuscitation by the mouth-to-mouth method.
5. Describe the sequence of events that occurs during inspiration and expiration of air.

CHAPTER 18

THE RESPIRATORY GASES AND BAROMETRIC PHYSIOLOGY

Objectives

A. Compare the composition of respired air and alveolar air
B. Discuss the role of partial pressure in exchange of gases in the lungs and tissues
C. Discuss the role of hemoglobin in the transport of gases
D. Describe the three ways in which carbon dioxide is carried in the blood
E. Describe the effects of high altitude and deep-sea diving on respiration

GAS EXCHANGE AND TRANSPORT

In Chapters 16 and 17 the respiratory system was discussed as a basis for understanding the physiological principles that regulate air distribution and gas exchange. This chapter deals with the processes that play a critical role in maintaining the constancy of the internal environment (homeostasis). The proper functioning of the respiratory system ensures that tissues receive an adequate supply of oxygen and that carbon dioxide is removed promptly throughout a wide range of environmental conditions and body activities. An adequate and efficient regulation of gas exchange between the body cells and circulating blood under changing conditions is the essence of respiratory function.

The mechanics of breathing in respiration involve pulmonary ventilation—the constant replacement of fresh air into the lungs (discussed in Chapter 17). The first phase in the respiratory cycle is the movement of oxygen from the alveolar air to the blood and the movement of carbon dioxide in the reverse direction. The blood leaves the lungs low in carbon dioxide and saturated with oxygen. The second phase occurs in the capillary walls throughout the body. Again, there is the movmeent of the respiratory gases: Oxygen diffuses out of the blood into the cells and carbon dioxide passes from the cells into the blood.

Composition of Respired and Alveolar Air

Air is composed chiefly of nitrogen (79 percent by volume), oxygen

(21 percent), and a small amount of carbon dioxide (0.04 percent). The air we breathe does not change in its percentage composition. That is, whatever the atmospheric pressure, nitrogen would still occupy four-fifths, oxygen one-fifth, and carbon dioxide a small amount of the total volume of dry air. Of these gases, the volume of oxygen and carbon dioxide in the body is relevant to their rate of exchange from the atmosphere to the tissues. Although there are no known metabolic reactions in the human body involving molecular nitrogen (N_2), compounds containing nitrogen are found in cellular waste products, in the blood, and in tissues.

The composition of respiratory gases varies in **inspired, expired,** and **alveolar air,** as indicated in Table 18-1. Nitrogen and water vapor are not included in this table, and therefore the combined percentages do not equal 100 percent.

As would be expected, expired air contains less oxygen and more carbon dioxide than inspired air. However, note that expired air contains *more oxygen* and *less carbon dioxide* than does alveolar air. This occurs because expired air includes inspired air that remains in the larger air passages, where gas exchange cannot take place. The space containing the air left in the airways is known as **dead space,** a term applied to the passages of the pharynx, trachea, bronchi, and bronchioles. The air in the dead space is fresh, atmospheric air that does not reach the alveoli with each inspiration. Thus, expired air is a mixture of alveolar air and inspired air from the dead space. The composition of alveolar air is relatively constant, and only one fifth of the air in the lungs is renewed by each inspiration. In this way, sudden and marked changes in the composition of alveolar air are prevented.

In addition to differences in the percentage composition of oxygen and carbon dioxide, there are other physical differences among inspired, expired, and alveolar air. Expired and alveolar air are saturated with water vapor and warmed to body temperature. The amount of water vapor and the temperature of inspired air vary markedly from region to region and with seasonal changes. A loss of body heat is attributed to warming the inspired air whenever its temperature is below body temperature. Heat is also lost by saturating the inspired air with water vapor whenever its water-vapor content is low as it enters the respiratory passages.

Table 18-1
Concentrations of
Respiratory Gases

Type of Air	Oxygen	Carbon Dioxide
Inspired	20.93%	0.04%
Alveolar	14.00%	5.50%
Expired	15.70%	4.40%

Partial Pressures of Respiratory Gases

In the previous chapter we saw that the diffusion of gases takes place because of pressure gradients between the alveoli and the blood. The volume of oxygen in either air or blood is referred to as oxygen **tension,** or **partial pressure,** and is written as p_{O_2}. The tension, or partial pressure, exerted by a gas in a mixture of gases is equal to the number of molecules of that gas present. Thus total pressure of a mixture of gases is equal to the sum of the partial pressures of all the gases in the mixture (Dalton's law). The atmosphere exerts a pressure of 760 mm Hg at sea level and oxygen composes 20.93 percent of the gases in the atmosphere. Therefore, the partial pressure exerted by oxygen is one fifth of 760, or approximately 159.1 mm Hg under standard conditions. The partial pressure of carbon dioxide accounts for 0.04 percent, or 1/2500, or 760, which is 0.3 mm Hg. Thus the partial pressure of a gas is directly proportional to its concentration in the atmosphere.

Under ordinary conditions, inspired air contains very little water vapor, but as this air passes through the nose and pharynx, it becomes fully saturated with water. At normal body temperature (37°C), the partial pressure of water vapor is 47 mm Hg. Since the total gas pressure must equal atmospheric pressure, the pressure of air in the lungs measures 713 mm Hg (760 − 47) and is the sum of the partial pressures of oxygen, carbon dioxide, and nitrogen in the tracheobronchial tree and alveoli. Knowing the total gas pressure, the p_{O_2} of alveolar air would be 14.0 percent (see Table 18-1) of 713, or 100 mm Hg.

External respiration.   The blood entering the capillaries of the lungs has a lower p_{O_2} and a higher p_{CO_2} than the alveolar air. The rate of diffusion in the lungs involves the partial pressure of gases in the alveoli and in the blood. The pressure gradients then permit the diffusion of oxygen

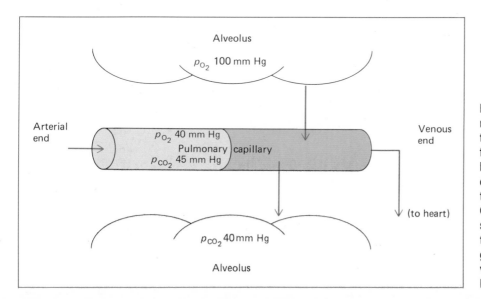

Figure 18-1   External respiration. Diagram of the diffusion of oxygen from the alveolus to the blood and of carbon dioxide from the blood to the alveolus. *Note:* Only in the pulmonary system do arteries transport poorly oxygenated blood (left) and veins highly oxygenated blood (right).

from the alveoli into the blood in the pulmonary capillaries and of carbon dioxide from the blood through the alveolar walls (Fig. 18-1). Under normal conditions the partial pressure of the gases in venous and arterial blood is basically similar. (See Table 17-2.)

Several factors can determine the amount of oxygen that diffuses into the blood of the capillaries of the lung. The rate of diffusion depends upon the oxygen pressure gradient between the alveoli and venous blood. Any situation that decreases the alveolar p_{O_2}, such as high altitude, will decrease the oxygen pressure gradient and thus the rate of oxygen entering the venous blood in the capillaries of the lungs. A second way in which oxygen diffusion may be reduced is by decreasing the surface area of the alveoli. For example, there is a decrease in the diffusion surface in *emphysema*, a disease that may be caused by smoking. This disease produces changes in the structure of the lung tissue that reduce the surface area available for respiration.

Internal respiration. The exchange of gases between the blood in the capillaries of the tissues and the tissue fluid depends upon the overall pressure gradients. The metabolic activities of cells consume oxygen and produce carbon dioxide. Thus the p_{O_2} of the surrounding tissue fluid will be lower than the p_{O_2} of the arterial blood, while the p_{CO_2} will be higher in the tissue fluid than in the arterial blood. Consequently oxygen will diffuse out of the blood and into the cells, while carbon dioxide will diffuse into the blood and be carried away (Fig. 18-2). As a result "arterial" blood becomes "venous" blood and returns to the heart for another trip to the lungs.

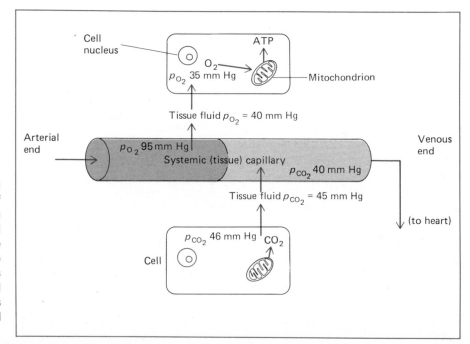

Figure 18-2 Internal respiration. Diagram of the diffusion of oxygen from the blood to the cell and of carbon dioxide from the cell to the blood. Systemic arteries carry highly oxygenated blood (left) and veins carry poorly oxygenated blood (right).

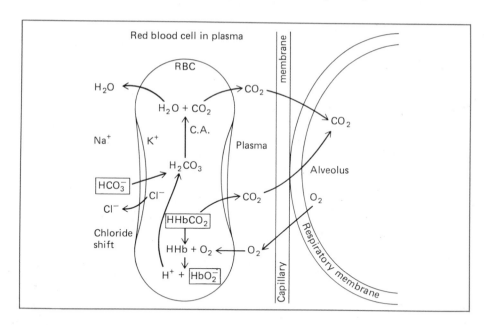

Figure 18-3 The exchange of gases between a red blood cell in a pulmonary capillary is an alveolus. Oxygen diffuses from the alveolus into the plasma and RBC. Carbon dioxide diffuses from the plasma and red blood cell into the alveolus. C.A. represents carbonic anhydrase, which speeds up the breakdown of bicarbonate ions into CO_2 and H_2O.

Diffusion of Gases into the Blood

On entering the blood, both oxygen and carbon dioxide form simple solutions in the plasma. Since there is a limit to the amount of gas that can dissolve in a given volume of a plasma (Henry's law), some oxygen and carbon dioxide combine chemically with other components of the blood.

Representations of these essential physical processes and chemical reactions are shown in Fig. 18-3 and 18-4. Several important facts are illustrated to assist in understanding the basic principles of gas transport. In Figure 18-3 the reactions proceed in a clockwise direction. As oxygen diffuses from its area of highest partial pressure (the alveolus) into an area of lower pressure (the plasma and the red blood cells), carbon dioxide simultaneously diffuses from its area of highest partial pressure (the blood) into an area of lower pressure (the alveolus) for exhalation.

In Fig. 18-4, carbon dioxide diffuses from its area of greatest partial pressure (the tissue cell) into an area of lower pressure (the plasma and red blood cells) for transport to the lung. If you follow the reactions in a counterclockwise direction, you will note the simultaneous diffusion of oxygen from its area of greatest partial pressure (the red blood cell) into an area of lower pressure (the tissue cell). In comparing the two figures, the conditions that have changed are the partial pressures of the gases and the place where gas exchange occurs. In the lung, oxygen is entering the blood while carbon dioxide is leaving. In the tissue cells, carbon dioxide moves into the blood while oxygen is leaving.

Oxygen transport.   The amount of oxygen that dissolves in the plasma is extremely small. Approximately 0.5 ml of oxygen will dissolve in

Figure 18-4 The exchange of gases between a red blood cell in a systemic (tissue) capillary and a tissue cell. CO_2 diffuses through the tissue fluid into the blood. A small amount is carried in solution in the plasma. CO_2 also enters the RBC, forming bicarbonate ions (HCO_{3-}). The majority of these ions leave the RBC and travel in the plasma as $NaHCO_3$; some ions remain in the RBC as $KHCO_3$. The remaining CO_2 combines with reduced hemoglobin (HHb) and is carried to the lungs as carbaminohemoglobin ($HHbCO_2$). Simultaneously, O_2 diffuses out of the bloodstream into the tissue cell. Negative chloride ions move into the RBC in a process called the chloride shift. (C.A. = carbonic anhydrase.)

hypo = below

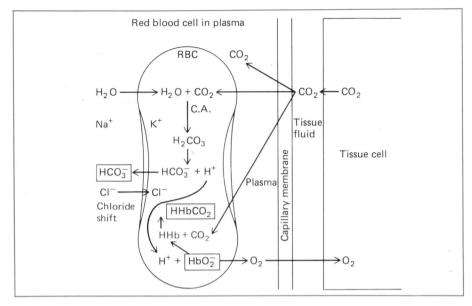

every 100 ml of blood. However, there is actually about 20 ml of oxygen in every 100 ml of blood. This additional amount of oxygen chemically combines with **hemoglobin (Hb),** the protein in red blood cells. Hemoglobin has the ability to combine chemically with oxygen to form **oxyhemoglobin:**

$$Hb + O_2 \rightleftharpoons HbO_2$$

One gram of hemoglobin can unite with approximately 1.3 ml of oxygen. With 15 grams of hemoglobin in 100 ml of blood, 20 ml of oxygen can be carried. This is more than the 0.5 ml of oxygen that is carried in the plasma. Hemoglobin that has lost its oxygen-carrying capacity is called *reduced hemoglobin* and can lead to a condition called **hypoxia.**

Carbon dioxide transport. Carbon dioxide diffuses from the tissues into the blood. Most of the carbon dioxide enters the red blood cells, with a small amount remaining in the plasma. Of this carbon dioxide in the plasma, some is carried in simple solution and some combines with water to form carbonic acid:

$$CO_2 + H_2O \rightleftharpoons H_2CO_3 \rightleftharpoons H^+ + HCO_3^-$$

| carbon dioxide | water | carbonic acid | hydrogen ion | bicarbonate ion |

Carbonic acid, if produced in large amounts, lowers the pH of the blood. **Buffers** are substances found in a solution that can prevent the accumulation of hydrogen ions (H^+). Since this reaction is not catalyzed by any enzyme, it is very slow and the free hydrogen ions are quickly *buffered* by the *plasma buffer systems.*

Acid-base balance. The major buffering of carbon dioxide occurs within the red blood cells. Within these cells is a special enzyme called **carbonic anhydrase.** This enzyme speeds up the reaction of carbon dioxide with water so that large quantities of carbonic acid are formed. Hemoglobin then serves as a buffer by combining with the excess hydrogen ions, thus preventing their accumulation in the blood plasma. The reaction can be written as follows:

$$\text{Hb} \quad + \quad \text{H}^+ \quad \longrightarrow \quad \text{HHb}$$

hemoglobin   hydrogen ions   reduced hemoglobin

The large numbers of negative bicarbonate ions (HCO_3^-) resulting from the rapid formation of carbonic acid diffuse rapidly through the cell membrane of the red blood cell. As a result, the interior of the cell loses its electrical neutrality. To reestablish the electroneutrality of the interior of the cell, negative chloride ions (Cl^-) in the plasma move to the interior of the red cells. This exchange of Cl^- for HCO_3^- is called the **chloride shift.** The process is reversed when blood reaches the lungs. (See Figs. 18-3 and 18-4.)

Carbon dioxide is also carried in combination with hemoglobin and forms a compound called **carbaminohemoglobin** ($HHbCO_2$). Thus hemoglobin functions to carry oxygen to the tissues and aids in the removal of carbon dioxide from the tissues.

hem = blood

BAROMETRIC PHYSIOLOGY

With the increasing ability to ascend to higher and higher altitudes in mountain climbing or by means of airplanes and space vehicles, it has become progressively more important to understand the effects of high altitude and low gas pressure on the human body. In contrast, descending beneath the sea or working in areas of high atmospheric pressures causes problems associated with excessively high gas pressures in the lungs.

Different organ systems of the body work together to maintain homeostasis as barometric conditions change. This is a good example of the interdependence of all the organ systems of the human body.

Effects of Low Pressure

Hyperventilation. As individuals are exposed to conditions found above 3 kilometers (10,000 feet), a series of physiological reactions occurs as the body tries to maintain homeostasis under new and, in this case, adverse conditions. The first response is *hyperventilation*, or an increase in the rate and depth of breathing, as the body tries to acquire more oxygen. Hyperventilation eliminates (by exhalation) an excessive amount of carbon dioxide and thus decreases the level of carbonic acid, which means that the hydrogen ion concentration of the plasma is also reduced. When the arterial p_{CO_2} is lowered, ventilation decreases. The decrease in breathing rate (lowering the H^+ concentration) raises the

hyper = excess

pH of the blood, but the kidneys restore the plasma pH to its normal level.

Chemical regulation.   The kidney responds to hypoxia by releasing a substance known as the **renal erythropoietic factor** into the blood. This substance causes a plasma transport protein to release a compound called **erythropoietin**. Erythropoietin then causes a more rapid production of red blood cells, and there is a significant rise in the number of red blood cells in the circulating blood. There is also an increase in the number of red blood cells released from storage areas such as the spleen. This increases the total hemoglobin content and thus the oxygen-carrying capacity of the blood. A slower reaction occurs in the muscles, where the oxygen-binding **heme protein** is **myoglobin**. In the muscles there is an increase in the production of myoglobin, which increases oxygen transport to the contracting muscle.

erythro = red

myo = muscle

There is some evidence that an extended exposure to high altitude results in an increased production of blood vessels in the lungs and other tissues of the body. Such an increase in the actual number of blood vessels allows the blood to deliver more oxygen to the tissue cells.

Effects of High Pressure

The three gases that a deep-sea diver breathes are usually nitrogen, oxygen, and carbon dioxide. However, helium is often substituted for nitrogen in the air mixture given to divers. Nitrogen in air breathed under high pressure can cause problems.

Breathing gases under pressure.   Approximately four-fifths of the air is nitrogen. At sea-level pressure the nitrogen has no known effect on the functions of the body. At high pressures, however, dissolved nitrogen becomes a problem. Divers must breathe air that is under greatly increased pressure to compensate for the high pressure of their surroundings. Under such conditions, excessive amounts of nitrogen dissolve in the plasma and in the tissue fluid. Such levels may produce giddiness and other effects similar to those produced by the consumption of alcohol. This condition, called **nitrogen narcosis**, can lead to unconsciousness. This effect has been seen not only in deep-sea divers but also in tunnel workers who work under high pressure.

osis = condition

Decompression sickness, also known as **caisson disease** or the **bends**, may follow too rapid an ascent or return to normal pressure. Nitrogen dissolved in the blood comes out of solution rapidly as pressure is lowered. It forms bubbles in the blood, which cause severe pain and discomfort in places where the bubbles become trapped. The most commonly affected areas are the muscles and joints. Treatment of decompression sickness depends upon *recompression*, a gradual rise in air pressure until excess nitrogen is slowly eliminated through the lungs.

Breathing oxygen under very high partial pressures can be detrimental to the central nervous system, causing acute oxygen poisoning. Exposure to 3 atmospheres of oxygen (p_{O_2} = 2280 mm Hg) will cause convulsions and coma in most individuals within one hour. These convulsions occur without warning and are likely to be lethal to a diver submerged beneath the sea.

PATHOLOGY OF THE RESPIRATORY SYSTEM

There are some common disorders that result from alterations in the transport or diffusion of the respiratory gases. These problems may be caused by the composition of inhaled air, improper ventilation, or a malfunction of a respiratory organ.

Cyanosis is characterized by a blue color of the skin and nail bed. The blueness of the skin occurs when there is too much reduced hemoglobin in the blood flowing through the capillaries. Cyanosis is a sign of failure of either respiration or circulation, or both. Since it is so readily observable, it is of great usefulness to the physician in diagnosing respiratory and circulatory disorders.

Hypoxia is the term used for inadequate oxygenation. It may refer to inadequate oxygenation of the air, the blood, or the cells. *Hypoxemia* specifically signifies a decreased oxygen saturation of the blood. A complete absence of oxygen is termed *anoxia;* absence of oxygen in the blood is called *anoxemia.* This extreme condition is rarely, if ever, encountered.

emia = blood

Polycythemia, a mild form of cyanosis, is a condition in which there is an abnormal increase in red blood cells. The cyanosis is caused by excessive amounts of reduced hemoglobin in the blood. There is so much hemoglobin that the amount not completely oxygenated is sufficient to give rise to a mild cyanosis. In addition, the great number of cells increases the viscosity of the blood and slows the flow of the blood, occasionally decreasing circulation.

SUMMARY

Respiratory gases are exchanged between alveolar air and venous blood in the lungs and between arterial blood and the cells in the tissues of the body. In the exchange of gases between alveolar air and venous blood, the air is exchanged in the lung capillaries. In the exchange of gases between arterial blood and cells, the air is exchanged in the tissue capillaries. The movement of the gases is caused by a diffusion.

The gases are transported by the blood as solutes and in combination with hemoglobin. About 0.5 ml of oxygen is transported dissolved in 100 ml of blood. About 19.5 ml oxygen per 100 ml blood is transported as oxyhemoglobin in the red blood cells. Carbon dioxide is transported in the blood in three ways: a small amount dissolves in plasma, most of the carbon dioxide is converted to bicarbonate ion by the enzyme carbonic anhydrase, and the remainder is carried in the red blood cells as carbaminohemoglobin.

VOCABULARY REVIEW

Match the statement in the left column with the correct word(s) in the right column. *Do not write in this book.*

1. A compound formed when hemoglobin unites with oxygen.
2. A chemical substance that helps to maintain the pH of the blood.
3. An enzyme concerned with the reaction of carbon dioxide with water.
4. Process maintaining the electrical neutrality of the red blood cell membrane.
5. The form in which carbon dioxide is carried by hemoglobin.
6. Causes rapid production of red blood cells in hypoxia.
7. When an inadequate amount of oxygen is available to the cells.

a. buffer
b. carbamino-hemoglobin
c. carbonic anhydrase
d. chloride shift
e. oxyhemoglobin
f. cyanosis
g. anoxia
h. hypoxia
i. polycythemia
j. reduced hemoglobin
k. erythropoietin
l. dead space

TEST YOUR KNOWLEDGE

Group A

Write the word or words that will correctly complete the sentence. *Do not write in this book.*

1. The process by which cells receive oxygen and dispose of carbon dioxide is called   (a) internal respiration   (b) external respiration   (c) inhalation   (d) exhalation.
2. Alveolar air, compared to inspired air, contains less   (a) carbon dioxide   (b) water vapor   (c) oxygen   (d) nitrogen.
3. Hemoglobin   in   contact   with   carbon   dioxide   forms   (a) oxyhemoglobin   (b) carbaminohemoglobin   (c) carbonic anhydrase   (d) carbonic acid.
4. The pH of the blood is normally about   (a) 3   (b) 6.6   (c) 7.4   (d) 10.
5. The acidity of the blood is held within a narrow range by a chemical reaction called   (a) chloride shift   (b) oxidation   (c) buffering   (d) neutralization.

Group B

Answer the following. Briefly explain your answer. *Do not write in this book.*

1. Briefly outline the chemical changes that occur in the blood when it is in contact with oxygen and carbon dioxide.
2. What is the significance of the chloride shift in CO_2 transport?
3. How does the body adapt to exposure to high altitude?

THE TRANSPORT SYSTEMS

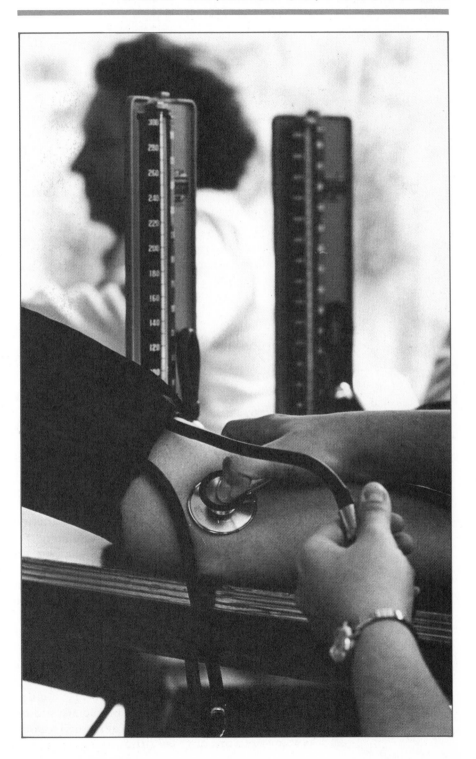

CHAPTER 19

THE BLOOD

Objectives

A. Describe the identifying features of erythrocytes and note their functions

B. Identify the various types of leukocytes and the role each plays in the body

C. List each of the components found in blood plasma

D. Explain the events that take place in hemostasis

E. Distinguish between blood types and identify blood groups

COMPOSITION OF BLOOD

The major role of the circulatory system is to maintain the most favorable conditions in the body's internal environment for optimal functioning of all activity. Through the distribution of heat, food, and oxygen, the removal of metabolic waste products, and the regulation of water content and the chemical composition of body fluids, the circulatory system maintains a relatively stable internal environment. This self-regulation and maintenance of a state of dynamic equilibrium is another example of **homeostasis**.

The cells of the human body are organized into tissues and organs that together carry on all of the body's functions. In order for the body to function as a unit, there must be a system to transport materials between different organs and tissues. This function is served by the circulation of *blood* and by the mechanism that moves it—the **cardiovascular system**. In this chapter the cellular components of blood—the red blood cells, or **erythrocytes**, white blood cells, or **leukocytes**, and platelets, or **thrombocytes**—will be described as well as their transporting liquid, the **plasma**. In Chapter 20, the anatomy and physiology of the heart will be examined. The heart functions as the pump in the cardiovascular system. The intricate system of passageways that makes up the vascular portion of the cardiovascular system is considered in Chapter 21.

The system of vessels that connects with the veins of the cardiovascular system is the **lymphatic system.** This system is a one-way transporting network of vessels that returns excess interstitial fluid from the

leuko = white
thrombo = clot
cyto = cell

tissues to the bloodstream and that plays a role in the body's defense against infection (discussed in Chapter 22).

Blood Volume

Blood is a type of *liquid connective tissue* in which 55 percent of the total volume is the fluid plasma, with the remainder consisting of the formed elements. This fact can be easily demonstrated by placing a specimen of whole blood in a test tube with a small amount of oxalate to prevent clotting. If the sample is centrifuged or allowed to stand for a sufficient length of time, it will be found that the blood cells will settle toward the bottom of the test tube while the plasma remains on top. By this means, the percentage of red cells in whole blood, known as the *hematocrit,* can be determined. The normal hematocrit is approximately 45 percent. (See Fig. 19-1.) The total circulating blood volume makes up about 8 percent of the body's weight. Accordingly, a 70-kilogram (152-pound) person will have 5 to 6 liters (10.5 to 12.6 pints) of blood. Blood is a somewhat viscous and sticky liquid. Its viscosity is about five times that of water. It has a characteristic odor, a salty taste, and a pH value of from approximately 7.35 to 7.45 (slightly alkaline).

Formed elements.   The most common formed elements in blood can be grouped according to the following list.

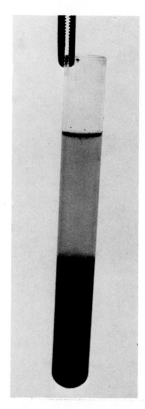

Figure 19-1 **When a blood sample is centrifuged, it separates and its different components can be identified.**

- Erythrocytes (red blood cells)
- Leukocytes (white blood cells)
    Granulocytes (polymorphonuclear granular leukocytes)
       Neutrophils
       Eosinophils
       Basophils
    Agranulocytes (monomorphonuclear nongranular leukocytes)
       Lymphocytes
       Monocytes
- Thrombocytes (platelets)

Formation of Blood Cells

The erythrocytes, the leukocytes, and the thrombocytes are formed in the red marrow of many bones. The lymphocytes and monocytes, moreover, are also formed in lymphatic tissue. Because the formed elements of the blood are constantly being worn out, they must be replaced by the body at a corresponding rate. The formation and development of blood cells is called *hemopoiesis.*

Two different types of marrow are found in bones. The **red marrow,** found in spongy bone, owes its name to the presence of a large amount of blood. It is the tissue responsible for the formation of erythrocytes, granulocytes, and thrombocytes. The *yellow marrow* is primarily composed of fat and is not capable of blood-cell formation and development. Before birth, blood cells are formed in the yolk sac, liver, spleen, thymus

hemo = blood

gland, lymph nodes, and red bone marrow of the fetus. All bones contain red marrow at birth. At about age 7, some of the red marrow becomes fatty and incapable of forming blood cells, especially in the bones of the extremities and to a certain extent in other bones. After age 20, blood-cell production only occurs in the red marrow of the upper ends of the humerus and femur and in the bones of the cranium, ribs, sternum, scapulae, clavicles, vertebrae, and ilia.

Blood cells arise in the red marrow from **mesenchymal stem cells,** which undergo a process of cell division and differentiation. In successive steps the stem cells give rise to the **hemocytoblasts,** which eventually form the various cellular elements of the blood. The development of the various types of blood cells is illustrated in Fig. 19-2. **Erythroblasts** eventually develop into erythrocytes; **myeloblasts** differentiate into the various kinds of granulocytes; and **megakaryoblasts** develop into megakaryocytes, the cytoplasm of which breaks up into fragments forming platelets. Some stem cells migrate to lymphatic tissue, where they form **lymphoblasts.** In turn, these primitive cells develop into lymphocytes and give rise to **monocytes.**

erythro = red

During the changes that occur in the cells of the red marrow, the future erythrocytes lose their nuclei, and the leukocytes develop cytoplasmic structures that will later differentiate them into their various types. When blood-cell formation is occurring in the bone marrow, the blood vessels in which the cells form are plugged by the new blood cells. Within these tubes the immature cells pass through their various stages. When the blood cells are mature, the vessels open and discharge their contents into the general circulation.

Erythrocytes

Mature red blood cells, also known as **red corpuscles,** normally lack nuclei and have a life span of about 120 days. Just before they reach maturity, erythrocytes are known as **reticulocytes** (see Fig. 19-2); some of these may escape into circulation. The number of reticulocytes in the adult human averages about 0.8 percent of the total number of erythrocytes in the circulating blood. Their percentage in a blood smear is a dependable index of the red-blood-cell formation rate.

The number of red blood cells circulating through the body remains remarkably constant. Every second of every day, millions of red corpuscles wear out and are destroyed, and millions of new ones are formed to take their place. The number of red blood cells per cubic millimeter (mm^3) of blood can be determined by counting a limited number of cells spread on a **hemocytometer,** a ruled microscopic slide. The red-blood-cell count is approximately 5,450,000/mm^3 for healthy males and 4,750,000/mm^3 for healthy females. The higher value for males is due to their higher metabolic rate. The total number of erythrocytes in an adult is estimated at over 30 trillion.

The process of erythrocyte production is called **erythropoiesis.** Under certain conditions the body may suddenly need more erythrocytes. For

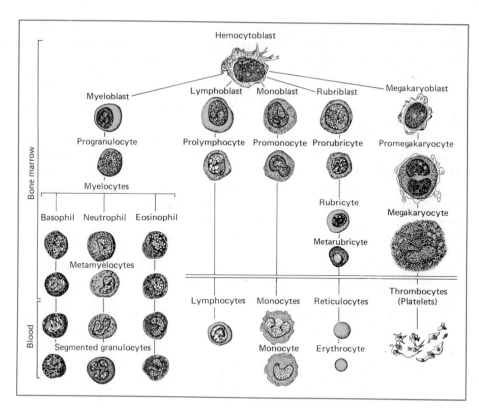

Figure 19-2 Origin, development, and structure of blood cells.

example, muscular exercise, ascending to high altitudes, and a rise in the environmental temperature may require a homeostatic mechanism to step up erythropoiesis. The stimulus for such a mechanism is oxygen deficiency in kidney cells, which release a hormone called **erythropoietin.** This hormone circulates through the bloodstream and activates the hematocytoblasts in red bone marrow to initiate the production of erythrocytes.

poiesis = to make

The normal red blood corpuscle looks round when viewed from above, but biconcave when seen from the side. As seen in Fig. 19-3, the red blood cell has a diameter of between 6.5 and 8 μm, or micrometers, and averaging 7.5 μm (1 micrometer = 0.000,001 meter). At the margin, the cell measures about 2.2 micrometers in thickness, but the center is only approximately half as thick. This inequality in thickness causes red blood cells to appear doughnut-shaped. The shape of these cells is of great importance from the standpoint of their function. It has been calculated that the erythrocyte has a surface area of about 140 square micrometers, which presents to the plasma a surface 20 to 30 percent greater than the corpuscle would were it a sphere.

The membrane that surrounds the corpuscle is complex. It is made up of protein and lipid molecules and is readily permeable to water, urea, and salt. From high-speed motion picture studies made of red blood cells flowing through capillaries, it has been found that the envelope of the corpuscle is highly elastic. A capillary is the smallest vessel in the body both in diameter and length and is the link between arteries that carry blood from the heart and veins that return blood to the heart. Many of the capillaries in the tissues are smaller in diameter than the corpuscles themselves. The result is that when they pass through these minute tubes, they assume a conelike shape, with the apex of the cone pointing forward; the trailing side is then concave. Just what effect this change in shape has on the exchange of gases between the tissues outside the capillaries and the corpuscles inside is still a question. It may, however, hasten the exchange by increasing the surface area that is in direct contact with the capillary wall.

Hemoglobin.   The red, oxygen-carrying pigment in red blood cells is **hemoglobin.** It is composed of the protein *globin* a *polypeptide*, and the pigment *heme*. A molecule of hemoglobin has a molecular weight of 64,450. The ability of hemoglobin to combine with oxygen is due to the four iron atoms associated with each heme group within the molecule. Each hemoglobin molecule can readily combine with four molecules of oxygen, attaching one oxygen molecule to each of the iron atoms. Oxygen combines weakly with the hemoglobin molecule. When saturated with oxygen, hemoglobin is called *oxyhemoglobin.* When this oxygen is released to the tissues of the body, the hemoglobin is called *reduced hemoglobin* or *deoxyhemoglobin.* It has been estimated that one erythrocyte contains approximately 280 million hemoglobin molecules.

About 20 different types of hemoglobin have been identified in human beings. Some of these will be discussed later, but mention should be made now of a change that normally occurs during a person's life.

Figure 19-3 (a) Electron micrograph of a RBC. (b) RBC's can change their shape as they pass through vessels.

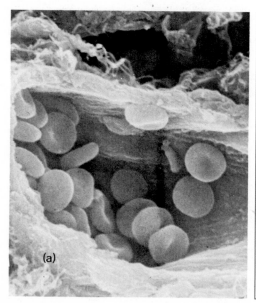

(a)

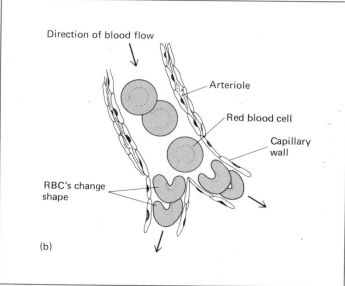

Direction of blood flow

Arteriole

Red blood cell

Capillary wall

RBC's change shape

(b)

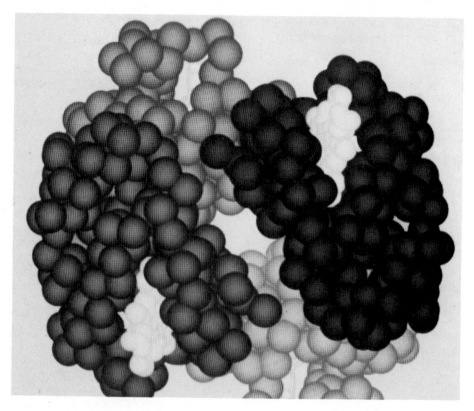

Figure 19-4 **A structural model of hemoglobin.**

Before birth, the fetus has a type of hemoglobin that is different from that of the adult. This fetal hemoglobin (HbF) differs from the adult type in that it can be saturated with oxygen at a lower oxygen tension. This is a special adaptation of the fetus, which must receive all of its oxygen through the maternal circulation. The HbF is replaced by adult hemoglobin during the first few months after birth.

Leukocytes

White blood cells, or leukocytes, are less numerous than red blood cells and number from 5,000 to 9,000 per cubic millimeter of blood. On the average, red blood cells outnumber the leukocytes by about 700 to 1. The size and shape of leukocytes vary. All of them are nucleated. Two kinds of leukocytes, the neutrophils and monocytes, exhibit ameboid movement and can actively leave the blood vessels. These cells can squeeze through the walls of capillaries and enter the surrounding tissues by a process called *diapedesis*. This process occurs normally but is greatly increased when stimulated by various pathological conditions, such as infections. On arriving in an area of infection, white cells move toward the invading bacteria and send out projections that surround and engulf the bacteria. Once inside the leukocyte, the bacterial cell is digested and rendered harmless. This process is known as *phagocytosis*, and cells that perform this function are called *phagocytes*.

phago = eat

Table 19-1 The Cellular Elements of the Blood

Name	No./mm^3	Diameter (micrometers)	Characteristics and Function
Erythrocytes		7.2–7.7	No nuclei, biconcave disk. Transport O_2.
Male	5,450,000		
Female	4,750,000		
Leukocytes	5,000–9,000		
Granulocytes			
Neutrophils	3000–6000	7–9	Fine granules that stain pale lavender. Multilobed nucleus. Phagocytic.
Eosinophils	300	9–10	Large granules that stain red with eosin. Bilobed nucleus. Phagocytosis of antigen-antibody complexes.
Basophils	25–100	7–9	Granules stain blue with basic dye. Exact function unknown, but store histamine.
Agranulocytes			
Lymphocytes	1500–3000	6–8	Large single nucleus with scant cytoplasm. Synthesis of antibodies.
Monocytes	500	14–20	Large single nucleus with considerable cytoplasm. Very motile and phagocytic—form macrophages.
Thrombocytes	250,000–500,000	2–4	Nonnucleated fragments of megakaryocytes, disc-shaped. Contain serotonin and several clotting factors. Aid in coagulation of blood.

The types of white corpuscles can be identified by different methods of staining. The most commonly used stain for blood smears is *Wright's stain*. When a blood smear is stained with Wright's stain, the nuclear materials of the cells are colored a deep blue or purple, any acidic cytoplasmic granules are stained red, and basic granules stain blue. In some leukocytes, cytoplasm doesn't stain and is nearly transparent.

On the basis of their staining reactions, white blood cells are classified into two major groups. Those with cytoplasmic granules and lobed nuclei are called **granulocytes** or *polymorphonuclear cells*. The granulocytes are formed in red bone marrow and are subdivided into three types on the basis of the staining reaction of their granules. These are the *neutrophils, eosinophils,* and *basophils.* The leukocytes that are free of cytoplasmic granules and have but a single, large nucleus are called **agranulocytes.** The agranulocytes are subdivided into two groups. The small white blood cells are the *lymphocytes* and are formed in lymphatic tissue. The *monocytes* are the large white blood cells and are formed in bone marrow.

Table 19-2 Chart of Plasma Composition

Component	Normal Range	Description
Water	90–92%	Water acts as solvent and suspending agent for plasma solids.
Plasma proteins	6–8%	
Albumins		Most abundant plasma protein. Helps maintain osmotic pressure. Maintains viscosity.
Globulins		May help in osmotic balance. A group of important antibodies.
Fibrinogen		A small fraction of plasma proteins but is essential in clotting of blood.
(The remaining solutes make up 1.5%)		
Inorganic salts (electrolytes)		Anions: Cl^-, PO_4^{---}, SO_4^{--}, HCO_3^-. Cations: Na^+, K^+, Ca^{++}, Mg^{++}. Salts help to maintain normal pH balance.
Nutrients		Transport of foods in solution.
Glucose		
Amino acids		
Fats (lipids)		
Waste materials		Breakdown products of cellular metabolism. They are carried to organs of excretion.
Urea		
Uric acid		
Creatine		
Creatinine		
Ammonium salts		
Respiratory gases		A trace dissolved in plasma, primarily carried by hemoglobin.
Oxygen		
Carbon dioxide		
Regulatory substances		Organic catalysts for cellular metabolism.
Enzymes		
Hormones		Produced by endocrine glands for distribution.
Vitamins		

Thrombocytes

Blood *platelets,* or thrombocytes, are the smallest of the formed elements in blood. They are small, granular, nonnucleated, irregularly shaped objects. Their size varies from 2 to 4 μm in diameter and they number between 250,000 and 500,000 platelets in each cubic millimeter of blood. As Fig. 19-2 shows, they arise from a fragmentation of the cytoplasm of megakaryocytes, which are formed in the red bone marrow. As considered in a later section, platelets are important in preventing loss of blood by initiating a chain of reactions that results in blood clotting. Table 19-1 will help to identify all of the different formed elements of blood.

Plasma

The liquid portion of the blood, the *plasma*, is a complex solution that performs many of the functions of the circulatory system. The normal volume of plasma in a 70-kilogram adult male is about 5 percent of the body weight, which amounts to approximately 3.5 liters. This fluid is transparent and amber in color. Basically, plasma is 90 to 92 percent water, in which is dissolved a variety of inorganic and organic substances. A number of ions constitute the inorganic salts in plasma. They include the cations of sodium (Na^+), potassium (K^+), calcium (Ca^{++}), and magnesium (Mg^{++}) and such anions as chloride (Cl^-), phosphate (PO_4^{---}), bicarbonate (HCO_3^-), and others.

Plasma proteins are found in the plasma and include **albumin, globulins, fibrinogen,** and **lipoproteins.** In addition, plasma normally contains varying amounts of hormones, enzymes, pigments, and vitamins. Some of the major constituents of plasma are shown in Table 19-2. The composition of plasma varies with the body's activity and different physiological states. For example, following a meal plasma contains an increased amount of nutrient materials. After a period of strenuous exercise, plasma contains wastes that the blood has removed from active tissues. Thus from one moment to the next, plasma reflects the physiological changes in the body.

PHYSIOLOGY OF THE BLOOD

The functions of the blood are closely tied to the functions of all the body's organ systems. Blood transports all materials that are required by the cells or discarded by them.

Transport Functions

The gases involved in respiration are carried by the blood. Oxygen is carried from the lungs to the cells in chemical combination with hemoglobin. Carbon dioxide is carried from the cells to the lungs in various chemical combinations. Some of the CO_2 dissolves directly in the plasma; a larger percentage changes into bicarbonate ions (HCO_3^-), which diffuse into the plasma, and hydrogen ions (H^+), which are buffered in the red blood cells. About 25 percent of the CO_2 reacts with the amino groups of hemoglobin to form carbamino compounds. In the lungs, carbon dioxide is discharged into the alveoli and is exhaled.

Foods that have been rendered soluble by digestion are absorbed into the plasma and are transported throughout the body. In a similar manner, soluble wastes are removed from the cells and carried by the plasma to those organs that can either eliminate them or change them into compounds that are useful for other purposes. The glands that manufacture chemical regulators of bodily functions—hormones—empty their secretions directly into the blood for transportation around the body.

Maintenance of water content.   Controlling the water level in tissues is one of the principal functions of the circulatory system. If the water content of a particular region of the body is lowered as a result of increased chemical activity, the blood makes up this deficit. The circulatory system then draws on reserve supplies of water from other regions to maintain its own fluid level.

Maintenance of body temperature.   The blood as a whole helps in stabilizing body temperature. If some tissue, such as muscle, is very active, the increased rate of oxidation occurring in it raises its temperature. Continued exposure to high temperature has a serious effect on enzymatic reactions in the cells. It is therefore essential to bring the temperature of the tissue to a normal point as rapidly as possible. Thus any increase in the activity of a tissue is also accompanied by an increase in the circulatory rate, and the blood passing through it is heated. This excess heat in the blood is then eliminated through the surface of the body, the lungs, or excretions.

Protection and acid-base balance.   Protection against disease is another important function of the blood. This is accomplished by the white blood cells in one of two ways. First, certain granulocytes, primarily neutrophils, act as phagocytes and engulf bacteria. Second, still other types of white corpuscles have the ability to manufacture chemical substances (antibodies) to destroy bacteria or to neutralize the toxins (poisons) produced by invading organisms.

The blood also plays an important part in maintaining the *acid-base balance* of the tissues. This is achieved by the buffering action of bicarbonates, phosphates, and hemoglobin, which neutralize the usually small amounts of acids or alkalis that the blood absorbs (see Chapter 18).

Hemostasis

In the event of an injury involving blood loss, a number of events take place to arrest the blood flow. Specific reactions must occur to control and maintain **hemostasis.**

The reaction of small vessels that have been broken or injured is to constrict immediately. The smooth muscles in their walls go into spasm, which may be intense enough to close the injured vessel completely. As a result of the initial spasm, the endothelial cells lining the vessel stick together. The vessel remains closed as if "glued together" even after the muscular wall begins to relax.

While they do not adhere to normal smooth endothelial lining cells, platelets will adhere to a rough surface, such as a broken vessel. Undoubtedly the broken vessel exposes some collagen, onto which the platelets immediately stick. These clumped platelets attract others until an aggregate, or plug, of platelets is formed. The platelets that form

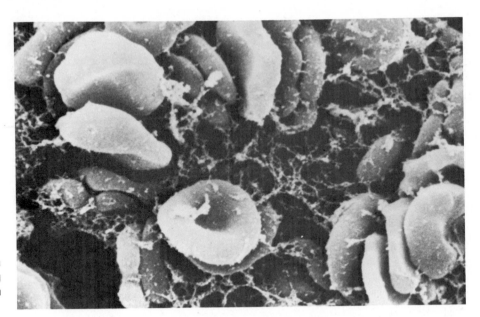

Figure 19-5 Electron micrograph showing blood cells enmeshed in fibrin threads.

the plug release two powerful vasoconstrictors called **serotonin** and **epinephrine.** These chemicals prolong the constriction of the injured vessels.

Blood clotting. Coagulation of blood is the most effective and complex of the hemostatic mechanisms. In blood-clot formation, blood loses its fluid properties and becomes a semisolid mass similar in consistency to gelatin. In the process of blood coagulation, soluble **fibrinogen** is converted into an insoluble protein, **fibrin,** by the enzyme **thrombin.** Fibrin threads form a meshwork that traps blood cells and plasma—the clot. The clotting mechanism that leads to the formation of fibrin involves a complex series, or "cascade," of reactions that involve at least twelve factors.

Formation of the clot. There are three stages in clot formation.

1. Thromboplastin triggers the clotting mechanism by converting prothrombin into thrombin. Some thromboplastin is released when platelets disintegrate at the site of vascular injury—the **intrinsic pathway** for thromboplastin release. Additional thromboplastin is released by the injured tissues themselves, especially vascular walls, in the presence of calcium ions (Ca^{++})—the **extrinsic pathway.**
2. In this stage prothrombin is converted to thrombin by both intrinsic and extrinsic thromboplastin and several plasma coagulating factors. Again, calcium ions are necessary, along with vitamin K.

$$\text{prothrombin} \xrightarrow{\substack{\text{thromboplastin} \\ + \text{ calcium ions}}} \text{thrombin}$$

3. In this stage fibrinogen, which is soluble, is converted to fibrin, which is insoluble, by the thrombin formed in stage 2. Additional plasma coagulation factors are also involved.

$$\text{fibrinogen} \xrightarrow{\quad\text{thrombin}\quad} \text{fibrin}$$
$$\text{(soluble)} \qquad\qquad\qquad \text{(insoluble)}$$

The insoluble fibrin forms a meshwork of fine protein threads that traps erythrocytes—the clot (Fig. 19-5). Since red blood cells have nothing to do with the chemical activities involved, clotting will occur in plasma even when the blood cells have been removed. The formation of the clot occurs in 5 to 8 minutes. Shortly thereafter, the clot appears to become denser and stronger by the pulling back, or retraction, of the fibers. Retraction occurs because platelets contain certain contractile proteins. The platelets pull the fibrin fibrils together and squeeze out the remaining fluid—the **serum.** This has the advantage of pulling the vessel walls closer together at the site of the clot. Once hemorrhaging is stopped, tissue repair can begin.

Physical Factors of Coagulation

Clotting of blood is affected by physical factors. A surface that can be wetted by water hastens the process, while one that prevents molecules of water from clinging to it retards clotting. For example, if blood is carefully removed from a vessel, so that it does not come in contact with injured tissues, and then placed in a clean glass tube, a clot will appear in a short time. If, however, the inside of the tube is coated with a layer of paraffin or wax (which water will not cling to), the clotting time will be greatly prolonged.

There appears to be a rapid destruction of the blood platelets when they contact a rough surface. Since all tissues are in contact with blood, a cut tissue surface itself aids in the formation of a clot. The process can be hastened by placing a piece of gauze or other cloth with a relatively rough surface over a wound. This further stimulates the formation of a clot.

Temperature also affects the rate of clotting. At a relatively high temperature, blood will clot easily, while under carefully controlled refrigeration, blood will remain fluid for a longer period of time.

The fact that blood does not normally clot in the blood vessels has been attributed to **anticoagulants** in the blood as well as to the smooth surface of healthy endothelial cells lining the blood vessels. One naturally occurring anticoagulant is **heparin.**

Heparin acts as an anticoagulant by interfering with the ability of thrombin to catalyze fibrinogen into fibrin. However, under normal conditions, heparin is not present in the blood in sufficient quantities to be detectable as an anticoagulant. Therapeutic doses of heparin reduce clot formation and are therefore very valuable in treating patients after intravascular clotting has already occurred.

Plasma Proteins

Plasma proteins (other than fibrinogen) are classified as either **globulins** or **albumins.** These terms are used to designate certain types of protein molecules. Both of these types of proteins are widely distributed in plant and animal tissue. A good example of an albumin that is not a constituent of blood is that in the white of an egg, while a globulin is present in the egg yolk. When fibrinogen is removed, the plasma, minus its clotting proteins, is called *serum.* From the serum, the albumins and globulins can be chemically separated and are found to constitute approximately 7 percent of the sample.

Serum albumin. Serum albumin is more plentiful. Its principal function is to control the **osmotic pressure of the blood** . As blood flows through the capillaries within the tissues, its osmotic pressure is highest just after the capillary branches from an arteriole. From that point onward through the capillary, the osmotic pressure slowly decreases. The result of this gradient in pressure is that food and oxygen leave the blood early in its passage through the capillary. Then waste materials pass into the blood as the capillary joins others to form a venule. Normally, the blood proteins do not pass through the capillary walls because of their relatively large size. However, since they are colloidal materials, they can give up, or take up, water-soluble substances.

If there is a decrease in total serum albumin, fluids leave the blood throughout the entire length of the capillaries and the surrounding tissues fill with fluid and become enlarged. Such a condition is known as *edema.* Certain types of kidney disorder result in the abnormal removal of serum proteins from the blood, with resulting edema.

Gamma globulin. The serum globulin helps in the maintenance of osmotic pressure and has other functions to perform. Among the substances that have been extracted from serum globulin is **gamma globulin.** Associated with this substance are the *antibodies* that protect against invasion by bacteria and viruses. For example, if a person has had measles, certain chemical materials in the blood will combat another attack of this disease. These materials are present in the serum globulin as definite antibodies that will neutralize the effects of a future invasion of the body by the same organism.

Acid-Base Equilibrium

One factor in maintaining a constant internal environment in which the cells can function normally is to control the amount of acids and bases present. Metabolic processes are constantly forming acid waste materials, primarily in the form of CO_2. The blood normally keeps the amount of these substances within limits that will permit the cells to function properly. Except under abnormal conditions, such as those arising during disease, the hydrogen ion concentration (pH) of the blood remains at an almost constant level.

There are three factors that help the blood to maintain its pH value:

1. the action of buffers in the blood
2. the excretion of carbon dioxide by the lungs
3. the excretion of fixed acids by the kidneys.

A buffer system is any solution that can adjust to large changes in pH when an acid or base is added. If a solution contains chemical compounds that form weak acids or salts, these compounds "cushion" or buffer the effects of acids or bases that enter the solution. In Table 19-2, there is a list that includes some of the more common inorganic substances in plasma. They include cations of such chemical elements as sodium and potassium, which may combine with bicarbonate ions to form such compounds as sodium bicarbonate ($NaHCO_3$) or compounds of sodium and carbonates and phosphates. These compounds act as buffers by changing their composition rapidly in response to the amounts of hydrogen (H^+) or hydroxyl (OH^-) ions in the blood.

Bicarbonates in the blood will react to neutralize such acids as lactic acid, which is produced by muscular exertion.

lactic acid   sodium bicarbonate   sodium lactate   carbonic acid

$$HLa \quad + \quad NaHCO_3 \quad \longrightarrow \quad NaLa \quad + \quad H_2CO_3$$

Carbonic acid is a weak acid and therefore is a more stable compound. Under normal conditions, therefore, the pH of the blood is maintained at a level of 7.4 by these chemical reactions.

In addition to the inorganic materials present in the plasma, the blood proteins also act as buffers. As previously pointed out, proteins are highly complex compounds containing different components that can ionize in aqueous solutions. These components react to shifts in the blood pH in much the same way as the inorganic compounds. These reactions result in the return of the pH of the blood to its normal level.

The exchange of oxygen and carbon dioxide in the tissues involves constant changes in the acidity of hemoglobin. The compound **oxyhemoglobin** is relatively acid. When blood containing a large amount of oxygen in the form of oxyhemoglobin passes through a tissue that is deficient in oxygen, hemoglobin changes chemically to release oxygen, allowing it to diffuse into the tissue. The hemoglobin is therefore more alkaline and accepts the acidic carbon dioxide that the tissue has formed. The carbon dioxide is then carried to the lungs, where the conditions are reversed. Here the blood encounters a region that is rich in oxygen, and hemoglobin once more becomes more acid as it accepts the oxygen while giving up the carbon dioxide it has taken on in the tissues. These reactions are buffered principally by the presence of bicarbonates in the plasma and in the red blood cells.

Blood in active tissue absorbs an extra amount of carbon dioxide resulting from oxidation in the tissue. Carbon dioxide combines with water in the plasma to form carbonic acid. When blood carrying excessive amounts of carbon dioxide passes through the respiratory center of the medulla, the center is stimulated and the rate of respiration increases. The excess carbon dioxide is thus removed and does not greatly affect the pH of the blood.

Blood Types

The most successful method of combating excessive loss of blood is transfusing whole blood into the circulatory system of the victim. Usually, whole blood is superior to any blood substitute or component part of blood. Recovery from blood loss is hastened to a marked degree when whole blood is used.

To transfuse blood from one individual to another, the blood types of both must be known. The blood of the **recipient,** the patient who receives blood, and the **donor,** the person from whom blood is taken, must be able to mix safely, that is, to be *compatible.* The mixing of unsafe, or *incompatible,* blood may lead to very serious or possibly fatal transfusion reactions. The problem occurs when the protein outer layer of the red corpuscles of the donor becomes "sticky" when introduced into an antagonistic plasma of the recipient. The "sticky" red blood cells clump together and block the vessels of the patient. This clumping of blood cells is called **agglutination.** The plasma proteins concerned in agglutination are antibodies called **agglutinins,** while the proteins in the walls of the corpuscles are antigens, called *agglutinogens.*

ABO groups.   The blood group of each person is designated as A, B, AB, or O, depending on the kind of agglutinogen (antigen) present on the red blood corpuscles. Antigen A is present in group A, antigen B is present in group B, both antigens A and B are present in group AB, and neither antigen is present in group O individuals. This blood trait is inherited, with A and B being dominant, while O is recessive (see Chapter 30). This system of nomenclature was suggested by Dr. Landsteiner of the Rockefeller Institute of Medical Research in New York. He also found that there were two types for agglutinins (antibodies) in plasma. These he called **alpha** and **beta,** or more simply *a* and *b.* If a person, for example, belongs to group A, the erythrocytes contain antigen A and the plasma automatically contains antibody b. Since b is a protein that is antagonistic to the erythrocyte protein antigen B, transfusion from a B donor to an A recipient would result in the agglutination of the B type of red cells. Likewise, a B individual has antibody a, and the anti-A protein in the plasma prevents the person from receiving blood from anyone in group A.

Table 19-3 summarizes the distribution of the A and B agglutinogens (antigens) and the a and b agglutinins (antibodies). It also shows what types of blood are compatible and may be used in transfusions.

Transfusion reactions.   The terms *universal donor* and *universal recipient* are frequently applied to members of groups O and AB, respectively.

In cases of dire emergency, group O blood can be transfused into most people without ill effects. Although the plasma of this type of blood contains both a and b agglutinins, it does not cause agglutination. When transfused these antibodies become so dilute in the recipient's plasma that they are ineffective. It is the destruction of the transfused cells that produces the transfusion reaction. Since group O corpuscles lack agglutinogens, they will not clump in the presence of either a or b agglutinins. In like manner, a person with group AB is a universal recipient because the blood lacks any agglutinins that affect the corpuscles of the donor.

To prevent transfusion of blood that is not compatible, the blood of the donor and the recipient must be properly classified. As a safeguard, since there are a number of other incompatible erythrocyte antigens and plasma antibodies besides those of the ABO groups, the blood of the recipient and the donor should be **crossmatched** before transfusion. In crossmatching the donor's and recipient's bloods are mixed together. If the bloods are compatible, no agglutination will occur.

The distribution of blood groups varies considerably among the peoples of the world. Blood group is an inherited characteristic. Table 19-3 shows how the different groups are distributed in one population.

Table 19-3
Blood Groups and Corresponding Antigens and Antibodies

Blood Group	Antigens of Red Cells	Antibody in Plasma or Serum
A	A	Anti-B
B	B	Anti-A
AB	AB	None
O	None	Anti-A and anti-B

Table 19-4
Blood Grouping

Type	Percentage of Population	Red Cell Antigens or Agglutinogens	Plasma Antibodies or Agglutinins
ABO			
A	41%	A	Anti-B
B	10%	B	Anti-A
AB	4%	A, B	
O	45%		Anti-A, anti-B
Rh(D)			
Positive	85%	Rh	
Negative	15%		

Rh factor.   There are several other proteins in the blood that may bring about agglutination under certain conditions. The most important of these is the **Rh factor,** which, like the blood type, is an inherited characteristic. The letters "Rh" are used since this factor was first studied in the rhesus monkey. A number of different kinds of Rh antigens have been discovered, but by far the most important is antigen D. In the United States, approximately 85 percent of the people have the Rh factor, and are said to be Rh positive (Rh^+). This means that these individuals have antigen D on the surfaces of their red blood cells, as well as one of the antigens of the ABO system. The 15 percent or so of the population who do not have antigen D on their erythrocytes are said to be Rh negative (Rh^-). These individuals will produce antibodies against that antigen if given a transfusion of blood which is Rh^+. These antibodies remain in the Rh^- individual so that a later transfusion of Rh^+ blood results in a severe reaction. Testing for the presence of antigen D is done in much the same manner as for the general blood type, as described earlier.

Hemolysis.   Occasionally, damage to an infant's blood occurs before birth as the result of the presence or absence of the Rh factor. If the father of a child is Rh^+ and the mother Rh^-, the child has a genetic possibility of being Rh^+. If the baby is Rh^+ during the mother's first pregnancy, there are no complications. However, the mother has now developed antibodies against the Rh factor. Should she become pregnant again and again carry an Rh-positive child, her antibodies can cross the placenta into the fetus, whose erythrocytes they attack, causing **hemolysis,** which is the breakdown of erythrocytes and the release of hemoglobin into the plasma. The released hemoglobin may damage many organs, including the brain. This serious effect on the child is called **erythroblastosis fetalis,** a condition characterized by severe anemia and jaundice. If the child is alive at birth, the condition is treated by means of numerous transfusions by which the child's blood is replaced with Rh^- blood. The antibodies are thus removed.

blast = a sprout

BLOOD DISORDERS

A deficiency in the number of erythrocytes or a reduced hemoglobin level is known as *anemia.* From a physiological standpoint, anemia results from one of the following principal causes:

1. an acute or chronic loss of blood (hemorrhage) that reduces the number of red corpuscles
2. the excessive destruction of red corpuscles or the presence of abnormal hemoglobin—for example, as in sickle-cell anemia
3. a decrease in red-blood-cell production because bone marrow fails to function normally or because of nutritional deficiency.

Types of Anemia

Regardless of the immediate cause of anemia, there is always a lack of hemoglobin, with a corresponding deficiency in the oxygen supply to the body. The typical person suffering from anemia appears pale, weak, and has little energy. There is labored breathing, rapid heart rate, and intolerance to cold.

Hemorrhagic anemia.   Excessive loss of red blood cells—whether from large wounds in a short period of time (acute), from prolonged bleeding from gastric ulcers, from abnormally heavy menstrual bleeding, or from other chronic conditions—is called *hemorrhagic anemia.*

Immediately following sudden loss of blood, the arterial vessels constrict and thus reduce the flow of blood through them. Fluid from the tissues then passes into the blood vessels, restoring the normal blood volume. Approximately 48 hours after the loss, many immature red blood cells appear in the bloodstream along with an excess of leukocytes. This indicates that the red marrow has been stimulated to increase its production of blood cells.

Hemolytic anemia.   When erythrocytes become misshapen, their outer membrane tends to rupture prematurely. This condition may be inherited and involves abnormal hemoglobin, red-blood-cell enzymes, or cell membranes. Other possible causes are toxins, parasites, or Rh incompatibility as seen in erythroblastosis fetalis.

lysis = destruction

One common type of hemolytic anemia is *sickle-cell anemia.* The replacement of one single amino acid in the hemoglobin molecule by another (out of 287 amino acids, the amino acid valine is substituted for glutamic acid at one position in the molecule) is responsible for this condition. The resulting hemoglobin is referred to as hemoglobin S and apparently causes red blood cells to become deformed (sickle shaped) when exposed to low concentrations of oxygen. (See Fig. 19-6). Deformed cells are very fragile and tend to rupture. Because of their shape, they tend to block small blood vessels. If the number of affected cells is relatively low, there are no outward symptoms evident, although the cells can be identified in blood smears. In severe cases, however, the red cells are rapidly destroyed and jaundice develops as a result of the accumulation of decomposed hemoglobin in the bloodstream. Sickle-cell anemia occurs almost exclusively among Blacks.

Figure 19-6 **The shape of the RBC in a person who has sickle-cell anemia.**

Aplastic anemia.   Faulty bone marrow results in *aplastic anemia.* Typically, red bone marrow is replaced by fatty tissue and fibrous tissue. The activity of the marrow can also be reduced by exposure to ionizing radiation (X rays), drugs used in cancer therapy, and poisoning with benzene, bacterial toxins, radioactive isotopes, and excessive amounts of certain antibiotics.

Nutritional blood deficiencies. A nutritional deficiency frequently occurs in growing children if iron is in inadequate supply, leading to *iron deficiency anemia*. The result is that there is insufficient hemoglobin in the red corpuscles, which appear pale (hypochromic) and are usually smaller than normal (microcytic).

Pernicious anemia is the result of a lack of vitamin B_{12}. If the lining of the stomach does not secrete enough of the glycoprotein produced by the parietal cells of the stomach, vitamin B_{12} (cyanocobalamin) is not absorbed. Since vitamin B_{12} is required for hemopoiesis, its absence results in cells that are too large (macrocytic) and often misshapen. Since fewer large cells can occupy the same space as many small cells, there is less total surface area for absorption of oxygen. In addition, the large cell's membrane is fragile and ruptures easily.

Hemophilia

Hemophilia is a sex-linked inherited condition in which the ability to produce a sufficient amount of thromboplastin is greatly reduced. Females are ordinarily not affected, but if one is a carrier, half of her sons may be affected. If a male carries the trait, his blood will fail to clot within a normal period of time, or not at all. External bleeding may be controlled but internal bleeding is extremely difficult to stop. Bleeding may occur in the muscles and joints and blood may appear in the urine. Recent work has indicated that there is present in the serum globulin of normal blood a factor called *antihemophilic globulin* (AHG). An absence or even an abnormal amount of antihemophilic globulin decreases the production of thromboplastin, and in turn of thrombin. Transfusions are a helpful but only temporary treatment. There is at present no cure for hemophilia.

Thrombosis

A blood clot, called a *thrombus,* occurs occasionally in blood vessels. If blood clots become too large, they seriously hamper the flow of blood through a vessel. If the clot becomes dislodged from its place of origin, it is called an *embolus.* Thus if a clot moves into and blocks a coronary artery in the heart, it would be called a coronary embolism, and if it moves into the lungs, pulmonary embolism can result. If the patient survives the initial attack, doctors frequently use an anticoagulant such as heparin to reduce the thrombus. Another drug that is being currently used to reduce a thrombus is *Dicumarol*. This compound was first isolated from sour clover fodder that was found to be the cause of hemorrhages in cattle who suffered minor cuts. Vitamin K, which is essential in the formation of prothrombin, counteracts the action of both heparin and Dicumarol.

Hemorrhage

Any undue loss of blood from the circulatory system is a **hemorrhage.** This loss may be *external*, as a result of a wound, or *internal*, with blood escaping into the surrounding tissues or into the body cavities.

The amount of blood lost in a hemorrhage determines the effect on the body. A person may lose one-half liter (approximately 10 percent of the total volume of blood) without any ill effects beyond a general weakness for a day or two. On the other hand, if the loss amounts to about 35 percent, serious damage may result; while a 50 percent reduction in volume is usually fatal. If the loss is small, as is the case in donating a pint of blood to a blood bank, the body restores the fluid in a period of 6 to 8 hours, and the level of red cells is brought to normal within about 2 weeks. White cells are restored more quickly. Following the loss of blood there is a slight leukocytosis, but this disappears rapidly and the number of white blood cells returns to normal within a short time.

If the loss of blood is sudden and extensive, the quantity of fluid is reduced to a point where the heart can no longer pump it through the body efficiently. This loss is indicated by a sudden drop in blood pressure. Because of this decrease in blood volume, the tissues begin to feel the effects of the lack of oxygen. The nervous system responds to the emergency by activating vasoconstrictor mechanisms. As a result, vessels in the skin and abdominal viscera constrict, and at the same time blood flow to vital organs is increased.

With the decrease in the blood pressure comes an increase in the rate of the heart beat. Since there is less blood flowing, the pulse appears to be weak but in fact the pulse rate increases. There is also an increase in the rate of respiration as the body attempts to compensate for the lowered supply of oxygen going to the tissues. Increases in heart and respiratory rates are important indications of internal hemorrhages. There may be few signs other than these of the loss of blood into a body cavity.

SUMMARY

Blood is considered a type of connective tissue composed of approximately 55 percent intercellular fluid (the plasma) and 45 percent formed elements. The contents of the blood maintain the water content of the body, protect against disease, and maintain an acid-base equilibrium in the internal environment.

Special features of blood include hemostasis (blood clotting), the control of osmotic balance by plasma proteins, the capacity of the blood to buffer acids or bases, and the ability for blood from different individuals of the same blood type to be mixed.

Disorders of the blood include different forms of anemia, clotting within vessels (thrombus), hemophilia, and hemorrhaging.

VOCABULARY REVIEW

Match the statement in the left column with the correct word(s) in the right column. *Do not write in this book.*

1. An abnormal condition resulting from a reduction in the amount of hemoglobin.
2. The process by which leukocytes engulf bacteria and other foreign bodies.
3. An immature red blood cell found circulating in the peripheral blood.
4. The principal blood-forming tissue.
5. The protein part of the hemoglobin molecule.
6. The smallest formed element of the blood.
7. An iron-containing compound.

a. anemia
b. thrombocyte
c. fibrin
d. gamma globulin
e. globin
f. heme
g. hemorrhage
h. reticulocyte
i. phagocytosis
j. red bone marrow

TEST YOUR KNOWLEDGE

Group A

Write the letter of the word(s) that correctly complete(s) the statement. *Do not write in this book.*

1. The most abundant compound in blood plasma is (a) fibrinogen (b) serum albumin (c) urea (d) serum globulin (e) water.
2. Of the following, the one *not* concerned with blood clotting is (a) calcium (b) thrombin (c) vitamin K (d) iron (e) thromboplastin.
3. The most important function of erythrocytes is to carry (a) carbon dioxide (b) oxygen (c) nitrogen (d) iron.
4. Of the following, the one that is *not* a plasma protein is (a) pepsin (b) prothrombin (c) fibrinogen (d) globulin (e) albumin.
5. A universal donor has blood type (a) A (b) B (c) AB (d) O (e) Rh⁻.
6. The blood cells that engulf bacteria are (a) erythrocytes (b) platelets (c) neutrophils (d) thrombocytes (e) red blood cells.
7. The most abundant white cells in the blood are (a) basophils (b) monocytes (c) lymphocytes (d) neutrophils (e) eosinophils.

Group B

Write the word or words that will correctly complete the sentence. *Do not write in this book.*

1. Hemoglobin is composed of an iron-containing pigment, _____.
2. White blood cells that engulf bacteria are called _____.
3. In the event of an injury involving blood loss, a number of events take place to stop the blood flow. This is called _____.
4. The process of red-blood-cell production is _____.

CHAPTER 20

THE HEART

Objectives

A. Describe the structure and functions of the heart
B. Describe the path of blood through the heart, naming its various chambers, valves, and connecting vessels
C. Describe the cardiac cycle and the mechanism of the conducting system
D. Compare the influence of various chemical and physical factors on the rate of the heartbeat

THE CARDIOVASCULAR SYSTEM

The blood provides for the metabolic requirements of all body cells. While at rest, these requirements remain at a basal, or minimum, level. If body activity is increased, there is greater demand for oxygen and nutrients that supply energy, and at the same time there is an increase in the formation of carbon dioxide and waste products. The cardiovascular system meets these demands by increasing blood flow to the areas of greater activity or to the entire body, if all of it is involved.

cardio = heart

Structure of the Heart

The heart is a hollow, muscular organ located in the thoracic cavity between the lungs, above the diaphragm. It is shaped like a blunt cone lying obliquely in a space called the **mediastinum,** with two thirds of its mass located to the left of the midsagittal plane. The size of the human heart varies. A large person can be expected to have a proportionately larger heart; the size of a person's heart generally corresponds to the size of the person's clenched fist. Normally, the heart is about 12 centimeters (4¾ in.) long, 9 centimeters (3½ in.) wide at its broadest point, and 6 centimeters (2½ in.) thick. The top, or base, is at the level of the second intercostal space (between the second and third ribs), and the **apex** (pointed end) projects downward and to the left between the fifth and sixth ribs. The weight of the heart ranges from 229 to 340 grams; the weight is dependent on the individual's size, sex, and age.

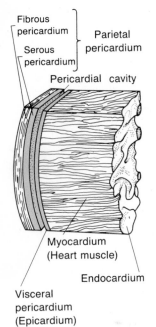

Fibrous pericardium
Parietal pericardium
Serous pericardium
Pericardial cavity

Myocardium (Heart muscle)

Endocardium

Visceral pericardium (Epicardium)

Figure 20-1 Structural view of the layers of the pericardial sac

myo = muscle

endo = within

In embryonic development, the heart has a fixed number of cardiac muscle cells that are present at birth and remain throughout life. The actual growth of the heart is due to an increase in the size of these cells, not to an increase in their number. The heart's diameter increases 2.6 times during the growth period.

Pericardial sac.   The heart and the base of the heart's major blood vessels are enclosed in a slippery, loose-fitting sac called the **pericardium.** This invaginated sac consists of two layers. The external fibrous layer that is attached to the diaphragm, to the sternum, and to the large blood vessels that enter and leave the heart is called the **parietal pericardium.** The internal, serous layer, or **visceral pericardium,** adheres to the heart wall and is known as the **epicardium.** The inner surface of the parietal pericardium has a layer of serous cells that lie opposite the outer serous surface of the visceral pericardium. Between these two layers is a narrow space called the **pericardial cavity,** filled with 10 to 15 milliliters of serous **pericardial fluid.** With every beat of the heart, this slippery fluid lubricates and prevents friction between the membranes. (See Fig. 20-1.)

The heart wall.   The heart has three distinct layers that make up its wall: the external layer (epicardium), the middle layer (myocardium), and the inner layer (endocardium). The **epicardium,** as already mentioned, is the *visceral pericardium.* Fat frequently infiltrates the epicardium layer. The **myocardium,** which comprises the bulk of the heart, consists of branching cardiac muscle fibers that appear striated and are involuntary. It is the myocardium that is responsible for the contracting action of the heart. The **endocardium** is a thin membrane of

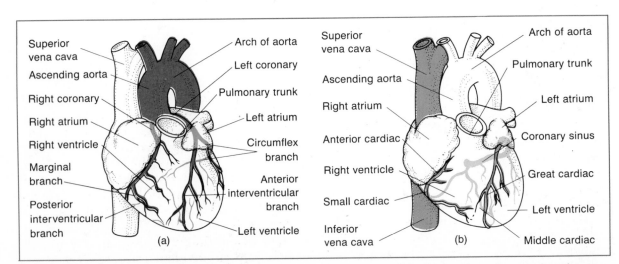

Figure 20-2 Coronary circulation: (a) anterior view of coronary *arteries*; (b) anterior view of coronary *veins*

endothelial and connective-tissue cells. It lines the inner surface of the myocardial layer and covers the heart valves and tendons that are attached to them. The endocardium is continuous with the endothelial lining of the major blood vessels that enter and leave the heart.

Heart chambers and valves.   The heart is divided into right and left halves, with each side further divided into upper and lower chambers. The upper chambers, the **atria,** are separated by the **interatrial septum;** the lower chambers, the **ventricles,** are separated by the **interventric-ular septum.** On the surface of each atrium is an external flap, or appendage, called an **auricle.** The auricle probably increases the atrium's surface area. On the surface of the interatrial septum is an oval depression, the **fossa ovalis,** a remnant of an oval hole, the **foramen ovale,** through which blood passes in the fetal heart. In the fetus blood flows directly from the right side to the left side without passing through the lungs to become oxygenated. The lungs only begin to function as an oxygen source after birth, and the opening then closes. From this time on, the right half receives only deoxygenated blood while the left half receives oxygenated blood from the lungs.

The heart contains two types of valves: the **atrioventricular valves,** which include the **tricuspid** and the **bicuspid,** or **mitral, valve;** and the **semilunar valves,** which include the **pulmonary** and **aortic semilunar valve.**   *pulmo = lung*

The atrioventricular valve between the right atrium and right ventricle is called the tricuspid valve. It consists of three irregularly shaped flaps, or **cusps,** formed mainly by fibrous tissue and covered by a thin endocardium. The pointed ends of the flaps point in toward the ventricle. They are attached by cords called **chordae tendineae** to small muscular projections, the **papillary muscles,** found along the inner surface of the ventricles. The left atrioventricular opening is guarded by the *bicuspid,* or *mitral, valve.* As its name indicates, it consists of two flaps, or cusps, instead of three, as is the case with the tricuspid valve. The bicuspid valve is heavier and stronger than the tricuspid, since the left ventricle exerts greater force in its contraction. Its cusps are also attached to chordae tendineae and papillary muscles.

A second type of valve is found where the two major arteries—the pulmonary artery and the aorta—leave the ventricles. These semilunar valves are crescent-moon–shaped and consist of three flaps. The *pulmonary semilunar valve* prevents blood in the pulmonary artery from reentering the right ventricle. The *aortic semilunar valve* prevents blood from reentering the left ventricle as it enters the aorta.

Blood Supply to the Heart

The heart, like any other muscle, must be supplied with nutrients and oxygen and must have metabolic wastes removed. Consequently, the heart has its own extensive blood supply. The **coronary arteries**   *cor = heart* supply the myocardium with blood.

Figure 20-3 Diagram of the heart showing the anterior internal areas. The solid and broken arrows indicate the flow of blood through the heart. Blood enters the right atrium, then passes through the tricuspid valve into the right ventricle. It then moves up through the pulmonary semilunar valve into the pulmonary artery that goes to the lungs. It returns to the heart by way of the pulmonary veins and enters the left atrium, thence through the bicuspid valve into the left ventricle. From there it passes through the aortic semilunar valve into the aorta and out into the body.

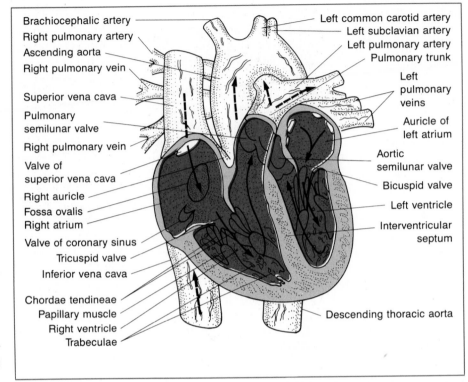

Coronary circulation.   The two vessels, the left coronary artery and the right coronary artery, are the first branches of the aorta. They encircle the heart, with the left coronary running under the left atrium and then dividing into the anterior interventricular and circumflex branches. The right coronary artery runs under the right atrium and branches into the posterior interventricular and the marginal branches. From the coronary arterial system, blood flows into the coronary veins. The main coronary veins empty into the coronary sinus, which in turn empties into the right atrium.

PHYSIOLOGY OF THE HEART

The fetal heart muscle begins to beat rhythmically some time before the central nervous system has innervated the heart. In fact, a heart will continue to beat for a number of hours even if removed from the body if it is supplied with appropriate nutrients, oxygen, and salts. How is it possible for a largely muscular organ to contract without direct control by the brain or spinal cord? The answer is that a specialized **conducting system** within the heart facilitates the rapid and coordinated spread of excitation.

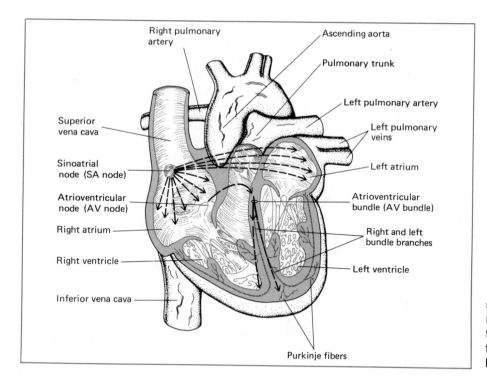

Figure 20-4 **Conduction system of the heart. Arrows indicate the path of the electrical impulse through the heart chambers.**

The Conducting System

Within the heart the conducting system consists of three structures: the **sinoatrial node** (*SA node*), the **atrioventricular node** (*AV node*), and the **atrioventricular bundle** (*AV bundle*), or the *bundle of His,* with its connecting *Purkinje network.* The SA node is a mass of specialized myocardial (heart muscle) cells embedded in the wall of the right atrium, near where the superior vena cava enters the heart. This structure initiates each heart beat, and thus sets the pace for the heart rate—as a result, the SA node is often called the **pacemaker.** The electrical activity of the SA node occurs on the average of 70 to 72 times per minute and spreads rapidly over the atria, exciting them almost at the same time so that both atria contract simultaneously. The impulse—a wave of depolarization from the SA node—reaches the atrioventricular node (AV node) at about the conclusion of atrial contraction. The AV node is located at the base of the right atrium very near the interatrial septum. Once the AV node becomes depolarized, a delay occurs for approximately 0.1 second, allowing the atria to contract and empty their contents into the ventricles.

Following this interval, the impulse then travels over the connecting link, the atrioventricular bundle (AV bundle). The AV bundle consists of a group of specialized cardiac muscle fibers adapted for conduction rather than contraction. The bundle divides and passes down each side of the interventricular septum to form the Purkinje network—the conducting system within the walls of the ventricles. Transmission over the AV bundle and Purkinje fibers is six times faster than if the impulse only traveled over the surface of the myocardium. Thus contraction proceeds from the inferior portions of the ventricles upward, ensuring that blood leaves the ventricles through the arteries.

The Cardiac Cycle

The sequence of heart action, called the **cardiac cycle,** may be said to have three stages, as shown in Fig. 20-5. Blood enters the atria through the *pulmonary veins* and the *superior* and *inferior venae cavae.* The pulmonary circuit brings blood from the lungs, while the systemic circuit returns blood from the head, neck, arms, thorax, trunk, and legs. In addition, the right atrium receives blood from the heart wall itself through the *coronary sinus.*

Pressure changes of the heart chambers. The walls of the atria are relatively thin and weak because these chambers simply receive blood from the great veins and pass it into the ventricles. While the atria are filling, the atrioventricular valves remain open so that there is a continuous flow from the veins through the atria into the ventricles. As the ventricles fill, pressure builds up within them. When the blood in the ventricles begins to close the tricuspid and bicuspid valves, the muscles in the walls of the atria contract to force all the blood that remains in them into the ventricles. For the brief time of the contraction, there is very little blood in the atria, since the pressure within the atria is greater than the pressure of the veinous blood returning to them.

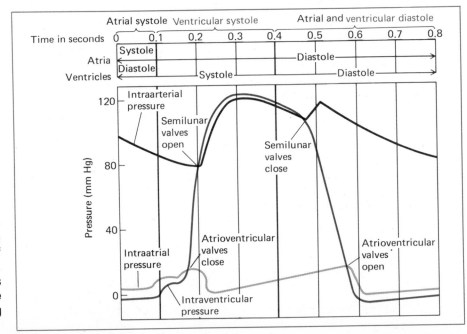

Figure 20-5 The cardiac cycle showing the systole and diastole of the atria and ventricles. The pressure changes during the cycle cause the opening and closing of the valves.

As the ventricles fill with blood, they become distended. The AV node is activated and the AV bundle with its connected Purkinje fibers causes a wave of muscular contraction. The contraction of the ventricles begins at their lower ends and moves upward toward the atria. This action closes the atrioventricular valves and also forces blood out of the ventricles through both the pulmonary artery and the aorta. The blood flows from the right ventricle through the pulmonary artery and its branches to the capillaries in the lungs. Here the blood gives up its carbon dioxide and takes up oxygen. The blood, containing oxyhemoglobin, then returns through the four pulmonary veins to the heart, entering the left atrium. The blood passes the bicuspid (mitral) valve and flows into the left ventricle. When the ventricle contracts, the blood is forced out through the aorta to all parts of the body except the lungs.

Under the pressure of ventricular contraction, the atrioventricular valves are prevented from being forced back into the atria by strong, elastic tendons, the chordae tendineae. During contraction, since the ventricles shorten slightly, the action of the papillary muscles compensates for this decrease in length. The atrioventricular valves are tough fibrous strands. With each normal contraction of the ventricles the valves must withstand a pressure of approximately 120 millimeters of mercury. During violent exercise, this pressure may exceed 180 millimeters of mercury.

When blood leaves the right ventricle, it passes through the pulmonary semilunar valve into the pulmonary artery. The cusps are pressed against the sides of the artery by the force of the ventricle's contraction. As the ventricle relaxes, the back pressure of the blood in the pulmonary artery forces the three flaps closed. This prevents blood from passing back into the right ventricle. A similar action takes place in the left ventricle with the aortic semilunar valve.

Diastolic and systolic phases.   The period of the filling of the atria and the ventricles is referred to as the *diastole.* This is a period of relaxation in which the chambers become distended. Relaxation is followed by a period of active contraction, or *systole,* which is marked by a shortening of the muscle bundles. If a heart beats at the rate of 70 times per minute, the duration of these phases is as shown in Table 20-1.

It should be noted that the total elapsed time for a complete cycle of both the atria and the ventricles is 0.8 second. During these cycles there

Table 20-1
The Cardiac Cycle

Intracardiac Pressures	Time	Events
Atrial diastole	0.7 sec.	Period of filling
Atrial systole	0.1 sec.	Active contraction
Ventricular diastole	0.5 sec.	Period of filling
Ventricular systole	0.3 sec.	Active contraction
Quiescent period for entire heart	0.4 sec.	

is a period of 0.4 second when both the atria and ventricles are filling. This results in a period when the heart muscles are not contracting.

The amount of blood leaving the heart with each contraction is about 160 milliliters when the body is relaxing; 80 milliliters are forced out of the left ventricle into the aorta, and 80 milliliters enter the pulmonary artery from the right ventricle. This is known as the **stroke volume** of the heart. Assuming a rate of 70 beats per minute, a total of approximately 5.6 liters (70 × 80) of blood is leaving each ventricle. This is known as the **cardiac output,** or *minute volume* of the heart.

Control of the Heart

The origin of the heart beat, as has been discussed, is a built-in mechanism, but the action of the heart can be modified by a number of other influences. For example, the heart rate is affected by the secretions of endocrine, or ductless, glands. Hormones secreted by the thyroid (thyroxine) and adrenal glands (epinephrine) increase the heart rate. This is quite evident when one is frightened or under other conditions of stress.

The heart is also under the control of the nervous system. Nerves coming from the parasympathetic division of the autonomic nervous system act to slow down the heart, while nerves from the sympathetic division speed it up. By their mutually antagonistic effect, they keep the heart rate steady. The parasympathetic nerves originate in the medulla and travel to the heart by way of the vagus nerves. Vagus fibers innervate both the SA and AV nodes. When stimulated, the vagus nerve releases acetylcholine, which *inhibits* the action of the heart by slowing the pacemaker, resulting in a weak systole and prolonged diastole.

Sympathetic impulses that accelerate the heart originate in the *cardioacceleratory center* of the medulla. They travel down the spinal cord to the upper thoracic region, where they pass over accelerator nerves to various cervical sympathetic ganglia, which in turn stimulate the SA and AV nodes by releasing norepinephrine, a chemical that increases the heart rate and strength of ventricular contraction.

Pressoreceptors.   Within the heart itself and also in the walls of some of the large vessels are nerve endings that help to control the rate of the heart beat. These nerves, depending on their location, may be grouped into three categories: the **right atrial reflex** (called the *Bainbridge reflex*), **the aortic reflex,** and the **carotid sinus reflex.** Nerve endings in the walls of the venae cavae and the right atrium are stimulated by increased venous pressure. Whenever a group of skeletal muscles becomes active, they press against the walls of the veins, which increases the flow of blood back to the heart. This increased flow results in increased pressure within the right atrium and the venae cavae. Whenever venous pressure increases, the pressoreceptors send impulses that stimulate the cardioacceleratory center in the medulla. This action causes the heart rate to increase. This burst of speed is known as the Bainbridge reflex.

Another reflex originates in the walls of the aortic arch, where there are nerve endings called the **stretch receptors.** If the pressure of blood in the aorta becomes excessively high, these receptors are stimulated by the stretching of the walls of the vessels. Impulses from them pass along the vagus nerve to the cardioinhibitory center, and the rate of heart beat is reduced, with an accompanying drop in blood pressure.

If the blood pressure drops to a very low point, as it does following severe hemorrhage, these pressoreceptors are no longer stimulated and their inhibiting action is removed. The cardioacceleratory center is free to dominate, resulting in a rise in heart rate and increased pressure.

The *carotid sinus reflex* is concerned with supplying normal blood pressure to the brain. On the right and left side of the neck, the common carotid artery divides into an internal and external branch. The external carotid supplies blood to the outer surfaces of the head, while the internal carotid carries blood to the brain. At the point of their separation, there is a small swelling in the internal carotid artery, known as the **carotid sinus.** Within its walls are stretch receptors from which nerve fibers pass along the ninth cranial nerves (the glossopharyngeal nerves), to the cardioinhibitory center. These nerve fibers form the third group of controlling nerves. If the carotid sinus becomes distended with blood, as might occur with high blood pressure, impulses stimulate the center in the medulla and the heart rate is decreased with an accompanying fall in pressure. On the other hand, if there is a decrease in blood pressure, the receptors are not stimulated and the rate of the heart beat accelerates.

The name *carotid* comes from the Greek word *karos*, which means "sleep." The ancients found that certain people could be made to sleep by pressure on the carotid arteries below the region of the sinus. This effect is caused by reduction in the amount of blood in the sinus, with the accompanying drop in blood pressure in the vessels of the brain. Today we would call this loss of consciousness a fainting spell. Fainting is the result of a drop in blood pressure in the vessels supplying the brain, due to any of a number of causes, including psychological disturbances. One of the common methods of preventing a fainting spell (syncope) is to lower the head below the waistline of the body. This allows more blood to pass to the brain, but it does not have an effect on the sinus receptors, thus resulting in a rise in blood pressure within the vessels of the brain.

Chemical influence.   In addition to neural control of the heart rate, there are many chemical compounds that affect the heart. For example, **atropine,** the active substance obtained from the deadly plant *nightshade,* has an action comparable to that obtained by cutting both vagus nerves—the heart beat is greatly accelerated. **Muscarine,** the poisonous substance found in some mushrooms, acts like *acetylcholine* and will inhibit heart action to the point where the beat stops entirely. The action of both of these inhibitors can be neutralized by atropine.

Nicotine, which is a highly poisonous alkaloid, evidently has a most unusual effect on the post-synaptic receptors. First, nicotine stimulates these receptors, and second, it seems to paralyze them. One of the most noticeable effects of nicotine is on the blood pressure. Even among those people who are accustomed to heavy smoking, the systolic pressure is raised about 19 millimeters of mercury, and the diastolic pressure is raised about 14 millimeters. The pulse rate is increased and there is a constriction of the coronary arteries. Therefore, it is quite unwise for a person who suffers from high blood pressure even to think of smoking. The action of nicotine on the nervous system is shown by its effect on the temperature of the different regions of the body. By the use of thermocouples, physiologists have found that smoking lowers the temperature of the skin on the toes by approximately 2°C while that of the fingers is reduced by just about 3°C. This decrease in temperature is due to the constriction of the blood vessels in these extremities.

Chart 20-1
The Relation of Age to Heart Rate

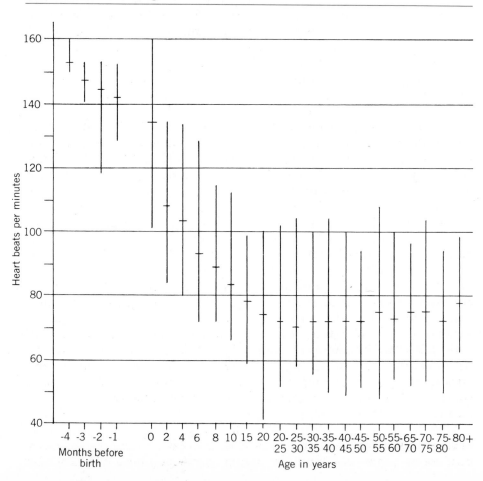

Rate of Heart Beat

The rate at which the heart normally beats is determined by many factors. Age, sex, position of the body, amount of physical activity, temperature of the surroundings, and thought processes are all reflected in the rate of the heart beat. Chart 20-1 compares the average rate of the heart beat as it relates to age. Before birth the rate is high, but there is a steady decline after birth until a fairly constant average is reached. The vertical bars show the limits that can be expected among healthy individuals, while the short horizontal bars indicate the average for the group. Thus a person 15 years old may have a normal resting rate between 59 and 98 beats per minute; the average rate for the entire group is 72 beats per minute. This data has not been compiled on the basis of the sex of the individuals included. Thus it does not indicate that women have a slightly higher average rate than men. During sleep, a man may average about 59 beats per minute, whereas a woman has a heart rate of approximately 78 beats per minute. When waking, both sexes show an increase, but the average number of beats is greater in women.

The electrocardiogram.   In common with all muscles, the heart develops an electric current when it contracts. This arises within the muscle itself as a result of the movements of ions across cell membranes. Known as the **action current,** it flows between any active tissue and an inactive region. The production of current in the heart is especially important from a medical standpoint. It allows the various stages in the heart beat to be gauged with a great deal of precision.

The instrument that measures and records this current is called an **electrocardiograph** and the resulting record is called an **electrocardiogram (EKG).** By amplifying the electrical current produced by the heart, a stylus can trace the pattern of heart waves on a sheet of moving paper. In order to pick up the current, various leads (electrodes) are placed on the subject's body, usually on the right arm, left arm, right leg, and left leg, as well as on the surface of the chest itself.

In a normal EKG, the electrical activity with each complete beat is designated by waves that have been labeled P, Q, R, S, and T. (See Figure 20-6.) Each wave corresponds to a particular part of the cycle. The P wave shows the passage of a positive current over the atria and corresponds to its depolarization and resulting contraction. Normal P waves indicate that the current originated in the SA node, while a distorted wave might indicate that the impulse originated outside the SA node. QRS waves occur as the ventricles are depolarized and reflect the time required for the impulse to travel through the AV bundle and Purkinje fibers. Following the contraction of the ventricles is the T wave, which represents ventricular recovery and repolarization of the tissue.

Heart sounds.   A physician, using a **stethoscope,** listens to the heart beat for two reasons. First, by noting the beginning and end of the ventricular systole, one can time other events in the cardiac cycle. Second, the sounds produced by the heart indicate the condition of the atrioventricular and semilunar valves.

There are two heart sounds, each of which is related to a phase in the cardiac cycle. The first sound is the result of the closing of the atrioventricular valves and the contraction of the muscles in the walls of the ventricles. This is a low-pitched sound of some duration that coincides with the ventricular systole. Actually, it starts a fraction of a second before there is visible evidence of contraction in the ventricles and represents the instant at which the atrioventricular valves close. In relation to the electrocardiogram, the beginning of this sound corresponds to the upstroke of the QRS wave and continues to a point between the S and T waves, as seen in Fig. 20-6. It sounds like *lub* and can be heard best over the region of the apex of the heart, between the fifth and sixth ribs to the left of the sternum.

The second sound is much shorter in duration and of a higher pitch. This sounds like the word *dub* and represents the sudden closing of the semilunar valves. It is heard most distinctly in the second intercostal space, between the second and third ribs, near the sternum. When it is correlated with the electrocardiogram, it is heard just before the appearance of the T wave.

HEART DISORDERS

Every heart has the ability to perform work. Whenever there is an increased demand, the heart can utilize its own cardiac reserve. However, if a heart is abnormal at birth, or if the individual develops certain risk factors, the frequent demands placed on the cardiac reserve tax the heart muscle to the point of shortening its life span. The risk factors considered to be dangerous to normal heart activity are: a high level of blood cholesterol, high blood pressure, cigarette smoking, obesity, lack of exercise, and diabetes mellitus (sugar diabetes). If one of these factors, either directly or indirectly, reduces the coronary blood supply

Figure 20-6 (left) Recordings of a normal EKG and correlated sounds of the heart.

Figure 20-7 (right) An EKG of a person suffering from auricular fibrillation. Note the series of P waves before the appearance of the R wave.

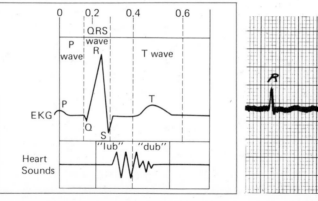

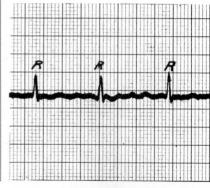

to the heart muscle, the action of the heart is reduced. The lack of blood supply to a restricted area is called *ischemia.* If this restriction results in destroying tissues of the heart, it is then called *infarction.* This is one of many causes of "heart attack."

Fibrillation

In Fig. 20-7 is seen an electrocardiogram of a person afflicted with *atrial fibrillation.* This is the most common of all serious heart irregularities. In this condition the atria are never completely emptied of blood, and their walls quiver instead of giving the pronounced contraction typical of a normal heart beat. The beat may be very rapid (300 to 500 beats per minute) and is completely irregular and disorganized. As a result, the AV node is activated at irregular intervals, resulting in a ventricular irregularity. A rapid heart beat is called *tachycardia.*

If fibrillation should occur in the ventricles, little or no blood is pumped out of the heart through the pulmonary artery or aorta, and circulation of the blood stops. Therefore, without emergency treatment, ventricular fibrillation, if it lasts for more than a few minutes, is fatal.

Heart Murmurs and Valve Defects

The major cause of heart murmurs is defective or diseased heart valves. If a valve becomes narrow—called *stenosis*—the blood forced through this restriction becomes turbulent, which generates a sound. If the valve does not close properly, blood leaks through it in the wrong direction, again producing a characteristic sound. Most murmurs can only be heard with the aid of a stethoscope.

Atrial septal defect.   If the fetal foramen ovale fails to close completely after birth, since pressure is lower in the right atrium than the left, blood will move from the left to the right atrium. This puts an added strain on pulmonary circulation, resulting in fatigue.

Ventricular septal defect.   If the interventricular septum is defective in development, deoxygenated blood may mix with oxygenated blood in the left ventricle. Thus the victim's blood does not take on a bright red appearance and the skin, fingernails, and lips appear blue, or *cyanotic.* Through open heart surgery, the defective septum can usually be repaired.

SUMMARY

The heart, a hollow muscular organ that lies in the mediastinum, is housed in the pericardium. The wall of the heart has three layers: the epicardium, myocardium, and endocardium. The heart acts as a double pump, with the right and left sides further divided into upper chambers (atria) and lower chambers (ventricles). The partition between the atria is called the interatrial septum; that between the ventricles, the interventricular septum. Guarding the passage between the upper and lower chambers are atrioventricular valves. A coronary system of vessels supplies the heart wall with oxygen and nutrients and removes waste materials.

The pacemaker, or SA node, initiates the impulse; the AV node receives and transmits the impulse over the AV bundle to the Purkinje network. The contraction phase is called systole and the relaxation phase is called diastole.

Nerve innervation to change the rate of the beat involves the parasympathetic division of the autonomic nervous system, which inhibits heart action, and the sympathetic division, which accelerates it. In addition, pressoreceptors influence the heart rate, and certain chemicals may activate or depress heart action.

The electrocardiogram (EKG) measures the electrical activity of the heart and transfers some of it into a wave pattern of each complete beat. Normal and abnormal waves are used as diagnostic tools for the physician. Heart sounds also assist the physician in identifying abnormal blood flow through the various valves, which produce a *lub-dub* sound.

The risk of heart attack may be directly related to certain factors that hinder normal heart action, such as cholesterol level, high blood pressure, smoking, obesity, lack of exercise, and diabetes mellitus. These factors can cause the coronary blood supply to a specific area to be reduced and eventually can lead to an infarction.

VOCABULARY REVIEW

Match the statement in the left column with the correct word(s) in the right column. *Do not write in this book.*

1. The first chamber to receive deoxygenated blood.
2. The major artery carrying oxygenated blood.
3. Prevents the backflow of blood to a ventricle.
4. The space between the lungs, above the diaphragm, in which the heart is found.
5. Another name for the sinoatrial node.
6. Prevents an atrioventricular valve from bulging into an atrium.
7. The period in which a heart chamber is filled.
8. A device that measures the movement of current across the heart.
9. A rapid heart beat.
10. The inner lining of the chambers and valves within the heart.

a. abdominal cavity
b. aorta
c. chordae tendineae
d. diastole
e. electro-cardiograph
f. endocardium
g. mediastinum
h. pacemaker
i. pericardium
j. pulmonary artery
k. right atrium
l. semilunar valve
m. systole
n. tachycardia

TEST YOUR KNOWLEDGE
Group A

Write the letter of the word(s) that correctly completes the statement. *Do not write in this book.*

1. The double-walled membrane surrounding the heart is called the   (a) periosteum   (b) perilymph   (c) pericardium   (d) peritoneum   (e) endocardium.
2. The coronary arteries supply blood to the (a) heart   (b) stomach   (c) pancreas   (d) spleen   (e) lungs.
3. The heart is divided into two halves by the (a) foramen ovale   (b) foramen magnum   (c) septum   (d) sacrum   (e) AV bundle.
4. The valve between the right atrium and the right ventricle is the   (a) bicuspid   (b) aortic semilunar   (c) pulmonary semilunar   (d) mitral   (e) tricuspid.
5. Blood enters the left atrium of the heart through the (a) venae cavae   (b) pulmonary veins   (c) sinus venosus   (d) atrial appendage   (e) coronary sinus.
6. The muscular wall of the heart is called the (a) pericardium   (b) endocardium   (c) epicardium   (d) myocardium   (e) periosteum.
7. Papillary muscles are found in the (a) ventricles   (b) atria   (c) skin   (d) chordae tendineae   (e) abdomen.
8. The period of emptying the atria or ventricles is called   (a) diastole   (b) diastase   (c) systemic   (d) systole   (e) dilation.
9. Oxygenated blood enters the heart in the (a) right atrium   (b) left atrium   (c) right ventricle   (d) left ventricle   (e) none of these.
10. The normal period of time for one complete cardiac cycle in an adult is about (a) 0.08 second   (b) 0.8 second   (c) 1.8 seconds   (d) 8 seconds   (e) 80 seconds.

Group B

Write the word or words that will correctly complete the sentence. *Do not write in this book.*

1. The _____ comprises the bulk of the heart.
2. Another name for the left atrioventricular or bicuspid, valve is the _____ valve.
3. The SA node, AV node, and AV bundle all belong to the _____ which is responsible for activating the heart muscle.
4. The _____ nervous system is responsible for inhibiting or accelerating the heart rate through nerve impulses.
5. The tracing on paper of the electrical activity of the heart is called an _____.

CHAPTER 21

THE VASCULAR SYSTEM

OBJECTIVES

A. Identify the various vessels that form the vascular system
B. Compare the structure and function of each of the major blood vessels
C. Distinguish between the pulmonary and systemic circuits of the circulatory system
D. List five ways the body maintains normal blood pressure
E. Discuss the mechanisms that control the distribution of blood throughout the vascular system

ANATOMY OF BLOOD VESSELS

In simple forms of life, the requirements for obtaining oxygen and nutrients and eliminating carbon dioxide and other metabolic wastes are met by direct contact with the environment, usually water, through the body surface. In larger more complex animals, humans included, deep-lying cells do not have direct contact with the environment. Their needs must be met with a system for transporting materials from the outside to each cell and for removing wastes efficiently from all cells to the outside. The method used involves a fluid transport system and a complex network of passageways, which bring the fluids in contact with all the cells of the body.

The blood vessels form the network of **veins, arteries,** and **capillaries**—the vascular portion of the cardiovascular system. This system of passageways transports blood to all parts of the body through the action of the heart. The arteries carry blood *away* from the heart and toward the tissues. The great arteries leave the heart and divide into small arteries and then into smaller-size vessels called the **arterioles.** Progressively, the arterioles branch into the capillaries found in the tissues. Blood passing through these microscopic vessels releases oxygen and nutrients to the tissue cells and removes carbon dioxide and other metabolic wastes. The capillaries have the thinnest walls because they are constructed of a single layer of endothelial cells (modified simple squamous epithelium). The capillaries emerge from the tissues as small veins, or **venules,** which in turn form larger veins. The largest of these vessels return blood to the heart.

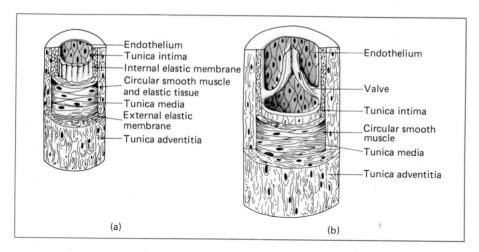

(a)

(b)

Figure 21-1 Diagram of blood vessels comparing the layers in (a) arteries and (b) veins.

The larger blood vessels, arteries and veins, are quite similar in structure. They both have three layers: (1) The outer layer, the **tunica adventitia,** is composed of connective tissue with varying amounts of elastic fibers. (2) The middle layer, the **tunica media,** is made up of smooth muscle fibers. (3) The inner layer, the **tunica intima,** contains endothelium that is continuous with the endothelial lining of all the blood vessels.

Arteries

The *tunica adventitia* is composed of varying amounts of elastic tissue and collagenous fibers that strengthen the wall of the artery. The fibers allow the large vessels to expand when the heart contracts, providing a temporary reservoir for the blood that leaves the heart. During ventricular diastole, the arteries spring back, forcing the blood toward the direction of the body tissues. The elasticity of the wall of the artery aids in maintaining blood pressure. In small arteries, the external layer is very thick, while in the larger arteries it is relatively thin.

The *tunica media,* the middle layer, contains smooth muscle fibers arranged circularly around the vessel. The thickness of the wall is due to the amount of smooth muscle present. Elastic fibers are present in the larger arteries, whereas smooth muscle is most predominant in the smaller arteries.

The innermost layer, the *tunica intima,* is composed of an endothelial lining and a layer of elastic fibers that aids in strengthening the wall when the internal pressure increases.

Many of the larger arteries have very thick walls containing their own system of blood vessels and a nerve network to control the amount of blood passing through them (Fig. 21-1).

Arterioles. Because of the extensive muscular wall of medium-size arteries, they can control the *distribution* of blood according to the needs

Figure 21-2 (left) Branching of an arteriole to form a capillary network. The precapillary sphincter, found between the arteriole and capillary, acts as a valve.

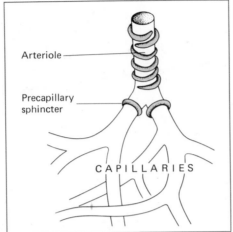

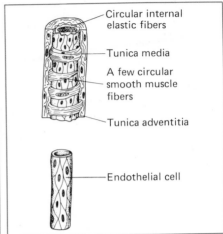

Figure 21-3 (right) The structure of a capillary.

of the various body organs. This is accomplished when the arteries undergo **vasoconstriction** (narrowing) or **vasodilation** (widening). On the other hand, the smaller arterioles *regulate* the flow of blood into the capillary network. The diameter of arterioles gets smaller the further they are from the artery to which they are connected. Gradually, their walls are replaced by a thin layer of endothelial tissue with isolated smooth muscle cells encircling the tubes. The muscle cells form **precapillary sphincters** that regulate the blood flow into the capillary network. (See Fig. 21-2.)

The Capillary Network

When the last traces of smooth muscle and connective tissue cells have disappeared in arterioles there remains only a single layer of endothelial cells (Fig. 21-3). These cells compose the walls of the capillaries and are continuous with the tunica intima of the large vessels. These flat cells join together like pieces of a jigsaw puzzle.

Rate of blood flow. The diameter of the capillaries is about 8 micrometers (μm) and some are 5 micrometers or less in diameter, while others may be 15 or 20 micrometers in diameter. The length of each capillary is between 0.4 and 1.0 millimeter. In these narrow channels the rate of the flow of blood decreases as the vessels become smaller and more numerous. This permits an exchange of materials between the blood and the cells.

It has been estimated that the total area of the capillary walls in human voluntary muscle is about 6000 square meters (1 square meter = 10.764 square feet). With this amount of surface exposed in this active tissue, the transfer of materials is greatly facilitated.

One of the features of the capillary network is that not all capillaries are open at the same time. This permits the flow of blood to those tissues that are most active. In the brain, for example, the majority of the capillaries remain open, but in a resting muscle only 1/20 to 1/50 of the capillaries carry blood to the fibers. To supply food and oxygen to actively moving muscles as many as 190 capillaries per square millimeter may be open. When the same muscles are resting, there may be only 5 capillaries per square millimeter in active use.

Some of these differences can be accounted for by the functions of two types of capillaries. One type connects the arterioles and venules directly. These capillaries are always open and offer little resistance to blood flow. The other type is formed by offshoots of the arteriole-venule system. The flow of blood through these capillaries can be regulated by the requirements of the particular tissue they supply.

Veins

The **venules** are the smallest part of the venous system. They are formed by the union of several capillaries. The difference between a capillary and a venule is in the nature of their vessel walls. The walls of venules (measuring 50 μm in size) contain smooth muscle cells, which are entirely lacking in capillaries. They differ very little from arterioles of corresponding size.

The venules rapidly converge to form small veins, which in turn join to form large veins. Gradually the diameters of the vessels increase, ultimately forming the great veins that enter the heart. The walls of the veins are thinner than those of the arteries. Veins have relatively thin

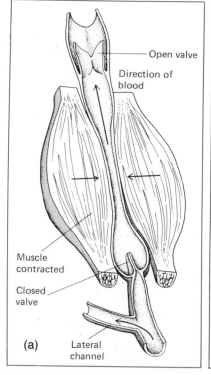

Figure 21-4 (a) Diagram of valves of a vein. (b) Closer view of an open and closed valve. In which direction does the blood move when the valve is open?

layers of muscles with fewer elastic fibers. Thus the structure of veins is not adapted to withstand the high internal pressures to which arteries are subjected.

Valves.   The inner lining of veins is made of endothelial tissue. It differs from the lining of arteries in that it forms **valves** that help to direct the flow of blood. These are composed of semilunar folds of tissue with their free edges pointing in the direction of blood flow. When blood flows toward the heart, they are pressed back against the walls of the vessels so that the blood can flow freely. If a back-pressure develops in the vein, the valves close. This prevents the blood from flowing back toward the capillaries. Valves in veins are of special importance in those regions of the body where blood has to flow against the force of gravity, as in the lower extremities. Review the major types of vessels in the Summary table at the end of this chapter.

The Circulatory Routes

The blood vessels of the human body are arranged in two circuits: (1) The less extensive of the two is the **pulmonary** circuit, which carries blood from the right ventricle to the lungs and back to the left atrium. (2) The more extensive **systemic** circuit carries blood from the left ventricle to all parts of the body and back to the right atrium. (See the middle of the text for diagrams of the systemic circulation.)

Pulmonary circulation.   Deoxygenated blood passes from the right ventricle through the pulmonary semilunar valve and into the **pulmonary trunk.** This short artery (about 5 centimeters long and 3 centimeters in diameter) extends upward and curves over the heart. It then divides into right and left **pulmonary arteries,** each of which goes to the right and left lungs. Within the lungs, the pulmonary arteries divide and subdivide, eventually forming **pulmonary capillaries.** These capillaries surround the alveoli of the lungs, in which blood takes up oxygen and loses carbon dioxide. The returning blood collects into venules, which successively get larger as they pass out of the lungs and form veins. Two **pulmonary veins** leave each lung and the four vessels pass to the posterior surface of the left atrium. These vessels carry oxygenated blood to be distributed from the heart by systemic arteries. Pulmonary veins have no valves (Fig. 21-5).

Systemic circulation.   The arteries of the systemic circuit arise from the aorta and lead either to one specific organ or divide into smaller branches that supply a more distant region. Several of these branches are described in the following section.

The **ascending aorta** arises from the left ventricle. Near its origin only two branches, the **right** and **left coronary arteries,** leave the ascending aorta. These branches form a crown around the heart and supply blood to the heart muscle. As the ascending aorta curves upward, it moves to the left of the midline and arches backward. It then descends within

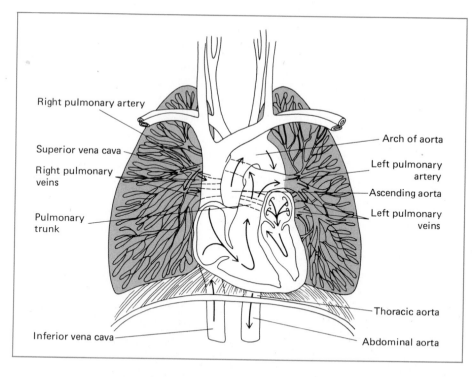

Right pulmonary artery

Superior vena cava

Right pulmonary veins

Pulmonary trunk

Inferior vena cava

Arch of aorta

Left pulmonary artery

Ascending aorta

Left pulmonary veins

Thoracic aorta

Abdominal aorta

Figure 21-5 **Pulmonary circulation. The arrows indicate the flow of blood between the heart and lungs.**

the thoracic cavity on the left side of the vertebral column and passes through the diaphragm. Three branches extend from the *arch of the aorta:* the **brachiocephalic artery,** the **left common carotid,** and the **left subclavian.**

brachio = arm
cephalic = head

The first branch, the *brachiocephalic* artery, is large in diameter but only 4 to 5 centimeters long. It divides into the **right common carotid artery** and the **right subclavian artery.** The right subclavian extends toward the right arm, where it forms the **axillary artery,** and then passes to the upper arm as the **brachial artery.** At the elbow it divides into the *ulnar* and *radial* branches, which eventually supply blood to the hands and fingers.

The second branch, the *left common carotid,* arises directly from the aortic arch. Both the right and the left common carotid arteries pass upward in the neck to the level of the larynx, where each artery divides into *external* and *internal carotid arteries.* The external carotid artery on each side gives off numerous branches that supply the neck, face, and scalp. The internal carotid artery ascends vertically and passes through the carotid canal in the temporal bone and enters the cranial cavity, where it supplies the brain.

The third branch of the aortic arch, the *left subclavian artery,* arches slightly above the clavicle and passes laterally toward the left upper extremity. It forms the same subdivisions and branches as the right subclavian.

The descending *abdominal aorta* is a continuation of the aorta that traverses the posterior mediastinum in the thorax. Thirteen arteries branch off the abdominal aorta but only four will be considered: **the celiac artery,** the **superior mesenteric artery,** the paired **renal arteries,** and the **inferior mesenteric artery.** The *celiac artery* is a short thick vessel that branches off the anterior surface of the abdominal aorta just below the diaphragm. It immediately divides into three branches: the *left gastric artery,* which supplies the esophagus and stomach; the *common hepatic artery,* which supplies the liver, the duodenum, and the pancreas; and the *splenic artery,* which conducts blood to the spleen, pancreas, and stomach. The *superior mesenteric* arises from the anterior surface of the aorta just below the celiac artery. It supplies all of the small intestine, the pancreas, and part of the large intestine. The paired *renal arteries* are short vessels that arise on the lateral sides of the aorta and supply the right and left kidney. The **inferior mesenteric artery** is an unpaired vessel that arises from the anterior surface of the aorta just before it divides. It supplies most of the large intestine and the rectum.

At the level of the fourth lumbar vertebra, the abdominal aorta divides into two descending lateral branches, the **common iliac arteries.** At the level of the sacroiliac joint, each common iliac artery divides into the smaller **internal iliac artery (hypogastric artery),** which supplies the pelvic region, and the larger **external iliac artery,** which enters each lower extremity. Upon entering the thigh, the artery is known as the **femoral artery.** Behind the knee it becomes the **popliteal artery,** and ends by dividing into the **anterior** and **posterior tibial arteries** of the leg.

Veins of the systemic circulation return blood to the right atrium of the heart. In general, the veins accompany the arteries and usually have the same names but are more numerous than the arteries and differ from them in the following ways: (1) Veins consist of two types—**superficial veins** that lie just under the skin's surface and **deep veins** that accompany the principal arteries. (2) Within the skull are found **cranial venous sinuses** that return blood from the brain. (3) Blood from the intestines is returned to circulation by way of the liver—called the *portal system.*

Blood is returned from the head and neck by the **jugular veins.** The **internal jugular vein** returns blood from the cranial sinuses and joins the subclavian vein of its own side to form the brachiocephalic vein. Lateral to the internal jugular is the **external jugular,** which drains blood from the scalp, facial muscles, and parotid salivary glands and enters the subclavian.

Deep veins run parallel to the arteries in the upper limb and enter the right and left subclavian veins, which unite with the internal jugulars to form the brachiocephalic veins. These unite to form the **superior vena cava** prior to its entrance into the right atrium of the heart.

Veins of the abdomen and pelvis include the largest vein in the body, the **inferior vena cava.** It is formed by the union of the two **common iliac veins,** which drain the lower extremities and pass upward anterior to the vertebral column and to the right of the aorta. The inferior vena cava perforates the diaphragm and ascends through the thorax to the right atrium. Along its route, numerous branches enter to return blood from various abdominal organs, such as the kidneys, testes or ovaries, diaphragm, and liver. In addition to the deep veins that run parallel to arteries, there are numerous superficial veins. Valves are present in both sets, but they are more numerous in deep veins.

Blood from each leg is returned by superficial and deep veins. The main superficial veins are the **great** and the **small saphenous veins.** The two great saphenous veins are the longest veins in the body, beginning at each foot and ascending to each thigh. They empty into the **femoral veins** in the groin region. The small saphenous veins also begin in the foot and move up the back of the leg, where they empty into the **popliteal veins** in back of the knee.

The deep veins that parallel the large arterial trunks include the **posterior** and **anterior tibial veins, popliteal vein,** the **femoral vein, the external** and **internal iliac,** and the **common iliac veins.** At the level of the fifth lumbar vertebra, the common iliacs unite to form the inferior vena cava.

Portal circulation. Blood is supplied to the intestines by the superior and inferior mesenteric arteries. From the intestines, nutrients are picked up by the capillaries in the villi, which eventually form the **superior mesenteric vein.** Along with the **splenic vein,** which receives

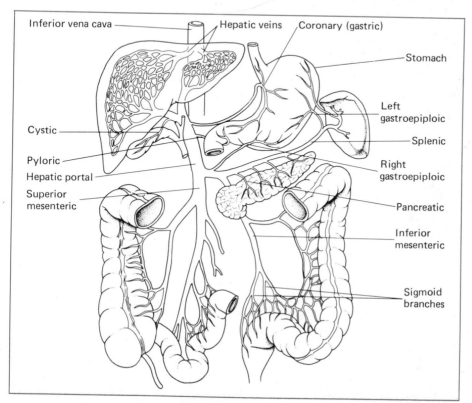

Figure 21-6 **The portal system.**

blood from the pancreas, stomach, a part of the large intestine, and gallbladder, the superior mesenteric vein and the splenic vein join to form the **portal vein,** which enters the liver. Within the liver, arterial blood from the hepatic artery mixes with portal blood and after passing through the liver eventually reaches the **hepatic veins,** which empty into the inferior vena cava. (See Fig. 21-6.) There are no valves in the portal system, so that blood can flow in either direction.

hepatic = liver

PHYSIOLOGY OF VESSELS

The human circulatory system, like that of all other vertebrates, is a closed system in which blood travels through the arteries under high pressure, and under decreased pressure in the veins. The average pressure in the aorta is about 100 millimeters of mercury, and by the time blood has traveled through arteries, arterioles, capillaries, venules and veins, the pressure has dropped to about 2 millimeters of mercury in the venae cavae. Blood pressure is the result of several factors. The five main ones are discussed below. Variations from the normal in any of these factors cause changes in the blood pressure.

Vessel Elasticity

The heart is a phasic pump; that is, the action of the ventricles causes the blood to spurt. This *phasic flow,* however, is converted to *constant flow* by the elastic vessels. If the walls of the arteries were rigid, the blood would continue to spurt with each beat of the ventricles. During the time when the ventricles are filling, there would be almost no flow of blood in the vessels. Because the walls of the arteries are elastic, they tend to expand as the blood pressure increases and contract as it decreases. This alternate expansion and contraction of the arterial walls slows or hastens the flow of blood, keeping it steady. In a child the walls of the vessels are highly elastic. With advanced age, the vessels slowly become more rigid and greater internal pressure develops.

Cardiac Output

A major factor in blood pressure is the pumping action of the heart. It is estimated that approximately 80 milliliters of blood are pumped from each ventricle at each beat. A heart beating at the rate of 70 beats per minute would discharge about 5.6 liters of blood per minute per ventricle, or about 8000 liters (over 2000 gallons) in the course of 24 hours. This volume increases in direct relation to the activity of the heart. It has been estimated that the left ventricle of an athlete competing in a strenuous race pumps about 25 to 30 liters of blood per minute. The **cardiac output,** or amount of blood expelled with each heart beat, depends on the force of that particular beat. Measurements taken in the aorta show that the pressure of the blood in the aorta as it leaves the heart may vary between 90 and 120 millimeters of mercury. The period in which the pressure of the ventricles exceeds that in the arteries is the **systolic pressure,** produced by the contraction of the ventricles. During the relaxation of the heart, the **diastolic pressure** in the aorta drops to between 60 and 90 millimeters of mercury. During this period, the ventricles relax and the pressure of the ventricles decreases until it becomes lower than the pressure within the arteries. The pressure within the ventricles may drop to between 2 and 8 millimeters of mercury.

Blood Volume

Blood pressure is partially determined by the amount of blood actually present in the system. Normally, this quantity varies only slightly, but conditions may arise in which the volume of the blood may be either increased or decreased. An extra amount of blood may be added to the arterial circulation when the spleen and large veins contract slightly and give up some of the stored or slowly moving blood. A fall in blood pressure may result from the loss of blood in hemorrhage (discussed in Chapter 20).

Peripheral Resistance

When blood travels from the larger arteries to several smaller arterioles, the pressure is greater in the smaller tubes because of friction. The slowing up of the flow builds up back-pressure in the larger tubes. Constriction of the arterioles and the precapillary sphincters under various stimuli also increases the blood pressure. An example is the response to the secretion of the adrenal glands. In moments of stress, these glands produce epinephrine, which enters the bloodstream and causes some arterioles to contract. The result is a reduced flow of blood to the capillaries and an increased pressure within the arteries.

Peripheral resistance is lowered when the arterioles and the sphincters expand, as under psychological stimuli. Thus the pain from an injury or an unpleasant experience may cause the capillaries to be suddenly filled with blood. Pressure in the brain may then fall to a point where the cells are not receiving a normal supply of blood and fainting occurs.

Blood Viscosity

The viscosity of a liquid is the degree to which it resists flow. Viscosity results from internal friction produced by the rubbing together of the tiny particles in the liquid. Blood is about six times more viscous than water. Normally, the viscosity of blood does not change and is a minor factor in determining resistance. If there is a pronounced increase in the number of red blood cells, as in the condition known as *polycythemia vera*, the viscosity of the blood increases because of friction among the red corpuscles as well as between them and the walls of vessels. This causes the blood pressure to rise.

Blood Pressure Measurement

In 1896, the modern method of measuring blood pressure was developed by Dr. Riva-Rocci. This method of measuring blood pressure in

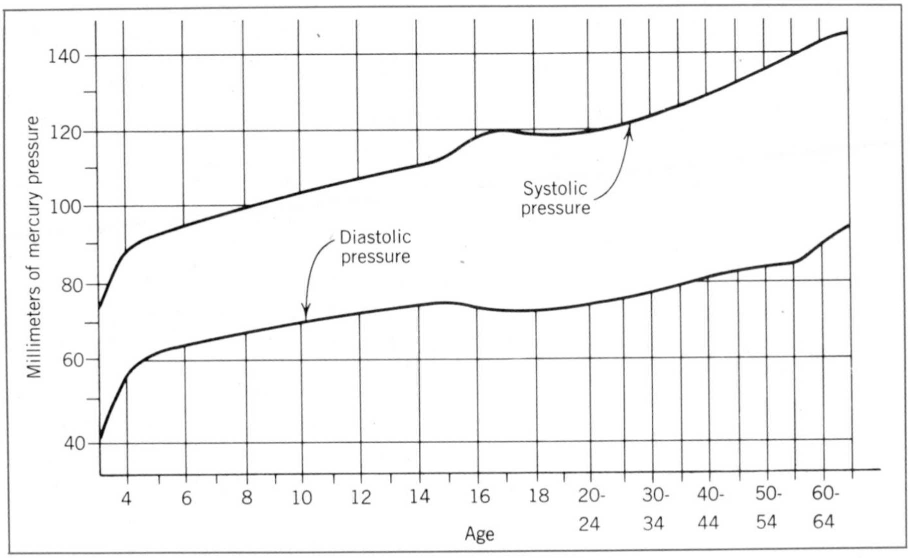

Figure 21-7 **The relation of blood pressure to age. The lines showing systolic and diastolic pressure are averages for each age group. Some variation from these averages is considered to be within the normal range.**

the arteries is based on the principle of applying just enough pressure to the artery to stop the flow of blood through it. The apparatus used is called a **sphygmomanometer.** It consists cf a cuff containing an inflatable rubber bladder which is wrapped around the arm. Connected to the rubber bladder are a bulb, by which it can be inflated, and a pressure gauge. A valve in the bulb allows air to escape slowly from the bladder. A column of mercury is used in some sphygmomanometers, while in others there is a modified aneroid manometer scaled to measure pressure in millimeters of mercury.

When blood pressure is measured, the cuff is adjusted around the upper arm and a stethoscope is placed under the edge of the cuff, above the brachial artery. Air is then pumped into the bladder until the sound of the pulse in the artery can no longer be heard through the stethoscope. Air is then released slowly. The point at which the sound reappears is noted on the pressure gauge. This gives the **systolic pressure,** because it is the pressure exerted by the wave of blood that has been ejected during the ventricular systole. As air is allowed to escape a change in the quality of the sound is heard just before the sound disappears. At this point the pressure recorded on the gauge is again noted. This gives the **diastolic pressure,** which is the pressure during the period when the ventricles are filling. Figure 21-7 shows the usual limits of the average blood pressure at varying ages. About a 5 percent variation from the averages is still considered normal.

The arterial pulse. You may be familiar with the fact that blood flows through the arteries in a series of waves. You have probably felt the pulse beat at your wrist and elsewhere in your body. However, the motion of the artery that can be felt is not the actual passage of a wave of blood that has started from the heart a short time before. There are two distinct processes involved in the pulsing of the arteries. First, there is a wave of muscular contraction and relaxation that constitutes the *pulse wave.* This is followed by the *wave of blood.*

The pulse wave originates in the aorta with each systole of the heart and spreads to all the arteries. This is a wavelike contraction that passes along the muscular coat of the arteries at a rate of between 6 and 9 meters per second. The wave of blood, expelled from the heart at the same instant, travels at a rate of only 1 to 5 meters per second. The pulse you feel does not, therefore, correspond to the wave of blood that has just left the heart. A pulse wave will travel from the heart to the arterioles in the sole of a foot in about 0.25 second, while about 7 seconds are required for the wave of blood that left the heart at the same instant to reach this region.

The venous pulse.   A venous pulse similar to the arterial pulse cannot be detected in veins. The reason for this is that the pulse wave is lost when the arteries lose their muscular coat as they merge into capillaries. Veins arise from the convergence of the capillaries. Although they contain a small amount of smooth muscle tissue, the pulse wave is not transmitted to them through the capillary network because of the lack of muscle tissue in capillaries. There is one region where a pulse exists in veins. This occurs in the large veins leading to the heart, but the origin of this pulse is quite different from that of the arterial pulse. Its cause is purely mechanical and depends on the alternate flow of blood into the atria and its sudden stoppage during systolic periods of these chambers.

Control Mechanisms

The flow of blood to different parts of the body is controlled mainly by nervous impulses to the vessels. If there is an increase in activity in a certain region, more blood passes to that region and other regions are deprived of their maximum supply. Since there is only a certain amount of blood to supply the entire body, its distribution must be carefully controlled. It might be of interest to note the following about quantities of blood in the different parts of the peripheral circulatory system:

Kind of Vessel	Percentage of Total Blood Volume
Large Arteries (down to 7 mm)	8
Small Arteries	5
Arterioles	2
Capillaries	5
Small Veins, Venules, Venous Sinuses	26
Large Veins	30

Thus the veins represent the main storage area in the body for blood, and therefore can deliver or remove blood from the remainder of the system as needed. It should be noted that 25 percent of the total blood volume is found in the heart and lungs.

A good example of controlled distribution is seen when a person goes swimming in cold water too soon after a heavy meal. The digestive processes demand a large supply of blood, with the result that other regions of the body are deprived of an adequate supply to meet emergency conditions. In such a case, the muscles may not be supplied with enough blood to maintain the temperatures they require for oxidation. Under these conditions muscles will contract violently and a cramp will result.

Vasoconstriction.   The normal distribution of blood to various regions of the body is controlled by the action of two nerve centers in the medulla of the brain. One of these, the *vasoconstrictor center*, is constantly sending out a series of impulses that keep the arterioles in a state of slight contraction, or tonus. These impulses reach the arterioles along paths that involve the sympathetic ganglia. Impulses that result in the further constriction of the arterioles are generally stimulated by such emotional states as anger or fear, or by changes in normal requirements of some part of the body for an added supply of blood. The other center is concerned with vasodilation.

Vasodilation.   The relaxation of muscle fibers in the blood vessels may be controlled by the *vasodilator center*, which lies 2 or 3 millimeters distant from the vasoconstrictor center. We cannot say definitely that impulses arising in this center are the sole cause for the dilation of the arterioles. There appears to be a reciprocal action between the constrictor and dilator centers, so that when one is stimulated the other is inhibited. At present there is no easy method of distinguishing between the active role of the constrictor center and the more passive part played by the dilator center.

The influence of the cerebral cortex.   From many parts of the cerebral cortex nerve pathways converge in the hypothalamus, which in turn influence the cardiovascular centers. For example, if one meets with an embarrassing situation, the person's face may become flushed. The steps leading to this reaction are extremely complicated because they involve memory patterns, previous training, and peculiarly individual reactions to certain situations. The flush is caused by inhibition of the vasoconstrictor center, with the result that the surface of the skin becomes filled with blood as the skin arterioles dilate. Other emotional stresses may have the opposite effect.

The effect of carbon dioxide.   In Chapter 18, you studied the effect of an increased amount of carbon dioxide on the rate of respiration. While there is some debate about the effect of this gas on the vasomotor system, there are certain well-recognized facts connected with carbon dioxide's appearance in the blood in greater than normal quantities. It has been found, for example, that in early stages of asphyxiation, blood pressure rises. On the other hand, if the carbon dioxide content of blood is lowered as a result of forced breathing, a feeling of dizziness results, which may be attributed to a decrease in blood supply to the cerebrum. The mechanism involved in these cases appears to be the presence of *chemoreceptors* in the carotid sinus and the aortic bodies. Impulses arising in these regions result in the stimulation of the vasoconstrictor center. If the activities of these centers decrease as the result of too little carbon dioxide in the blood, the blood pressure is lowered and fainting follows. Conversely, an increase in the amount of carbon dioxide in the blood results in rising blood pressure, which has a reflex action on the heart and inhibits its rate of beating. (Refer to Chapter 9 for a review of the nervous system.)

The effect of drugs.   *Epinephrine,* a product from the adrenal glands and of sympathetic effector nerve endings, has a profound effect on the action of the arterioles and precapillary sphincters. Under the influence of this hormone, the heart is stimulated and the blood pressure may rise sharply but remains elevated for only a short time. The vasoconstrictor effect is used by physicians, who may apply a dilute solution of epinephrine (about 1:10,000 parts of water) locally when performing minor operations on the eye, nasal membranes, or some other local area. This causes constriction of the arterioles in the region and helps to reduce hemorrhage.

Acetylcholine is produced at the junction of motor nerves and skeletal muscles. In skeletal muscles, it has a stimulating action that results in contraction. In the heart and blood vessels it acts as a relaxing agent and causes dilation of vessels.

Histamine is a compound that is found in all tissues. If formed in larger than normal quantities, it dilates the arterioles and capillaries. The congestion of mucous membranes, a common symptom of asthma, the common cold, and certain allergies is the result of this action. *Antihistamine* drugs have been developed to combat this congestion. These drugs do not affect the underlying causes of the symptoms, and therefore should not be considered a cure.

Alcohol is essentially a vasodilator that may produce a feeling of warmth if taken on a cold day. This effect is caused by dilation of the skin capillaries, and may result in loss of heat over the surface of the body. The sensation of warmth may be misleading if the body is later subjected to extremely low external temperatures. Some extremely low internal body temperatures [13 to 18.5°C (55 to 65°F)] have been reported following a rather excessive use of alcohol and later severe exposure to cold. This action is the result of the depressant effect that alcohol has on the vasomotor center.

Nicotine, on the other hand, has an effect that is opposite to that of alcohol. It is what physiologists call a **vasoconstrictor.** Its action, however, occurs at the postsynaptic receptors, which is the region where acetylcholine (ACh) plays its important role.

Ephedrine, a drug obtained from plants, is frequently used in nose drops. It reduces the swelling of nasal membranes by the constriction of blood vessels locally. It is doubtful whether this affects the vasoconstrictor center. The effect seems to be mainly local, confined to stimulation of the constrictor nerves in the region.

External Pressure Changes

Since the heart is a muscular pump that forces blood through the body under normal pressure, any pronounced variation from the normal may have a profound effect on the flow of blood. The blood vessels, also, are designed to withstand pressures at sea level, or within a reasonable distance above or below it. If the body is subjected to forces that exceed those normally encountered, the circulatory system has no means of adjusting to these changes. The result may be disastrous unless proper precautions are taken. For instance, a pilot may be subjected to abnormal forces when pulling the plane out of a dive or when banking it sharply around a curve. In these maneuvers, the blood is forced into the lower parts of the body with a corresponding deficiency in circulation through the brain. This may cause loss of consciousness, commonly known as *blackout*. If the maneuver forces blood away from the lower organs (as in an outside loop), the pilot may experience a condition called *redout*, in which everything is seen through a red haze. This effect is probably caused by excess blood in the vessels of the eye, and may be rapidly followed by unconsciousness. The reactions may be prevented to some degree by the use of pressure suits that can be inflated with air to produce counter-pressures that tend to neutralize the effects of these unusual forces.

DISORDERS OF THE VASCULAR SYSTEM

Of the nearly 2 million individuals that die each year, more than half of them (51 percent) die from one or more diseases of the cardiovascular system. The American Heart Association indicates that nearly 40 million Americans have some form of heart and blood vessel disease. Of these, by far the most prevalent is hypertension (high blood pressure), followed by coronary heart disease, rheumatic heart disease, and stroke.

Figure 21-8 In atherosclerosis, the passageways in arteries become narrowed by fatty deposits that form into plaque around the vessel wall. As shown in (a), (b), and (c), plaque progressively builds up to block the arterial passageway.

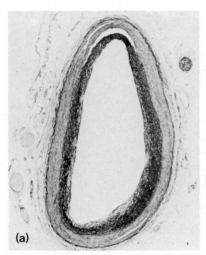
(a)

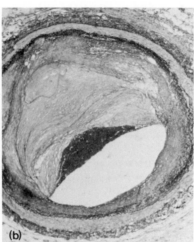
(b)

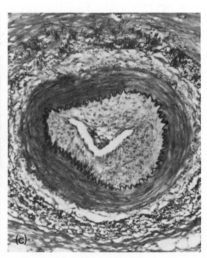
(c)

Hypertension

Nearly 25 million individuals in America suffer from **hypertension.** Less than one-third of these individuals have the condition adequately under control. However, recent data suggests that an increasing number of those with high blood pressure are aware that they have the condition and are receiving treatment. The condition may damage the kidneys, weaken blood vessels, and cause stroke and heart disease.

Conditions of hypertension. The most common kind of hypertension, accounting for nearly 90 percent of the cases, is called *primary high blood pressure* (essential hypertension). It is a hereditary type of unknown cause, for it does not seem to be related to any other disease. The remaining cases of high blood pressure are called *secondary high blood pressure.*

Arteriosclerosis

Arteriosclerosis, or hardening of the arteries, is the underlying cause of more deaths in the United States than any other disease. It is usually a slow, progressive disease and may start in early childhood. Early symptoms are usually absent even in advanced stages.

Atherosclerosis. In this disease, the linings of the arteries become thickened and roughened by deposits of lipids, especially cholesterol, proteins, and calcium, which together form *plaques.* (See Fig. 21-8.) As this plaque buildup occurs, the walls lose their ability to contract and expand. Because of increased friction in the movement of blood, it is easier for an abnormal clot or *thrombus* to form within the lumen and block the channel. This in turn raises blood pressure and may deprive the heart, brain, or other vital organs of a normal supply of oxygen. In addition, the clot may dislodge or break off and move through the vessels to become an **embolus.** If the blockage occurs in a coronary artery, the result is **coronary thrombosis,** one form of *heart attack*; or if it moves to the brain, the result is a *cerebral thrombosis,* one form of **stroke.**

Summary Table   Comparison of Blood Vessels

	Arteries	Arterioles	Capillaries	Veins
Direction of flow	Away from heart	Away from heart	Between arterioles and venules	Toward the heart
Walls:				
Tunica externa	Thick—consisting of collagen and elastic fibers	Thin	Absent	Thin
Tunica media	Smooth muscle and elastic fibers	Mainly smooth muscle	Absent	Thin—little smooth muscle or elastic fibers
Tunica intima	One layer of endothelial cells resting on connective tissue	Prominent layer of endothelial cells	Single layer of endothelial cells	Layer of endothelial cells with little connective tissue

VOCABULARY REVIEW

Match the statement in the left column with the correct word(s) in the right column. *Do not write in this book.*

1. A ring of muscle that controls the flow of blood through a capillary.
2. The smallest division of the venous system that contains smooth muscle fibers.
3. A tiny vessel that offers little resistance to the flow of blood.
4. A characteristic of blood that affects blood pressure.
5. The reduction of blood flow to a region of the body.

a. arteriole
b. capillary
c. diastolic pressure
d. precapillary sphincter
e. pulse
f. systolic pressure
g. vasoconstriction
h. venule
i. viscosity

TEST YOUR KNOWLEDGE

Group A

Write the letter of the word(s) that correctly completes the statement. *Do not write in this book.*

1. The blood vessels with the smallest diameter are called   (a) capillaries   (b) arterioles   (c) venules   (d) lymphatics.
2. Enlargement of a blood vessel is called   (a) vasodilation   (b) vasoconstriction   (c) vasopressin   (d) edema.
3. A venous blood vessel that carries nutrients to the liver is called the   (a) carotid   (b) jugular   (c) hepatic   (d) portal.
4. The muscular layer of an artery is called the   (a) tunica intima   (b) tunica media   (c) tunica externa   (d) tunica albuginea.
5. With aging, blood pressure generally   (a) increases   (b) decreases   (c) remains the same   (d) varies widely.

Group B

Write the word or words that will correctly complete the sentence. *Do not write in this book.*

1. The heart and the connecting blood vessels form the _____.
2. Veins differ from arteries in that _____.
3. The largest vein that enters the heart is the _____.
4. The vessels of the body that contain the major portion of the blood are the _____.
5. The part of the brain that controls the size of vessels is the _____.

CHAPTER 22

THE LYMPH SYSTEM

OBJECTIVES

A. Describe the structure and principal functions of the lymph system
B. List the major kinds of lymphatic vessels
C. Describe how lymph is formed and distinguish between lymph and plasma fluid
D. Name three lymphatic organs and explain the functions of each
E. Compare cellular immunity and humoral immunity and give an example of each

CIRCULATION OF THE LYMPH SYSTEM

The *lymph system* is related to the cardiovascular system in origin, structure, and function. It consists of networks of *lymphatic capillaries*, an elaborate system of collecting vessels or *lymphatics*, and two large *lymph ducts.* These ducts connect with the venous circulation and empty into the bloodstream at the junction of the jugular and subclavian veins. Interspaced along the length of the lymph vessels are numerous clusters of cells called *lymph nodes.* The nodes act as filters and producers of lymphocytes, which prevent harmful substances from entering the body tissues. Interstitial fluid that does not return directly to the capillaries but enters the *lymphatics* instead is called *lymph.* Lymph acts as a carrier of large protein molecules and also of fats, which enter the lymph in the small intestine and are carried into the bloodstream as lymph enters the veins.

Structure of Lymphatic Vessels

Fluids are transported to the tissues of the body by the circulatory system. The transporting vessels of the lymphatic system assists in returning this tissue fluid to the bloodstream.

Lymphatic capillaries. Lymph capillaries are microscopic tubes with walls composed of thin, flat endothelial cells. Their structure is similar to the blood capillaries in that both derive from the same venous en-

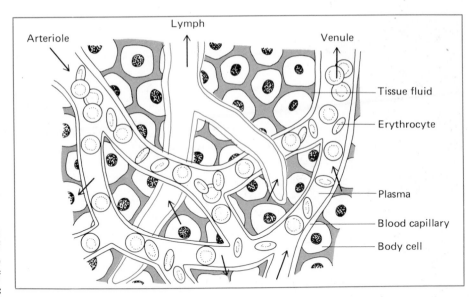

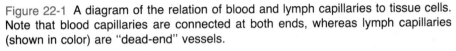

Figure 22-1 A diagram of the relation of blood and lymph capillaries to tissue cells. Note that blood capillaries are connected at both ends, whereas lymph capillaries (shown in color) are "dead-end" vessels.

Figure 22-2 (a) Valve in a lymphatic. The flaps of the valve are closed to prevent the backflow of lymph. (b) Lymphatic vessel laid open to show the valves.

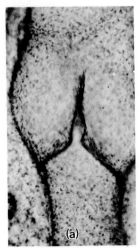

(a)

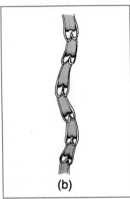

(b)

dothelium. However, the walls of the lymph capillaries are more permeable than blood capillaries; thus such large particles as free plasma proteins and large foreign substances can pass into the lymphatic system. Lymphatic capillaries are larger in diameter than blood capillaries and form vast branching networks throughout most of the body. The lymphatic capillaries found in the villi of the small intestine are called **lacteals.** These vessels are responsible for the absorption of fats from the intestine and their distribution to the vascular circulation.

Lymphatic veins.   The various networks of lymphatic capillaries connect and drain into a system of lymphatic veins (Fig. 22-1). These lymphatic vessels are similar to those of blood veins in that both consist of three similar tissue layers: (1) an internal coat of thin, transparent, and slightly elastic elongated endothelial cells; (2) a middle coat of smooth muscle with a few elastic fibers; and (3) an external coat of connective tissue with few smooth muscular fibers. The walls of lymphatics are so thin and transparent that their tissue fluid is visible. Collecting lymphatics contain *valves* similar to the valves found in veins. The valves of lymph veins occur so frequently that they give the lymphatic surface a beaded or knotted appearance (Fig. 22-2).

The lymph nodes.   Situated along the route of the lymphatics are many lymph nodes (Fig. 22-3). Each node consists of lymphatic tissue enclosed in a capsule of fibrous connective tissue. These bean-shaped structures range from 1 to 25 millimeters in length. The slight depression on the *concave* side, called the **hilum,** is where *efferent* lymphatics

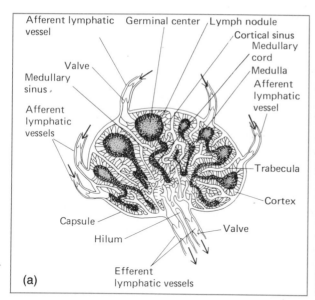

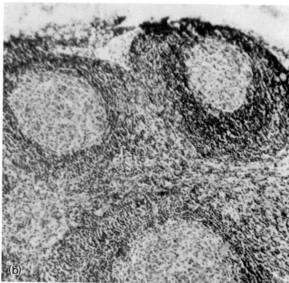

(a) Labels: Afferent lymphatic vessel, Germinal center, Lymph nodule, Cortical sinus, Medullary cord, Medulla, Afferent lymphatic vessel, Valve, Medullary sinus, Afferent lymphatic vessels, Trabecula, Cortex, Capsule, Hilum, Valve, Efferent lymphatic vessels

Figure 22-3 (a) Structure of a lymph node. The route of lymph is shown by the arrows. (b) Photomicrograph of a section of a lymph node.

leave the node. *Afferent* lymphatics enter the node along the convex side and are more numerous than efferent vessels.

In examining a node cross section we find that the node is divided into compartments by a framework of partitions called ***trabeculae.*** The filtering portion of the node consists of lymph sinuses, a system of microscopic channels lined with *phagocytic* cells. The movement of lymph is slow as it passes through these tiny channels, thus permitting bacteria and other foreign materials to be engulfed by the phagocytes. In addition, lymphocytes are added to the lymph before it passes from the node into the efferent lymphatics.

While lymph nodes are found throughout the body, they are more prevalent in certain areas of the body, such as the cervical region, or neck; in and around the armpit (axilla); the inguinal, or groin, area; deep within the iliac, lumbar, and thoracic regions; and in the mesentery of the small intestine.

The lymphatic ducts.   Ultimately the lymphatics join to form two large terminal ducts: the ***right lymphatic duct*** and the ***thoracic duct.*** Several lymph vessels join the right lymphatic duct prior to its entrance into the right subclavian vein of the blood vascular system. These lymphatics drain lymph from the right side of the head, neck, and upper extremity; the right side of the thorax and of the heart; the right lung; and part of the upper surface of the liver. The right lymphatic duct is a short tube about 1.25 centimeters in length.

The thoracic duct, or *left lymphatic duct*, receives lymph from a more extensive area of the body. Lymph coming from the lower extremities, the intestine, the pelvic region, and the kidneys collects at the level of the second lumbar vertebra and drains into the lymphatic called the *cisterna chyli.* From this point this portion of the thoracic duct extends upward through the diaphragm and along the posterior mediastinal cavity. At the level of the fifth thoracic vertebra the duct empties into the left subclavian vein at an angle formed by its junction with the left internal jugular vein.

The left lymphatic duct receives lymph from the left side of the head, neck, and chest, the left upper extremity, and the entire lower body. In the adult, the duct is about 38 to 45 centimeters in length. It contains valves that allow lymph to move in only one direction and that deliver lymph to the bloodstream. Figure 22-4 shows the location of the right and left ducts and the distribution of the lymphatic vessels.

Organs of the Lymph System

The lymph nodes are composed of lymphatic tissue. The **tonsils, thymus,** and **spleen** are also organs consisting of lymphatic tissue. They like lymph nodes produce lymphocytes.

Tonsils.   The tonsils are masses of lymphoid tissue embedded in mucous membrane. The **palatine tonsils** are two prominent masses with one located on each side of the palatine arches in the back of the mouth. There are also **pharyngeal tonsils** (also called *adenoids*) located on the posterior wall of the nasopharynx. The surface of these structures is marked with pits called **crypts,** which are surrounded by lymphoid tissue. Blood vessels and nerves supply each tonsil. The tonsils are usually larger in children, but after puberty they become smaller in size and assume a disclike shape.

Thymus.   The *thymus gland* is located in the mediastinum behind the sternum. In children it appears as two lobes and is pinkish gray in color. The thymus reaches its maximum size during puberty. By age 25 it begins to decrease in size, and by old age it may diminish or disappear entirely. As an individual grows older, the thymus' lymphoid cells are replaced by fat and connective tissue.

Spleen.   The spleen is the largest organ of lymhatic tissue in the human body. It may vary in size, but in the adult years it is about 12 centimeters in length, 7 centimeters in width, and 3 to 4 centimeters in thickness. The spleen is located just below the diaphragm, lying to the left of and slightly posterior to the stomach. In general the shape of the spleen is oblong and flattened, it is soft in texture, and it has a dark purplish color. In front and on the midsurface is an indentation called the **hilum** through which blood vessels, lymphatic vessels, and nerves enter.

The surface of the spleen is covered by a capsule consisting of two layers: an *outer serous* and an *internal fibroelastic* coat. In the area of the hilum, the fibroelastic connective tissue forms an internal framework of trabeculae. Between the trabeculae are small spaces, or **areolae,** that contain **splenic pulp,** a highly vascular lymphatic tissue.

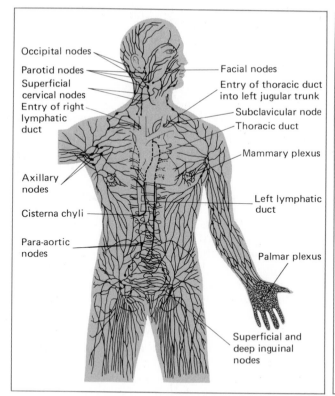

Figure 22-4 The lymphatic system with its distribution of vessels.

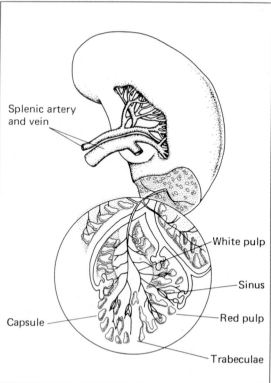

Figure 22-5 The splenic vessels and microscopic view of a cross section of the internal structure of the spleen.

The splenic pulp is composed of **white pulp** and **red pulp.** The white pulp is primarily lymphatic tissue that forms the lymphatic nodules. They are the source of lymphocyte production in the spleen. In addition, the white pulp forms **plasma cells** when bacteria or other foreign proteins enter the bloodstream. The red pulp contains large numbers of red blood cells, reticular cells, and leukocytes (neutrophils and monocytes).

The blood supply passes into the large splenic artery and through its several branches to enter the hilum of the spleen. After passing through the red pulp of the spleen, the blood collects in veins and travels to the splenic vein, which goes to the portal system.

The spleen is not considered an essential organ since a person can continue a normal life after it has been surgically removed. Apparently other organs in the body can take over splenic function. See Fig. 22-5.

FUNCTIONS OF THE LYMPH SYSTEM

The lymphatic system is closely associated with the circulatory system in its principal functions: (1) It collects and transports fluid from the intercellular spaces of the tissues and returns it to the blood by way of lymph vessels. (2) Through its immune mechanism, it provides protection from invading bacteria and other foreign materials. (3) The lymph system absorbs fats from the villi of the intestinal tract and transports them to the bloodstream.

Formation of the Lymph

As part of the body's homeostasis, the distribution of fluid volumes must be carefully maintained. The total quantity of fluids in *all* the cells of the body is commonly known as **intracellular fluid.** The remaining **extracellular fluid** is distributed in blood plasma; the fluid found between the cells, called **interstitial fluid**; and the fluid that enters lymph capillaries as lymph. The volume of lymph that returns to the bloodstream is about 120 milliliters per hour.

Intracellular fluid.   The intracellular fluid contains approximately 25 percent protein materials by volume; the remaining 75 percent is an aqueous solution of both organic materials and inorganic salts. Plasma, on the other hand, contains about 8.0 percent protein materials, while there is only about 2 percent protein in the interstitial fluid. The reason for these differences is that protein molecules are generally too large to pass easily through cell membranes. The concentration of protein material within cells is a contributing factor in maintaining **osmotic pressure** in the cells.

Interstitial fluid.   The interstitial fluid bathes the cells and serves as an intermediary between the blood and the cells. Since blood capillaries do not come into direct contact with all cells, it is necessary for some fluid to act as a carrier between the blood and cells. This is the primary function of interstitial fluid. It comes from the portion of the blood plasma that diffuses through the walls of the capillaries. This process is called **transudation.** The plasma may pass through the limiting membrane of the endothelial cells, or through small openings or pores between these cells. The pressure exerted by fluids within the capillaries (hydrostatic pressure) is greater at the arteriole end of the capillary bed. As a result a filtration gradient causes some plasma proteins to be forced out of the bloodstream to enter the interstitial fluid. The pressure is less at the venule end of the capillary bed and only some of the interstitial fluid containing protein re-enters the capillaries.

Medium for transport.   The fluid that leaves the blood gradually filters through the spaces among the cells. It serves as the vehicle for exchange of those materials that are essential for maintaining cellular activities. Oxygen, for example, is released into the plasma by hemoglobin as it reaches an area of low oxygen tension. The gas, carried by the plasma solution, diffuses through the capillary wall and enters the cells that lack oxygen. Other materials are distributed to cells in the same way.

In addition to the required materials that pass into cells, various metabolic wastes diffuse out of the cells and into the interstitial fluid. Some of the wastes enter directly into the bloodstream because of the difference in osmotic pressure between the tissue fluid and plasma.

The osmotic pressure of plasma is lower at the beginning of the arteriole portion of the capillary and becomes higher toward the venule end. The osmotic gradient that exists at the arteriole end of the capillary forces materials to exit from the blood. The opposing osmotic gradient at the venule end permits the reabsorption of other substances. Other materials, including proteins that have escaped from the plasma and excess interstitial fluid, can enter the more porous dead-end lymphatic capillaries as lymph.

Flow of lymph. The lymph capillaries are found in all the tissues. Gradually they join together forming larger tubes, the lymphatics. These vessels contain one-way valves that control the direction of flow. Lymph is pushed through these tubes by the massaging or milking action of the tissues in which they lie. The flow is aided by the contractions of skeletal muscle and the smooth muscle layer of the abdominal walls. Lymph flows much more slowly than blood.

Immunity

Immunity involves physiological activities that alert the body to the entrance of foreign materials and destroy or render them harmless. Foreign substances may include: bacteria, viruses, various unicellular and multicellular organisms, or specific proteins introduced into the body as organ transplants. Organisms may produce harmful toxins. The ability of the body to distinguish its own substances (self) from foreign materials (not self) that enter the body and to dispose of them is a key factor in an immune response.

Immune response. Besides the various protective barriers (the layer of skin, the lacrimal, sebaceous, and sweat glands, and the sticky mucous membrane lining of passages), the body uses two specific immune cells for defending itself. These are known as **T lymphocytes** (**T cells**) and **B lymphocytes** (**B cells**). All lymphocytes are originally formed from stem cells in bone marrow, from which they later migrate. Some of them enter the *thymus*, and a number travel to lymph nodes and other lymphatic tissue throughout the body. The T lymphocytes enter the thymus and change in such a way as to be able to recognize and destroy foreign material. They leave the thymus and enter the lymph nodes, where they carry on **cellular immunity** involving allergic reactions, rejection of foreign tissue transplants, viruses, and certain cancer cells.

The foreign material, called an **antigen,** is attacked by the T lymphocytes. The T cells secrete toxins and enzymes that dissolve the cell membrane of the antigen and ingest the contents, destroying the cell. T cells, once sensitized to a specific antigen, begin to divide and produce numbers of similar T cells. These cells may remain in the blood for a long period of time, thus maintaining a ready defense against the same type of antigen.

The B lymphocytes enter the lymph nodes, spleen, and other lymphatic tissues. The B cells change into *plasma cells* when they are exposed to a specific antigen (often a virus or bacterium). Plasma cells are responsible for producing **antibodies** or **immunoglobulins,** which both destroy a specific kind of antigen. The antibodies are released into the bloodstream and may remain there for a great length of time. Some of the B lymphocytes may remain in the lymph nodes, where they will change into antibody-releasing plasma cells when the same antigen reappears. This type of immunity is called **humoral immunity** and involves the plasma or fluid part (humor) of blood.

Viruses or bacteria, called **pathogens,** cause acute infections. Antibodies may prevent a virus from attaching to a host cell. In other instances, antibodies may cause invading bacteria to clump together, thus enabling them to be phagocytized (engulfed) by macrophages. Certain antibodies neutralize the toxins produced by pathogenic organisms, thus rendering them harmless.

Functions of the spleen. The spleen has four main functions: (1) It clears the blood of damaged blood cells and of foreign substances in plasma. (2) It is one of the sites for the transformation of B lymphocytes and monocytes into plasma cells and macrophages. (3) It reacts to blood-borne antigens by producing specific antibodies. (4) It normally contributes new lymphocytes to the blood. If the spleen is surgically removed, the liver and bone marrow take over some of its functions.

Absorption of Lipids

In the center of each villus, found in the inner lining of the small intestine, is a lymphatic capillary called a **lacteal.** These vessels are responsible for the absorption and transport of digested fats into the bloodstream. The emulsifying action of bile salts makes it possible for the lacteals to absorb the products of fat digestion, namely fatty acids and monoglycerides. These products are further changed so that they can diffuse through the epithelial lining of the small intestine. Within the epithelial cells monoglycerides are converted to glycerol and fatty acids, which form triglycerides. They pass out of the epithelial cell with other fats as tiny lipid droplets called **chylomicrons** and enter into the lacteal. The chylomicrons flow into larger lymphatics and into the circulation of the bloodstream.

DISORDERS OF THE LYMPH SYSTEM

Normally the body's immune system is able to recognize and dispose of harmful substances. Individuals seldom have lymphocytes that are activated to produce antibodies against their own cells. Such a recognition is called **tolerance.** Sometimes, and for no apparent reason, the immune mechanism breaks down and the T cells and B cells start to reject the body's own constituents. Several diseases are thought to be caused by this **autoimmunity.** Some of these diseases are rheumatic fever, myasthenia gravis, multiple sclerosis, and systemic lupus.

Hodgkin's Disease

Hodgkin's disease is sometimes called cancer of the lymphoid tissue. This disease affects the lymph nodes and spleen, which become greatly enlarged due to an increase in cell production. The symptoms are similar to many cancers in that there is a loss of weight, fever, anemia, and possible cough if lymph nodes in the thoracic cavity are involved.

Edema

Edema is not a disease in itself, but is a symptom in a number of illnesses. The condition results from an abnormal accumulation of interstitial fluid within the tissues of the body. There are a number of possible causes of edema: (1) cardiac failure, which is the most common cause and affects the entire body; (2) kidney disease; (3) an increase in capillary permeability; and (4) some blockage of lymphatic drainage, as seen with inflammation, injury, surgery, or parasitic infection. Since the lymphatic system is closed, any blockage will cause an accumulation of lymph.

Tonsillitis

A bacterial infection of the tonsils results in tonsillitis. Many organisms may be involved but the most common is streptococcus. It may represent a secondary infection accompanying the common cold. Because bacteria may be the causative agent, the infection may spread to other individuals and thus cause an epidemic in a close community. Tonsillitis may be symptomatic of such other diseases as diphtheria, measles, and scarlet fever. If tonsillitis becomes chronic, a tonsillectomy, or adenoidectomy, or both, may be performed.

SUMMARY

The lymph system is closely related to the cardiovascular system. It is a closed system of vessels that drains lymph in one direction, eventually joining with the veins of the blood. The smallest lymph vessels are the lymph capillaries. The lymphatic capillaries in the villi are called lacteals. Lymphatics are larger vessels, containing valves similar to those of the blood system. Groups of lymph nodes are found in the neck, under the arms in the axillary area, in the groin, and in the abdomen and chest. After passing through the lymph nodes, the lymph enters the blood.

The primary functions of the lymph system are to: (1) return interstitial fluid to the bloodstream; (2) provide an immune mechanism to remove anything that is foreign from the body; and (3) absorb fats from the digestive tract and transfer them to the bloodstream via the lymphatics.

Disorders of the lymph system include autoimmune diseases, or abnormal immunity where antibodies produced in the body work against the body's own cells.

VOCABULARY REVIEW

Match the statement in the left column with the correct word(s) in the right column. *Do not write in this book.*

1. Blind-end microscopic lymphatic tubes.
2. Lymph vessels found in the villi of the digestive system.
3. Vessels that are thinner, but have the same structure as veins.
4. Numerous lymph filters.
5. Lymphatics converge to form these structures.
6. The major lymph drainage enters this passageway.
7. The largest lymphatic organ.
8. A gland that is related to immunity.

a. adenoids
b. ducts
c. lacteals
d. dead-end lymph capillaries
e. lymphatics
f. nodes
g. right lymphatic duct
h. spleen
i. thoracic duct
j. thymus

TEST YOUR KNOWLEDGE

Group A

Write the letter of the word(s) that correctly completes the statement. *Do not write in this book.*

1. In a lymph node, efferent lymphatics leave by way of a slight depression called a   (a) trabecula   (b) lacteal   (c) hilum   (d) cortical sinus   (e) germinal center.
2. Lymph moves through lymphatics by   (a) the action of the heart   (b) contraction of skeletal muscles   (c) suction   (d) hydrostatic pressure   (e) blood pressure.
3. Cell-mediated immunity directly involves   (a) phages   (b) toxins   (c) antibodies   (d) B lymphocytes   (e) T lymphocytes.
4. The removal of dead erythrocytes is the responsibility of the (a) lymph nodes   (b) heart   (c) spleen   (d) thymus   (e) tonsils.
5. The body reacts against its own cells in   (a) immunity   (b) awareness   (c) autoimmunity   (d) tolerance   (e) conceit.
6. Lymphatic tissue found along the palatine arch makes up the (a) spleen   (b) tonsils   (c) adenoids   (d) thymus   (e) tongue.

Group B

Write the word or words that will correctly complete the sentence. *Do not write in this book.*

1. The lymph system is closely related to the _____.
2. The lymph system collects _____ and returns it to the blood by way of the lymph vessels.
3. Collecting lymphatics, like veins, contain _____.
4. Fluid found in all of the cells of the body is commonly called _____ fluid.

UNIT 7

THE REGULATORY SYSTEMS AND METABOLISM

CHAPTER 23

THE SKIN AND ITS APPENDAGES

OBJECTIVES

A. Describe the gross anatomy of the skin, including the layers of the skin
B. List and describe the functions of the skin
C. Describe the appendages of the skin and their functions
D. Compare apocrine and eccrine glands

GROSS ANATOMY OF THE SKIN

The skin is the body's largest and one of its most important organs. Its role in the human body can be described as vital, diverse, complex, and extensive. Its surface area is as large as the body itself, in average-sized adults roughly 10 to 17 square feet. Its thickness varies from slightly less than 1/50 of an inch to slightly more than 1/8 of an inch. The functions of skin are crucial to survival. They are diverse because they include protection, excretion, and sensation. In addition, the skin plays a part in maintaining water balance in the body, as well as regulating electrolytes necessary for the proper functioning of all body organs. Maintaining normal body temperature is another essential role of the skin.

As a protective organ the skin prevents entry of most microorganisms. It also lessens mechanical injury of underlying structures when tissue is damaged. Ordinarily the skin does not allow penetration of water, most chemicals, and excessive sunlight.

Millions of microscopic nerve endings are distributed throughout the skin. They are sensors that keep the body informed of changes in the environment, providing information that is essential for health and at times vital for survival.

The skin is also called the **integument** and **cutaneous membrane.** As a membrane it is far more complex than other membranes of the body. Since the skin is exposed, it is subjected to many more dangers than are internal membranes, and its structure varies accordingly. Also, depending on its location in the body, skin structure can vary. For ex-

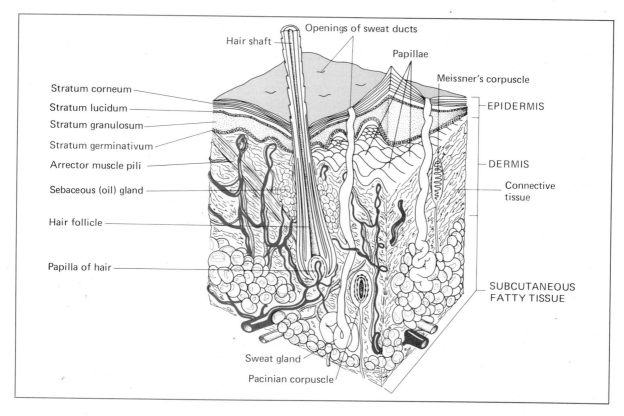

Hair shaft

Openings of sweat ducts

Papillae

Meissner's corpuscle

Stratum corneum

Stratum lucidum

Stratum granulosum

Stratum germinativum

Arrector muscle pili

Sebaceous (oil) gland

Hair follicle

Papilla of hair

EPIDERMIS

DERMIS

Connective tissue

SUBCUTANEOUS FATTY TISSUE

Sweat gland

Pacinian corpuscle

Figure 23-1 Microscopic view of the skin. The epidermis is raised at one corner to reveal the papillae.

ample, the skin is thicker on the palms of the hands and the soles of the feet than it is over the eyelids, because the hands and feet are subjected to more wear and tear.

The Layers of the Skin

The skin covers the entire surface of the body. It consists of two main layers that are quite different in structure and function: the **epidermis** and the underlying **dermis.** These layers are firmly bound together. However, excessive rubbing of the skin, as happens when wearing poorly fitted shoes, for example, may cause the epidermis to be separated from the dermis. Tissue fluid then accumulates in this area, further separating the two layers and causing a blister. Immediately under the dermis, but not part of the skin, is **subcutaneous tissue,** or **superficial fascia.**

epi = on the outside
dermis = skin

fascia = thin layer of connective tissue

The epidermis.   The epidermis is composed of stratified squamous epithelial tissue, which makes up the thin surface layer of skin. Since epithelial cells have no blood vessels, the only living cells are in the deepest layer of the epidermis, which is nourished by tissue fluid from the capillaries in the dermis. These cells are constantly undergoing mitosis, and the daughter cells are pushed toward the surface, away from the tissue fluid. As these cells die from lack of nourishment, they undergo a chemical transformation and their soft cytoplasm becomes filled with a waterproof protein known as **keratin.** When this occurs, the cells have become *keratinized,* or *cornified.* These cells are waterproof and prevent evaporation of water from the deeper layers of the skin.

The epidermis of thick skin has four layers. These layers are the *stratum corneum,* or *horny layer,* which consists of dead keratinized cells; the *stratum lucidum* and the *stratum granulosum,* which are middle layers where the epithelial cells gradually die and become keratinized; and the *stratum germinativum,* which is adjacent to the dermis and contains only epidermal cells that can reproduce themselves.

stratum = layer

The **stratum corneum** is one of the principal defenses of the body against invasion by bacteria and against mechanical injury to the delicate underlying tissues. So long as this outer layer remains unbroken, bacteria cannot easily invade the body. In fact, the protective action afforded by this keratinized layer is reportedly equal in value to the immunizing action of the blood in preventing infections. The thickness of the horny layer varies in different regions of the body. In some areas it is quite thin, while in others it becomes thick as the result of friction. When the thickening develops outwardly in a local region, such as the palms of the hands, the enlargement is known as a *callus.* If the thickening results from inward growth of the stratum corneum, as it does occasionally on the toes, the growth is called a *corn.*

Historically, the skin was considered to be impermeable to all substances. Today researchers know that gases and lipid-soluble substances pass relatively easily through the skin. Skin, however, remains essentially impermeable to electrolytes and water.

An initial barrier to the penetration of substances is formed by an electrically charged double-layered structure, located between the junction of the keratinized and nonkeratinized layers of cells. It is this electrical barrier that probably limits the passage of electrolytes. A second barrier exists at the basement membrane upon which the stratified squamous epithelium of the epidermis rests. Most substances that do not have an electrical charge appear to be held up at this point. If a substance passes these two barriers, no restriction is imposed on its entry into the blood. Substances that are thought to be able to pass these barriers are phenols, such hormones as testosterone and estrogen, vitamins A, D, and K, aspirin, gases (O_2, CO_2), and organic bases.

The **stratum lucidum** is present in thick skin only. It is named *stratum lucidum* because of the presence of a translucent compound called *eleidin,* from which the keratin forms.

The **stratum granulosum** layer is so named because of the granules visible in the cytoplasm of its cells.

The cells in the stratum germinativum produce **melanin,** the pigment responsible for differences in skin color. Skin color comes in a wide assortment of colors and hues. All peoples have some melanin in their

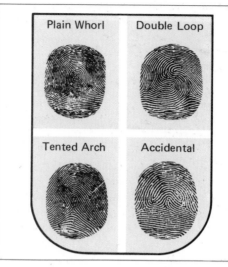

Figure 23-2 (left) The surface of the skin magnified 100 times. The ridges are papillae.

Figure 23-3 (right) Four typical fingerprint patterns.

skin. However, skin color can be modified by numerous other factors. Heredity is a crucial factor in determining skin color. Geneticists now believe that four to six pairs of genes exert the primary control over the amount of melanin formed by melanin-producing cells called **melanocytes.**

Thus heredity determines how dark or light one's basic skin color will be. But other factors can modify the genetic effect. Sunlight is the obvious example. Prolonged exposure of the skin to sunlight causes melanocytes to increase melanin production and darken skin color. Melanocytes will also increase melanin production under the influence of an excess of adrenocorticotropic hormone (ACTH) or an excess of melanocyte-stimulating hormone (MSH), two of the hormones secreted by the anterior pituitary gland. Freckles are merely irregular patches of melanin. An absence of melanin due to a genetic defect is called *albinism.*

The production of excess pigment is also stimulated by ultraviolet rays. It is these solar radiations that are responsible for sunburning and the subsequent tanning of the skin. Ultraviolet rays in the neighborhood of 3200 angstrom units (visible violet rays are about 4000 angstrom units in length) cause the familiar reddening of exposed skin. Following the sunburn, a moderate tanning will appear. Continued exposure will darken the skin color as more pigment is formed in the cells. It should be remembered that this effect is a defense mechanism the body has to protect the lower-lying tissues from injury by ultraviolet rays. The wisdom of prolonged exposure to sunlight is being questioned at present because of its possible connection with the development of skin cancers and premature aging of the skin.

The dermis.   The dermis is the deeper layer of the skin. It is composed of two layers of connective tissue. The upper layer, or **papillary layer,**

is composed primarily of ordinary loose connective tissue, fine elastic fibers, and an extensive supply of capillaries, which provide nourishment for the living cells of the epidermis.

The capillaries also play a role in temperature regulation. In thick skin the papillary layer is arranged in ridges. The epidermis follows the contours of the papillary layer, and this produces epidermal ridges that can be found on the fingers, palms of the hand, toes, and soles of the feet. These ridges develop during the third and fourth months of fetal development in a pattern that never changes except to enlarge. The unchanging pattern formed by these ridges is peculiar to each individual. There is even a difference between patterns in identical twins. Accordingly, fingerprints and footprints are excellent means of identification. The medical and legal use of skin ridges is called *dermatoglyphics*. See Fig. 23-3.

The lower layer, or **reticular layer,** is much thicker than the papillary layer, and it consists of dense, irregularly arranged connective tissue. There are interlacing bundles of collagenic fibers and coarse elastic fibers, which give strength, extensibility, and elasticity to the skin.

The Subcutaneous Tissue

Beneath the dermis lies the subcutaneous tissue, or superficial fascia. This tissue differs in various parts of the body and in different individuals. It may consist of loose adipose (fatty) tissue, or a dense type of connective tissue. The dermis is anchored to the subcutaneous tissue by collagenic fibers. The subcutaneous tissue then attaches the dermis to such underlying structures as muscles and bones.

In young people, the skin is extensible and elastic, but as they grow older certain changes occur. The skin becomes thinner, there are fewer elastic fibers, and fat disappears from the subcutaneous tissue. This combination of events results in skin that is wrinkled in appearance.

Appendages of the Skin

The cutaneous glands, the hair, and the nails comprise the appendages of the skin. During fetal development, they are derived from differentiating epidermal cells that grow down into the developing dermis.

Glands of the skin.   There are two major types of cutaneous glands associated with the skin: **sebaceous** and **sweat** (*sudoriferous*) glands.

Sebaceous glands are found almost everywhere on the surface of the body except for the palms of the hands and the soles of the feet. These glands lie in the dermis. Their excretory ducts usually open into the upper part of hair follicles. Sebaceous glands produce an oily secretion called *sebum* that prevents hair from becoming dry and brittle. Sebum also forms an oily layer on the skin surface, keeping it soft and helping it to be waterproof. Persons with underactive sebaceous glands tend to have dry hair and skin, while those with overactive glands have hair and skin that are more oily. Blackheads are discolored accumulations

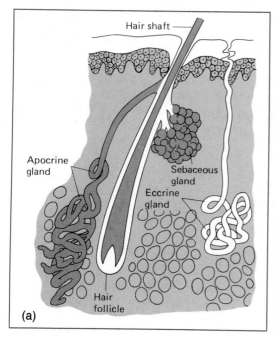

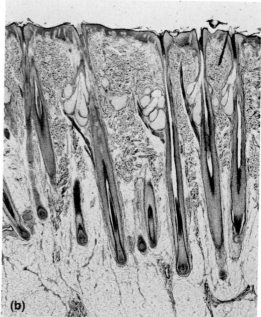

of dried sebum. Acne may be the result of an infection of the sebaceous glands.

Sweat glands are of two general types: **apocrine glands,** which are associated with hair follicles; and **eccrine glands,** which are not. The coiled secretory parts of both types are located in the subcutaneous tissue immediately below the dermis.

Apocrine sweat glands are not very numerous. They are particularly found under the arm (*axilla*) and surrounding the anal and genital regions. The ducts of these tubular glands empty into hair follicles just above the openings of the sebaceous glands. Upon reaching the body surface, the milky, sticky secretions of the apocrine glands are decomposed by the action of bacteria that are normally present on the skin. These products of decomposition give rise to the distinctive odor of sweat. Apocrine secretion is increased under conditions of emotional excitement.

Other apocrine glands believed to be related to the sweat glands are the compound tubular glands of the *areola* (the pigmented area surrounding the nipple) of the breasts, the relatively simple *ciliary glands* of the eyelids, and the *ceruminous* (*wax*) glands of the external auditory meatus of the ear.

Eccrine sweat glands are the most common type in humans. They are distributed over most of the body, except in a few areas, such as the lips. Their excretory ducts pass upward in a spiral manner, through the dermis and epidermis, and open as pores on the surface of the skin. Eccrine glands are the major source of sweat. They secrete a mixture of water, salts (mostly sodium chloride), and such waste products as

Figure 23-4   (a) The sweat glands. Note that the apocrine gland opens into the hair follicle, while the eccrine gland opens directly onto the surface of the skin. (b) Photomicrograph of sweat glands.

urea and uric acid. These glands are supplied with nerves that promote sweating when it is necessary for the body to lose heat by the evaporation of water from its surface.

Hair.   In the third month of fetal development, the epidermis begins to send downgrowths into the underlying dermis or even into the subcutaneous tissue. These epidermal downgrowths become hair follicles and give rise to hair. With certain exceptions, such as the palms and soles, hair is present over most of the body surface, but it is most abundant on the scalp.

Although a part of each hair lies within the dermis, it is actually surrounded by an inward projection of the epidermis, which forms a sloping tube, the *hair follicle.* At the lower end of this tube is an upward projection of the surrounding tissue, the **papilla** of the hair. The papilla contains capillaries to nourish the cells of the follicle, from which the hair is formed by the rapid division of cells at its lower end.

Microscopic examination of a hair shows that it is composed of three distinct parts. On the outside is the **cuticle,** a single layer composed of flat scalelike structures in which the cytoplasm has been replaced by keratin. These overlap each other slightly and give the surface a scaly appearance. Within the cuticle is a keratinized layer of elongated, nonliving cells that make up the **cortex** of the hair. The pigment of the hair is in the cortex and the central cavity, or **medulla,** if one is present.

The distribution of hair over the body, its type, and its color are the result of hereditary factors that are beyond the scope of this text. However, two common conditions should be discussed briefly: these are the graying of hair and the loss of hair. White hair has pigment only in the medulla, while blond hair has no medulla and just a small amount of color in the cortex. The period of life at which graying of the hair occurs appears to be genetically determined. There is no evidence that it is connected with any vitamin deficiency, as is the case in some animals. The loss of hair may be due to several causes. First, if the scalp is too tightly drawn over the bone framework of the skull, a temporary loss of hair may occur. This condition is the result of interference with the blood circulation to the hair follicles and can be relieved by massage. Illness, when accompanied by a very high fever, may result in the loss of hair. If the body temperature does not reach a point where there is destruction of the cells in the hair follicles, the hair will return. Local diseases of the scalp may also cause the loss of hair over small areas. A common infection of this type is caused by the growth of a fungus and is known as **ringworm** of the scalp. The most common type of baldness seems to be hereditary. Inheritance of certain factors determines not only the loss of hair but also the approximate time of life when this occurs and the pattern that it follows. Aging and the activities of the male sex hormones are also factors connected with baldness.

Attached to each hair follicle on the side toward which it slopes is a small bundle of smooth muscle fibers, the **arrector pili muscle.** This has its origin on the upper surface of the dermis and its insertion on the wall of the follicle. In some animals, the contraction of this muscle causes the hair to be erect, but in humans the hair is too weak to allow this action to occur. When this muscle of the skin is stimulated, as by sudden chilling, it contracts and causes the skin to pucker around the hair. This produces the condition commonly called *goose flesh*. Contraction of the muscle also results in pressure on the sebaceous glands and the secretion of a small amount of oil.

Nails.   The nails are also classified as appendages of the skin. These are hard structures, slightly convex on their upper surfaces, and concave on the lower surfaces. They are formed by the epidermis, where they first appear as elongated cells that then fuse together into plates. During this process their protoplasm is replaced by keratin. The area producing the nail is called the **nailbed,** or **matrix.** As long as any part of this region remains intact, it will produce the nail substance. It is possible, therefore, for a fingernail to be lost and later be replaced by a new one.

Sense Receptors

Receptors in the skin differ considerably in structure—from very highly complex to very simple. The simplest receptors are merely free

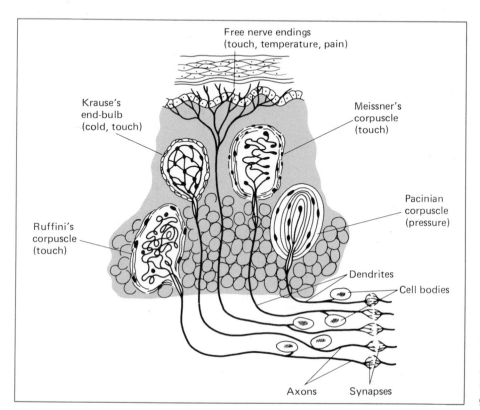

Figure 23-5 Diagram of various sense receptors. The cell bodies and synapses of these afferent neurons are located in the dorsal root ganglia and spinal cord.

nerve endings of sensory dendrites in the skin that give rise to sensations of touch, pain, and temperature. These free nerve endings are distributed in the dermis of the skin and the cornea of the eye and surround the root of a hair follicle.

Somewhat more complex skin receptors are those named **Meissner's corpuscles, Ruffini's corpuscles, pacinian corpuscles,** and **Krause's end bulbs.** Meissner's corpuscles (tactile corpuscles) are found in the papillae of the skin. They are encapsulated nerve endings found in hairless portions of the skin. Meissner's corpuscles are responsible for the sensation of touch. Ruffini's corpuscles were formerly thought to mediate the sensation of heat. Presently researchers think that Ruffini's corpuscles mediate the sensation of touch instead.

Pacinian corpuscles are encapsulated nerve endings widely distributed in subcutaneous tissue. They are known as pressure receptors because they mediate the sensation of deep touch and pressure. Other encapsulated nerve endings are Krause's end bulbs. These receptors lie near the surface of the skin. Opinions on the function of Krause's end bulbs are also undergoing change. They may mediate the sensation of cold, but evidence indicates that they are not the only type of receptor for cold. There is also some evidence they may be involved in the sensation of touch.

PHYSIOLOGY OF THE SKIN

The skin performs many essential functions, such as that of a protective barrier, temperature regulator, and sense organ. The skin also produces vitamin D and plays a minor role in the elimination of water and salts.

Protective Barrier

The most outstanding function of intact skin is protecting underlying tissues. The skin acts as a mechanical barrier to prevent the invasion of bacteria and other foreign matter and to prevent physical injury to the delicate tissues below the skin.

Although the outer epidermal cells function as a barrier to microorganisms invading the skin, they also provide a dwelling place for considerable numbers of bacteria, fungi, and one microscopic animal, the hair-follicle mite. All these tiny inhabitants live primarily on the secretions produced by the skin. The most abundant inhabitants are bacteria, which are found in greatest numbers on the skin of the face, armpits, and groin.

The skin is a waterproof covering that makes it possible for the body to maintain a high content of water, even in dry air. The importance of the skin in this respect is dramatically illustrated in the case of a person who suffers extensive burns over the entire body. This destruction of the skin results in such a rapid loss of fluid from the underlying tissues that the water balance of the entire body is upset. Burn victims frequently die from loss of water, which results in the collapse of the circulatory system.

Temperature Regulation

The maintenance of a constant body temperature within a fairly narrow range depends on a number of factors. The skin, however, plays an important role. Its many eccrine glands constitute one of the mechanisms for cooling the body. When the environment becomes too warm, the evaporation of sweat from the surface of the skin enhances a loss of heat, cooling the body. An even greater amount of body heat is lost through radiation. Excess heat is released through the amount of blood brought to the surface of the skin. This is an important factor in temperature regulation. When blood vessels are dilated, more heat is lost. When the blood vessels are constricted, less heat is lost. The regulation of body temperature is discussed in greater detail in Chapter 24.

Sensation

The skin is a very efficient sense organ, containing nerve endings that are responsive to touch, pain, pressure, and changes in temperature. It provides an important source of information that enables individuals to respond to changes in the environment.

Other Functions

When the skin is exposed to the ultraviolet rays in sunlight, vitamin D is produced from a precursor (forerunner) in the epidermis. This vitamin is then absorbed into the body, where it plays an important role in the absorption of calcium, a major constituent of bone tissue.

Relatively small amounts of water, salts, and nitrogenous wastes are excreted in sweat. The kidneys are far more important than the skin in this excretory role. However, excess sweating in extremely hot environments can seriously deplete sodium chloride from the extracellular fluids of the body. Individuals who work in extremely hot environments add extra salt to their diets.

PATHOLOGY

Infections are perhaps the most common disorders affecting the skin. Bacteria, mostly **staphylococci** and **streptococci,** are the most common infectious agents. **Boils** are the result of coccal infections. Viruses may produce **herpes** (small vesicles that itch), or **warts.** Fungi produce the cracks or fissures found in **athlete's foot.** The fungi thrive in warm, moist environments such as that provided by a tennis shoe and are easy to transfer from a tennis shoe to a locker room floor.

Common Skin Problems

Acne is a condition that affects more than 80 percent of teenagers. It is probably due to the effect of **androgenic hormones** (male sex hormones) on the sebaceous glands and hair follicles. It occurs in both sexes, because androgens are produced by the adrenal glands as well as the gonads.

Burns

Today nearly 1 million people are burned severely enough to be hospitalized annually. Of this million, it has been estimated that about 7000 die. Burns are one of the major categories included under accidents, one of the leading causes of death between the ages of 1 and 44.

Exposure to flames, scalding, hot objects, or some chemicals destroys the skin, and its protective functions are lost. The depth of the burn is described by *degrees*. First-degree burns involve only the surface layers of the epidermis. This skin is reddened and tender to the touch. A sunburn is a good example of a first-degree burn.

Second-degree burns are those in which much of the epidermis is destroyed. The burn also extends into the dermis. There is redness, and blisters are usually present. A burn resulting from briefly touching a hot object is usually a second-degree burn. This type of burn is usually very painful because the nerves of the dermis are being irritated by products of cell destruction.

Third-degree burns involve all skin layers, with no epidermal remnants present in the burned area. Charring of the skin is common. With third-degree burns the skin is insensitive to stimuli because of destruction of nerves in the dermis and subcutaneous tissues.

SUMMARY

The skin is referred to as the integument as well as cutaneous membrane. The skin is composed of three main layers: the epidermis, dermis, and subcutaneous tissue. In thick skin there are four layers in the epidermis. In thin skin the stratum lucidum is not present. The dermis contains fibrous connective tissue, sensory receptors, blood vessels, and glands. The subcutaneous tissue contains much fat.

Glands, hair, and nails are appendages (derivatives) of the skin. The sweat glands consist of apocrine and eccrine glands. The sweat glands are primarily concerned with heat loss in the body. The sebaceous glands lubricate the hair and skin. Hair grows from a follicle and is present over most of the body surface. Nails are plates of clear, hard keratin formed by epidermal cells.

The functions of the skin include protection, temperature regulation, sensation, production of vitamin D, and excretion. The skin is a sensitive indicator of whole-body physiology. It may develop infectious disorders that are quite common or it may be burned.

VOCABULARY REVIEW

Match the statement in the left column with the correct word(s) in the right column. *Do not write in this book.*

1. A minute opening on the surface of the skin through which secretions can leave.
2. The result of the action of the arrector pili muscles.
3. A nonliving material that replaces the protoplasm of the cells of the stratum corneum.
4. Produces a secretion that has a high lipid content.
5. A layer of the skin that can become thickened.
6. Has papillae that project into the stratum germinativum.

a. subcutaneous tissue
b. basement membrane
c. cuticle
d. dermis
e. epidermis
f. goose flesh
g. keratin
h. melanin
i. pore
j. sebaceous gland
k. sweat gland

TEST YOUR KNOWLEDGE

Group A

Write the letter of the word(s) that correctly completes the statement. *Do not write in this book.*

1. The outer layer of the skin is called the   (a) epidermis (b) ectoderm   (c) dermis   (d) endoderm.
2. One of the principal defenses of the body against bacteria is the (a) stratum granulosum   (b) stratum corneum   (c) papillary layer   (d) reticular layer.
3. The sweat glands that are associated with hair follicles are the (a) apocrine glands   (b) ciliary glands   (c) ceruminous glands (d) eccrine glands.
4. The appendages of the skin include all of the following *except* (a) hair   (b) fingernails   (c) sebaceous glands   (d) sublingual glands.
5. The hair follicle is made up of cells of the   (a) dermis   (b) epidermis   (c) boundary membrane   (d) subdermal layer.

Group B

Answer the following. Briefly explain your answer. *Do not write in this book.*

1. (a) What are the functions of the skin? (b) What structure, or structures, are mainly responsible for each of these?
2. (a) Briefly describe the structure of the epidermis and its method of growth. (b) What structures of epidermal origin are present in the dermis?

CHAPTER 24

REGULATION OF BODY TEMPERATURE

OBJECTIVES

A. Compare heat production and heat loss and identify the heat-regulating center

B. Describe the processes by which the body loses heat

C. Explain how body temperature is maintained and the factors affecting heat loss and gain

D. Describe the physiological variations in body temperature that occur normally and those that are caused by dysfunction of heat regulation

GENERAL PRINCIPLES OF BODY TEMPERATURE

The maintenance of a fairly constant body temperature in humans is an excellent example of **homeostasis.** A constant body temperature is of utmost importance, because healthy survival depends on biochemical reactions that occur only within a narrow range of temperatures. The amount of heat produced by the body varies greatly in accordance with the activity of skeletal muscles, the types of foods eaten, the temperature of the environment, and the condition of certain endocrine glands. It is estimated that a resting adult who fasts for 24 hours produces about 1,700 Calories (c) of heat. If the adult is moderately active during this 24-hour period but still fasts, the Calories produced could be 3,000. If the fasting adult does heavy work, the Caloric production could be close to 7,000. Whatever the amount of heat produced in the body, the circulatory system is needed to distribute the heat throughout the body.

Heat Loss

Ordinarily, the temperature of the human body is maintained within very narrow limits, close to 37° C. With a rise in temperature of 10° C the rate of activity of an enzyme approximately doubles. A rise in internal temperature therefore threatens the delicately balanced biochemical reactions taking place all through the body.

Since heat is produced continuously during life, there must be a mechanism for heat loss. If there were no heat loss, the temperature of the body at rest and in a fasting state (basal conditions) would rise by 1 C° (1.8 F°) per hour. With moderate activity it would rise by 2 C° (3.6 F°) per hour.

Not only must there be continuous heat loss, but the amount lost must vary in relationship to the changes in heat production. Most of the heat loss takes place through the skin and is regulated by the amount of blood flowing through the blood vessels of the skin and by the activity of the sweat glands.

In order to maintain a balance between heat production and heat loss, fine adjustments are needed. A device somewhat like a thermostat is required. The regulating mechanism of the body differs from a thermostat, however, in that heat production cannot be turned off completely, no matter how hot the external environment becomes.

The heat-regulating center.   Several factors influence body temperature. These include the nervous-controlled dilation and constriction of the blood vessels, changes in muscle tone, and the production of sweat. The centers controlling these activities lie in the lower part of the medulla, but their function in temperature regulation appears to be under the control of a ***heat-regulating center*** located in the hypothalamus. It has been found that if the hypothalamus is separated from the lower part of the medulla in an experimental animal, all control over heat regulation is lost.

There are two distinct areas in this heat-regulating center. One of these controls the loss of heat, and the other controls its production. If the anterior part of the center is destroyed or injured, the body temperature is adjusted normally in cool surroundings. However, in a warm place the body temperature rises because excess heat cannot be disposed of.

When the anterior region is warmed in an experimental animal, all of the outward signs of heat loss are shown: The animal pants, sweats, and its body temperature drops. If the posterior part of the center is injured, heat production lags behind heat loss and the body temperature falls. Under normal conditions the hypothalamus is influenced directly either by sense receptors in the skin or by the blood.

Heat Production

Heat production is the result of the oxidation of nutrients in all cells. It is a by-product in the production of energy for body activities.

Because much of the body is composed of skeletal muscle, most body heat is produced by the metabolic activity of these muscles. During rest, heat production is lowest because skeletal muscle activity is low. During strenuous exercise, heat production by the skeletal muscles increases greatly. If the body temperature falls, a reflex mechanism initiates rhythmic contraction of muscles, or shivering, to increase heat production.

Secretory activity of glands also produces heat. Since the liver is the largest and most active gland in the body, it accounts for about 20 percent of the heat production when the body is in a resting condition. It is a very important source of heat when skeletal muscles are relaxed, as during sleep.

Both epinephrine and norepinephrine have a direct effect on the metabolic rate and thus play a role in heat production. Heat production may be increased as much as 20 to 30 percent during maximum secretion of these substances.

The increased heat production that occurs after food is eaten is partially due to increased activity of the smooth muscle in the gastrointestinal tract and to activity of the glands that secrete digestive enzymes. In addition, the intake of foods produces what is known as a **specific dynamic action.** This is, the heat that is produced from the breakdown of food. The increase in heat varies with the type of food eaten, being greatest with protein foods. This increased heat production usually lasts for several hours after the food has been consumed. One of the requirements for basal metabolism tests is that no food should be eaten during the preceding 14 to 18 hours. Another minor source of heat is the temperature of the food when it is eaten.

Processes of Heat Loss

There are three principal means by which the body loses heat: through the skin, from the lungs, and through excretions. Of these three, the skin plays the most important role with 87.5 percent of body heat lost through its surface. Heat is lost from the lungs when air, warmed as the result of contact with heated respiratory tissue, is expired. The excretions of the body (urine and feces) are so small that they are taken into consideration only when making the most exact measurements of loss of body heat.

Loss of heat over the surface of the body may occur in any of several ways, depending on environmental conditions. (See Table 24-1.)

Radiation.   Body heat is lost by **radiation** in the form of invisible waves at the infrared region of the spectrum (Fig. 24-1). Radiation occurs whenever an object is warmer than its surroundings. A radiator in a cold room gives off infrared (heat) waves, which are absorbed by the walls and furniture. The human body is constantly radiating heat from the skin. The amount and type of clothing that people wear, as well as the amount of heat brought to the skin by the blood, will greatly influence the radiation process.

Vaporization.   *Vaporization*, or evaporation, of sweat from the surface of the skin is the principal means by which the body loses heat when subjected to high external temperature. The capacity of a fluid to lower the temperature of the surface from which it is evaporating depends on its ability to absorb heat from that surface. This heat is the energy which converts the fluid into a vapor, and is known as the **heat of vaporization.** It can be defined as the amount of heat needed to vaporize one gram of a liquid without changing the temperature of the liquid. The amount of heat absorbed is measured in terms of small calories. It varies with each fluid and also with the temperature of the fluid.

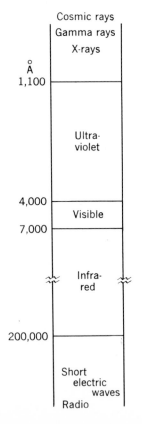

Figure 24-1   Divisions of the electromagnetic spectrum.

Cosmic rays
Gamma rays
X-rays
Å
1,100
Ultra-violet
4,000
Visible
7,000
Infra-red
200,000
Short electric waves
Radio

**Table 24-1
Heat Loss from the Body**

Processes	Percentage loss at 21°C (70°F)
Radiation and conduction	70.0
Vaporization	27.0
Respiration	2.0
Urination and defecation	1.8

Water has a very high heat of vaporization in comparison to other liquids. In effect, its rate of evaporation is sufficiently slow to prevent too rapid chilling. This also makes it very efficient for cooling. At a skin temperature of 30°C each gram of water that evaporates requires 579.5 calories of heat. However, the rate of evaporation of sweat depends on factors such as the rate of the movement of air, the temperature of the air, and the relative humidity of the air. Relative humidity refers to the amount of moisture present in the air at a given temperature in relation to the absolute amount of moisture it could hold at that same temperature. As air becomes warmer, it can hold more moisture. Therefore, on a hot and humid day the body is still able to lose heat by evaporation of sweat, although there is a general feeling of discomfort. A relative humidity between 40 and 60 percent is generally considered to be the most desirable for health and comfort.

Conduction.   *Conduction* of heat may occur when the skin comes in contact with a surface of lower temperature, as when a lightly dressed person lies on cold ground. In general, however, humans can be protected from heat loss by conduction by wearing proper clothing. For example, clothing made from materials that have numerous small pockets of air trapped within the fibers are poor conductors of heat.

Convection.   *Convection* is partly responsible for loss of body heat. When at rest in a cool room, a person loses heat as a result of convection air currents that are set up over the skin. Layers of warmed air next to the skin are constantly replaced by cooler air that flows in as the lighter warm air rises. The same process occurs when a person swims in cold water. The thin layer of warmed water that had been in contact with the skin is replaced by cold water. In these conditions there is a great loss of body heat by convection.

MAINTAINING BODY TEMPERATURE

The main heat-regulating centers are apparently controlled in two ways: by the temperature of the blood bathing these areas and by the receptors for temperature located in the skin. As previously indicated, these centers operate like a thermostat, responding to temperature changes by stimulating heat-producing or heat-conserving mechanisms when they are cold, or heat-loss mechanisms when they are warm.

Factors Affecting Heat Loss and Gain

The major mechanisms for heat gain and heat loss are summarized in Fig. 24-2. In a cold environment, the heat of metabolism is lost by radiation (R), convection (C), and evaporation of sweat (S). To maintain body temperature, heat production must equal heat loss. This can be accomplished by an increase in metabolism or a decrease in losses.

The first line of defense against a loss of body heat is *vasoconstriction*. The blood supply to the skin decreases, thereby preventing heat delivery to the surface of the skin. If heat loss continues to be greater than heat production, the second line of defense mechanisms comes into play. These include increased muscle tone, shivering, and increased muscle activity, all of which increase heat production. Also important is an increase in thyroid gland activity. Another mechanism for the conservation of heat is to add clothing and seek shelter.

In a hot environment heat is lost primarily by evaporation. In this circumstance, the first line of defense against a gain in body heat is *vasodilation*. The blood supply to the skin increases, thereby transferring large amounts of heat to the surface of the body, where it is lost

Figure 24-2 Schematic representation of how the body reacts when it is subjected to a cold or hot environment.

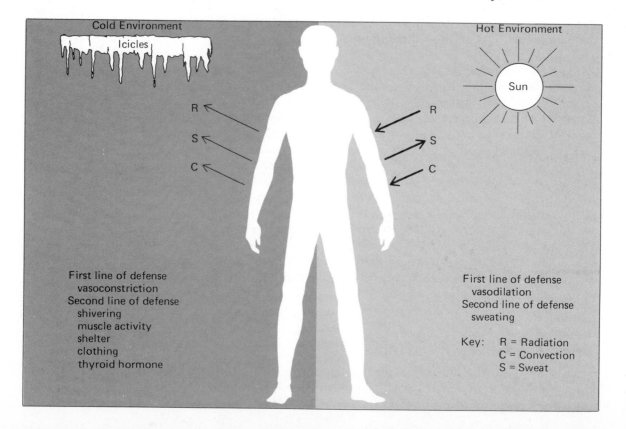

Cold Environment
Icicles
R
S
C

Hot Environment
Sun
R
S
C

First line of defense
 vasoconstriction
Second line of defense
 shivering
 muscle activity
 shelter
 clothing
 thyroid hormone

First line of defense
 vasodilation
Second line of defense
 sweating

Key: R = Radiation
 C = Convection
 S = Sweat

by radiation and convection. The second line of defense against a rise in body temperature is sweating. In a dry atmosphere, large amounts of sweat will evaporate, resulting in the loss of large amounts of heat and water. In a humid environment, sweat cannot evaporate sufficiently, so it pours off the body as a liquid.

Physiological Variations in Body Temperature

Newborn babies have an imperfectly developed heat-regulatory mechanism. Infants are, therefore, quite susceptible to cold air, which is apt to lower their body temperature to a dangerous level. Temperature changes in young children are also common and are usually due to strenuous activity, exercise, or emotional disturbance. The elderly have a similar inability to regulate their body temperature, with the result that they are easily affected by changes in temperature.

Normal body temperature.   The average oral normal temperature of a human is 37°C (98.6°F). Rectally, the normal temperature is 37.5°C (99.6°F). Most probably the oral temperature is closer to the real average of the whole body. It must be pointed out that this is only an average temperature. Among any large group of people, the temperatures probably range anywhere from 35.8° to 37.4°C.

An individual's oral temperature fluctuates to a greater degree than is generally realized during a 24-hour period. The body temperature drops below the normal average during sleep at night. As the day progresses, there is a buildup from the below-average temperature at night to the normal average and then even a rise above the average as the day progresses. In Fig. 24-3 a graph records data obtained from one person and undoubtedly shows some slight variations from similar data that might be obtained from another individual. Some interesting observations have been made on people who have changed their employment so that their working hours come during the night instead of the day. Such persons show a reversal of the normal temperature patterns after a few weeks so that their maxima are reached during the early hours of the morning.

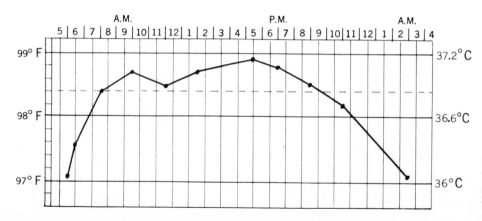

Figure 24-3 A graph of a person's body temperatures during a 24-hour period. Note how the temperatures deviate from the usually accepted average oral temperature.

Temperature measurement.   The idea of measuring body temperature as a part of the treatment of disease was first used in the early part of the seventeenth century, only a few years after Galileo had invented the thermometer. It was not, however, used in the same way as it is today. This later relationship between a specific disease and its effect on temperature did not appear until the middle of the nineteenth century, when standards for body temperatures were established. Since then, the use of the thermometer in medical procedures has become widespread. The oral thermometer was the first type used, and even today remains the most common method of measuring body temperature. Its reliability, however, is questioned because of the changing temperatures in the mouth. For more constant and reliable readings, rectal temperatures are preferred.

DYSFUNCTIONS OF HEAT REGULATION

Not only is heat produced by healthy active tissues, but there are certain types of diseases that produce excess heat in local areas. Since all heat is a type of radiation in which the wavelengths are longer than red in the visible spectrum, these radiations cannot be seen. They are called **infrared radiations.** It is possible, however, to photograph them if a special type of camera and film sensitive to infrared are used. This principle has been adapted recently to diagnose certain diseases, such as a cancer, which may be near the surface of the body, or an obstruction in a superficial blood vessel.

thermo = heat
graphy = record

If a sufficiently sensitive apparatus is used, the site of the disorder can be accurately located by the slight difference in temperature between it and the surrounding healthy tissue. Differences of as little as 2°C can be detected by this means. The term **thermography** is applied to this method.

High External Temperature

Heat exhaustion, or **heat prostration,** occurs when the circulatory system fails to function properly under adverse temperature conditions. In this condition the pulse is weak, the blood pressure is low, and the skin is moist and clammy. A person suffering from heat exhaustion should be kept warm. Complete rest is the best method of treating this condition.

Heat cramps.   Heat cramps develop when a person has been subjected to a high temperature and has been sweating profusely with the loss of an excessive amount of water and sodium chloride from the body. The cramps are not accompanied by any rise in body temperature. The main characteristic of this condition is the painful cramps in the muscles that have been used most.

Heat stroke.   A second type of disturbance of heat regulation is a *heat stroke* or *sun stroke*. Both these terms refer to the same condition, which is brought about by high external temperature and high humidity. The body is unable to lose heat normally and the temperature rises. Heat stroke is characterized by a dry, hot skin, a rapid pulse, and high blood pressure. The internal body temperature may rise to 43.3°C (110°F). At this temperature, destruction of brain cells results and death follows. Care for a victim of heat stroke may include external application of ice packs or wrapping the person in sheets soaked in cold water in order to rapidly reduce the body temperature. Being overweight, drinking alcohol, and wearing unsuitably heavy clothing are all factors that may contribute to heat stroke.

Low External Temperature

If the body is subjected to excessive cold, or if there is too great a loss of body heat through insufficient clothing or by immersion in cold water, the body temperature falls markedly. Along with this, there is a feeling of exhaustion and an uncontrollable desire to sleep. Slow respiration, low blood pressure, low blood CO_2 content, and, finally, coma are signs of **hypothermia.** This is a condition whereby the internal body temperature falls below 35°C (95°F).

Fever

One of the most common types of disturbance of the heat-regulating mechanism is *fever*. In the past, some people held the opinion that fever was the result of an increased rate of oxidation in the cells. Currently, it is thought that a fever is a result of the temporary failure of the "thermostat" in the hypothalamus to function properly due to the action of bacterial toxins or viral toxins, or due to serious emotional upsets. Nervous control of the heat-regulating mechanisms is not well established in babies and young children; therefore, a young child will develop a fever more readily than will an adult.

A fever is accompanied by lack of sweat and constriction of the skin capillaries. This latter condition may cause the victim to feel cold, even though the victim's temperature is higher than normal.

SUMMARY

Temperature regulation requires a balance between heat production and heat loss. The heat regulating center in the hypothalamus responds to stimuli from the environment to regulate heat loss and heat production. In a cold environment, vasoconstriction, increased muscle tone, shivering, increased muscle activity, and shelter and clothing produce or conserve heat, which helps regulate body temperature. In a hot environment, vasodilation and sweating are the major means of heat loss. Body temperature varies with age and with time periods.

VOCABULARY REVIEW

Match the statement in the left column with the correct word(s) in the right column. *Do not write in this book.*

1. The location of the thermostat of the human body.
2. If properly stimulated, makes an experimental animal sweat or shiver.
3. Process due to the function of sweat glands.
4. May cause air disturbances over a radiating surface in a cold room.
5. The feeling of cooling experienced while holding an ice cube is due to this.

   a. conduction
   b. convection
   c. heat-regulating center
   d. hypothalamus
   e. vaporization
   f. radiation

TEST YOUR KNOWLEDGE
Group A

Write the letter of the word(s) that correctly completes the sentence. *Do not write in this book.*

1. The normal internal temperature of the human body   (a) varies within narrow limits   (b) varies over a range of 6°C (11°F)   (c) is always the same   (d) usually decreases during daylight hours.
2. The human body loses heat mainly through   (a) the skin   (b) the lungs   (c) excretions   (d) secretions.
3. Heat is produced in the body by the process of   (a) evaporation   (b) condensation   (c) oxidation   (d) osmosis.
4. The largest quantity of heat is produced in   (a) nerve tissue   (b) epithelial tissue   (c) muscle tissue   (d) connective tissue.
5. The first line of defense against loss of body heat is   (a) vasodilation   (b) clothing   (c) vasoconstriction   (d) evaporation.
6. In a hot environment heat is lost primarily by   (a) vasoconstriction   (b) radiation   (c) evaporation   (d) convection.

Group B

1. Explain why running up a flight of stairs may make you sweat.
2. What role does each of the following play in regulating body temperature on a day when the air temperature is 33.9°C (93°F):   (a) radiation   (b) vaporization   (c) conduction   (d) convection?
3. Explain the concepts about the cause of fever.
4. Relate each of the following to the regulation of body temperature: (a) wearing clothing   (b) drinking salty water   (c) drinking alcohol   (d) staying in the shade when the air temperature is 43.3°C (110°F)   (e) swimming in water that is 18.3°C (65°F).

CHAPTER 25

METABOLISM

OBJECTIVES

A. Define metabolism and name the processes involved in biological oxidations

B. Define the term *basal metabolism* and discuss factors affecting the basal metabolic rate

C. Explain the process of carbohydrate metabolism, including glycolysis, the citric acid cycle, and the electron transport system

D. Describe the metabolism of lipids and proteins

ENERGY OF METABOLISM

Foods are first digested, then absorbed, and finally metabolized. Metabolism may be defined as all the changes foodstuff undergoes from the time it is absorbed from the small intestine until it is excreted from the body as waste products. It also refers to all the chemical changes that occur within the cells of the body. Metabolism consists of two phases. Processes involving the building up, or synthesis, of substances are called **anabolism.** The second phase, called **catabolism,** is the breaking down of substances. The processes involved in catabolism are primarily *oxidative reactions.* These reactions release energy that is used by the body to carry on its many functions. The energy which is not used is stored in the body in the form of the high-energy phosphate bonds of ATP.

Energy is involved in every activity of the human body. The action of the heart, the digestion and absorption of food, and the building and repairing of tissues cannot occur without energy. In addition, the vital functions of each living cell are dependent upon specific chemical reactions that require a constant supply of energy. It is ATP that supplies energy directly to the oxidative reactions of all cells.

Energy Sources

In order to supply the body's constant demand for energy, cells use the process of **biological oxidation.** In this process certain cellular reactions enable the cells to obtain energy from foodstuff without utilizing body tissue at the same time. Because of the catalytic action of intracellular enzymes, absorbed nutrients can be oxidized at the low temperatures that are compatible with the internal conditions of the cell. Biological oxidation is a metabolic process that can occur through a series of enzyme reactions in a cell. To supply the body's constant demand for energy, cells use the process of biological oxidation.

Biological oxidation.   Biological oxidations are similar to other types of oxidation in that there is a loss of electrons either by molecules that combine with oxygen or by molecules from which hydrogen is removed. In the cells of the body, the transfer of hydrogen atoms (electrons) through intermediary acceptors and finally to oxygen, generates the greatest amount of energy.

The agents involved in these reactions are **enzymes, coenzymes,** and **hydrogen acceptors.** The enzymes that catalyze the removal of hydrogen from a substrate are **dehydrogenases.** Enzymes that act on oxygen and eventually cause it to take part in the electron-transport system are called **oxygenases.** Certain coenzymes, such as nicotinamide adenine dinucleotide (*NAD*), can be alternately reduced and oxidized. Thus they act as hydrogen acceptors or hydrogen donors. NAD accepts hydrogens from molecules of a foodstuff (becoming $NADH_2$), transfers the hydrogens to other hydrogen acceptors (thus becoming oxidized), and returns to its original state, NAD.

The end products of digestion are completely oxidized by way of the **citric acid cycle** and **electron-transport system.** The citric acid cycle is also called the **tricarboxylic acid (TCA) cycle** or **Krebs cycle.** The latter was named after the scientist Sir Hans Krebs, who received the Nobel Prize for postulating the pathway for biological oxidation. It is from this cycle that a tremendous amount of energy is liberated, in addition to CO_2 and H_2O, which are excreted. Much of the energy produced is stored temporarily in the high-energy phosphate bonds of ATP. The remainder is converted into heat to maintain normal body temperature. ATP, found in every living cell, is a reservoir of energy that can be used for mechanical work, as when muscles contract, for chemical work, as in the building and repairing of tissue, and for electrical work, as in the transmission of nerve impulses.

Carbohydrates, fats, and proteins all contain stored energy. This energy can be converted to ATP following oxidation by way of the citric acid cycle and the electron-transport system. The potential energy value of food is known as the **caloric,** or **heat value. A large calorie (Cal)** is the amount of heat required to raise the temperature of 1 kilogram of water 1 degree centigrade. Each gram of carbohydrate releases 4 Cal of heat when it is oxidized. Fats have an even higher caloric value, yielding 9 Cal per gram when oxidized. Under normal conditions, only a relatively small amount of protein is used to supply energy; each gram of protein provides 4 Cal at the end of oxidation.

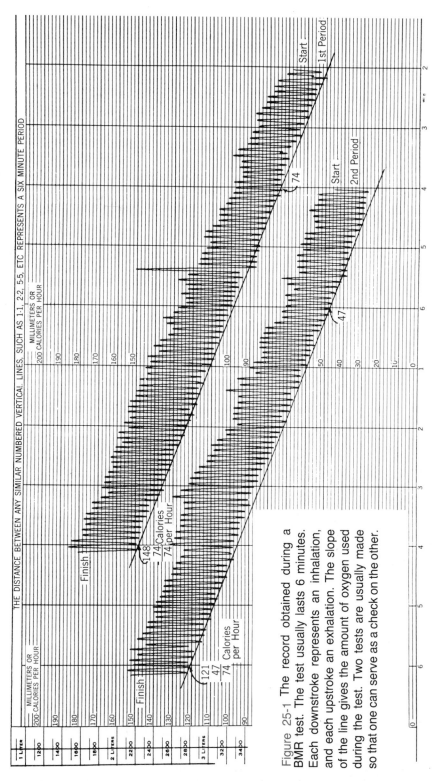

Figure 25-1 The record obtained during a BMR test. The test usually lasts 6 minutes. Each downstroke represents an inhalation, and each upstroke represents an exhalation. The slope of the line gives the amount of oxygen used during the test. Two tests are usually made so that one can serve as a check on the other.

Basal Metabolism

Basal metabolism refers to the energy output required to just maintain life activities, such as regulation of body temperature, respiration, and circulation, when the body is at rest (basal state). The amount of energy used (heat produced) is related to the surface area of the body. Metabolism at basal state is about 1000 Cal per square meter of body surface per day, or 38 to 40 Cal per square meter per hour for males and 36 to 38 Cal for females. Variations of plus or minus 10 percent are considered within normal range. This means that on the basis of 40 Cal for a normal young adult, the limits are 36 (40−4) or 44 (40+4) per square meter per hour.

Measuring metabolic rate. The **basal metabolic rate,** or **BMR,** is reported as a percentage of normal. For example, if the BMR is recorded as +30, it means that the basal metabolic rate is 30 percent higher than the normal under standard conditions, that is, for a person of the same size, age, and sex.

The basal metabolic rate is determined by measuring the heat produced by the body. In a research laboratory or medical clinic the basal metabolic rate is determined by an indirect method. This procedure involves a simpler method which measures the oxygen consumption. The subject breathes oxygen from a special chamber and expels carbon dioxide through a rubber tube placed into the mouth. After a period of time, the amount of oxygen used from the chamber is measured. Each liter of oxygen consumed is equivalent to a certain amount of calories. From this, combined with the surface area of the subject, the BMR can be calculated. During the test, the subject must be completely relaxed and must not have eaten for 12 to 14 hours since the previous meal. Figures 25-1 through 25-3 illustrate the BMR test.

Factors that affect BMR. Several factors influence the basal metabolic rate. For example, it is higher in children because of their rapid growth rate. It is usually higher in males and proportional to body size and amount of muscle tissue. It also varies with the body's state of nutrition. For example, when malnutrition is present, the BMR is low. Abnormal activity of the thyroid gland can affect the BMR since thyroid hormone plays an important role in controlling the rate of chemical reactions in the cells. If the thyroid gland is overactive, the basal metabolic rate increases above the normal limits. If the gland is underactive, the basal rate is decreased below the normal limits. Recording BMR has some disadvantages because it only measures oxygen consumption which can be affected by a vast variety of circumstances and may not be related to the thyroid hormone.

Protein-bound iodine. The iodine in the protein-bound thyroid hormone is called **protein-bound iodine (PBI).** Determining the amount of PBI in the blood is a laboratory test used to provide important infor-

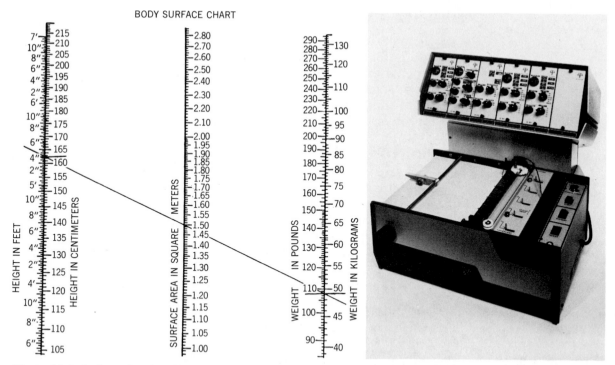

Figure 25-2 A chart showing the relationship between weight, height, and body surface area. To use the chart, lay a ruler across it so that one end intersects the weight line at body weight and the other end intersects the height line at body height. Then read the body surface area (in square meters) where the ruler intersects the middle line.

Figure 25-3 The equipment seen in the photo is used for measuring the BMR.

mation regarding thyroid function. The normal PBI range is between 4 and 8 milligrams per milliliters of plasma. Decreased levels are found in underactivity of the thyroid gland and increased levels are found in overactivity of the gland.

The **radioiodine uptake** is another test of thyroid function. The function of the thyroid gland can be evaluated more directly by measuring its ability to take up iodine from the blood. There are other tests that measure the amount of thyroid hormone by radioactive methods. However such techniques require facilities equipped to handle radioactive materials.

Energy requirements. The total caloric requirement varies with every individual depending primarily on the energy expended when eating, working, moving, and performing other basic body functions.

For example, a sedentary individual may require anywhere from 2000 to 2400 Cal per day for the total range of activities. However, individuals engaging in hard muscular work may require 6000 to 8000 Cal per day to meet energy needs. In any case caloric intake should be balanced by energy output.

FUNCTION OF FOOD METABOLISM

Carbohydrate metabolism includes all the oxidative reactions of carbohydrates for the main purpose of providing energy. Lipids are vital constituents of all living cells, and when oxidized, they also are an important source of energy. The primary function of protein metabolism is to supply the amino acids necessary for building new tissue, blood proteins, enzymes, and many hormones.

Carbohydrate Metabolism

Carbohydrate metabolism is essentially glucose utilization. There are various pathways that glucose takes following absorption from the small intestine into portal circulation and finally into systemic circulation to the cells.

Some glucose may enter the cells immediately after absorption from the small intestine and oxidize, forming carbon dioxide and water. The release of energy in this reaction is needed to maintain body temperature and to support muscular activity.

Figure 25-4 Glycolysis. Glycolysis essentially involves the conversion of a six-carbon molecule (glucose) into two three-carbon molecules (pyruvic acid). For each molecule of glucose broken down two molecules of $NADH_2$ and six of ATP are produced. Four molecules of ATP, however, are used up in this reaction to give a net energy gain of two molecules of ATP.

Figure 25-5 Citric acid cycle. Substances entering cycle are in colored boxes; products of the cycle are colored. The formation of acetyl-CoA from pyruvic acid yields 2 NADH (= 6 ATP) and one FADH (= 2 ATP) for each molecule of glucose. *Note:* Glucose produced two pyruvic acid molecules. Two molecules of carbon dioxide are released and eight hydrogen atoms are passed on to coenzyme molecules (NAD and FAD). One molecule of ATP is formed by phosphorylation during each turn of the cycle.

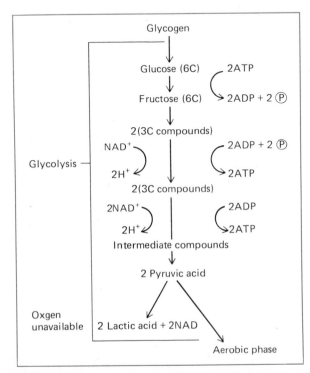

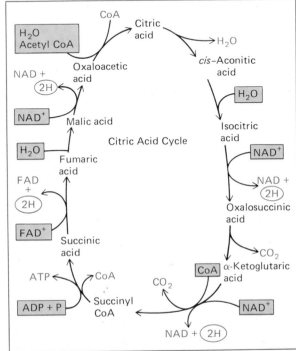

Basic metabolic processes.   Glucose that is not used immediately by the body is converted into glycogen (glycogenesis) for temporary storage in the liver and muscle cells. ATP provides the necessary energy for the synthesis of glycogen. Glucose metabolism provides energy for all the tissues of the body.

If there is a large intake of carbohydrate within a short period of time, some of the glucose may be eliminated by the kidneys. This is called **alimentary glycosuria,** because the glucose excess spills over into the urine. A similar spillage may occur in cases of uncontrolled *diabetes mellitus.* Because of a deficiency of insulin, diabetic patients are unable to metabolize carbohydrate efficiently. This causes an increase in the blood glucose level and a large percentage of glucose in the urine.

When the blood glucose level begins to fall, as it does between meals, the liver responds in two ways. First, it converts its stores of glycogen into glucose by means of the action of the hormone **glucagon** for release into the bloodstream (glycogenolysis). Second, liver cells form glucose from noncarbohydrates such as amino acids (gluconeogenesis).

All of these mechanisms maintain the blood glucose level within narrow limits. The average range for blood glucose is between 80 and 100 milligrams per 100 milliliters of blood. Normally, the blood glucose varies continuously, but it is kept within its narrow limit by the processes associated with glucose metabolism.

Catabolism of Carbohydrates

There are two phases of catabolism involved in the liberation of energy from carbohydrates. The first, an **anaerobic phase** is called **glycolysis.** This process results in the breakdown of glucose with the formation of *pyruvic* and *lactic acids.* Oxygen is not required in this phase, and only a few ATP molecules are formed. The second, an **aerobic phase,** involves the citric acid cycle and **oxidative phosphorylation,** which requires oxygen. In this phase pyruvic acid undergoes a series of changes so that large amounts of additional energy in the form of heat and ATP are produced.

Glycolysis.   The anaerobic phase of carbohydrate metabolism includes both the utilization and the production of ATP. This process involves the transfer of hydrogen atoms and their energy to the hydrogen acceptor, NAD. Energy becomes available for heat or for ATP production as the hydrogens are transported through the electron-transport system. Figure 25-4 summarizes anaerobic metabolism of glucose.

The symbol P is used to represent phosphate. Some of the energy of glucose is used to produce four molecules of ATP. However, two molecules of ATP are used in the initial stage of converting glucose to glucose-6-P, and fructose-6-P to fructose-1, 6-diphosphate (diP). The *net* yield is 2 molecules of ATP. Some of the energy of glucose contained in hydrogen is transferred to NAD, the hydrogen acceptor. If oxygen is available, this energy is released later when the hydrogens go through the electron-transport system. If no oxygen is available, $NADH_2$ is used to reduce pyruvic acid to lactic acid.

Citric acid cycle. The citric acid cycle is illustrated in Fig. 25-5. As pyruvic acid is oxidized to **acetyl-CoA,** carbon dioxide is given off and 2 hydrogens are transferred to a hydrogen acceptor. Acetyl-CoA then condenses with oxaloacetic acid to form citric acid and begins the citric acid cycle. During one complete turn of the cycle, two molecules of carbon dioxide are given off, one molecule of ATP is formed directly, and eight atoms of H are carried into the mitochondrial inner membrane for passage through the electron-transport system.

Electron-transport system. Note that in the citric acid cycle (Fig. 25-5) a total of eight hydrogens are released from intermediary molecules and attached to hydrogen acceptors with each turn of the cycle. It is believed that NAD and **flavin adenine dinucleotide (FAD)** are the principal hydrogen acceptors for this system. The hydrogens attach to NAD and FAD, and the energy contained in the hydrogens is liberated through the exchange of hydrogens, or their electrons (e), from one acceptor to another. This system has been called the **electron transport system,** the **cytochrome system,** and the **respiratory chain.** A proposed scheme for this system is shown in Fig. 25-6. The $NADH_2$ formed during oxidation reactions transfers its hydrogens to FAD, which then becomes $FADH_2$. The energy (E) released at this point, coupled with phosphate (P), is used to convert ADP to ATP. The hydrogens on $FADH_2$ lose their electrons (e) to cytochrome b and become hydrogen ions ($2H^+$). These electrons are now transferred from one cytochrome to another, alternately reducing and oxidizing the cytochromes as the electrons are transferred. Finally cytochrome oxidase transfers these electrons to oxygen, which then reacts with the $2H^+$ released from $FADH_2$ to produce water. During two phases in the transference of electrons by the cytochromes, energy is released and trapped in ATP. Since phosphate is also necessary to convert one molecule of ADP to ATP, this reaction is often referred to as **oxidative phosphorylation.**

The total number of ATP molecules ultimately produced by one turn of the citric acid cycle is 12. For each acetyl-CoA, regardless of the source (glucose, fatty acids, or amino acids), completely oxidized in the citric acid cycle, enough energy for the production of 12 molecules of ATP will be produced. However, if this same foodstuff enters the system as pyruvic acid, three additional ATP molecules will be produced. Thus complete oxidation of pyruvic acid yields 15 molecules of ATP.

A summary of the total ATP production from the complete oxidation of a molecule of glucose totals about 40 percent (Fig. 25-7), while the remaining 60 percent is liberated as heat.

Lipid Metabolism

When lipids are absorbed from the small intestine, they are transported in the blood as **chylomicrons,** and partly as **lipoprotein complexes.** Chylomicrons are composed primarily of *triglycerides* combined with a small amount of cholesterol, phospholipid, and protein.

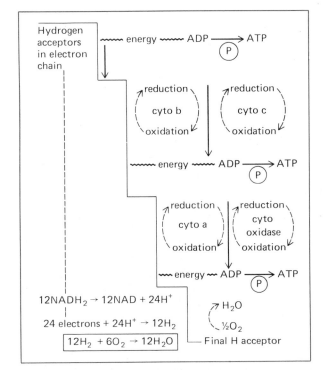

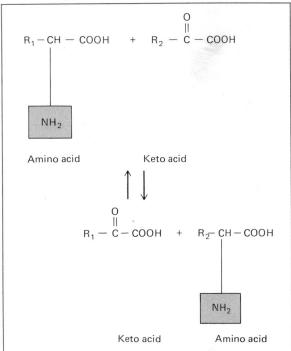

Figure 25-6 Electron transport system. Each molecule of glucose (which produced two molecules of acetyl-CoA) yields 2 ATP by oxidative phosphorylation, 6 NADH$_2$ (= 18 ATP), and 2 FADH$_2$ (= 4 ATP), for a total of 24 ATP.

At the end of the chain the final hydrogen acceptor combines the electrons with hydrogen ions and with oxygen to form 12 molecules of water. Thus the complete oxidation of one molecule of glucose by glycolysis and the citric acid cycle produce 38 molecules of ATP.

Figure 25-7 Transamination. Note that the keto acid to which the amino group is transferred becomes an amino acid.

Storage of triglycerides. The storage of lipids in the form of adipose tissue provides insulation and mechanical protection as well as energy reserves. The liver serves as a central clearing house for fats. It is the major organ for the synthesis of lipids as well as their breakdown into smaller substances.

The storage of fats involves the formation of adipose tissue in various parts of the body. Because a limited amount of carbohydrate can be stored in the liver and muscle, these fat deposits constitute the bulk of stored, energy-producing nutrient in the body. Storage fat is in a continuous state of change; it is constantly being deposited and used. However, deposits may exceed usage when there is an excessive intake of food.

Oxidation of fatty acids. The oxidation of fatty acids involves a complex series of reactions that result in the formation of a number of molecules of acetyl-CoA, component compounds in the citric acid cycle. The oxidation of fatty acids takes place mainly in liver cells, but it can also occur in other tissues. When the liver generates more acetyl-CoA than is required for its own metabolic needs, the liver condenses two molecules of acetyl-CoA to form one molecule of acetoacetic acid. This *keto acid* diffuses into the blood and is carried to active tissues throughout the body. Upon entering a tissue cell, the process is reversed and two molecules of acetyl-CoA are formed. These molecules then enter the citric acid cycle and are oxidized, releasing energy.

Although the major pathway for acetyl-CoA is its oxidation by means of the citric acid cycle, this compound also serves as the means of the synthesis of lipids in the body. Acetyl-CoA is the major precursor in the synthesis of fatty acids, acetylcholine, cholesterol, and other sterols.

Protein Metabolism

Proteins must be constantly supplied in the diet to replace that amount used for the synthesis of chemical substances and for growth and repair of body tissue. The building blocks of proteins are amino acids (refer to Chapter 12).

There are three sources of amino acids in the body: absorption from the small intestine at the end of digestion, synthesis by liver and tissue cells, and the catabolism of tissue protein. The amino acids in the blood form a pool that can be used for protein metabolism. In essence, the maintenance of the amino acid pool determines protein metabolism.

Amino acid synthesis.   All amino acids are required by the body, but only certain ones can be formed within the body. Those that cannot be synthesized are called *essential amino acids.* These must be obtained from food sources. On the other hand, *nonessential amino acids* can be synthesized in the liver if they are missing from the diet. For example, the liver cells can transfer the amino group from an amino acid to a keto acid and thus form a different amino acid. This is known as *trans-amination.* The liver cells can also form an amino group from some other nitrogen-containing compound (ammonia) and by combining it with a keto acid, a different amino acid can be formed.

The amino acid nitrogen in the body shifts continually from one amino acid to another, being added and then removed from the pool. It is then incorporated into tissue, liver, or plasma protein and then split again and returned to the pool. Urea is the chief nitrogenous waste product. Less important are creatinine and ammonium salts.

Protein synthesis.   Each tissue selects the specific amino acids from the amino pool that are needed to manufacture its particular proteins for growth, maintenance, and proper functioning. A rapidly growing individual requires great quantities of amino acids because new tissue is being formed throughout the body. As an individual gets older, amino acids are needed to replace the tissue that breaks down under the wear and tear of daily living. The production of enzymes and hormones are other examples of protein synthesis. Every cell in the body can produce the enzymes that are necessary to catalyze its own internal chemical reactions. However, only special cells in the gastrointestinal tract can synthesize the extracellular enzymes needed to digest food. The production of certain hormones also requires highly specialized cells, such as those in the islets of the pancreas, which produces insulin. For every type of protein there is a blueprint that indicates the exact sequence of amino acids required for the synthesis of a specific protein. Muscle cells build sarcoplasm, not hemoglobin. One sequence of amino acids will result in the blood protein, *albumin*, while another sequence of amino acids will form the blood protein *globulin*. Deoxyribonucleic acid (DNA) and ribonucleic acid (RNA) are of fundamental importance in all protein synthesis (discussed in Chapter 3).

Protein catabolism.   Protein molecules are catabolized when they are needed as a source of energy. The two methods whereby the amino acids are catabolized are *transamination* and **deamination** in the liver.

Deamination is the enzymatic removal of the amino group of an amino acid with the formation of ammonia and a keto acid. The ammonia is converted to urea and eliminated in the urine. The keto acids are converted by any of a variety of processes to **ketone bodies,** such as acetoacetic acid, some of which is converted to acetyl-CoA and converted into various intermediates in the citric acid cycle.

When energy needs must be met, these compounds transfer electrons, releasing energy along the citric acid cycle. However, if energy is not required and if structural protein supplies are adequate, these substances are converted to fat for storage.

DISORDERS IN METABOLISM

When no food is eaten, the sequence of events does not change concerning the interrelationship of all the processes involved in metabolism of carbohydrate, fat, and protein. Despite popular belief, it is possible to survive long periods of time without food as long as water is available. Consequently, starvation of body tissue takes a long time and survival depends upon the age of the individual and available fat stores of the body.

Effects of Starvation

Carbohydrates cannot be stored in sufficient amounts to meet glucose needs during starvation. The main storage of glucose is in the form of glycogen, but this store is not enough to supply the needs of the brain even for 24 hours. Therefore, except for the first few hours of starvation, the major effect on the body is the use of tissue fat and protein. Since fat is the prime source of energy, it will continue to be utilized until most of the fat stores of the body are gone.

When other sources have been depleted then protein is primarily used for glucose, two-thirds of which is used mainly to supply energy to the brain. The glucose formed by the protein is produced in the liver by the process of *gluconeogenesis*. Under normal circumstances the brain cannot utilize other metabolic substances for energy besides glucose.

Ketosis.   As starvation progresses, however, the protein stores are also depleted and the gluconeogenesis decreases markedly. This leads to *ketosis.* Fortunately, in this circumstance, the brain can utilize some of the ketone bodies for energy. However, this cannot go on indefinitely since proteins are necessary for the maintenance of cellular functions. Death occurs when the proteins of the body are depleted.

Children cannot withstand starvation as compared with adults. During starvation, growth of the child stops and if starvation continues, normal growth is never attained even after proper nutrition is restored. In addition, there are irreversible changes in the nervous system that may involve brain damage.

SUMMARY

All three classes of macronutrients can enter the citric acid cycle and electron-transport system and thus all three can be used as sources of chemical, electrical and mechanical energy. Hydrogen and oxygen provide the substrates for the oxidative phosphorylation. Some of these hydrogens are derived directly from the citric acid cycle and other hydrogens may come from glycolysis, or the breakdown of fatty acids. Glucose can be converted to fat or to amino acids by way of intermediate metabolites such as pyruvic acid and acetyl CoA. Similarly, amino acids can be converted into glucose and fat. The glycerol portion of triglycerides can be converted into glucose. Metabolism is thus a highly integrated process in which all classes of nutrients can be used to provide energy or to synthesize other nutrients.

VOCABULARY REVIEW

Match the statement in the left column with the correct word(s) in the right column. *Do not write in this book.*

1. A general term used to describe all the chemical changes food undergoes in the body.
2. The building-up process which results in tissue synthesis.
3. A chemical test for thyroid function.
4. Urea is formed through this process.
5. The break-down process resulting in the release of energy.
6. Involved in the formation of one amino acid from a keto acid.
7. The pathway for biological oxidation.
8. Intermediate metabolite for entrance of major nutrients into citric acid cycle.

a. PBI
b. catabolism
c. deamination
d. anabolism
e. transamination
f. metabolism
g. acetyl-CoA
h. citric acid cycle
i. glycolysis
j. basal metabolism

TEST YOUR KNOWLEDGE

Group A

Write the letter of the word(s) that correctly completes the statement. *Do not write in this book.*

1. The nutrient that provides the greatest amount of energy per unit of weight is   (a) starch   (b) protein   (c) sugar   (d) fat.
2. The caloric requirements of the body can be determined indirectly by measuring its intake of   (a) nitrogen   (b) carbon dioxide   (c) oxygen   (d) carbohydrates.
3. Of the following, the term that is *not* closely related to the chemical processes within living cells is   (a) catabolism   (b) anabolism   (c) embolism   (d) metabolism.
4. One product of oxidation of nutrients in the cells is   (a) water   (b) carbon monoxide   (c) glycerol   (d) amino acid.
5. The rate of metabolism is greatest in   (a) young children   (b) teenagers   (c) adult males   (d) adult females.
6. The rate at which a person performs the fundamental chemical reactions is abbreviated by the letters   (a) DNA   (b) BTI   (c) BMR   (d) ATP.
7. Of the following factors, the one that has *least* effect on the rate of metabolism is   (a) age   (b) sex   (c) altitude   (d) external temperature.
8. The hydrogen acceptors and hydrogen donors in the electron transport system are primarily   (a) enzymes   (b) oxidases   (c) coenzymes   (d) nutrients.
9. The citric acid cycle is also called the TCA and the   (a) electron-transport system   (b) BMR   (c) Krebs cycle   (d) oxidative phosphorylation cycle.
10. The oxidation of fatty acids takes place mainly in the   (a) stomach   (b) liver   (c) pancreas   (d) spleen.

Group B

Answer the following. Briefly explain your answers. *Do not write in this book.*

1. The doctor reports that an individual's BMR is +8. What does this mean?
2. Why is the citric acid cycle called a cycle? Describe how glucose, fatty acids, and amino acids can enter the citric acid cycle.
3. How many ATP molecules can be produced from an amino acid that is deaminized to pyruvic acid?
4. Can an individual get fat from a high protein diet? Explain.

CHAPTER 26

THE KIDNEY

OBJECTIVES

A. Describe the anatomy of the kidney
B. List the structures of the nephron and relate each of the parts to its function in the kidney
C. Describe filtration in the kidneys and identify the mechanisms that control the rate of filtration
D. Explain the processes of tubular reabsorption and secretion and how they affect the composition of urine

ANATOMY OF THE KIDNEY

The constancy of many materials in blood and in tissue fluid is determined by the activity of the kidneys. Since the normal function of all body cells depends upon this constancy, it is necessary to understand the functions of the kidneys. Not only do all living cells constantly take materials from the blood during metabolic activity, but they also release many metabolic waste products into the blood during such cellular activities as growth, repair, and maintenance. The lungs remove carbon dioxide and the kidneys remove a wide range of products that can dissolve in water, particularly those referred to as **nitrogenous wastes.** Water is eliminated from the body by several routes such as the skin, lungs, and large intestine, but most of it is released as urine, a product of kidney function. There is a tendency to focus on the elimination of waste products as the prime role of the kidneys, but their role in maintaining a chemical balance in the body is equally important.

It has been noted by William B. Snively that "... what matters most for our internal chemical equilibrium is, not what we eat and drink, but what our kidneys retain. They are truly master chemists of our sea within."

retro = behind

Gross Structure

The kidneys are located in the posterior part of the abdominal cavity, one on either side of the spinal column, behind the peritoneum (*retroperitoneal*). The kidneys are not fixed in a rigid position against the abdominal wall, since they move with the diaphragm during inspira-

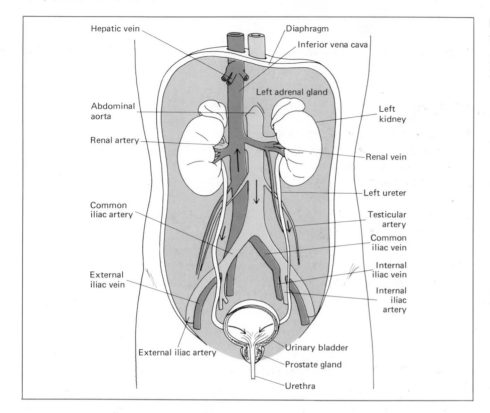

Figure 26-1 **Overview of the male urinary system and its major blood vessels.**

tion. However, they are embedded in a mass of adipose tissue around which is found a supporting layer of fibrous tissue called the **renal fascia.** The renal arteries and veins give added support to the kidney.

The kidney is a bean-shaped structure. Its inner medial border is concave, while the outer lateral border is convex (see Fig. 26-1). Vessels and nerves enter and leave the kidney through a longitudinal fissure called the **hilum,** located along the concave border. The ureter leaves the kidney through this fissure and descends toward the urinary bladder. The kidney is covered with a fibrous capsule, or tunic, which provides a firm, smooth covering.

Internal regions.   If a longitudinal section is made through the kidney as illustrated in Fig. 26-2, three general regions may be observed: the **renal cortex,** the **renal medulla,** and the **renal pelvis.**

The renal cortex, or outer portion of the kidney, is granular and reddish brown in appearance. It arches over the pyramids of the medulla and dips in between adjacent pyramids. These inward extensions of cortical substance are called the **renal columns** (*of Bertin*).

The renal medulla is darker in color and consists of striated, cone-shaped masses called the **renal pyramids.** The pyramids vary in number, but the average is 12. The base of each of these pyramids is directed toward the cortex, and the free end, or apex, of each projects centrally

renal = kidney

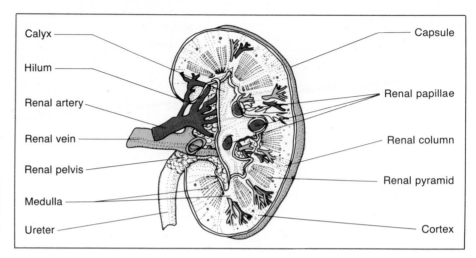

Figure 26-2 A drawing of a longitudinal section through the kidney showing the kidney's internal structure.

Labels in figure: Calyx, Hilum, Renal artery, Renal vein, Renal pelvis, Medulla, Ureter, Capsule, Renal papillae, Renal column, Renal pyramid, Cortex

toward the renal pelvis, where it forms a *papilla*. The summit of each papilla resembles a sieve, because it is studded with a variable number of openings. Urine flows through these openings into an extension (***calyx***) of the renal pelvis.

The renal pelvis is a funnel-shaped sac that forms the upper expanded end of the ureter. It receives urine from all parts of the kidney by way of the *calyces*, which are cup-shaped extensions of the sac. Each kidney has a variable number of calyces, since a single calyx may surround more than one papilla. **Kidney stones** (renal calculi) may form in the pelvis of the kidneys. They frequently cause pain and blood in the urine (*hematuria*).

Microscopic Anatomy

There are approximately 1 million highly specialized tubules, or ***nephrons,*** within each kidney. Each tubule is about 14 millimeters long and 0.055 millimeter in diameter. The upper end of the tubule is expanded into a saclike structure known as Bowman's capsule. It is composed of two layers of cells and its shape is like that obtained by pushing a blunt object against one side of a thin-walled rubber ball to form an indentation. The afferent blood vessel, a minute branch of the renal artery, enters at this indentation. The afferent vessel then forms a knot of about 50 separate capillaries, the ***glomerulus,*** which fills the cavity of the capsule. Although the diagram of the nephron in Fig. 26-3 does not show it, the inner wall of the Bowman's capsule dips between the capillaries to surround them. This brings each blood vessel into direct and intimate contact with the capsule wall.

Beyond the capsule, the nephron becomes tubular and quite convoluted (twisted) for a short distance. Both the capsule and the convoluted tubule lie in the cortex, but the tube suddenly dips into the medulla to form the **loop of Henle** and then returns to the cortex. It once more becomes convoluted before opening into a ***collecting tubule.*** This larger vessel is joined by other tubules that open into the pelvis from the surface of a pyramid. After the afferent vessel has formed the glomerulus, it continues out of the capsule as the efferent vessel and this again branches to form the capillaries that surround the tubules. These eventually flow together to form a small branch of renal vein, which takes the blood from the kidney.

PHYSIOLOGY OF THE KIDNEY

As already mentioned the renal artery branches into smaller vessels immediately upon entering the kidney. The result of the sudden decrease in the size of the vessels leads to a sudden increase in pressure, like that which would occur if many small rubber tubes were attached to the end of a garden hose. Thus when the blood flows into the glomerulus, the pressure rises as the artery branches into capillaries. In most of the capillaries throughout the body, the pressure is about 25 millimeters of mercury, but within the glomerulus it rises to between 60 and 70 millimeters. As a result of this high pressure, blood fluid is filtered through the capillary wall and passes into the capsule.

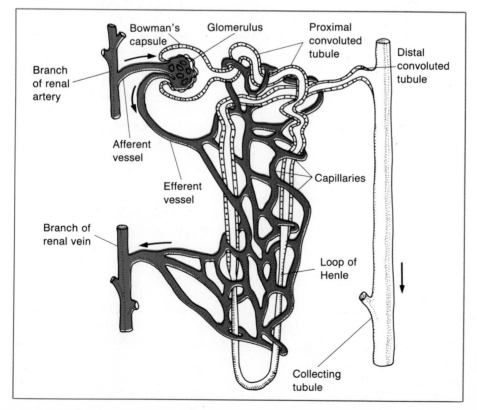

Figure 26-3 Diagram showing a nephron and the blood vessels associated with it.

Glomerular Filtration

The total amount of fluid passing from the capillaries is estimated to be approximately 125 milliliters per minute. In the course of 24 hours this amounts to about 180 liters. In spite of this tremendous withdrawal of fluid from the blood, the body loses only 1.0 to 1.5 liters of fluid per day in the form of urine. The reason may be found in the relationship of the blood capillaries to the tubular parts of the nephron.

Composition of filtrate. Because of the entry of large amounts of fluid into the capsule, the blood in the efferent vessel is highly concentrated and viscous. The blood remaining in this vessel contains blood cells and molecules, mostly proteins, that are too large to pass through the capillary membrane. The membrane of the glomerular capillaries has a special structure that makes it very permeable to all but the larger protein molecules. This means that the glomerular filtrate is similar in composition to plasma, but lacks most of the proteins. Proteins of molecular weight lower than about 60,000 may be present in the filtrate. Also present in the filtrate are metabolic wastes such as urea, plus compounds that are essential for body chemistry. These include glucose, and ions such as Na^+, K^+, Ca^{++}, Mg^{++}, Cl^-, and HCO_3^-. Amino acids and vitamins are also present in the filtrate.

Rate of filtration. Constriction of the *afferent* blood vessels decreases the rate of blood flow in the glomerulus. This leads to a decrease in glomerular pressure, and as a consequence the rate of filtration drops. Dilation of the afferent vessels in turn moderately increases the glomerular filtration rate.

Constriction of *efferent* blood vessels causes resistance to flow in the glomerular capillaries and increases pressure. Up to a point, the filtration rate increases. As more fluid is removed from the glomeruli, the concentration of plasma proteins increases. Eventually this tends to slow down the loss of fluid and the filtration rate drops.

Tubular Reabsorption

As the glomerular filtrate flows through the renal tubule many substances are reabsorbed into the bloodstream. This is vital since many of them are needed by the body. For example, water, glucose, amino acids, sodium chloride, bicarbonate, and other tiny particles are removed from the tubular fluid and returned to the blood as it flows through the capillaries surrounding the renal tubules. Over 80 percent of the reabsorption of water and solutes occurs in the proximal tubules. The rest of the absorption will take place in other parts of the nephron, particularly the distal tubule and the collecting ducts (Fig. 26-4). Substances are returned to the blood by means of diffusion, osmosis, and active transport.

Osmosis. Water reenters the bloodstream by osmosis, mainly in the proximal tubules. The tubules in this area are very permeable to water.

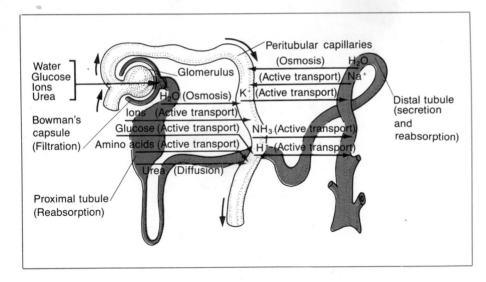

Water
Glucose
Ions
Urea

Bowman's
capsule
(Filtration)

Proximal tubule
(Reabsorption)

Glomerulus

Peritubular capillaries
(Osmosis) H_2O
(Active transport) Na^+
(Active transport)

H_2O (Osmosis) K^+ (Active transport)
Ions (Active transport)
Glucose (Active transport) NH_3 (Active transport)
Amino acids (Active transport) H^+ (Active transport)

Urea (Diffusion)

Distal tubule
(secretion
and
reabsorption)

Figure 26-4 Diagram showing filtration, reabsorption, and secretion in a nephron. Filtration occurs in Bowman's capsule as water, glucose, ions, urea, and other materials leave the blood. They are reabsorbed in varying degrees in the proximal tubule. Additional reabsorption and secretion take place in the distal tubule.

Passive transport.   The concentration of urea in the tubules is higher than in the efferent vessels. The tendency is for the urea to diffuse back into the bloodstream. The tubular membrane is only slightly permeable to urea, however, so that over half of it remains in the tubules and is excreted.

Active transport.   Many substances, such as glucose, amino acids, vitamins, and various ions, are transported across tubular membranes into the bloodstream by the expenditure of energy. Much of this reabsorption takes place in the proximal tubules.

The presence of glucose in the urine, a condition known as *glycosuria*, indicates either a loss of ability of the tubules to reabsorb glucose or an excess of glucose in the blood.

When glycosuria is not accompanied by an increase of glucose in the blood, it is called renal glycosuria. If there is an increase in glucose in both the blood and urine, the condition is called *diabetes*.

Tubular Secretion

The kidney plays an important role in regulating the acid-base balance of the body. Through the ability of its tubular cells to secrete varying amounts of hydrogen ions and ammonia, the kidney can increase or decrease the acidity of the urine. These secretory activities occur primarily in the distal portion of the tubules and are part of the final phase of urine formation.

The kidney should not be considered as a simple filter that removes the end products of protein metabolism in an automatically uniform manner. It is a highly discriminating organ that is able to remove wastes and foreign substances. At the same time the kidney makes ad-

Table 26-1
Percentage Composition of Plasma and Urine

Material	Plasma %	Urine %	Change in Concentration in the Kidney
Water	90-93	95	——
Proteins, other colloids, and fats	8.0	——	——
Glucose	0.1	——	——
Sodium	0.32	0.35	1
Potassium	0.02	0.15	7
Chlorine	0.37	0.6	2
Phosphates	0.009	0.15	90
Ammonia	0.001	0.04	40
Urea	0.03	2.0	60
Uric acid	0.004	0.05	12
Creatinine	0.001	0.075	75

justments that will conserve some materials that are in low supply or remove these same substances when they are in excess. There may be a great variability in the quantity and composition of urine samples from the same individual even within a short period of time. This indicates that the materials are handled individually in order to serve the body effectively.

The average concentration of various materials present in the urine as compared with their presence in the plasma is shown in Table 26-1 adapted from the work of Dr. A. R. Cushny.

Factors affecting secretion.   The blood supply to the kidneys is largely responsible for determining the amount of urine produced under normal conditions. When the nerves to the kidneys are cut, the kidney blood vessels dilate. If the sympathetic nerves are then stimulated, raising the blood pressure, more blood passes through the kidneys, resulting in greater production of urine by the kidney.

The amount and concentration of the urine are affected by ingestion of water or salt. Drinking large quantities of fluid increases the amount of urine. This urine is quite dilute because of the presence of an increased amount of water. On the other hand, if very salty substances are eaten, the quantity of urine is reduced, but the concentration of dissolved materials increases. The events that bring about these reactions involve both the nervous and endocrine systems. Intake of fluid or salt changes the osmotic pressure of the blood. This in turn affects certain receptors (*osmoreceptors*) of the hypothalamus of the brain. The hypothalamus then sends impulses to the posterior lobe of the pituitary gland, which produces an **antidiuretic hormone (ADH).** Diuretic refers to urine production. If too much fluid has been taken into the body, the production of this hormone is reduced and the liquid is excreted in order to maintain the correct osmotic state of the blood. On the other hand, if an excess of salt has been eaten, the osmoreceptors are stimulated and the amount of antidiuretic hormone is increased.

The effect of the osmotic pressure of the blood on the production of ADH and the reciprocal effect of ADH on the blood's osmotic pressure constitutes an excellent example of negative feedback maintaining a system in equilibrium. A rise in osmotic pressure stimulates production of ADH, which prevents further rise in osmotic pressure by inhibiting excretion of water in the urine. As the osmotic pressure falls further, the production of ADH is no longer stimulated. This is just one example of the many homeostatic (self-regulatory) mechanisms present in living organisms.

Aldosterone, sometimes called a salt and water hormone, is secreted by cells in the cortex of the adrenal glands. This hormone plays an important part in the active transport of both sodium and potassium. When the concentration of aldosterone is high, sodium is retained in excessive amounts in the body, and potassium is excreted in greater amounts than usual. On the other hand, when there is a deficiency of aldosterone, as in *Addison's disease,* potassium is retained while excessive sodium is lost from the body and with it an equivalent amount of water. As a consequence, the amount of fluid in the body decreases to seriously low levels.

Composition of Urine

Urine is a watery solution of nitrogenous wastes and inorganic salts that is removed from the plasma and eliminated by the kidneys. The color usually is amber, but varies with the amount produced. The pH of urine is acid on a mixed diet and alkaline on a vegetable diet. The specific gravity usually is between 1.016 and 1.020, but may vary from 1.002 to 1.040 in normal kidneys, according to whether the urine is very dilute or very concentrated. The quantity of urine is about 1.5 liters (1.6 qts.) daily. An increase in daily output that is persistent is called *polyuria,* and a temporary increase is called *diuresis.* A decrease below the average amount is called *oliguria.*

The composition of urine varies with the waste products to be removed and with the need for maintenance of homeostasis. About 95 percent of the urine is water. The average amount of solids excreted per day is about 60 grams. Of this, about 25 grams consist of inorganic salts and 35 grams consist of organic substances.

Urination.   The elimination of urine, urination or *micturition,* is under the control of the nervous system and is brought about by stimulation of bundles of smooth muscles in the walls of the ureters, bladder, and urethra.

The ureters enter the bladder at a sharp angle, which prevents the backflow of urine into them. Liquid flows along them in an almost continuous stream, although the quantity may vary greatly depending on the factors mentioned earlier. The normal capacity of the bladder is about 470 milliliters. Its opening into the urethra is controlled by two groups of sphincter muscles that are under voluntary control.

The process of voiding the urine is the result of a combination of involuntary and voluntary processes. In the first place, the presence of urine in the pelvis of the kidney stimulates peristaltic waves in the walls of the ureters, which pass urine to the bladder. The bladder becomes distended as the quantity of fluid in it increases, and receptors in its walls are stimulated. A feeling of fullness results when the bladder contains between 250 and 300 milliliters of liquid. In the infant, the flow of urine is purely reflex in nature and will continue to be until proper training (*conditioned reflex*) has been established. In an adult, the nerve impulses from the bladder pass to a center in the spinal cord just as they do in an infant. However, in the adult they continue upward to the cerebral cortex and result in the control of the expulsion of urine through the urethra.

KIDNEY DISORDERS

A variety of diseases can all render the kidney incapable of functioning normally. **Uremia,** or kidney failure, is the name given this state of inadequate renal function. *Uremia* implies urea in the blood. However, urea is a normal constituent of the blood and only becomes a serious concern when the level in the blood is increased due to kidney impairment. The measurement of blood urea is expressed as blood urea nitrogen (BUN), and is one way to check for the presence of renal failure. BUN can be elevated somewhat by other conditions, notably dehydration.

Hemodialysis—Artificial Kidney

hem = blood

Uremia is frequently treated by **hemodialysis.** The principle of dialysis is the same as the process described in Chapter 3. Dialysis is the process of separating crystalloids from colloids in solution by virtue of the difference in their rates of diffusion through a semipermeable or selectively permeable membrane.

The equipment used for hemodialysis is frequently referred to as the **artificial kidney.** Briefly, it consists of a long cellophane tube arranged in coils and submerged in a temperature-controlled dialyzing solution. Dialysis begins when blood from one of the patient's arteries is pumped through the tubing. Because of their large size, red blood cells and plasma proteins are unable to pass through the membrane; thus they remain in the bloodstream. Crystalloid substances that need to be removed from the blood, such as urea in the case of kidney failure, diffuse across the cellophane membrane from the area of higher concentration in the blood to the area of lower concentration in the dialyzing solution. After passing through the length of the tubing, the blood is returned to the patient's circulation by way of a vein.

SUMMARY

The kidneys are located in the posterior portion of the abdomen, retroperitoneally. They are bean-shaped structures with the ureter and large vessels entering and leaving the concave medial border. In longitudinal section, the kidney consists of the renal cortex, the outer portion of the kidney; renal medulla, composed of pyramids; and the renal pelvis, a funnel-shaped sac with extensions called calyces.

The unit of structure and function of the kidney is the nephron. The nephron consists of the glomerulus and Bowman's capsule, forming the renal corpuscle, and the renal tubule, composed of a proximal portion, a loop of Henle, and a distal portion. Urine formation occurs in the nephron by means of glomerular filtration, tubular absorption, and tubular secretion.

VOCABULARY REVIEW

Match the statement in the left column with the correct word(s) in the right column. *Do not write in this book.*

1. A vessel that branches over a short distance into about 50 smaller ones.
2. The structural unit of filtration in the kidney.
3. A knot of capillaries through the walls of which filtration occurs.
4. A small saclike structure with walls composed of two layers of cells.
5. The condition of having protein present in the urine.
6. A product that is formed in the liver and excreted by the kidneys.
7. A tube connecting a kidney and the urinary bladder.
8. The appearance of glucose in the urine.
9. A chemical material that increases the production of urine.

a. albuminuria
b. Bowman's capsule
c. diuretic
d. glomerulus
e. glycosuria
f. nephron
g. renal artery
h. renal vein
i. urea
j. ureter
k. urethra

TEST YOUR KNOWLEDGE

Group A

Write the letter of the word(s) that correctly completes the statement. *Do not write in this book.*

1. The pelvis of the kidney opens directly into the   (a) urinary bladder   (b) urea   (c) ureter   (d) urethra.

2. Over 80 percent of the reabsorption of water and solutes occurs in the   (a) proximal tubules   (b) glomerulus   (c) distal tubules   (d) ureter.

3. The plasma component that is normally *not* found in urine is   (a) uric acid   (b) ammonia compounds   (c) creatinine   (d) albumin.

4. The chief function of the kidneys is excretion of the decomposition products resulting from the catabolism of   (a) starches   (b) sugars   (c) proteins   (d) fats.

5. The region of the kidneys that is composed chiefly of blood capillaries and tubules is the   (a) tunic   (b) cortex   (c) apex   (d) pelvis.

6. The kidneys are a part of the   (a) portal circulation   (b) pulmonary circulation   (c) cardiac circulation   (d) renal circulation.

7. The saclike end of a kidney tubule is called the   (a) glomerulus   (b) collecting tubule   (c) Bowman's capsule   (d) alveolus.

8. Urea passes from the bloodstream to the tubules by means of   (a) active transport   (b) osmosis   (c) passive transport   (d) facilitated diffusion.

9. The principle of the artificial kidney is based on the process of   (a) filtration   (b) active transport   (c) diffusion   (d) pressure filtration.

10. The part of a nephron that is most intimately related to Bowman's capsule is the   (a) proximal tubule   (b) distal tubule   (c) glomerulus   (d) loop of Henle.

Group B

Answer the following. Briefly explain your answer. *Do not write in this book.*

1. What part does each of the following structures play in the process of excretion: (a) the liver   (b) the hilum of the kidney   (c) the pelvis of the kidney   (d) Bowman's capsule   (e) a glomerulus?

2. (a) How does the kidney remove waste materials from the blood? (b) Why are not all of the materials that are removed from the blood excreted in the urine?

3. (a) Why does the urine normally lack many proteins and carbohydrates? (b) Under what conditions may these substances infrequently appear in the urine?

4. (a) In what respects does the composition of the plasma differ from that of the urine? (b) How is the balance between these two fluids maintained?

THE ENDOCRINE SYSTEM

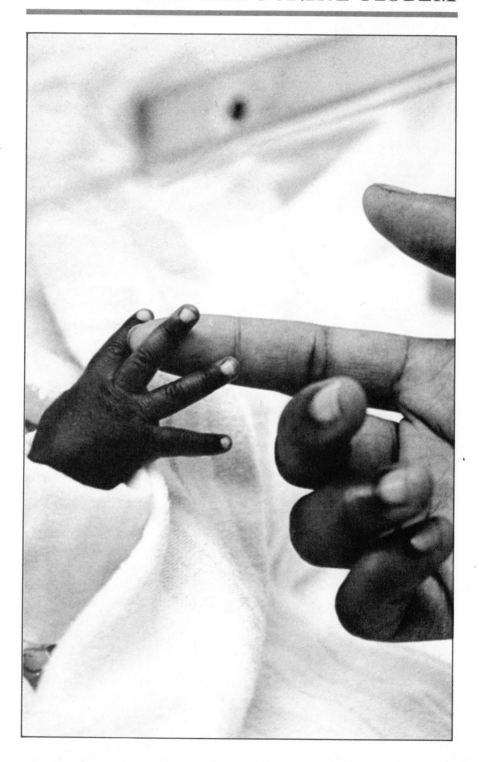

CHAPTER 27

ENDOCRINE GLANDS AND THE FUNCTION OF HORMONES

OBJECTIVES

A. Identify the locations of the major endocrine glands
B. Describe the physiological effects of the hormones produced by the major endocrine glands
C. Compare the effects of steroid and amino hormones and protein hormones on their target cells
D. Describe the negative feedback mechanism in hormone production and give an example

THE ENDOCRINE GLANDS

The survival of all living organisms is dependent upon the degree to which they can maintain their internal environment in a relatively constant state known as *homeostasis.* Of paramount importance in maintaining homeostasis are the two great communication systems in the body, the nervous system and the *endocrine system.* The two systems act together to respond to the many external and internal conditions that would upset this constant state. Certain parts of the central nervous system may activate or inhibit endocrine function and, in turn, the endocrine system is capable of stimulating or inhibiting the flow of nerve impulses. The nervous system enables the body to adjust quickly as changes take place in the environment. To these activities, especially in more complex organisms, is added a special integrating mechanism of chemical regulation. The glands of the endocrine system provide this second communication system of the body. Their chemical products, carried to cells by the bloodstream, regulate continuing processes of long duration such as growth, sexual maturation and reproduction, and other cellular functions.

All endocrine glands act by secreting *hormones* that are transported throughout the body by the blood circulatory system. These glands work independently of one another, yet share the responsibility of controlling the body's activities. Other organs that are not considered endocrine glands have the endocrine function of producing hormones. For example, *secretin,* produced by the small intestine, has the endocrine function of activating the pancreas to secrete its digestive juice.

endo = inside

Types of Glands

By definition, a gland is any group of cells or any organ that secretes a substance. The body contains several kinds of glands that are classified according to the way they secrete. The endocrine glands differ from other glands that were previously discussed in that their secretions diffuse directly into the blood of capillaries associated with the endocrine tissue. The secretions are not carried to a definite internal or external surface through ducts. Endocrine glands are thus known as **ductless glands,** or **glands of internal secretion.** Since the bloodstream transports the endocrine products in minute quantities all over the body, the primary effect of these products may be produced on specific **target cells** in an organ far removed from the site of secretion, or they may have a general effect on the activities of many cells.

Glands, such as the lacrimal glands of the eye that secrete tears, salivary glands in the mouth that produce saliva, or sweat and sebaceous glands, whose respective secretions of water and oil are evident on the skin, all have ducts and are called **exocrine glands.** Prototypes of exocrine and endocrine glands are illustrated in Fig. 27-1.

A few glands have the ability to carry on both exocrine and endocrine functions. Such **heterocrine,** or **mixed, glands** are represented by the pancreas and the gonads (the testes and ovaries).

hetero = mixed

As seen in Fig. 27-2, the endocrine glands are distributed throughout the body. These glandular organs are so diverse in structure and embryological development that it is difficult to consider them as an organ system. Since all of these organs secrete their products into the bloodstream and all function to regulate the activities of other structures, grouping these glandular organs as a system can be justified.

The hypophysis or pituitary gland. The **hypophysis** (which comes from the Greek meaning "outgrowth") is a small gland measuring about 1 cm in diameter (see Fig. 27-2). It hangs from the base of the brain on a short stalk, the **infundibulum.** It sits in a protective housing of the sphenoid bone and is covered by the dura mater. The hypophysis is divided anatomically and embryologically into two primary parts: the anterior lobe is called the **adenohypophysis,** and the posterior lobe is called the **neurohypophysis.** The two lobes release at least nine different secretions of great physiological significance. In addition to regulating the action of many other endocrine glands, these secretions directly influence growth, water absorption by the kidneys, milk secretion by the breasts, and uterine contractions during labor. For this reason, the hypophysis is often considered the "master gland" of the human body.

hypo = under

adeno = gland

The thyroid. The largest gland that is totally endocrine in function is the **thyroid gland.** The thyroid takes its name from its rather prominent "shieldlike" shape, which consists of two lateral lobes and a connecting anterior band of tissue, the **thyroid isthmus** (see Fig. 27-2). This gland is located in the neck at the junction of the larynx and the trachea. Normally, the thyroid weighs about 30 grams in the adult, but its size may vary due to age, pregnancy, diet, during stress, and when under

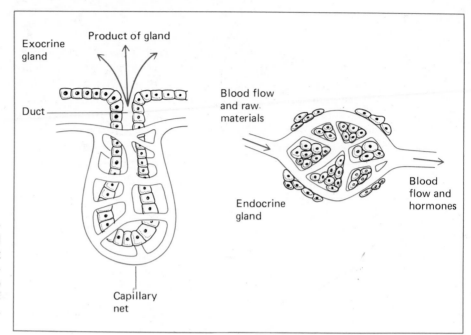

Figure 27-1 **The product of an exocrine gland leaves by way of a duct to a surface. The hormone of an endocrine gland is absorbed directly into the bloodstream for transport.**

the influence of other glands. The primary responsibility of the thyroid gland is to control the metabolic rate of the body, that is, the rate at which oxidative processes occur in cells and the resulting production of heat. Since the rate of metabolism is directly related to growth, the normal function of this gland is essential for development of the body, both mental and physical.

The parathyroids. The *parathyroid glands,* usually four in number, are small reddish bodies, situated on the posterior side of each thyroid lobe (see Fig. 27-2). The two pairs of flattened oval discs are either attached to or embedded in the thyroid tissue in a superior and inferior position. The primary function of these glands is to maintain the homeostasis of calcium salts. The parathyroids are essential to life since their secretion helps to maintain normal irritability of muscle and nervous tissue.

The adrenal glands. The *adrenal glands* are paired, caplike organs located one on top of each kidney; thus they are sometimes called *suprarenal glands* (see Fig. 27-2). They vary in size in different individuals but average about 5 cm in length, about 3 cm in width, and about 1 cm in thickness. The usual weight of each gland is between 3.5 and 5.0 grams. Each adrenal consists of two distinct parts. The outer portion is the *adrenal cortex* and accounts for 80 percent of the gland's weight. The adrenal cortex is derived from mesodermal cells closely associated with those that form the reproductive glands. The inner portion is the *adrenal medulla* and is derived embryologically from the same kind of cells that form the nervous system. Functionally, the adrenal medulla

supra = above

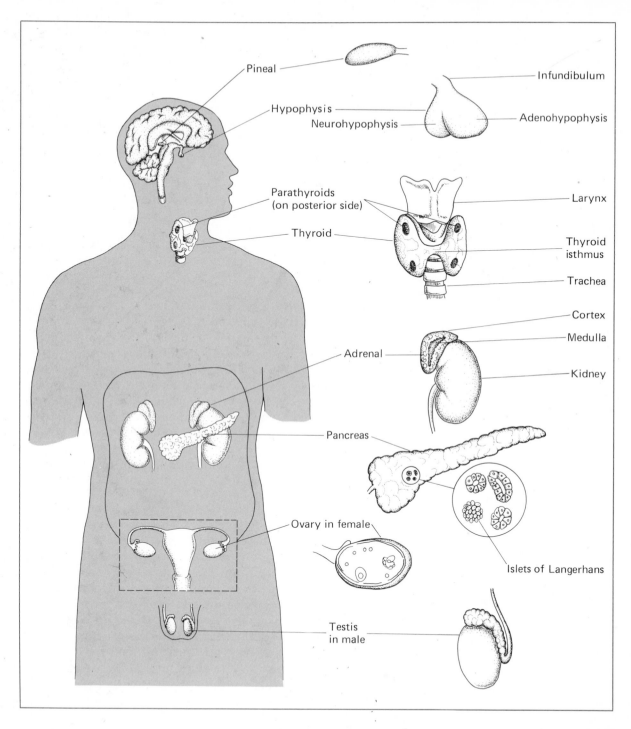

enables the body to cope with emergency situations. The activities of the adrenal cortex are more complex and diversified. They range from regulating electrolytes and water balance in tissues to increasing blood

Figure 27-2 General location of the major endocrine glands and associated structures.

sugar and promoting anti-inflammatory responses. The adrenal cortex also produces minute amounts of sex-related secretions.

The islets of Langerhans.   As previously mentioned, the pancreas is a mixed gland. The major portion of the gland produces a digestive juice concerned with the breakdown of nutrients. The endocrine portion, which amounts to about 1 or 2 percent of the weight of the pancreas, consists of minute islands of secreting cells. They are named *islets of Langerhans* for Paul Langerhans, a German anatomist who first noticed them in 1869. The islets of Langerhans may number as many as 1 million tiny clusters of isolated cells (see Fig. 27-2). The entire pancreas is set in a loop of the duodenum, posterior and slightly inferior to the stomach on the right side of the body. The islets of Langerhans produce two secretions that affect the metabolism of glucose. One secretion causes blood sugar to be transported across cell membranes. The other secretion causes the conversion of glycogen in the liver to glucose, thus elevating the level of blood sugar.

The gonads—ovaries and testes.   The gonads might also be considered mixed glands for their exocrine function is the production of gametes, or sex cells, which are carried by ducts to a surface. The female gonads, called **ovaries,** are paired glands located in the pelvic portion of the abdominal cavity (see Fig. 27-2). The development and maintenance of secondary sex characteristics of the female, such as the development of breasts, the changes of the body contour during puberty, and the onset and control of menstruation, are determined by secretions of each ovary. The male gonads are called **testes** and are paired glands located in a skin covered pouch called the **scrotum** (see Fig. 27-2). The secretions of the testes stimulate the development and maintenance of secondary sex characteristics in the male.

Other endocrine glands.   During pregnancy, the organ of metabolic exchange of nutrients and wastes between mother and developing child is the **placenta.** In addition to this vital role, the placenta carries on an endocrine function. A number of its secretions regulate a variety of activities. One prevents menstruation during pregnancy while others help to maintain and support pregnancy as well as to increase the development of the breasts in preparation for feeding the newborn.

The least understood endocrine structure is the **pineal gland** (see Fig. 27-2). It is a small cone-shaped body that develops as an outgrowth of the third ventricle of the brain. The pineal is 5 to 8 mm long, about 9 mm wide, and is attached to the roof of the ventricle by a short stalk. In early childhood it is glandular, attaining its maximum development by the age of seven. After seven it begins to atrophy (to become nonfunctional) and changes into fibrous tissue in the adult. While many anatomical facts are known about the pineal gland, its function has been the focus of much debate. Recent studies indicate that the gland may produce secretions that affect the physiology of the reproductive system.

CHARACTERISTICS OF ENDOCRINE SECRETIONS

The secretions of endocrine glands are called **hormones** (Greek *hormon*, "to excite or arouse"; "to set in motion"). In terms of chemical structure, a hormone may be a *protein*, as produced by the islets of Langerhans in the pancreas, or the parathyroids; an **amine** (a derivative of amino acids), as produced by the thyroid or the adrenal medulla; or a **steroid** (lipid) as produced by the adrenal cortex and the reproductive glands.

Hormones act as chemical regulators in three important ways: (1) A hormone may activate an immediate chemical reaction, as in an emergency. (2) A hormone may regulate a long-term process such as growth, bone development, maturation of reproductive organs, or the development of secondary sex characteristics. (3) A hormone may regulate secretion of another hormone or substance, usually by a *negative-feedback mechanism*. In this homeostatic mechanism, hormone *A* stimulates the production of hormone *B* while *B* inhibits secretion of *A* as *B*'s own level in the blood rises. By means of negative feedback hormones are secreted only when they are needed. A thermostat and furnace work together much the same way. A furnace is turned on by a thermostat if the temperature falls below a preset level. When the temperature rises to the preset level, the thermostat turns the furnace off. (See Fig. 27-3.)

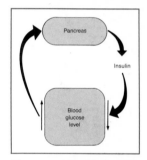

Figure 27-3 Negative-feedback control of insulin by the blood glucose level. The overall effect of insulin is to lower the blood glucose level.

Physiological Effects of Hormones

In order for a hormone to influence a cell, it must first be recognized by the cell, and then must signal the cell to change a function. It may alter rates of transport across the cell membrane, increase the synthesis of an enzyme, or convert an enzyme from an inactive to an active form. There are two general ways in which hormones act: (1) Steroid hormones and some amine hormones pass directly through the cell membrane and enter the cytoplasm, where they are recognized by protein receptors. These receptors in turn transport the hormones into the nucleus, where they interact with DNA to alter the rate of synthesis of one or more messenger RNAs to allow production of specific enzymes (Fig. 27-4a). (2) Protein hormones and some amines do not enter the cell, but rather bind to a receptor on the plasma membrane. Then the cell may release a new substance, called a second messenger, which causes changes in some cell function. In the most common mechanism, the bound hormone activates an enzyme in the membrane, adenyl cyclase, which converts ATP in the cell to cyclic adenosine monophosphate, or **cyclic AMP.** Cyclic AMP can influence many processes in a cell, including protein synthesis, membrane permeability, enzyme activation or inactivation, or secretion of hormones (Fig. 27-4b).

In both types of hormone action, the receptor must recognize the hormone and distinguish it from other substances in the circulation. The subsequent responses depend on the number of receptor molecules which contain hormone, which in turn depends on the amount of hormone in the blood.

Hormones may be continuously secreted in small amounts by endocrine glands. As a result, a great variety of hormones may be present in the blood at any one time. These minute amounts, some as little as one millionth of a milligram per milliliter, are transported in two ways: (1) Steroid hormones, because of their lipid nature, must be attached to plasma globular proteins. (2) Protein hormones are water-soluble and can be transported in the blood plasma without being bound to protein. Amine hormones may require both methods of transport. The body does not store hormones to any known extent after they have been secreted. They are removed from the bloodstream by target cells, by the liver, or by the kidneys, which excrete them in the urine. Thus, the rate of steroid hormone production can be determined by assays of urine.

Hormone Balance

The entire field of endocrinology, the study of endocrine glands, is very complex. The endocrine glands form an "interlocking directorate" over the activities of the whole body. Although each hormone has a specific function, that function can be modified by the effects of hormones from other glands. This interdependence of one gland on another results in the endocrine balance of the individual. Hormonal balance is generally achieved by negative feedback. Exceptions to this are certain neurosecretions, hormones called *releasing hormones*, produced by the hypothalamus of the brain. An excellent example of the workings

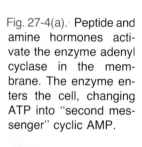

Fig. 27-4(a). Peptide and amine hormones activate the enzyme adenyl cyclase in the membrane. The enzyme enters the cell, changing ATP into "second messenger" cyclic AMP.

Fig. 27-4(b) Steroid hormone passes through plasma membrane and attaches to receptor protein, which enters the nucleus where specific genes are activated.

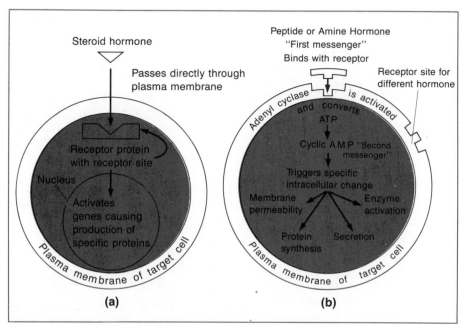

of this homeostatic system is shown by the action of the pituitary gland. This information will be discussed in detail in Chapter 28.

INVESTIGATION OF ENDOCRINE DYSFUNCTIONS

Various methods have been employed by research workers to collect information about the endocrine glands. Most of the basic investigations involved laboratory animals. First, in order to determine its function, the endocrine tissue was removed and any changes in the animal were noted. Second, an extract from the removed tissue or the tissue itself was administered to see if the animal returned to normal. Finally, the researchers tried to isolate the active chemical in pure form. With the passage of time, these methods have become sophisticated and precise.

Recent Techniques

New methods and techniques have evolved to aid research scientists in their quest for knowledge about the endocrine system. Clinical observations of patients suffering from some endocrine disorder correlated with the results of pathological studies provide knowledge in understanding endocrine physiology.

Great strides have been made in recent years in the identification of the molecular structure and the synthesis of artificial endocrine products. The use of radioactive isotopes and chromatography has made possible the measurement of hormones and products of metabolism in small amounts of blood and urine. These are but a few of the techniques now employed by research laboratories.

SUMMARY

Endocrine Gland	Major Action
Pituitary gland (hypophysis)	
Anterior lobe (adenohypophysis)	Regulates growth of body
Posterior lobe (neurohypophysis)	Controls absorption of water
Adrenal gland	
Adrenal medulla	Controls body mechanisms for emergency situations
Adrenal cortex	Regulates various metabolic processes
Thyroid gland	Controls rate of metabolism
Parathyroid gland	Maintains the body's balance of calcium salts
Islets of Langerhans (pancreas)	Controls rate of sugar metabolism
Gonads (ovaries and testes)	Stimulates development and maintains sex characteristics
Placenta	Regulates hormone secretion during pregnancy
Pineal gland	Possibly controls reproductive processes

VOCABULARY REVIEW

Match the statement in the left column with the correct word(s) in the right column. *Do not write in this book.*

1. Name for a gland of internal secretion.
2. An organic chemical produced in one part of the body and transported by the bloodstream to target cells away from the site of its origin.
3. A group of specialized cells forming secretions that are transported by ducts.
4. An example of a heterocrine gland.
5. The largest totally endocrine gland.
6. Another name for the adrenal gland.
7. Secondary sex characteristics are related to these glands.
8. Maintain normal irritability of muscle and nervous tissue.

a. endocrine gland
b. exocrine gland
c. gonads
d. hormone
e. hypophysis
f. pancreas
g. parathyroids
h. suprarenal gland
i. thyroid

TEST YOUR KNOWLEDGE

Group A

Write the letter of the word(s) that correctly completes the statement. *Do not write in this book.*

1. The hypophysis is located in the (a) head (b) neck (c) thorax (d) abdomen.
2. Many protein hormones cause the target cell to form (a) mRNA (b) ATP (c) steroid hormones (d) cyclic AMP.
3. Secondary sex characteristics are influenced by the (a) adrenal medulla (b) pancreas (c) gonads (d) thyroid.
4. As the blood glucose level rises, (a) insulin production stops (b) the endocrine glands shut down (c) insulin production increases (d) there is no effect on hormone production.
5. The gland that is considered the "master gland" is the (a) adrenal (b) thyroid (c) testis (d) hypophysis.

Group B

Answer the following. Briefly explain your answers. *Do not write in this book.*

1. For each of the following bodily functions, name the gland involved: (a) sugar metabolism, (b) metabolic rate, (c) homeostasis of calcium and phosphorus, (d) emergency reactions, (e) successful maintenance of pregnancy, and (f) contraction of the uterus during delivery of a child.
2. How does the "negative feedback system" work? Give an example.
3. (a) What is the difference between an exocrine gland and an endocrine gland? (b) How do the secretions of these two kinds of glands differ?

CHAPTER 28

SPECIFIC ENDOCRINE GLANDS

OBJECTIVES

A. Compare the structure and functions of the anterior and posterior lobes of the pituitary gland

B. Compare the structure and functions of the adrenal medulla and the adrenal cortex

C. Describe the endocrine functions of the thyroid and parathyroid glands, the islets of Langerhans, and the gonads

D. Describe the metabolic disorders caused by hyper- and hypoactivity of the endocrine glands

THE ANATOMY OF ENDOCRINE GLANDS

In Chapter 27, the general anatomy and physiology of the endocrine glands, including the characteristics of hormones, was introduced. This exposure should be of assistance in the in-depth study of the individual endocrine glands. While some of the glands appear to be independent structures, the majority of the endocrine glands greatly influence each other. They effectively share with the nervous system the role of maintaining and coordinating the activities of the body. The nervous system is essential in order for changes in the environment to be sensed, or recognized. Hormones, in turn, enable the body to adjust its chemical balance in response to these changes in the environment.

Terminology

It should be noted that certain terms are frequently used in endocrinology. The suffix *-tropic* is used to indicate a hormone that activates the gland indicated by the portion of the word that precedes the suffix. For example, *thyrotropic hormone* is one that stimulates the thyroid gland to secrete its hormones. In addition, two prefixes are commonly used to refer to the relative amount of hormone produced by an endocrine gland. *Hypo-* refers to a subnormal production of a hormone by a gland, the term being derived from the similar Greek word meaning "under." The prefix *hyper-* means "over" and is used to describe overproduction of a hormone. For example, undersecretion of the thyroid gland is called *hypothyroidism;* oversecretion is called *hyperthyroidism.*

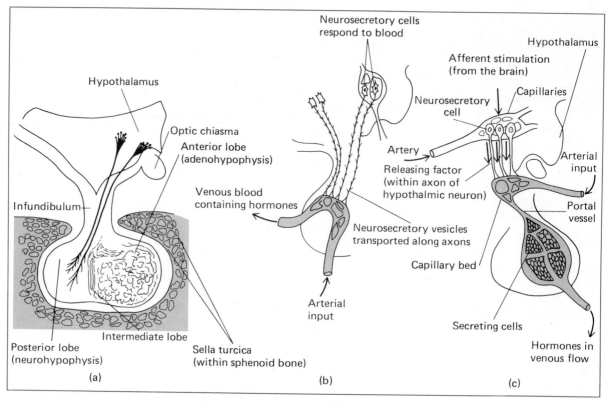

Figure 28-1 (a) Diagram showing pituitary gland in the sella turcica of the sphenoid bone. Infundibulum, optic chiasma, and hypothalamus are also shown. (b) Hormonal secretion of the neurohypophysis is regulated by the bloodstream. (c) Activity of the adenohypophysis is controlled by the hypothalamus.

Hypophysis (Pituitary Gland)

The hypophysis (pituitary) is a small gland, weighing about 0.6 gram, and is attached to the lower surface of the brain just behind the optic chiasma and below the hypothalamus. It is connected to the hypothalamus by a short stalk, the **infundibulum.** The pituitary sits protected within a saddle in the sphenoid bone called the **sella turcica.** (See Fig. 28-1.) The name "pituitary" (from the Latin *pituita*, or phlegm) was given this gland by the great anatomist Vesalius (1514–1564) because he thought that it produced a lubricating mucus.

The gland has three lobes. The largest is the anterior portion, called the **adenohypophysis.** This part develops embryonically from the roof of the mouth. It has a glandular appearance, loosely interspersed with different kinds of cells. The posterior portion, or **neurohypophysis,** is a downgrowth of the hypothalamus that remains attached to it by the short infundibulum. This lobe consists of modified nerve cells called **pituicytes,** numerous neuroglial cells, and a network of capillaries and nerve fibers. The cell bodies to which these fibers are attached originate in the hypothalamus. The **intermediate** lobe is the area where the anterior and posterior regions of the pituitary come in contact with one another. At present, this part of the pituitary is still not yet completely understood. However, it is known to control the production of skin pigment in some of the lower vertebrates. Therefore the discussion of the pituitary primarily concerns its anterior and posterior lobes.

In view of its many regulatory functions, the hypophysis has sometimes been called the "master gland" of the body.

The anterior lobe.   The anterior lobe, the adenohypophysis, releases several different hormones that regulate a whole range of body functions. This is accomplished in two ways: by regulating the metabolic activities of cells, and by stimulating other endocrine glands. The major control over the production of these anterior-lobe hormones is exerted by the hypothalamus of the brain (see Fig. 28-1). Special neuron cell bodies in the base of the hypothalamus produce **neurohormones,** or **releasing factors,** which move down the axons for storage at the nerve fiber endings in the stalk. Associated with these endings are capillaries formed from branches of the internal carotid arteries. These form a number of small venules that descend into the adenohypophysis.

When stimulated, the fiber endings discharge one or more of the releasing factors into the small venules so that the anterior lobe is stimulated to produce its secretion of hormones.

The posterior lobe.   The posterior lobe, or neurohypophysis, does not secrete hormones but functions as a storage area for two hormones produced in the hypothalamus. Two well-defined clusters of neuron cell bodies send out axons that descend through the connecting stalk and terminate near the posterior lobe capillaries (see Fig. 28-1). **Antidiuretic hormone (ADH),** also known as **vasopressin,** and **oxytocin,** are synthesized by these neurosecretory cells and are enclosed in small vesicles that move slowly (3 millimeters a day) down the neuroplasm of the axons to accumulate at the nerve endings. Upon nervous stimulation by the hypothalamus, a specific hormone is released into the capillaries to be distributed by the bloodstream.

Thyroid Gland

The thyroid gland is a dark red structure in the neck, lying along the sides of the larynx (Fig. 28-2). Its general shape is that of the letter H with the upright lobes about 5 centimeters long and about 3 centimeters broad. These are joined by a bridge of tissue, the **thyroid isthmus,** which is situated in front of the second and third cartilage rings of the trachea and is about 1.3 centimeters wide. In the adult, the thyroid gland weighs about 30 grams, being slightly larger in females than in males. It is covered by a connective capsule and the surface of the gland is divided into many lobules by partitions, or septa, which extend inward. The thyroid is richly supplied with blood, receiving approximately 100 milliliters per minute. For its weight, the thyroid has the highest rate of blood flow of any organ in the body.

Follicular cells.   The structural and functional unit of the thyroid gland is a **follicle.** It is composed of a spherical group of single layered cu-

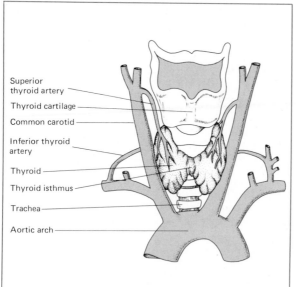

Figure 28-2 **The thyroid gland with its blood supply and associated structures—the trachea and larynx.**

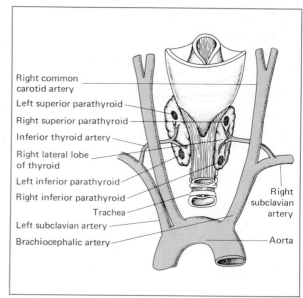

Figure 28-3 Posterior view of neck region showing the parathyroid gland and its associated blood supply on lateral lobes of thyroid.

boidal epithelial cells that surround an open cavity, or lumen. These cells secrete a viscous material of uniform texture known as **thyroid colloid,** which is deposited in the lumen of the follicle. These follicles do not have external openings but the spaces between them are richly supplied with blood and lymph vessels. In this way, the follicular cells are able to obtain raw materials to synthesize their products. One of the main products is a large protein called **thyroglobulin,** containing preformed thyroid hormones, which is stored in the colloid. Scattered around the follicles within the gland are parafollicular cells, called **C-cells,** that produce a different hormone associated with calcium metabolism, called *calcitonin.*

Parathyroid Glands

Attached to or embedded in the posterior surface of the two lateral lobes of the thyroid gland are small oval masses of tissue called the *parathyroid glands* (see Fig. 28-3). These reddish tan bodies are probably the smallest of the endocrine glands, each measuring about 3 to 8 millimeters in length, 2 to 5 millimeters in width, and 0.5 to 2.0 millimeters in thickness. They are usually four in number and are located at the four quadrants of the thyroid lobes, but their number and position are variable. The combined weight of the parathyroid glands varies from 0.05 to 0.3 gram.

The internal microscopic anatomy of the parathyroids consists of two kinds of epithelial cells. The most numerous are called **principal,** or **chief, cells** and are probably responsible for the production of the parathyroid hormone. The second type are called **oxyphil cells.**

Adrenal Glands (Suprarenals)

The paired **adrenal glands** are two masses of yellowish tissue that resemble caps and are located superior to each kidney (see Fig. 28-6). Because of this location, they are also known as **suprarenal glands.** In humans, the right adrenal gland resembles an admiral's cocked hat, while the left gland is more semilunar in shape. Their dimensions, in the adult, are about 4.5 centimeters to 5.0 centimeters in length, 3 centimeters in width, and from 1.0 to 1.5 centimeters in thickness. Each gland weighs approximately 4 grams. Like the thyroid gland in the neck, the adrenals are highly vascular, receiving blood from branches of the renal, pelvic, and gonadic arteries.

The human adrenal glands appear early in embryonic life and at one stage of development are the most conspicuous organs in the abdominal cavity. At birth they are approximately one-third the size of the kidneys, but in early infancy they become smaller. Each gland is composed of two distinct regions: an outer area that makes up the bulk of the gland is called the **adrenal cortex,** and the smaller central portion is called the **adrenal medulla.** These two parts have very different and distinct secretions, and their origins in the embryo are different.

The adrenal cortex.   This part of the adrenal gland is absolutely essential to life. Its removal or extensive injury in experimental animals invariably leads to death unless its secretions are replaced artificially. Microscopically, the cortex is subdivided into three zones, each producing different hormones. The outer zone produces various **mineralocorticoid hormones** and is called the *zona glomerulosa.* The larger middle zone forms various **glucocorticoid** hormones and is called the *zona fasiculata.* The inner zone, the *zona reticularis*, produces various sex hormones called **gonadocorticoids**. Fig. 28-4(c) shows a molecule of *cortisol*, one of the major corticoid hormones in humans. (See page 434.)

The adrenal medulla.   Unlike the adrenal cortex, which secretes several steroid hormones, the adrenal medulla produces two hormones derived from amino acids—**epinephrine** (also known as **adrenalin**) and **norepinephrine** (**noradrenalin**). Epinephrine is more potent in its action than norepinephrine and constitutes about 80 percent of the total secretion of the medulla, although neither hormone is essential for life.

Islets of Langerhans (Pancreas)

The exocrine activities of the pancreas were previously described in Chapter 15. At present we will consider the endocrine function, that is, the secretions produced by the pancreatic islets, known as the **islets of Langerhans.** Scattered throughout the pancreas are many small clusters of cells segregated from those of the digestive pancreas by reticular fibers. It is estimated that there may be as many as 1 million islets in the pancreas, collectively weighing about 1 gram of the total pancreatic weight of 60 grams.

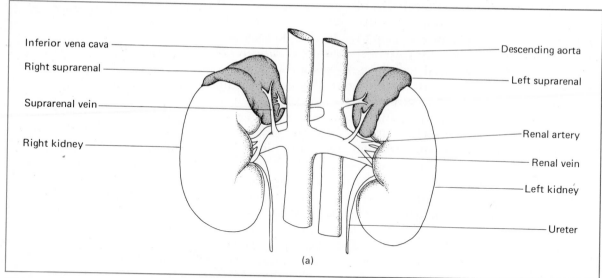

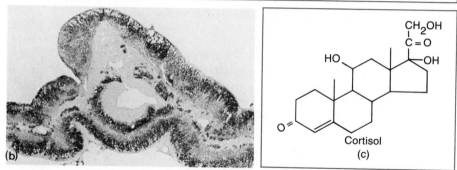

Figure 28-4 (a) Location of the suprarenal glands and their associated blood supply. (b) Cross section of adrenal gland (photo) showing the two regions. (c) Molecular structure of cortisol.

Islet cells.   Each islet is approximately 150 μm (microns) in diameter and contains two principal kinds of cells: **beta cells,** which represent three-quarters of the total number of cells and secrete the hormone **insulin;** and **alpha cells,** which are responsible for the production of the hormone **glucagon.** A third kind of cell, found in minute numbers, **delta cells,** produces somatostatin, which inhibits both insulin and glucagon (see Fig. 28-5).

The Gonads

The primary sex characteristics of any individual are established when the sperm nucleus unites with the ovum nucleus at fertilization. As will be explained in Chapter 29, the presence of certain chromosome combinations will determine whether the individual will be male or female. The chromosomal type that an individual inherits controls the development of certain primitive and essentially similar embryonic tissues into the reproductive organs (testes in the male and ovaries in the female). The term **gonads** is used to identify the reproductive organs. The testes produce sperm and characteristic male sex hormones. The ovaries are female gonads and produce ova and female sex hormones.

The sex hormones.   The human sex hormones control the development of the **secondary sex characteristics,** which are indicative of the sex of the individual. The effects of human sex hormones usually become noticeable between the ages of 12 and 15 years, when a person is becoming physically mature. At that time a young person begins to show the secondary sex characteristics, which so definitely distinguish males from females. This period is called **puberty.** Prior to this developmental stage, there is little structural difference between the two sexes other than the type of gonad and genital organs present.

With the onset of puberty and the increased production of sex hormones, the similarities in bodily structure disappear. The pitch of a male's voice becomes lower as a result of an enlargement of the larynx, while that of a female retains its higher register. Hair grows in the pubic region (above the genitals) and axillary region (in the armpit) in both, but a beard begins to grow in the male while the female's face remains relatively hairless. A female begins to develop a layer of *subdermal fat,* while a male does not and therefore keeps a more angular and muscular frame. The skeleton of a male increases in size and weight to a greater degree than that of a female. The hips of a female grow wider between the crests of the ilia, while those of a male continue their growth pattern without marked change. A female's *breasts* develop, while their growth in a male is inhibited.

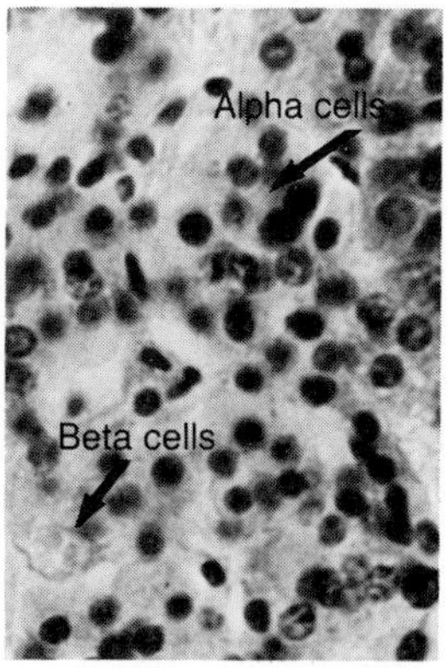

Figure 28-5 **Alpha cells and beta cells of the islet of Langerhans.**

Other Glands

During pregnancy, the development of the **placenta** involves a combination of interlocking maternal and embryonic tissues. This selectively permeable structure performs many important activities.

Embryonic circulation begins with nutrients, oxygen and carbon dioxide, and metabolic wastes being exchanged. The placenta also acts as a temporary endocrine gland by producing several hormones of crucial importance to the successful maintenance of pregnancy.

The **pineal body** is a small organ attached to the posterior commissure of the brain by a short stalk. It arises from the roof of the third ventricle under the posterior end of the corpus callosum. The pineal body is cone-shaped, measuring about 8 millimeters long and 9 millimeters at the base. It is reddish gray in color. The stalk contains some nerve fibers that are largely lacking in the pineal body itself. There are neuroglial cells present in the stroma of the gland. The pineal reaches its maximum size by age 7 and by puberty becomes nonfunctional.

PHYSIOLOGY OF ENDOCRINE GLANDS

The integration of the various body processes is regulated by both endocrine secretion and nerve action. In turn, the interaction of endocrine glands affects their individual activity. For example, the hyperactivity or hypoactivity of one gland will influence the activity of other glands.

Function of Anterior Pituitary Hormones

Each lobe of the hypophysis releases different hormones. At least six major hormones are produced by cells in the anterior lobe. These include human growth hormone (HGH); thyroid-stimulating hormone (TSH); adrenocorticotropic hormone (ACTH); two gonadotropic hormones, follicle-stimulating hormone (FSH) and luteinizing hormone (LH); and prolactin. (See Fig. 28-6.)

soma = body
tropic = changing

Human growth hormone is also known as **somatotropic hormone (STH).** A number of factors seem to influence when an increase or decrease in the production of HGH will occur. Low blood sugar, increased intake of proteins, stress, and exercise stimulate the hypothalamus to secrete a regulating hormone called **growth releasing hormone (GRH).** This releasing hormone is carried by the blood to the adenohypophysis, activating the lobe to increase HGH output. When conditions return to normal, GRH secretion decreases and the release of HGH slows down. An abnormally high level of glucose in the blood results in the hypothalamus producing another hormone called **growth inhibiting hormone (GIH)** or **somatostatin.** This substance inhibits the release of GRH with the subsequent decrease in HGH.

The primary function of human growth hormone is to increase protein synthesis in hard and soft tissues. Increasing the rate of active transport of amino acids into the cells and the synthesis of amino acids into proteins cause cells to grow and multiply. Growth of the skeleton and other connective tissues is indirectly influenced by HGH, which promotes the production of a peptide molecule called *somatomedin* in the liver. This hormone acts as a mediator, or "second messenger," that promotes cartilage and bone development. Human growth hormone also affects fat and carbohydrate metabolism.

The **thyroid-stimulating hormone (TSH),** also known as *thyrotropin,* promotes the synthesis and secretion of hormones produced by the thyroid gland. The rate at which TSH is produced is influenced by **thyrotropin releasing hormone (TRH),** a product of the hypothalamus.

The amount of TSH released by the anterior lobe of the pituitary depends upon the level of thyroid hormones in the blood. As the level of thyroid hormones increases, there is a slowing down in the production of TSH. As the amount of TSH drops, the production of thyroid hormones is diminished, allowing the TSH to once again increase. This is another example of a negative-feedback system.

Adrenocorticotropic hormone (ACTH) is also known as *corticotropin.* It controls the growth and secretions of the adrenal cortex. A low concentration of corticoid hormones in the blood, as well as the presence of stress (pain and extremes in temperature), causes the hypothalamus to produce **corticotropic releasing hormone (CRH).** This in turn stimulates the anterior lobe to release ACTH, resulting in an increase in adrenal cortical hormones.

Gonadotropic hormones, also known as *gonadotropins,* control the growth, development, and functions of the gonads (the testes and ova-

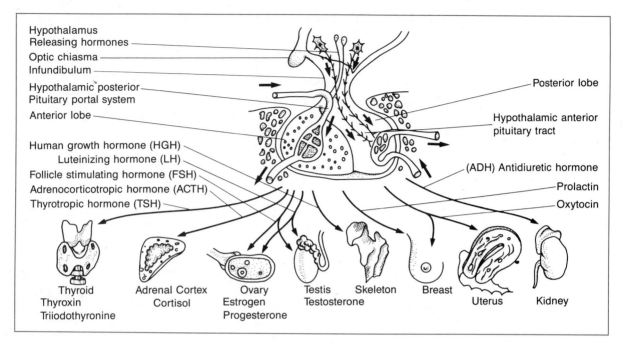

Hypothalamus
Releasing hormones
Optic chiasma
Infundibulum
Hypothalamic posterior
Pituitary portal system
Anterior lobe

Human growth hormone (HGH)
Luteinizing hormone (LH)
Follicle stimulating hormone (FSH)
Adrenocorticotropic hormone (ACTH)
Thyrotropic hormone (TSH)

Posterior lobe

Hypothalamic anterior
pituitary tract

(ADH) Antidiuretic hormone
Prolactin
Oxytocin

Thyroid
Thyroxin
Triiodothyronine

Adrenal Cortex
Cortisol

Ovary
Estrogen
Progesterone

Testis
Testosterone

Skeleton

Breast

Uterus

Kidney

Figure 28-6   Relation-
ship of the pituitary
lobes and their hor-
monal effects on the
other endocrine glands
and body structures.

ries). In the female, the **follicle-stimulating hormone (FSH)** stimulates
the development of follicle cells in the ovary, which surround each
developing egg (ovum). In the male, it stimulates the testes to produce
sperm. The production of FSH is under the influence of the hypothalamus, which produces **gonadotropin releasing hormone (GnRH).** GnRH
is released in response to sex hormones produced by the gonads and
involves a negative-feedback system. Another gonadotropin is the **luteinizing hormone (LH).** In the female, LH is essential for ovulation (the
discharge of the egg from the ovary) and the conversion of the follicle,
after ovulation, into a separate glandular structure, the *corpus luteum.*
LH in the male, sometimes referred to as the *interstitial cell–stimulating
hormone (ICSH),* stimulates the interstitial cells within the testes to
produce testosterone, the male sex hormone.

During pregnancy, female sex hormones (estrogen and progesterone)
stimulate the development of the breasts in preparation for feeding the
child. **Prolactin,** or lactogenic hormone, activates the breasts in producing milk. Prolactin also inhibits the secretion of LH and FSH following delivery of a child and prevents menstruation during nursing.

Function of Posterior Pituitary Hormones

There are two hormones released by the posterior lobe. One of the
hormones is the **antidiuretic hormone (ADH)** and the other is known
as **oxytocin.**

The principal physiological effect of *ADH* is the retention of water by
the kidneys. When osmotic receptors in the hypothalamus are activated
by changes in the fluid content of the blood, impulses are sent to the
posterior lobe, which releases the hormone. ADH acts on the distal

muscle in the uterus, which brings about the birth of the child. Oxytocin, along with prolactin, causes the release of milk by the breasts of a nursing mother.

Functions of Thyroid Hormones

Physiologically, the thyroid follicles absorb iodine from the bloodstream as sodium or potassium iodide, and tyrosine, one of the essential amino acids. These raw materials are synthesized into two hormones, **thyroxine,** also known as **tetraiodothyronine** or T_4, and **triiodothyronine** or T_3. Figure 28-7 shows the structural formula for these hormones. These products are stored in the colloid bound to the large protein molecule **thyroglobulin.** They remain in this state until needed by the body. When under the influence of thyroid-stimulating hormone (TSH), the bonds are broken and the hormones are secreted into the capillaries. Upon entering the bloodstream, the hormones are attached to a plasma protein for transport to target cells.

Thyroxine (T_4) is the principal thyroid hormone, making up about two-thirds of the circulating thyroid hormones, with triiodothyronine (T_3) making up about one-third. They both perform similar functions.

T_3 has a stronger and more rapid effect on cellular activity. The thyroid is primarily concerned with cellular metabolism. It regulates the metabolic rate, heat production, and oxidation rate of all cells, with the possible exception of those of the adult brain. In addition, it has the ability to convert glycogen into glucose as well as to increase the rate at which carbohydrates are absorbed from the small intestine, resulting in a rise in blood sugar. Thyroid hormones increase the rate of fat metabolism and decrease the levels of cholesterol in the bloodstream. Throughout the body, the secretions control the growth and differentiation of tissues, thereby influencing both physical and mental growth in a child.

A third thyroid hormone—**calcitonin** or **thyrocalcitonin**—is produced by cells found around the follicles (parafollicular cells). Calcitonin aids in maintaining the homeostatic level of calcium in the blood. It inhibits the reabsorption of calcium from the bones, thereby lowering blood calcium levels. The exact role of this hormone is uncertain. It seems to have little effect on adults but is more active in young individuals, where it may play a role in skeletal development.

Control of thyroid secretion. If the metabolic rate should decrease or the amount of thyroid hormones fall below a normal level, chemical sensors in the hypothalamus are activated to produce *thyroid-stimulating hormone releasing factor* (TSHRF). TSHRF stimulates the adenohypophysis to secrete thyroid-stimulating hormone (TSH), which in turn activates the thyroid gland to release thyroxine. This activity continues until the metabolic rate is in balance once again. As the level of thyroxine increases, the negative-feedback system is activated to depress the quantity of TSHRF. There is some evidence to indicate that the drop in thyroxine in the blood directly affects the adenohypophysis

creases, the negative-feedback system is activated to depress the quantity of TRH. There is some evidence to indicate that the drop in thyroxine in the blood directly causes the adenohypophysis to release thyroid-stimulating hormone during certain stressful situations. For example, extremely cold weather conditions may activate the hypothalamus to form TRH and reinforce the production of TSH.

Function of Parathormone

The hormone produced by the parathyroids is called **parathormone (PTH),** a peptide chain of 83 amino acids, which gives the molecule an atomic weight of 9500 units. Parathormone plays a part in regulating the calcium and phosphate ions that are present in blood plasma. These two ions are of importance in promoting proper nerve and muscle function as well as in maintaining bone structure.

The concentration of calcium and phosphorus in the blood acts as a control over the parathyroid hormone. When the total calcium drops below the normal level of about 10 mg per 100 ml of blood, the parathyroids are stimulated and the production of PTH is increased. If adequate amounts of vitamin D are present, PTH increases the level of calcium ions in the blood by promoting the absorption of these ions from the intestines and kidney tubules and by stimulating the activity of osteoclasts (bone-destroying cells). When bone tissue is destroyed, calcium and phosphate ions are released into the blood. As the level of calcium and phosphate ions increase, there is a decrease in the production of parathormone. If there is a deficient supply of the hormone in the blood plasma, bone formation ceases and the nervous system and muscles are adversely affected. (See Fig. 28-8.)

Vitamin D also plays a part in the retention of calcium and phosphate ions by keeping them at a constant level. If vitamin D is absent from the diet, an increased excretion of calcium and phosphate occurs. Without these, bones cannot develop; they become soft and bend.

Secretion of Adrenal Cortical Hormones

There are three types of corticoids produced by different layers of the adrenal cortex: *mineralocorticoids, glucocorticoids,* and sex hormones, primarily *androgens* (male hormones) and *estrogens* (female hormones). (The sex hormones are discussed later in the chapter.)

Action of corticoids.   **Aldosterone** is the principal and most potent hormone produced in the outer zone of the cortex. Its primary function is to regulate the electrolytic minerals by the reabsorption of sodium by the cells of the renal tubules of the kidneys and at the same time eliminate potassium. In fact, the production of aldosterone is controlled by the amount of sodium taken into the plasma of the blood. For example, if the body is deprived of sodium, or if there is an increase in the presence of potassium, there is an increase in the secretion of aldosterone. Water loss from the kidneys is decreased, causing a corresponding increase in blood plasma volume.

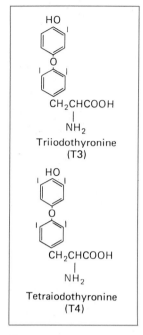

Figure 28-7 **Structural formula of the hormones of the thyroid gland.**

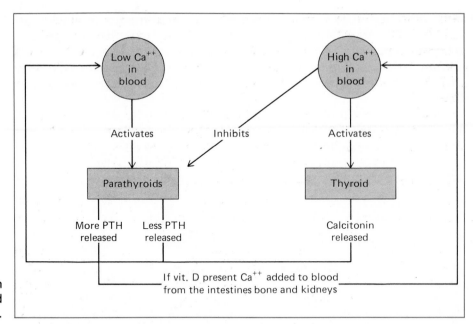

Figure 28-8 **Regulation of calcium ions in blood with PTH and calcitonin.**

Cortisol (hydrocortisone) is the principal glucocorticoid produced by the middle layer of the cortex. While *corticosterone* and *cortisone* are also produced in this region, cortisol represents 95 percent of its output.

The hormones of this group are concerned with the metabolism of carbohydrates, proteins, and fats. In addition, they enable a person to cope with stress. They contribute to the catabolism of proteins and stimulate the liver to take up amino acids and convert them into glucose. In addition to promoting *gluconeogenesis* (the production of new glucose), cortisol together with insulin facilitates the production of glycogen that is stored in the liver.

The production of glucocorticoids is primarily under the control of adrenocorticotropic hormone (ACTH).

Medullary Secretion

The release of epinephrine and norepinephrine is directly related to activities of the nervous system rather than to the influence of some other hormone. Within the adrenal medulla, the hormone-producing cells are directly innervated by neurons of the sympathetic division of the autonomic nervous system.

Emergency controls.   When the body is confronted with a threatening "flight" or "fight" situation, impulses received by the hypothalamus are transmitted to sympathetic neurons, which enter and stimulate the cells of the medulla. The result is a sudden increase in the output of the two hormones, initiating numerous physiological changes in the body. There is a rise in blood pressure due to an increasing heart rate and constriction of visceral blood vessels. Respiration is more rapid and bronchioles dilate. The digestive process is suppressed and urinary output decreases. Heart and skeletal muscles receive an increased blood supply while vessels in the skin are constricted. There is an increase in the conversion of glycogen to glucose and cellular metabolism is stimulated.

In all, the action of the adrenal medulla is to reinforce the activities of the sympathetic division of the autonomic nervous system and prolong the response the body makes during an emergency.

Secretory Function of the Pancreas

The beta cells in the pancreatic islets produce *insulin*. Insulin promotes the storage and utilization of all macronutrients (fatty acids, glucose, and amino acids), but the effect on glucose is most dramatic. The only insulin-sensitive tissues or target tissues are muscle (skeletal and cardiac), adipose tissue, and liver. In muscle cells, insulin facilitates the absorption of glucose and its conversion into glycogen. In the liver, while blood sugar enters the cells freely, insulin enhances the formation of glycogen, or *glycogenesis*. Insulin enables adipose cells to absorb glucose, which is then used as a raw material for the synthesis of neutral fats. All of these activities tend to lower the level of glucose in the blood. The regulation of insulin production is directly related to the level of glucose in the blood through a typical feedback mechanism.

The second hormone, produced by the alpha cells in the pancreatic islets, is **glucagon.** This secretion has the opposite effect on blood sugar levels as insulin by accelerating the conversion of liver glycogen into glucose. The production of glucagon is also controlled by the level of sugar in the blood. When the blood sugar level falls below normal, the alpha cells are stimulated into secreting glucagon. As the sugar level rises, glucagon production diminishes.

Secretion of the Gonads

All the secondary sex characteristics fail to develop if the gonadotropic hormones, the follicle-stimulating hormone (FSH), and the luteinizing hormone (LH) are missing or greatly deficient in quantity. These hormones are responsible for the development of the gonads, which in turn produce male and female hormones. (See Fig. 28-9.)

Androgens.   The male sex hormones are considerably less complex than those of the female since they are not concerned with the menstrual cycle, a developing child, or childbirth. The hormones produced by the interstitial cells of the testes are known by the general name of **androgenic hormones.** Of these steroids, the principal one is testosterone, which is by far the most potent and most abundant. It is a 19-carbon steroid synthesized from cholesterol.

andro = male

Testosterone is responsible for the development of adult primary sexual characteristics. During development the penis, scrotum, and testes increase in size. The internal prostate and bulbourethral glands enlarge and become active. As mentioned before, testosterone is responsible for the development of the male secondary sex characteristics.

Estrogens. The female gonads, paired ovaries, produce ova and the female sex hormones estrogen and progesterone. Each ovary, about the size and shape of a large almond, consists of an outer covering layer of columnar epithelial cells called the **germinal epithelium.** Beneath this covering is connective tissue called **stroma,** which is divided into a cortex and inner medulla. It is within the cortex that one finds thousands of immature ova, each surrounded by a layer of flattened epithelial cells forming a *follicle.* With the onset of puberty, under the influence of FSH, approximately 400 follicles will mature in the ovaries during the reproductive years of the female. It is from these follicular cells that the first group of female sex hormones is produced, the *estrogens.* One of these is **beta-estradiol,** the most potent and the principal female sex hormone. When an ovum is matured and released (ovulation) under the influence of LH, the follicular cells, now void of the ovum, form a new glandular structure, the *corpus luteum,* which produces the second ovarian hormone, **progesterone.** It is the balance that exists between the concentration level of two gonadotropic hormones of the adenohypophysis (FSH and LH) and the resulting formation of estrogen and progesterone that controls the various phases associated with the menstrual cycle (discussed in Chapter 29, Reproduction and Development).

Activity of the Pineal Gland

Because of its location in humans, investigators have found it difficult to ascertain the true physiology of the pineal gland. Most of the work has been done on experimental animals with little correlation to human function. In 1958, a hormone was isolated and named **melatonin** because of the blanching action it had on frog skin. Melatonin has no known function in humans, but current research suggests that it may participate in regulation of reproductive physiology.

ENDOCRINE DYSFUNCTION

It is beyond the scope of this textbook to consider clinical details about endocrine pathology. As a consequence, we will present in a nontechnical way some of the disorders based on hyposecretion or hypersecretion of hormones.

Hypophysis

Since the adenohypophysis produces so many hormones, the overproduction or underproduction of these secretions has widespread and

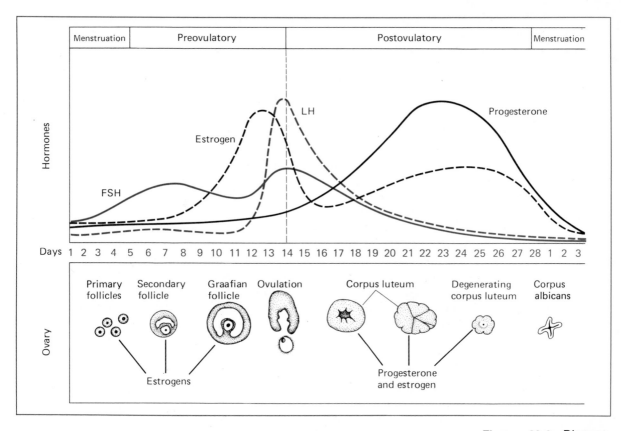

complicated consequences. One of the most striking evidences of a disturbance in this portion of the hypophysis is the effect of HGH on physical development. If this hormone is secreted in greater than skeletal quantities before puberty has been completed and before the epiphyseal junctions have closed in a normal manner, the child shows continued skeletal development. The victim of this disease may have a normal weight at birth but quickly shows signs of excessive growth. This accelerated growth results in a disorder called ***pituitary gigantism.*** One of the tallest giants reported in medical literature was nearly 3 meters (9.9 feet) tall. Gigantism is accompanied by an increase in protein metabolism, so there is corresponding growth of the internal organs as well (Fig. 28-10).

If hypersecretion of human growth hormone occurs after adult size is reached, the long bones are no longer affected since the epiphyseal plates are closed. There are, however, marked changes in the bones of the face, hands, and feet. The lower jaw enlarges and spaces appear between the teeth. Ridges develop over the eyes and the person's expression is greatly changed. The bones of the hands also become larger, especially in the region of the knuckles. This condition is called ***acromegaly.***

Figure 28-9 Diagram showing relationship between the ovarian hormones and those of the anterior pituitary.

acron = extremity
mega = large

Deficiency of the growth hormone in a child leads to **pituitary dwarfism.** In this condition the skeletal structures reach their maximum development early in life and are unable to grow any larger. The result of this stoppage of growth is a miniature person whose bodily proportions and mental ability are generally normal. Since the growth hormone from other animals has no effect on humans, therapy with human growth hormone is administered before the epiphyseal plates close.

Hyposecretion of HGH in the adult produces no physical changes. However, if there is dysfunction of the entire adenohypophysis, a condition called *panhypopituitarism* results. There is increased sensitivity of body cells to glucose, resulting in low blood sugar, low thyroid activity, adrenal insufficiency, and lowered sexual activity. There is loss of body hair and diminished muscular development.

The principal disorder associated with the neurohypophysis occurs when there is insufficient production of antidiuretic hormone (ADH). This condition is characterized by the daily output of huge volumes of dilute, sugar-free urine, which may be as great as 15 to 20 liters (15.9 to 21.2 quarts) depending upon water intake. This disease is called **diabetes insipidus.** Diabetes insipidus should not be confused with sugar diabetes (diabetes mellitus), which is associated with malfunction of the pancreas.

Thyroid

Disorders of the thyroid gland are either associated with a deficiency of iodine in the diet, which is the essential element in thyroid hormones, or with an underproduction or overproduction of the hormones themselves. The former, lack of iodine, results in the development of **simple goiter.** At the onset of this disease, the thyroid gland is unable to produce a normal supply of thyroxine. This deficiency activates the negative-feedback system, causing an increase in TSH. The thyroid enlarges under this stimulus, apparently trying to utilize the iodine available. Because of this efficient feedback mechanism, nearly normal levels of thyroxine may be produced by the enlarged gland. This disease is illustrated in Fig. 28-11.

Hypothyroidism.   The deficiency of thyroid hormones is characterized by a reduction in metabolic rate. If this deficiency should occur during the growing years, the skeleton and brain fail to develop normally. The resulting disease is called **cretinism** and is characterized by dwarfism and mental retardation. Cretins are frequently bow-legged and have dry, scaly skin. Their abdomens usually protrude and their tongues project slightly between the characteristically open lips. Their faces are puffy due to the formation of an excess of subcutaneous fat. Cretins' mental abilities are retarded so that they are hardly able to take care of themselves. Cretinism can be prevented by early treatment (before six months of age) with thyroid hormones.

myxo = mucous

Hypothyroidism during the adult years causes **myxedema.** Lack of thyroid hormones results in a lowered metabolic rate. This results in

Figure 28-10 The photograph of a giant shows the effect of abnormal functioning of the pituitary gland.

Figure 28-11 Villagers from a town in Paraguay, many of them suffering from goiter caused by a deficiency of iodine in their diets.

a drop in bodily temperature so that the hands and feet always feel cold. There is a decrease in heart rate, muscular weakness, general lethargy, and an increase in body weight. A classic symptom is *edema*, the retention of water by the body. The skin becomes puffy especially about the face and has a rather dry and waxlike texture. Myxedema is far more common in females than males. The symptoms are alleviated by the administration of thyroxine.

edema = swelling

Hyperthyroidism. When the thyroid gland produces an excess of thyroxine and triiodothyronine, a condition called **exophthalmic goiter** or *Graves' disease* results. The individual may show many symptoms: a basal metabolic rate that may exceed the normal by 50 to 75 percent; a tendency for increased appetite but loss in weight; possible increase in thyroid size; a rise in the respiratory rate; profuse sweating; an increase in heart rate and blood pressure; and increased blood sugar and irritability. One of the classic symptoms is protruding eyes. Why this condition occurs has not been fully explained.

Treatment of hyperthyroidism usually centers on reducing the amount of circulating thyroid hormones. Certain drugs such as *propylthiouracil* block the synthesis of thyroid hormones. Radioactive iodine may be effective in partially destroying the gland. Finally, partial removal of the thyroid gland (subtotal thyroidectomy) may be effective in lowering the hormone level.

Parathyroids

If the parathyroid glands are accidentally removed during thyroidectomy (the surgical removal of the thyroid) or if for some reason they become nonfunctional, there is a dramatic drop in blood calcium level. The result is a marked increase in the excitability of neuron membranes, causing them to discharge spontaneously. In turn, muscles begin to twitch, go into spasm (tetany), and general convulsions may occur leading to respiratory failure and death. The symptoms of parathyroidism can be relieved by administering large amounts of calciferol (vitamin D_2), and parathormone and by injections of calcium.

oma = tumor

Adenoma. Occasionally, a benign tumor called an **adenoma** may involve a parathyroid gland, stimulating the gland to produce excess amounts of parathormone. The result is a loss of muscle tone and excessive removal of calcium from bone causing the skeleton to become fragile and prone to spontaneous fractures. Treatment involves the surgical removal of the gland. Adenomas may also involve the pituitary, thyroid, adrenal cortex, and other endocrine glands.

Adrenals

The adrenal medulla is remarkably stable and few problems occur. Secretory deficiency presents no serious symptoms. Hyperfunction manifests itself in either intermittent or continuous elevated blood pressure. Occasionally a tumor develops in the medulla, stimulating it to oversecrete. Blood pressure usually returns to normal after the tumor is surgically removed.

A dysfunction of the adrenal cortex with accompanied insufficient production of both glucocorticoids and mineralocorticoids results in a rare condition called **Addison's disease.** Symptoms include: a disruption of electrolyte balance between sodium and potassium ions; a pronounced fatigue and muscular weakness; hypoglycemia (low blood sugar); kidney dysfunction and a fall in blood pressure; an inability to cope with stress, often resulting in emotional distress; and dark pigmenting of the skin and mucous membranes, called *bronzing*. Addison's disease can be controlled by administering cortisol and a synthetic mineralocorticoid and increasing salt intake.

Excessive production of ACTH or a tumor within the adrenal cortex results in an excessive production of cortical hormones. The result is **Cushing's syndrome.** Symptoms involve: obesity; high blood pressure; a rounded "moon" face because of both a redistribution of fat and edema; muscular weakness; a tendency to bruise easily; and poor healing of skin lesions. Therapy involves some means of removing the hyperactivity of the hormone involved.

Diabetes

If the beta cells of the islets of Langerhans fail to produce sufficient insulin, hypofunction results in a disease called **diabetes mellitus.** Tissues of the body rely on insulin for the absorption and metabolism of sugar. Without insulin, sugar is not absorbed and it accumulates in the blood plasma, raising its level to a point where the excess spills over into the urine. Excess sugar in urine produces an osmotic shift in the kidney's tubules so that less water is reabsorbed into the body. This water must be replaced, causing a diabetic to suffer from extreme thirst. Since glucose is not available for cellular oxidation, there is a shift in metabolism to fats and proteins for energy. This causes a loss in body weight. Diabetes may result in two types of vascular problems: large vessel involvement, such as that which causes heart attacks or gangrene; and small vessel involvement, which causes retinal and kidney dysfunction. The former is caused by atherosclerosis (fatty deposits) in the inner lining of blood vessels and the latter is related to diabetic control and may be caused by abnormal proteins in the linings of the vessels. In addition, many fats are incompletely oxidized to **ketone bodies,** which produce **acidosis.** This may result in *diabetic coma* (unconsciousness) and possible death.

It is estimated that 11 million people in the United States have diabetes mellitus. Ironically, half of these people do not know they have it since most individuals do not undergo routine physicals. The disease appears more often in the same family; thus there is strong evidence that inheritance plays a part in the incidence of diabetes. Approximately 10% have type I diabetes, which occurs in young people and is called *juvenile-onset* diabetes. Because it is characterized by severe insulin deficiency, this type requires replacement therapy and strict diet. Most diabetics (90%) have type II diabetes, which is often called *maturity-onset* diabetes and in which old age and obesity play a part. Frequently, dietary management and weight reduction are sufficient to establish control. When required, insulin must be given by hypodermic injection. It cannot be given orally because insulin, being a protein, is destroyed by digestive enzymes. If the case is mild, certain drugs which stimulate the pancreas to increase its output of insulin can be used. However, in severe cases, since injections only replace the missing insulin and are not a cure, the injections must be given for life.

Summary Table

Gland	Location	Secretions	Activity of Hormone	Dysfunctions
Hypophysis	Base of brain (sella turcica)			
Adenohypophysis	Anterior lobe	Growth hormone	Stimulates growth by promoting protein synthesis	Hypo- 　Child—dwarfism 　Adult—No physical 　symptoms Hyper- 　Child—gigantism 　Adult—acromegaly
		Thyroid-stimulating hormone (TSH)	Stimulates production of thyroxine and tri-iodothyronine	

Gland	Location	Secretions	Activity of Hormone	Dysfunctions
		Adrenocorticotropic hormone (ACTH)	Stimulates production of cortex hormones	
		Gonadotropic FSH	Stimulates formation of Graafian follicle and maturation of ovum	
		LH	Converts follicle after ovulation into corpus luteum	
		Prolactin	Development of breast during pregnancy	
Neurohy-pophysis	Posterior lobe	Antidiuretic hormone (ADH)	Promotes reabsorption of water	Hypo- Diabetes insipidus
		Oxytocin	Labor and production of milk after delivery	
Thyroid	Laryngeal-tracheal region	Thyroxine—T_4 Triiodothyronine—T_3 }	Metabolic rate; normal growth	Hypo- Child—cretinism Adult—myxedema Hyper- Exophthalmic goiter
		Calcitonin	Calcium balance; more active in children	
Parathyroids	Posterior surface of thyroid	Parathormone	Calcium and phosphate balance	Hypo- Tetany Hyper- Spontaneous fracture
Adrenals	Superior to kidney			
Medulla	Center of adrenal	Epinephrine Norepinephrine }	Emergency reactions	Hyper- Elevated blood pressure
Cortex	Periphery of adrenal	Aldosterone	Maintains water balance regulation of Na^+ and K^+	Hypo- Addison's disease Hyper- Cushing's syndrome
		Cortisol	Metabolism of carbohydrates, fats, and proteins; cope with stress	
Islets of Langerhans (Pancreas)	Abdominal cavity	Insulin	Lowers blood sugar; converts glucose to glycogen	Hypo- Diabetes mellitus
Gonads				
Testes in male	Scrotum	Testosterone	Secondary sex characteristics; production of sperm	
Ovaries in female	Abdominal cavity	Estrogen (follicle)	Secondary sex characteristics	
Pineal	Brain	Melatonin		

VOCABULARY REVIEW

Match the statement in the left column with the correct word(s) in the right column. *Do not write in this book.*

1. The stalk that connects the hypophysis with the hypothalamus.
2. The "master gland" of the body.
3. That portion of the brain closely associated with endocrine function.
4. The gland primarily concerned with regulating the rate of metabolism.
5. Produces epinephrine and norepinephrine.
6. A general term used to indicate the reproductive glands of male and female.
7. The discharge of the ovum from the follicular cells.
8. The hormone that is believed to induce labor by causing the uterus to contract during childbirth.
9. Probably the smallest of the endocrine glands.
10. The site of the manufacture of insulin.
11. The hormone that activates the body during an emergency.

a. adrenal cortex
b. adrenal medulla
c. alpha cells
d. antidiuretic hormone
e. beta cells
f. epinephrine
g. gonads
h. hypophysis
i. hypothalamus
j. infundibulum
k. insulin
l. norepinephrine
m. ovulation
n. oxytocin
o. parathormone
p. parathyroids
q. progesterone
r. testes
s. testosterone
t. thyroid

TEST YOUR KNOWLEDGE

Group A

Write the letter of the word(s) that correctly completes the statement. *Do not write in this book.*

1. The hormone that signals the ovary to produce beta-estradiol is (a) estrogen   (b) chorionic gonadotropin   (c) ACTH   (d) insulin.
2. Parathormone is chiefly concerned with the presence in the blood of   (a) calcium   (b) sodium   (c) nitrogen   (d) iron.
3. The adrenal glands are physically associated with the   (a) liver   (b) pancreas   (c) spleen   (d) kidney.
4. Of the following, the hormone that is *not* related to the function of the adrenal cortex is   (a) cortisol   (b) epinephrine   (c) cortisone   (d) testosterone.
5. Another name for epinephrine is   (a) ACTH   (b) adrenalin   (c) antidiuretic hormone   (d) noradrenalin.
6. Insulin is associated with all of the following terms *except* (a) goiter   (b) pancreas   (c) diabetes   (d) islets.
7. The human growth hormone is known as   (a) GRH   (b) TSH   (c) GIH   (d) STH.

EMINENT SURGEON

Louis Tompkins Wright (1891–1952) was a black doctor who made significant contributions to medical science. Dr. Wright received his education at Clark University in Atlanta, Georgia, where his widowed mother worked. He worked summers as a field hand to pay his way through medical school. He graduated from Harvard Medical School, *cum laude*, in 1915.

During World War I Dr. Wright was the youngest surgeon to be placed in charge of a base hospital in France. After the war he became the first black doctor to be appointed to the surgical staff of Harlem Hospital, and 23 years later he became the director of surgery there. After taking the civil service examination in 1928, he was appointed a police surgeon in New York City. Dr. Wright's speciality was surgery associated with fractures and head injuries. Later in his career he experimented with the antibiotics Aureomycin and Terramycin and was the first physician to use Aureomycin in humans. He did work in cancer research on the effectiveness of drugs used for chemotherapy. He wrote many scientific articles during his career, publishing a total of 89 papers on topics ranging from the validity of the Schick test for blacks to the treatment of injuries and the effectiveness of antibiotics.

8. Deficiency of thyroid hormones during the growing years can result in (a) diabetes (b) cretinism (c) low blood calcium (d) Addison's disease.

9. The principal effect of ADH is to (a) lower blood sugar (b) cause labor to begin (c) cause the kidneys to retain water (d) increase heart rate.

Group B

Answer the following. Briefly explain your answer. *Do not write in this book.*

1. (a) What are the main physical and psychological changes which take place in boys and girls at puberty? (b) What are the causes for these changes?

2. What are the differences between endocrine dwarfism and cretinism? Consider the causes, structural differences, and psychological aspects.

3. For each of the following diseases, name the gland and hormone involved, general symptoms, and possible treatment: (a) acromegaly, (b) myxedema, (c) Addison's disease, (d) diabetes mellitus, and (e) tetany.

UNIT 9

REPRODUCTION AND HEREDITY

CHAPTER 29

REPRODUCTION AND DEVELOPMENT

OBJECTIVES

A. Identify the structures of both the male and female reproductive systems
B. Explain the functions of the reproductive organs of both sexes
C. List the changes that occur during the ovarian and menstrual cycles
D. Describe the processes involved in the development of the fetus
E. Identify the stages involved in labor and parturition and the interaction of hormones involved in lactation

ANATOMY OF THE REPRODUCTIVE ORGANS

Reproduction is one of the most fundamental of all the functions of living things. On it depend the creation of all individual organisms and the continued existence of all complex, multicellular organisms. The survival of the human species and all other species depends upon reproduction.

In humans and in many other forms of life, a new individual is produced by the union of a female ovum and a male spermatozoan. The union of the ovum and spermatozoan is called **fertilization** and takes place in the body of the female. The fertilized ovum remains in the uterus of the female, where it develops until it is ready for independent existence.

The importance of normal reproductive system function is notably different from the end result of "normal function" as measured in any other organ system of the body. The proper functioning of the reproductive system and of its enormously complex control mechanisms ensures survival not only of the individual but also of the species. In both sexes organs of the reproductive system are adapted for the specific sequence of functions that are concerned primarily with propagation of the species. In addition, production of hormones that permit development of the secondary sex characteristics occurs as a result of normal reproductive system activity. In humans sexual maturity and the ability to reproduce occur at puberty. The organs involved in the process of reproduction may be grouped together as the **genitals,** or **genitalia.**

In both the male and female, the organs of the reproductive system are divided into two groups: the primary organs and the accessory organs. The primary organs are the **gonads:** the testes in the male and ovaries in the female. These organs produce the germ cells, spermatozoa and ova, and sex hormones. The accessory organs are associated with a series of ducts for the transportation of the germ cells. In the male, an organ that contains a part of the duct system, the **penis,** is modified for the transfer of germ cells into the body of the female. In the female a part of the duct system, the **uterus,** is modified to support the growth and development of the new individual.

Female Reproductive System

The female reproductive organs consist of the paired gonads, or **ovaries,** the paired **uterine tubes,** a single **uterus,** and a **vagina.** Associated structures of the system include the **external genitalia** and the **mammary glands.**

Uterus.   The uterus is a hollow, thick-walled, muscular organ. Its cavity communicates with those of the uterine tubes above and with that of the vagina below. The uterus resembles an inverted pear that is somewhat flattened from front to back. This means that its cavity is slitlike with the anterior and posterior walls close together. The upper portion of the uterus is called the **body,** and the lower constricted portion is called the **cervix.** The portion of the body above the entrance of the uterine tubes is referred to as the **fundus.**

The lining of the uterus covered with ciliated epithelial cells, and contains numerous glandular structures, the *uterine glands,* and many capillaries. In the uterus the ovum disintegrates unless it has been fertilized. The *cervix* opens into the vagina. The relation of these parts to other abdominal structures is shown in Fig. 29-1. The external opening of the vagina lies between folds of skin called the **labia.**

The wall of the uterus is composed of an outer serosal layer called the **perimetrium,** a middle smooth muscle layer called the **myometrium,** and an inner special mucous layer called the **endometrium.** The size of the uterus is capable of great change due to the muscle fibers in the myometrium. The endometrium is the layer that is shed during menstruation. The initial function of the uterus is to retain and to sustain the new individual during the first 40 weeks of growth and development. The final function of the uterus is to expel the newborn at the end of pregnancy.

Uterine tubes.   The uterine tubes are known also as the **Fallopian tubes** and **oviducts.** Each tube is about 11 to 12 centimeters long and opens into the uterine cavity. The trumpet-shaped free end is in intimate contact with the ovary. The end of the tube is also called the fimbriated end because of the presence of fingerlike projections, or **fimbriae.** Fertilization normally occurs in the uterine tube.

ovi = egg

The lining of the uterine tube is mucous membrane that is continuous with that of the uterus and the vagina. Because of this, it is possible for harmful bacteria to enter the vaginal opening and cause infection.

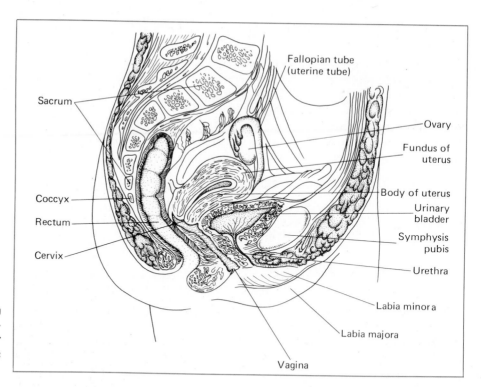

Figure 29-1 Diagram showing female reproductive organs and their placement in the pelvic cavity.

Ovaries. The ovaries are small, almond-shaped bodies that are about 4 to 5 centimeters in length. They are located with one on either side of the uterus. Structurally, each ovary consists of an outer zone, or *cortex,* and an inner zone, or *medulla.* It is in the inner zone that the ovarian follicles are formed. The ovum begins as a **primary follicle,** which consists of a germ cell called an **oogonium** and follicle cells.

At birth several hundred thousand primary follicles are present in the ovary. However, this number decreases steadily throughout life, as most of them degenerate and die. During a female's reproductive years, only 300 to 400 follicles reach maturity and release their oocytes.

Vagina. The vagina is a collapsible, musculomembranous tube, about 9 centimeters long, lying between the bladder and the urethra anteriorly, and the rectum posteriorly. As shown in Fig. 29-1, it extends downward and forward from the uterus to the external opening. The opening of the vagina is surrounded by a thin fold of mucous membrane called the **hymen.** The vagina serves as the excretory duct of the uterus and provides a sheath for the penis to deposit sperm.

External genitalia. The external genitalia are also known as the **vulva.** They consist of the following structures: The **mons pubis,** or **mons veneris,** is the rounded pad of fat in front of the pubic symphysis. (See Fig. 29-2). It is covered by dense skin and, at puberty, becomes covered with hair. The **labia majora** are two prominent longitudinal folds of skin and underlying fat that extend backward from the mons pubis

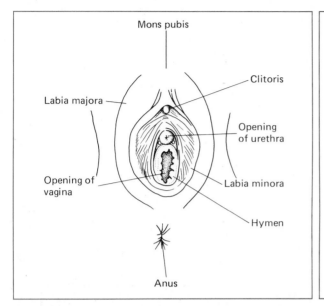

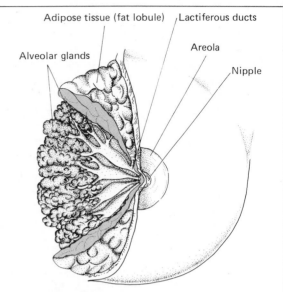

Figure 29-2 (left) External female genitalia.

Figure 29-3 (right) Structure of the mammary gland.

toward the anus. The **labia minora** are two small folds of skin that are situated between the labia majora. The **clitoris** is a small structure that is composed of erectile tissue, corresponding in structure and in origin with the penis in the male.

Mammary glands.   The mammary glands of the breasts are functionally related to the reproductive system since they secrete milk for nourishment of the young, but structurally they are related to the skin as modified *apocrine glands.*

The breast is composed of 15 to 20 lobes, each having its own individual excretory, or **lactiferous** duct. Each lobe consists of lobules of glandular tissue supported by a connective tissue framework that contains a variable amount of adipose tissue. The lobules are drained by intralobular ducts that empty into the main lactiferous ducts.

On the ventral surface of each breast, slightly below center, there is a cylindrical projection or **nipple.** In its rounded tip are 15 to 20 perforations, the openings of the lactiferous ducts. The nipple is surrounded by a pigmented area called the areola, which becomes darker during pregnancy (Fig. 29-3).

Male Reproductive System

The male reproductive system consists of a pair of male gonads, or testes, and a system of excretory ducts with their accessory structures. (See Fig. 29-4.) The ducts are the **epididymis,** the **ductus deferens,** and the **ejaculatory ducts.** The accessory structures are the **seminal vesicles,** the **prostate gland,** the **bulbourethral glands,** and the **penis.**

Testes.   The male gonads (testes) develop in the region of the embryonic kidneys. However, during the development of the embryo, they move through the body cavity to occupy a specialized fold of the skin, the **scrotum,** which hangs from the body in the ventral abdominal region between the thighs. During this movement, each testis passes through an opening in the muscular wall of the lower abdomen called the **inguinal canal.** This canal then normally closes up, although it may leave a weakened area in the muscular wall. Later in life, severe straining of this weakened area may result in a *hernia* (rupture) in which a fold of the intestine is pushed through the weakened spot. A hernia of this type is generally corrected by a surgical operation that draws the muscular walls together.

The testes are glandular organs that are divided internally into many small subdivisions called **lobules,** each containing one or more convoluted tubules. A cross section of one of these lobules and its tubule (Fig. 29-5a) shows that they are composed of several different types of cells. The **interstitial cells** produce the male sex hormones. The **spermatogonia** are the cells from which the mature sperm develop. The function of the **Sertoli cells** is apparently to supply nourishment for the sperm cells that occupy the cavity of the lobule at some distance from the nearest blood vessels. Compared with most of the other cells of the body, the sperm cells are very minute. The total average length of a sperm cell is about 0.05 millimeter and the **head,** which contains the nuclear materials, is only 0.003 millimeter in length. Behind the head region is a part usually referred to as the **middle piece,** which contains the mitochondria and the centrosome. Behind the middle piece is the **tail,** which serves as a locomotor organ (Fig. 29-5). There is, in fact, very

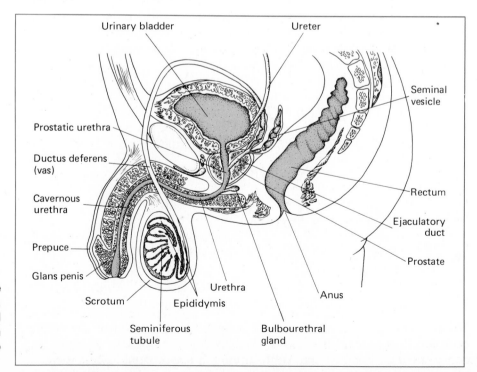

Figure 29-4 **The male reproductive system. Spermatozoa travel through the duct system from the epididymis to the urethra.**

little cytoplasmic material in a sperm cell. The functions of the sperm are to add its chromatin material to the ovum at the time of fertilization and to initiate the development of the ovum into a new individual.

Ducts of the testes.   The cavity shown in the lobule in Fig. 29-5a is a cross section of one of the long convoluted **seminiferous tubules** that occupy a large volume of the testis. These tubules, whose combined length may extend 20 feet, empty through other tiny tubules, the **vasa efferentia,** into the coiled epididymis, which in turn empties into the **vas deferens.** All of these tubules (like the oviducts) are lined with ciliated epithelium whose cilia constantly sweep the mature sperm toward the seminal vesicle, which secretes a thick fluid that is vital to sperm viability. The relationship of these various structures is shown in Figs. 29-4–5. Beyond the seminal vesicle, other fluids are added to the sperm by the **prostate gland** and the small **Cowper's glands.**

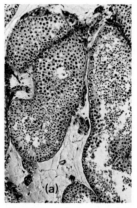

Penis.   Beyond the point of entrance of the prostate and Cowper's glands, the vas deferens opens into the urethra, which has its external opening at the end of the penis. Contractions of the epididymis and vas deferens expel semen into the urethra of the penis. The penis is the organ by which sperm cells are introduced into the vagina of the female. The outer surface of the penis is covered by a layer of skin, the lower end of which is more or less free from the body of the penis. This is the **prepuce,** which is sometimes removed from the enlarged end of the penis by a surgical operation called circumcision.

The body of the penis consists of very spongy tissues containing many blood sinuses. These erectile tissues are arranged in three groups: one is the **corpus spongiosum,** which lies below the urethra, while the others form the **corpora cavernosa** and lie above and to the sides of the urethra.

Semen.   The spermatozoa, plus the secretions of the seminal vesicles, prostate gland, and Cowper's glands, make up the semen, which is ejaculated during the sexual act. This thick, grayish-white fluid has an average pH of about 7.5.

The mucosal cells of the seminal vesicles secrete a thick, yellow, alkaline material that contains large quantities of fructose. This material is emptied into the ejaculatory ducts at the same time that the ductus deferens delivers the spermatozoa. The seminal vesicle secretion serves to provide nourishment for the spermatozoa and to enhance their motility by reducing acidity in their environment.

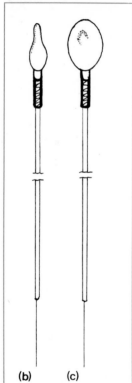

The bulk of the prostate glandular secretion is a thin, milky, alkaline fluid that aids motility of the sperm by helping to maintain an optimum pH. Cowper's glands secrete an alkaline, mucoid substance that coats the lining of the urethra prior to ejaculation. This probably neutralizes the acidity of the vaginal tract.

In the average male, each cubic centimeter of semen contains about 100 million spermatozoa, and the average amount of ejaculate varies between 2 and 4 milliliters. Spermatogenesis begins at puberty and continues throughout life. With advancing age, the seminiferous tu-

Figure 29-5 (a) Photomicrograph of a section through a lobule of the human testis. Human sperm as seen from the side (b) and top (c).

bules undergo a slow decline in their ability to produce mature spermatozoa but apparently never completely cease to function until death. Evidence indicates that many spermatozoa are required for the process of fertilization even though only one sperm can fertilize the ovum.

PHYSIOLOGY OF THE REPRODUCTIVE ORGANS

All body functions serve to ensure and maintain survival of the individual. Only the function of reproduction serves the important purpose of ensuring the survival of the whole species. Male functions in reproduction consist of the production of spermatozoa (male gametes) by means of *spermatogenesis* and the introduction of the spermatozoa into the female body. The introduction of the semen into the vagina is called *coitus, copulation,* or *sexual intercourse.* In order for coitus to take place, erection of the penis must first occur. In order for sperm to enter the female tract, both spermatozoa and secretions from the accessory glands must be introduced into the urethra.

Functions of the Testes

Erection is a parasympathetic reflex initiated mainly by certain tactile, visual, and mental stimuli. It consists of dilation of the arteries and arterioles of the penis, which in turn flood and distend spaces in its erectile tissue and compress its veins. Therefore more blood enters the penis through the dilated arteries than leaves it through the constricted veins. Hence it becomes larger and rigid and an erection occurs.

Emission is the reflex movement of spermatozoa and secretions from the genital ducts and accessory glands into the prostatic urethra. Once emission has occurred, ejaculation can then occur.

Ejaculation of semen is also a reflex response. It is the usual outcome of the same stimuli that initiate erection. Ejaculation and various other responses, such as an accelerated heart rate, increased blood pressure, increased respiration, dilated skin blood vessels, and intense sexual excitement, characterize the climax of coitus known as *orgasm.*

Functions of the Ovaries

Many changes occur periodically in women between the time when menstruation begins, called *menarche,* and the cessation of menstruation, called *menopause.* The most obvious cycle is menstruation, in which there is an outward sign of changes in the endometrium. In addition, however, many women notice periodic changes in other parts of their bodies. These changes are rhythmic and occur at fairly uniform intervals during the reproductive years of the female.

Ovarian cycle.   Once each month, on about the first day of menstruation, several *primary follicles* and their enclosed ova begin to grow and develop (Fig. 29-6). The follicular cells proliferate and start to secrete estrogens. Usually, only one follicle matures and migrates to the surface of the ovary. The surface of the follicle degenerates, causing expulsion of the ovum into the pelvic cavity. This is the process of **ovulation.**

Shortly before ovulation, the ovum undergoes meiosis, the process by which its number of chromosomes is reduced by half. Immediately after ovulation, cells of the ruptured follicle enlarge and become filled with a gold-colored lipoid material. This gold body is called the **corpus luteum.** The corpus luteum grows for 7 or 8 days, secreting **progesterone** in increasing amounts. If fertilization occurs, the corpus luteum will continue to maintain its size and secrete progesterone. If fertilization does not occur, the size of the corpus luteum decreases and the amount of progesterone decreases.

Menstrual cycle.   During menstruation, the endometrium slowly begins to die (*necrosis*) and sloughs off bits of its compact and spongy layers, leaving small bleeding areas. Following menstruation the cells of these layers begin to grow, causing the endometrium to reach a thickness of 2 or 3 millimeters by the time of ovulation. During this period the endometrium grows thicker and more vascular, until after ovulation it reaches a maximum of about 4 to 6 millimeters' thickness. The swelling of the endometrium is due to fluid retention as well as proliferation of **endometrial cells.** The day before menstruation starts again, the blood supply to the endometrium is diminished, leading to necrosis, sloughing, and once again menstrual bleeding.

The menstrual cycle is customarily divided into phases in which major processes occur. These phases are **menses;** the **postmenstrual,** or

post = after

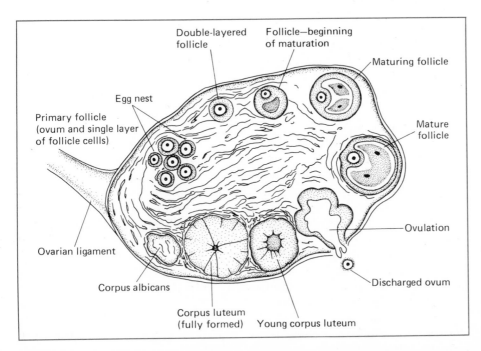

Figure 29-6 Diagram of an ovary showing successive stages of ovarian follicle and ovum development.

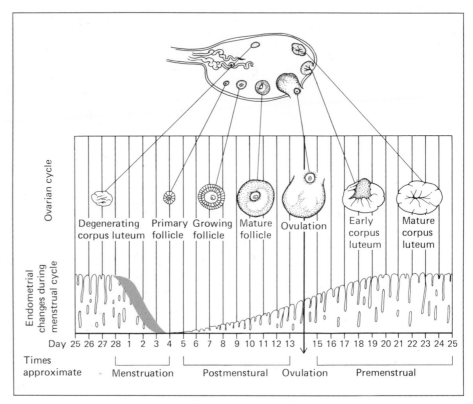

Figure 29-7 Schematic representation of one ovarian cycle and the corresponding changes in thickness of the endometrium. It is thickest just before the onset of menstruation and thinnest just as menstruation ceases.

preovulatory, phase; ovulation; and the **postovulatory,** or **premenstrual, phase.** (Refer to Fig. 29-7.)

The menses, or menstrual period, occurs on cycle days 1 to 5. There is some individual variation, however, that falls within the normal range. The postmenstrual, or preovulatory, phase occurs between the end of the menses and ovulation. In a 28-day cycle, it usually includes cycle days 6 to 13 or 14. The length of this phase varies more than do the others. The preovulatory phase is also called the **estrogenic** or **follicular phase** because of the high blood estrogen level resulting from secretion by the developing follicle.

Ovulation, that is, the rupture of the mature follicle with expulsion of its ovum into the pelvic cavity, generally occurs on cycle day 14 or 15 in a 28-day cycle. In a 32-day cycle the preovulatory phase would probably last until cycle day 17 or 18 and ovulation would then occur on cycle day 18 or 19. It is during this period that fertilization of the ovum can occur.

The premenstrual (or postovulatory) phase occurs between ovulation and the onset of the next menses. This phase is called the *luteal phase* or progesterone phase because the corpus luteum secretes the hormone progesterone during this time.

Gonadotropic cycle. The anterior pituitary gland secretes two hormones called **gonadotropins** that influence the female reproductive

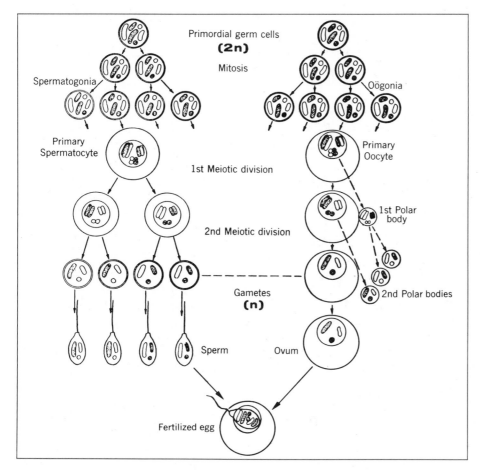

Figure 29-8 The process of meiosis in an animal having a diploid number of six chromosomes. Note the reduction in the number of chromosomes in the gametes.

cycles. Their names are *follicle-stimulating hormone (FSH)* and *luteinizing hormone (LH).* The amount of each gonadotropin secreted varies with a rhythmic regularity that can be related to the rhythmic ovarian and uterine changes (discussed in Chapter 28).

Formation of the Embryo

The science of development of the individual before birth is called *embryology.* It concerns the means by which a single microscopic cell is transformed into a complex human being.

An individual is formed by the union of two cells, a sperm and an ovum. These cells are called *gametes.* If sperm and ova all contained 46 chromosomes, the total number found in humans cells, then the fusion of one sperm and one ovum would result in a cell with 92 chromosomes. This obviously does not occur; therefore, some process of redirection must accompany the formation of gametes if the number of chromosomes is to remain constant from one generation to the next. This process is called *meiosis.* Since the lack of a chromosome, or the presence of an extra chromosome, can cause an abnormality, the

di = two

process of reducing chromosomal number must be very precise. It is essential to ensure that every individual contains the requisite 23 pairs of chromosomes, called the **diploid number** (2n). The requirement is satisfied if each individual receives one member of each pair from the mother and one member from the father. Therefore when a gamete forms, be it sperm or ovum, the members of each chromosome pair must separate in such a way that one particular sperm or ovum receives only one member of each pair of chromosomes. It does not matter which, and therefore the probability that from a particular pair a sperm receives one or the other of the two chromosomes is exactly 50 percent. The same is true for an ovum. Thus both sperm and ova have 23 chromosomes each, one of each pair, called the **haploid number** (n). When the sperm and ovum fuse in the process called **fertilization,** the cell that results will again have 46 chromosomes. This cell is called a **zygote.**

Meiosis is a modified form of mitosis and is illustrated in Fig. 29-8. There are two features of meiosis that should receive special attention. During the formation of the sperm, called *spermatogenesis*, most of the cytoplasm disappears from the cell, and the sperm contributes only chromatin material to the ovum at the time of fertilization. The cytoplasm of the ovum, on the other hand, supplies the early stages of the developing embryo with nourishment. It is therefore of extreme importance that the mature ovum retain as much cytoplasm as possible. During the formation of the ovum, called *oogenesis*, the meiotic divisions of the chromosomes are accompanied by unequal division of cytoplasm. As a result, miniature cells (polar bodies) are formed that contain the same number of chromosomes as the ovum but have a minimum amount of cytoplasm. These small polar bodies cannot normally be fertilized.

Fertilization.   Fertilization is the union of male and female gametes to produce a new one-celled individual called a zygote. After the semen is deposited in the female tract, the sperm are propelled by their motility and contractions in the female tract toward the uterine tube. The ovum normally awaits the sperm in the upper outer third of the uterine tube. Although numerous sperm surround the ovum, only one penetrates it. As soon as the head and neck of one spermatozoan enters the ovum, leaving the tail behind, the remaining sperm seem to be repulsed. The sperm head then forms itself into a nucleus that approaches and eventually fuses with the nucleus of the ovum, producing a zygote.

Cleavage and implantation.   *Cleavage* consists of repeated mitotic divisions, first of the zygote to form two cells, then of those two cells to form four cells, then of those four cells to form eight cells, etc., resulting in about 3 days' time in the formation of a solid, spherical mass of cells known as a **morula.** A day or so later the embryo reaches the uterus, where it starts to implant itself in the endometrium. Occasionally, implantation occurs in the tube or pelvic cavity instead of the uterus. The condition is known as an **ectopic pregnancy.**

ecto = outside

As the cells of the morula continue to divide, a hollow ball of cells or *blastocyst,* consisting of an outer layer of cells and an inner cell mass, is formed. Implantation in the uterine lining is now complete. About 10 days have elapsed since fertilization. The cells that compose the outer wall of the blastocyst are known as *trophoblasts.* They eventually become part of the *placenta,* the structure that nourishes the fetus. The placenta is derived in part from both the developing embryo and from maternal tissues. The relationship of maternal and fetal blood vessels is a close one, but maternal and fetal blood do not mix. Exchange of nutrients occurs across a fetal membrane called the *chorion.*

Another fetal membrane, the *amnion,* develops around the growing embryo, which is connected with the chorion by the body stalk. The body stalk ultimately becomes the umbilical cord of the fetus. The umbilical cord is composed of two umbilical arteries, which carry blood to the placenta, and one umbilical vein, which carries blood to the fetus.

Fetal circulation.   Circulation in the fetus differs from circulation after birth because the fetus must secure oxygen and food from the maternal blood instead of from its own lungs and gastrointestinal tract. Therefore, there are additional blood vessels in the fetus to carry the fetal blood into close proximity with the maternal blood and return it to the fetal blood. These structures are the two **umbilical arteries,** the **umbilical vein,** and the **ductus venosus.** The placenta functions as the structure where the interchange of gases, foods, and wastes between the fetal and maternal blood can take place. The exchange of substances occurs without any actual mixing of maternal and fetal bloods, since each flows in its own capillaries.

In addition to the placenta and umbilical vessels, three structures located within the fetus' own body play an important part in fetal circulation. The ductus venosus serves as a detour by which most of the blood returning from the placenta bypasses the fetal liver. The other two, the **foramen ovale** and **ductus arteriosus,** provide detours by which blood bypasses the lungs.

Differentiation

As the cells composing the inner mass of the blastocyst continue to divide, they arrange themselves into a structure containing two cavities separated by a double layer of cells known as the *embryonic disc.* The cells that form the cavity above the embryonic disc eventually form a fluid-filled, shock-absorbing sac called the **amniotic sac** in which the fetus floats. The cells of the lower cavity form the yolk sac, a small vesicle attached to the belly of the embryo until about the middle of the second month, when it breaks away. Only the double layer of cells that compose the embryonic disc is destined to form the new individual. The upper layer of cells is called the **ectoderm** and the lower layer the **endoderm.** Refer to Fig. 29-9, which follows the zygote from fertilization to implantation and development of the yolk sac. A third layer of cells, the **mesoderm,** develops between the ectoderm and endoderm.

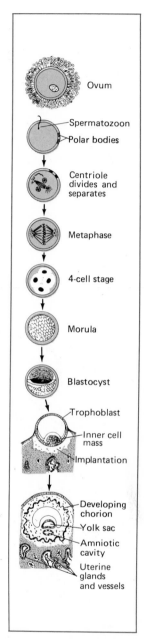

Ovum

Spermatozoon
Polar bodies

Centriole divides and separates

Metaphase

4-cell stage

Morula

Blastocyst

Trophoblast
Inner cell mass
Implantation

Developing chorion
Yolk sac
Amniotic cavity
Uterine glands and vessels

Figure 29-9 Diagram showing fertilization to implantation and development of the yolk sac. Rapid growth of uterine glands and vessels covers the developing blastocyst at the time of implantation.

histo = tissue

Histogenesis and organogenesis. The way in which the primary cell layers develop into many different kinds of tissues is called *histogenesis* and their arrangement into organs is called *organogenesis.* When two sex cells unite to form a single cell a new human body evolves by a series of processes consisting of cell multiplication, cell growth, cell differentiation, and cell rearrangements, all of which take place in an orderly sequence. Figure 29-10 shows development of the embryo from the yolk sac stage to 4 months of gestation. By the third week of development the third cell layer, the mesoderm, appears between the two initial layers of the embryonic disc. The developing heart is beating by the fifth week, and the fetus, only about 8 millimeters in length, is showing rapid development of every major organ system. By the end of the fourth month, about 120 days, development is a matter of growth.

Because the organ systems are rapidly developing during the first 4 months of pregnancy, the embryo-fetus is especially sensitive to events that take place in the mother's body. An *embryo* arbitrarily becomes a *fetus* after the first 8 weeks of life. Detrimental events that can cause major disruptions are German measles, influenza, polio, syphilis and infectious hepatitis. Radiation can also cause problems in development and X-rays are avoided.

Labor (Parturition)

At the end of pregnancy, about 280 days after the last menstrual period, the uterus begins to undergo rhythmic contractions that ultimately lead to the birth of the infant. The sequence of events involved in this process is referred to as **labor,** or **parturition.**

The *first stage* of labor consists of dilation of the cervix. With each uterine contraction the amniotic sac of fluid that surrounds the fetus

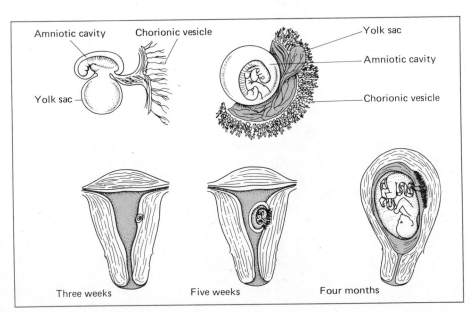

Amniotic cavity　Chorionic vesicle　Yolk sac
Yolk sac　Amniotic cavity
Chorionic vesicle

Three weeks　Five weeks　Four months

Figure 29-10 Development of the embryonic yolk sac to 4 months of gestation.

is forced into the canal of the cervix, where it acts as a wedge to dilate the canal. In a typical delivery the amniotic sac ruptures just at the close of this stage, and the fluid escapes.

When the dilation of the cervix is completed, the *second stage* of labor begins. This refers to the **delivery** of the infant, normally head first, through the cervix and vagina.

The third stage of labor consists of the **delivery of the placenta,** or **afterbirth.** After this the uterus rapidly decreases in size as a result of the contraction of smooth muscle fibers of the myometrium.

While it is believed that many hormones play a significant role in initiating labor, not a great deal is presently known about their essential action. On the basis of available information it seems that labor is initiated by genetically determined changes in the amnion or chorion. Also, prostaglandins may influence uterine contraction once labor is initiated. The hormone **oxytocin** apparently does not play a role in initiation of labor because it is rarely detected in maternal blood during the first stages of labor. It does appear in the second stage of labor and throughout to reinforce labor once it has begun.

Lactation

Many final changes take place during pregnancy through the interaction of many hormones. The mammary glands reach their final stage of development at the end of pregnancy. **Prolactin,** or **lactogenic hormone,** of the anterior lobe of the pituitary gland stimulates the production of the milk in the mammary glands, which have been prepared by estrogen and progesterone. The mammary glands secrete milk for the nourishment of the newborn infant. The process is called **lactation.**

During the latter half of pregnancy the mammary glands begin to produce a secretion that is called **colostrum.** This is the fluid obtained from the breasts before the secretion of true milk begins. It is ingested by the newborn for the first 2 days of life after birth. Colostrum has a

Figure 29-11   Photographs of models showing two stages in the birth of a child. In (a), the head is passing through the cervix into the vagina. In (b), the head is emerging from the opening of the vagina. At this stage the child is still attached to the placenta by the umbilical cord (not shown in the model).

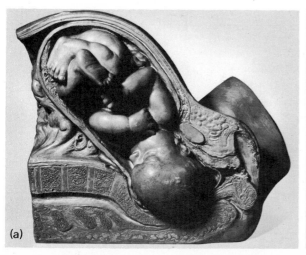
(a)

(b)

very high concentration of immune bodies that play a significant protective role in the newborn.

For the maintenance of secretion, removal of milk is absolutely essential. If milk is not removed by breast feeding, the swelling of the breasts eventually disappears. Milk ceases to be produced and the mammary glands eventually undergo **involution.** Under normal conditions, milk production continues for 6 to 9 months after the infant is born. During milk secretion and suckling by the infant there may be an inhibition of ovulation. However, the inhibition is short-lived in many women, and approximately 50 percent begin to ovulate despite continued nursing. Pregnancy is common in females by the mistaken belief that failure to ovulate is always associated with nursing the newborn.

PATHOLOGY

Teratology is the study of **abnormal development.** Some malformations occur in at least 10 percent of all live births. These may appear in any part of the body. The most frequent sites of occurrence are the face with the result of *harelip* and *cleft palate*. Septal defects can occur in the heart, such as failure of the foramen ovale to close between the right and left atria at birth. This allows for the mixing of venous with arterial blood and results in a *blue* baby. If a malformation in the brain occurs, it may produce a condition called **hydrocephalus. Pyloric stenosis,** or narrowed stomach opening, is an abnormal development of the digestive tract. Other **congenital anomalies** can also occur in the skin, such as *birthmarks*, and the skeletal system, such as *talus equinus*, or *clubfoot*. Most malformations have their origin during the first 4 months of development.

cephalo = head

SUMMARY

The male and female reproductive organs consist of the gonads, ovaries in the female, testes in the male, a system of reproductive ducts and accessory organs. The female produces a limited number of gametes, called ova, during her reproductive span. The male produces hundreds of thousands of spermatozoa during his entire life span. The union of the ovum and sperm results in the production of a new individual, called a zygote.

Reproduction involves maturation of the gametes by means of meiosis, fertilization, implantation, development of the placenta and membranes, the intrauterine development of the embryo and fetus, and development of the mammary glands for nourishment of the newborn.

VOCABULARY REVIEW

Match the statement in the left column with the correct word(s) in the right column. *Do not write in this book.*

1. Produced by the seminiferous tubules.
2. Surgical removal of the prepuce.
3. Occurs about 14 days before menstruation.
4. An organ important in the nourishment of the fetus.
5. Production of haploid number of chromosomes.
6. A germ layer between the ectoderm and endoderm.
7. A function of the mammary gland.

a. circumcision
b. semen
c. endoderm
d. follicle
e. lactation
f. meiosis
g. implantation
h. mesoderm
i. ovulation
j. ovum
k. placenta

TEST YOUR KNOWLEDGE

Group A

Write the letter of the word(s) that correctly completes the statement. *Do not write in this book.*

1. The cyclical shedding of the endometrium every 28 days is called (a) oogenesis (b) menstruation (c) parturition (d) ovulation.
2. One type of gonad is the (a) ovum (b) egg (c) ovary (d) sperm.
3. The primary sex organ of a female is the (a) oviduct (b) ovary (c) uterus (d) vagina.
4. The blood of a fetus bypasses its lungs through the *ductus arteriosus* and the (a) amnion (b) morula (c) foramen ovale (d) *ductus venosus.*
5. The first stage of labor consists of (a) placental delivery (b) production of oxytocin (c) start of histogenesis (d) dilation of the cervix.

Group B

Answer the following. Briefly explain your answer. *Do not write in this book.*

1. Describe the general development of the zygote through the formation of the primary germ layers.
2. Briefly describe the phases of the menstrual cycle.
3. (a) Describe the path taken by the sperm from the moment of its introduction into the vagina of the female through fertilization. (b) Relate the differences in structure between sperm and ova to their different functions.

CHAPTER 30

HUMAN GENETICS

OBJECTIVES

A. Explain the chromosomal basis of sex determination and the mechanism of sex-linked inheritance

B. Describe the roles of DNA, mRNA, tRNA, and rRNA in protein synthesis

C. Explain the principles of Mendelian inheritance in humans

D. Explain the basis for mutations and their consequences

THE BASIS OF HEREDITY

Heredity concerns the transmission of genetic material from parents to offspring. Geneticists study how characteristics are inherited, the mechanism involved in their transmission and expression, and the way in which inherited traits relate to all aspects of life. The science that deals with understanding the processes of heredity is called *genetics.*

Heredity and environment are interdependent. The interactions between heredity and environment are frequently difficult to determine. The environment influences the development of an individual, although heredity governs what an individual can become.

The science of genetics was started a century ago by Gregor Mendel through his studies of garden peas. By investigating a variety of traits of these plants, Mendel was able to make some important generalizations known as Mendel's laws. Later research showed that these laws applied to all plants and animals, including humans.

With better microscopes and staining techniques, it was found that the genetic material is carried on the chromosomes. Mendel's results were finally explained when it was recognized that traits are controlled by one or more segments of a chromosome. These segments are known as genes.

It is important to recognize that characteristics, abnormalities and diseases that are hereditary in nature are not all expressions of identical genetic mechanisms and, more importantly, that the pattern of transmission is not identical in all cases. Therefore, to understand human heredity and to appreciate its implications for the future, it is necessary to know *how* genes react and interact to produce their final expression

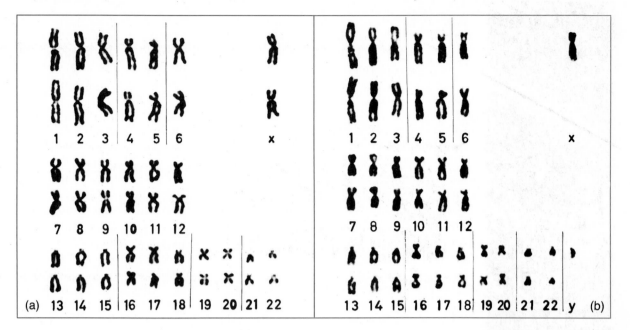

in the body. This chapter deals with the various modes of transmitting human traits.

Structure of Chromosomes

In spite of the great variety of cells in the body each one except the sex cells, or gametes, contains 46 chromosomes. Chromosomes consist chiefly of long spiral molecules of *DNA* (deoxyribonucleic acid). If the molecule of DNA in a human chromosome were uncoiled, it would be about 5 cm long. Since its diameter is 2 millionths of a millimeter, the DNA in a chromosome can be seen only with an electron microscope.

In cell division, the set of chromosomes in a cell is duplicated, and one of these sets is passed on to each of the daughter cells. This is accomplished by the process of **mitosis** (see pages 46 and 47).

Sex Determination

Microscopic examination of the chromosomes in a human cell getting ready to divide reveals that the 46 chromosomes are arranged in pairs. The two chromosomes of most of the pairs appear to be practically identical to each other. They are similar in size and shape and carry genes for the same traits. In females, the twenty-third pair is also made up of almost identical chromosomes. In males, the twenty-third pair is made up of two chromosomes that do not look alike. See Fig. 30-1. The first 22 matching pairs of chromosomes are called *autosomal chromosomes*. The chromosomes of the twenty-third pair, in males and females, are the *sex chromosomes*. This particular pair of chromosomes

Figure 30-2 Dr. Elizabeth S. Russell is a geneticist whose research has been in the study of mammalian genetics. She has contributed to the genetic understanding of hereditary anemias.

is of special importance for the determination of sex. In males this pair is called **XY** and the corresponding pair in females is called **XX.** The X in the male is identical to the two Xs in the female and is a relatively long chromosome (see Fig. 30-1); but the Y chromosome is a very short chromosome and in a few rare cases may even be missing. Mapping of human chromosomes is called **karyotyping.**

Deviation in sex determination.   Today it is believed that the Y chromosome is essential for the development of male traits. There are some rare individuals who lack the Y chromosome and therefore have 45 chromosomes, only one of which is an X. They are called XO individuals, where O refers to the fact that the Y chromosome is missing. These individuals are females. They tend to be abnormally short and often have a webbed neck. They are sterile, since they do not develop normal reproductive organs. The XO condition is called **Turner's syndrome.** This abnormality is considered to be the result of an accident that occurred when the particular gamete (usually the sperm) that gave rise to the abnormal individual was formed. There are some individuals with 47 chromosomes in whom there are clearly two normal X's, but also one Y. The name for this abnormal condition is **Klinefelter's syndrome.** These individuals (XXY) are males which shows the strong influence of the Y chromosome over the two X chromosomes. Abnormalities of the X chromosomes thus suggest that the Y chromosome is responsible for determining the male sex.

Nondisjunction of Chromosomes

Since the lack of a chromosome or the presence of an extra chromosome can cause an abnormality, the process of **meiosis** must be very precise. It is essential to ensure that every individual contains the requisite 23 pairs, or **diploid (2n) number,** of chromosomes. The requirement is satisfied if the offspring receives one of the pairs of chromosomes from each parent. Therefore when a gamete forms, the members of each chromosome pair must separate in such a way that a sperm or ovum receives half the number of chromosomes. The probability that from a particular pair of chromosomes a sperm receives one or the other of the two chromosomes is exactly 50 percent. The same is true for an ovum. Thus both sperm and ova have 23 chromosomes each, one of each pair, called the **haploid (n) number.** When the sperm and ovum fuse in the process called **fertilization,** the cell that results will again have 46 chromosomes.

It is important to note that meiosis is the time at which errors in the distribution of chromosomes are most likely to have serious consequences. Thus if the two members of one particular chromosome pair do not separate at the time of formation of gametes, a gamete may result that has two members of one pair, and another that has no member of that pair. This occurrence is known as **nondisjunction.**

Down's syndrome is usually accompanied by the presence of 47 chromosomes in the body cells. The abnormality most likely occurred be-

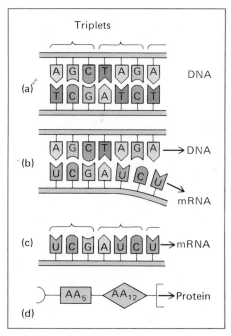

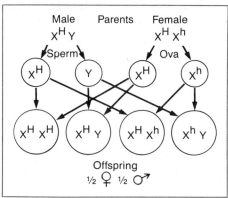

Figure 30-4 Cross between a $X^H Y$ male and a heterozygous $X^H X^h$ female. The probability is that one half of the females will be $X^H X^H$ and one half will be $X^H X^h$, and that one half of the males will be $X^H Y$ and one half will be $X^h Y$.

Figure 30-3 A gene determines what kind of protein will be made. (a) DNA triplets located on a chromosome. (b) An active strand of DNA sends messenger RNA to the cytoplasm. (c) The triplets UCG and AUC on the messenger code for the amino acids AA-5 and AUC AA-12. (d) A portion of a protein, a long chain of amino acids.

cause of errors in the distribution of chromosomes at meiosis. The extra chromosome in these individuals belongs to type 21. This condition is also called **trisomy 21;** the number identifying each of the three chromosomes. By means of a medical procedure called **amniocentesis,** it is possible to diagnose Down's syndrome before birth.

tri = three

DNA to Protein Synthesis

The structure of DNA and the genetic code have by now appeared in practically all biology books. (Refer to Chapter 3.) Chromosomes are chiefly threads of double-stranded DNA. Each strand is made of a long sequence of relatively simple compounds, the **nucleotides,** which are attached to one another. A nucleotide is made of the sugar **deoxyribose,** of **phosphate,** and of *one* of the following **four bases: adenine, A; thymine, T; cytosine, C;** and **guanine, G.** Since nucleotides are distinguished from one another only by the base, the four different types can be identified by the letters A, T, C, and G, corresponding to the initial letter of each base's name. The two strands forming DNA are perfectly complementary (Fig. 30-3): whenever there is an A in one strand, in the other strand at the corresponding site there is a T; opposite T is A; opposite G is C; and opposite C is G. Knowledge of the sequence of nucleotides in one strand can indicate the sequence in the other.

Mutations

A special mechanism in cells enables proteins to be made from amino acids by a process of "translation." In this process DNA is "interpreted" as protein in accordance with a "dictionary" called the genetic code.

mutare = change

Figure 30-5 **Cross be-tween two heterozy-gotes for galactosemia. The dark areas in the circles represent the normal (gal $^+$) gene; the light areas the galacto-semia (gal$^-$) gene.**

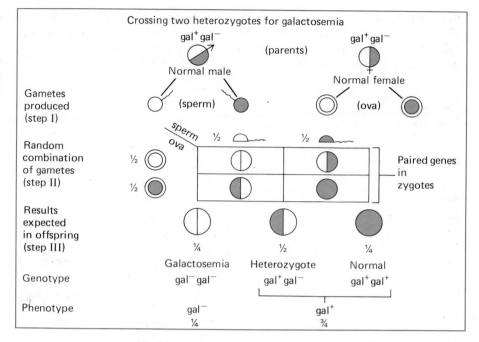

Before this phase of translation, however, there is another phase, "transcription," in which one of the two strands of a DNA segment is "replicated" (copied by simple complementarity rules) into a somewhat similar substance called ***messenger ribonucleic acid (mRNA).*** Messenger RNA is single and complementary to the DNA strand being transcribed (see Fig. 30-3). The DNA segment being transcribed corresponds to one gene or a set of neighboring genes and includes thousands of nucleotides. Complementarity is the rule for transcription as for DNA copying, but in RNA there is a different base, U (for uracil), which takes the place of T and is therefore found opposite A in the DNA strand (Fig. 30-3). The process of making protein consists of building protein chains out of sequences of amino acids, chosen from among the 20 possible amino acids. The genetic code is such that three nucleotides in a row correspond to a given amino acid (see Table 30-1). Thus the triplet in the DNA strand adenine-adenine-adenine (AAA) commands, according to the code, the introduction of the triplet uracil-uracil-uracil (UUU) in messenger RNA, and the introduction of ***phenylalanine*** (one of the 20 amino acids) into the protein chain. If the next three nucleotides in DNA are cytosine-adenine-guanine, then according to the code the next amino acid in the protein thread is always valine, and so on. Thus the sequence of nucleotides in DNA determines the properties of the protein. If a change occurs in DNA—for example, if one of the nucleotides is accidentally replaced by another—the code determines if and how the amino acid that goes into the protein at that site is changed. This is an example of a ***mutation,*** and it may happen that the altered protein has different properties from the original one even if only one amino acid is changed. The change may be so great as to make the entire molecule nonfunctional, or less functional, or to make it function in a substantially different way. On the other hand, the mutation may cause a difference that is slight or even unnoticeable.

Table 30-1
The Genetic Code

DNA	Messenger RNA	Amino Acid	DNA	Messenger RNA	Amino Acid
AAA	UUU	Phenylalanine	ATT	UAA	Chain end
AAG	UUC		ATC	UAG	
AAT	UUA		GTA	CAU	Histidine
AAC	UUG		GTG	CAC	
GAA	CUU	Leucine	GTT	CAA	Glutamine
GAG	CUC		GTC	CAG	
GAT	CUA		TTA	AAU	Asparagine
GAC	CUG		TTG	AAC	
TAA	AUU	Isoleucine	TTT	AAA	Lysine
TAG	AUC		TTC	AAG	
TAT	AUA		CTA	GAU	Aspartic acid
TAC	AUG	Methionine	CTG	GAC	
CAA	GUU	Valine	CTT	GAA	Glutamic acid
CAG	GUC		CTC	GAG	
CAT	GUA		ACA	UGU	Cysteine
CAC	GUG		AGG	UGC	
AGA	UCU	Serine	ACT	UGA	Chain end
AGG	UCC		ACC	UGG	Tryptophan
AGT	UCA		GCA	CGU	Arginine
AGC	UCG		GCG	CGC	
GGA	CCU	Proline	GCT	CGA	
GGG	CCC		GCC	CGG	
GGT	CCA		TCA	AGU	Serine
GGC	CCG		TCG	AGC	
TGA	ACU	Threonine	TCT	AGA	Arginine
TGG	ACC		TCC	AGG	
TGT	ACA		CCA	GGU	Glycine
TGC	ACG		CCG	GGC	
CGA	GCU	Alanine	CCT	GGA	
CGC	GCC		CCC	GGG	
CGT	GCA				
CGC	GCG				
ATA	UAU	Tyrosine			
ATG	UAC				

A = adenine, C = cytosine, G = guanine, T = thymine, U = uracil

Effect of DNA

The significance of a sequence of nucleotides and the type of proteins that it causes the body to produce is enormous. The structural materials and enzymes that are produced by the cells are determined by these sequences. A segment of a chromosome that codes for a specific protein is called a *gene.* Genes are responsible for the traits that appear in an individual. Since genes are locations on chromosomes and chromosomes occur in pairs, most genes also occur in pairs in a cell.

Mendel's studies with pea plants showed that one gene of a pair may be dominant over the other. For example, a pair of genes in pea plants determines flower color. If a plant with red flowers has a gene for red color and one for white, then the gene for red color is said to be *dominant* over the white. The gene for white color is said to be *recessive;* it is present but not expressed. There may be more than two genes that affect a characteristic, but only two are present in one individual. A group of two or more genes that have contrasting effects on the same trait are called *alleles.* An individual that receives identical genes for a trait from both parents is a *homozygote* for that trait. When the paired genes are not identical, the individual is a *heterozygote.* A male with a gene in his single X chromosome that is not paired with a gene on his Y chromosome is a *hemizygote.*

homo = same
hetero = different

genos = race

phen = to show

The term *genotype* refers to the genetic composition of an individual with respect to the genes for a particular characteristic. The expression of the genetic composition, or the *appearance,* of an individual is termed the *phenotype.* Because of the phenomenon of dominance, the genotype of an individual must be distinguished from the phenotype. This can be determined by karyotyping or chemically testing the blood. In some instances, there is no dominance or recessiveness because both alleles are expressed in the heterozygote. This is called *codominance* or *incomplete dominance.*

Sex-Linked Inheritance. In the early 1900's T. H. Morgan, working at Columbia University on inheritance of various traits in the fruit fly, *Drosophila,* observed that eye color was related to the sex of the fly. Experiments showed that the gene for eye color in *Drosophila* is located on the X chromosome. Morgan explained that for a recessive eye color to appear in a female fruit fly, both X chromosomes must carry the recessive genes. Since a male has only one X chromosome, if that X chromosome has the recessive gene, the male has the recessive eye color. Apparently there is no corresponding gene for eye color on the Y chromosome. As a result of its location on the sex chromosome, a male can transmit the recessive gene only to his female offspring. A female, however, can transmit the gene to male *and* female offspring.

Patterns of inheritance. For a recessive disease the female may be a homozygous dominant, heterozygous, or an affected homozygous recessive (has the disease). The heterozygous female, a carrier, has a one-in-two (½) chance of giving one particular chromosome to her son. If the son receives the X chromosome that carries a defective gene, he will pass it on to every daughter he fathers.

Because the male has only one X chromosome and the female has two, a particular gene that has nothing to do with the individual's sex but happens to be located on the X chromosome is transmitted in a crisscross pattern of inheritance, mother to son to grand-daughter. Never, under ordinary circumstances, does it go from father to son.

The presence of a dominant X-linked allele produces its effect in females as well as males. The inheritance pattern is quite different from that of a recessive X-linked gene. A dominant allele in the hemizygous male ensures that he will pass the condition to each of his daughters but to none of his sons. A female who has the disease because she is a heterozygote will transmit the condition to half her children of either sex. If a female is homozygous for the dominant allele *all* her children will be affected.

Hemophilia: A Recessive Trait

A famous example of a recessive gene located on the X chromosomes is the gene that causes the disease called **hemophilia.** There are at least two types of this disease; the more severe is called *hemophilia A*, and the other, less severe, *hemophilia B*. They are both rare. Hemophiliacs are usually males. Hemophilia is a disease in which blood clotting is slowed because of lack of protein normally present in blood. A hemophiliac who has a wound may die from loss of blood because of the lack of clotting substance. Today these individuals can be saved by transfusions of blood containing the necessary protein. Until a short time ago this treatment did not exist, and most hemophiliacs died before reproducing.

hemo = blood
philia = tendency toward

Using the symbol **H** for the normal allele and the small **h** for the hemophilia allele, the possible genotypes for the female are homozygous dominant ($X^H X^H$); heterozygous carrier ($X^H X^h$); and homozygous recessive affected ($X^h X^h$). The possible genotypes for the male are normal hemizygous ($X^H Y$) and affected hemizygous ($X^h Y$). Since there are three female genotypes and two male genotypes, altogether there are six (3 × 2) possible crosses. In order to predict the outcome of any cross, the same principles are followed:

1. Identify the genotype of both parents.
2. Determine which gametes (sperm or ovum) each parent can form and how many of each gamete type can be formed.

3. Combine the gametes of both parents *at random* and obtain the proportion of each new zygote thus generated as the *product* of the proportions of the gametes that have formed the zygote.

These principles can be applied to one of the possible crosses. A female who is heterozygous, $X^H X^h$, mates with a male who is $X^H Y$. The female parent can form two types of ovum, X^H and X^h, in equal proportions (½ X^H and ½ X^h). The male can form only one type of X carrying sperm, which is X^H. When the sperm joins with ovum X^H, an $X^H X^H$ homozygote dominant is produced; if the sperm joins an ovum X^h, a heterozygote $X^H X^h$ is produced (Fig. 30-4). These combinations occur if the sperm carries an X chromosome; however, an equal number of sperm carry a Y chromosome. Gametes receiving the Y chromosome will give rise to male offspring. Since the Y chromosome does not have the gene for this enzyme, the genotype of the male offspring will depend entirely on the gene contained in the particular ovum. A cross of this type will result in half the male offspring being of type $X^H Y$ and half of type $X^h Y$. The expectations of all six possible matings or crosses are given in Table 30-2.

THE PRINCIPLES OF HEREDITY

auto = by oneself
soma = body

The chromosomes other than X and Y are called **autosomes.** The laws of inheritance for autosomal genes were discovered by Mendel in 1865. It was only in 1900 that three different scientists rediscovered Mendel's laws. Once chromosomes were known and the chromosomal theory of inheritance had been developed, it became much easier to understand the laws for the inheritance of autosomal genes.

Table 30-2
Inheritance of Hemophilia*

Gametes Formed	Female (Genotype)			
	Type I (HH)	Type II (hh)	Type III (Hh)	
Ova	H	h	½H	½h
Male Type I				
Sperm				
X(H) normal	**HH**	**Hh**	½ **HH**	½ **Hh**
Y	H	h	½ H	½ h
Male Type II				
Sperm				
X(h) hemophiliac	**Hh**	**hh**	½ **Hh**	½ **hh**
Y	H	h	½ H	½ h

*Results of crosses are shown in bold type. Color indicates genotype of male offspring; black indicates genotype of female offspring.

Mendel's Laws and Autosomes

Mendel expressed the results of his observation of crosses of garden peas in terms of numbers of the various types of plants he found in the offspring, called **progeny,** and he interpreted the observed proportions of types in terms of simple probabilities, ½, ¼, etc. His main conclusions are summarized in the form of hypotheses.

1. If "pure lines" which are called *homozygotes* (AA and aa for two alleles A and a) are crossed (mated), all offspring (progeny) will be entirely *heterozygous* (Aa). They will be similar phenotypically between themselves. If neither parental trait shows dominance, they will be phenotypically intermediate in character between the two parents (*codominance*). If the offspring are all like one parent, that parental trait is referred to as *dominant* and the other unexpressed trait as *recessive.*

2. In a cross between two heterozygotes (intercross), Aa x Aa, there is a segregation (separation) into three classes: AA, Aa, aa in the proportions ¼:½:¼. In the case of dominance, there are only two phenotypes, and they segregate in the proportion ¾ dominant and ¼ recessive.

3. In a cross between a heterozygous and a homozygous parent (called a backcross), say Aa × AA (or Aa × aa), the two parental genotypes reappear in the progeny in equal proportions.

4. Different genes (e.g., the pairs of alleles A, a and B, b) segregate independently; i.e., the combinations of genotypes obtained for each pair of alleles considered by itself are simply obtained by multiplying the respective proportions. If in a given cross AA is expected in a proportion of ¼ and Bb in a proportion of ½, the genotype AABb is expected in a proportion ¼ × ½ = ⅛.

Galactosemia: an autosomal disorder.   Mendel's conclusions can be illustrated by the example of an unusual condition called **galactosemia.** One may wonder why, here and elsewhere, traits that are unusual have been used. Why not use eye color, nose shape, skin color, etc.? The answer is that the above-mentioned characteristics have a complicated inheritance; that is, they may be determined by many genes. Therefore, they would not be suitable examples for understanding the basic rules of genetics. It is simpler to study one gene at a time. This strategy was the basis of Mendel's success.

Fortunately, very few children are born with galactosemia. The main problem of victims with galactosemia is that they lack an enzyme called galactose 1-phosphate uridyl transferase, which is necessary for metabolizing the lactose in milk. The absence of this enzyme in children can cause a very serious reaction when drinking ordinary milk. Only substituting a different sugar in the milk makes these children healthy.

Children who come under observation for this disease usually have normal parents, and frequently there is no history of the disease among their relatives. Geneticists consider the disease to be inherited by observing that more than one child born to the same parents may show the disease. On this basis a geneticist can make more exact predictions. If the disease is due to a defective gene, incapable or markedly deficient in producing the enzyme, exact expectations can be predicted for these events. The two allelic forms of this gene can be designated *gal* + and *gal* −. The allele that cannot form the normal enzyme will be gal −. Offspring can therefore be of three possible genotypes: gal + gal + (homozygous for the normal gene), gal + gal − (heterozygous), or gal − gal − (homozygous for the abnormal gene).

When the quantity of the enzyme galactose 1-phosphate uridyl transferase is measured, it is found that homozygous normals manufacture twice as much as do heterozygotes. This is an indication that each of the two chromosomes of the pair works independently. The amount of enzyme produced by the heterozygote, although half the normal level, is sufficient for normal functioning. Thus the heterozygote (gal + gal −) does not suffer from galactosemia and is indistinguishable from homozygous normals (gal + gal +). Homozygotes for the gal − allele produce either no enzyme, or a very small amount.

Illustration of segregation. Figure 30-5 shows what happens in a mating between two heterozygotes. The circle indicating each individual is divided into two halves: Dark areas represent the normal allele of the gene for the galactose enzyme, while light areas represent the abnormal, galactosemic allele. Because both parents are heterozygotes, each is represented by a circle that is half light and half dark. To find what proportion of their children will be galactosemic, the same procedure is followed as in crosses involving sex-linked traits. First determine the gametes of each parent and the number of corresponding gene traits that can be expected. Then make random combinations of the possible gametes and, as the final step, count the progeny.

It is also useful to keep in mind that to compute the probability of union of one particular type of sperm and one particular type of ovum, simply multiply the relative proportions. Thus the proportions of zygotes given in Step 2 are computed simply as the products of the proportions (that is, the fractions) of the corresponding gametes. In this simple example, they all turn out to be $\frac{1}{2} \times \frac{1}{2} = \frac{1}{4}$. As shown in Step 3, it can be expected that one-quarter of the children will be normal, half will be heterozygous, and one-quarter galactosemic. This refers to the genotypes of the progeny. The phenotypes of the progeny cannot be determined without making an enzyme assay. Phenotypically three-quarters of the progeny will be normal and one-quarter galactosemic. It should be kept in mind that expected proportions will be observed only if the number of offspring is large. In a cross such as the one shown in Figure 20-5, for example, each child will have one chance in four of being homozygous normal.

These are the standard proportions expected in a cross between two heterozygotes. They are often expressed as 1:2:1 for homozygote normal: heterozygote: homozygote abnormal, and also as 3:1; 3 for the dominant phenotype and 1 for the recessive phenotype.

Independent Inheritance of Genes

Hemoglobin is one of the many proteins in our body, but its great importance is shown by the large amount present in the blood: approximately 14 grams in every 100 ml of blood. It serves as a carrier of oxygen from the lungs to the tissues. Its molecular structure is well known. Each molecule consists of two pairs of subunits, called alpha and beta, that are made up of chains of amino acids.

Several hundred different mutations in the two chains of hemoglobin are known today; most are rare and do not cause serious conditions. One of these mutations is well known and is referred to as hemoglobin S, where S stands for *sickle-cell*. Individuals homozygous for this trait suffer from **sickle cell anemia**, a frequently fatal disease. Hemoglobin S differs from hemoglobin A, which is the normal type, in the kind of amino acid at position 6 in the beta chain. The amino acid valine is in hemoglobin S whereas glutamic acid is in normal hemoglobin.

As a result of this difference the molecule of hemoglobin becomes elongated and relatively inflexible. This distorts the shape of the red blood cell and results in a sickle shape. The S gene frequently appears in various tropical areas of the world such as Africa and, to a lesser extent, India. It is also found in Greece and various other Mediterranean countries.

It is estimated that among African blacks transported to the United States the frequency of the sickle cell trait was approximately 16 in 100 black slaves, or 15.5%. The reason for such a high incidence appears to be that the individual who is heterozygous for the sickle cell trait has an advantage in an area where malaria is prevalent. The presence of one gene for sickle cells protects an individual from the serious infectious disease. Thus, the relatively high rate of sickle cell children born to black parents today is indirectly a consequence of the protective advantage offered against another disease, malaria. In areas where malaria is under control, the sickle cell gene is no longer useful, and causes a serious health problem in the black population.

Rh factor inheritance. Another inherited difference in properties of the red blood cells is the **Rh factor.** The mechanism of inheritance of the Rh factor is thought to be a complex one involving many linked genes. It may be assumed, for our purposes, that an r allele is recessive and gives rise to the phenotype called Rh-. All other alleles are dominant and give rise to the Rh+ phenotype. The genes controlling hemoglobin and Rh factor appear to function independently both physiologically and genetically. They are not on the X chromosome; otherwise, they would show the typical sex linkage in their pattern of inheritance. Thus the two genes could be on any of the other 22 pairs of chromosomes, and the odds are roughly 21 to 1 that they are on different sites.

Illustration of independent assortment. In determining the expectations for the offspring of crosses between individuals who differ with respect to two genes located on different chromosomes, the following cross demonstrates the probable results. The genotypes for the individuals are as follows:

Male	*Female*
Hb^SHb^A	Hb^AHb^A
Rr	rr

The male parent is heterozygous for both hemoglobin S and the Rh factor. The male parent carries the S (sickle-cell trait) gene and is Rh^+. The female parent is homozygous for normal hemoglobin and is an Rh^- individual. What progeny are expected from such a cross? The female parent can form only one type of gamete, Hb^Ar, being a homozygote for each gene. But the formation of gametes in the male is a little more complicated. The principles of gamete formation:

1. Each gamete contains **only one member of each chromosome pair,** chosen at random, and hence there is a probability of ½ that a gamete contains a given member of a pair. Thus half the gametes will contain one of the two alleles present in the heterozygote and the other half will contain the other.

2. **Each gamete normally contains one member of all pairs of chromosomes.**

Figure 30-6 represents the formation of the gametes of the male parent in this cross. Figure 30-6, for reasons of simplicity, does not represent all 23 pairs of chromosomes, but just those two pairs that are marked by the two genes under consideration. In case A in Figure 30-6 the random assortment of the members of each pair of chromosomes has put together in one sperm the gene Hb^S with the gene r and in the other the gene Hb^A with the gene R; these two types of sperm will be in equal proportions. However, case A is not the only possible partition of chromosomes when a gamete is formed. There is no reason why the black chromosome of one pair must go with the black chromosome of another pair (here black and white refer to the representation of the chromosomes in the figures). Thus the distribution of the chromosomes can also be as shown in case B. Here two types of sperm are formed again in equal proportions, one of them Hb^SR and the other Hb^Ar. To determine the relative proportions of the two possible cases, A and B, keep in mind that chromosome pairs do not influence each other when their members separate in the formation of gametes.

The ova produced by the Hb^AHb^Arr female parent are always of the type Hb^Ar. To find the genotypes of the offspring of this mating, put together the four types of sperm with the one type of ovum (Hb^Ar):

Zygotes formed

Hb^SHb^Arr
Hb^SHb^ARr
Hb^AHb^ARr
Hb^AHb^Arr

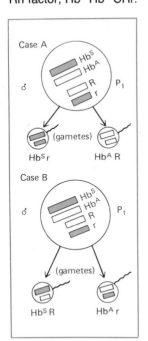

Figure 30-6 Illustration of the law of independent assortment using the genes that determine hemoglobin and Rh factor, Hb^S Hb^A CRr.

The chance of finding each of these four types is the same, because it depends directly on the proportions of the four types of sperm, which are all equal. Thus all four types of zygote are expected in equal proportions. This is an application of the *law of independent assortment* of genes, which applies whenever two genes are located on different chromosomes.

Autosomal Linkage and Crossing-Over

Early in this century it was found in observations of experimental crosses in plants and animals that the law of independent assortment was sometimes violated. The chromosomal theory of inheritance explained these exceptions. This theory predicted that genes located on the same chromosome must show a different behavior from that of genes located on different chromosomes. What happens to genes located on the X chromosomes (sex-linked genes) has already been noted. The situation for genes located on the same autosome is similar to the one we have already encountered with pairs of genes on the X chromosome, though slightly more complex.

It is clear that if the chromosomes were passed as an entire unit from parent to child, all the genes that are on one X chromosome or one autosome would tend to stay together and would be passed as a block (linkage). But at the time of gamete formation there is a phenomenon called **crossing-over,** which permits a reciprocal exchange of parts between two homologous (alike) chromosomes. Remember there are 23 pair of chromosomes. Crossing-over occurs between each pair of chromosomes with the exception of the X and Y chromosome, which are not a pair. Crossing-over takes place during the first division of meiosis. At one stage in this division, chromosome pairs become entwined, one chromosome around the other. The chromosomes may break due to the

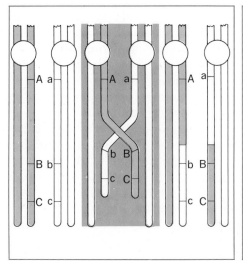

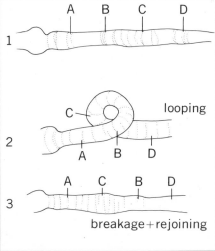

Figure 30-7 (left) In a crossover, the homologous chromosomes become entwined. When they separate, a new arrangement of genes results.

Figure 30-8 (right) Mutation in a chromosome. The gene sequence has been altered.

strains developed and exchange material as they rejoin (Fig. 30-7). Crossing-over and the recombination that accompanies it is the usual process by which new groupings of characteristics arise (Fig. 30-8).

GENETIC DISORDERS

It has already been mentioned that errors in the process of copying may occur, especially when cells and consequently their DNA reproduce. Such an error is usually the insertion of one nucleotide in the place normally taken by another. The result may be the substitution of one amino acid in the protein made by the segment of DNA concerned, the exact consequences being predictable by the genetic code (see Table 30-1). Another possibility is the loss or addition of a base (adenine, thymine, cytosine, or guanine) on the DNA molecule. Since the genetic code operates in terms of triplets of bases, as noted before, the loss or addition of *one* or *two* bases will give the sequence of bases an entirely different meaning from that point on. The sequence of amino acids, therefore, will be different from the correct sequence, and the protein will likely be nonfunctional. A third possibility is the loss or addition of three consecutive bases. In this case one amino acid will be deleted from or added to the protein chain.

Other types of mutation may also occur. Chromosomes may break and rejoin in different positions or they may be lost or sometimes duplicated. Mutations of these types, called **chromosome aberrations,** usually have more serious consequences than do the smaller mutational changes mentioned above (gene mutations).

Frequency of Mutations

It is theoretically expected that mutations are not very frequent. The human organism is like a complex piece of machinery (much more complicated than a rocket missile), all of whose parts, at least all vital ones, must function. In some cases the substitution of even a single amino acid in one protein may be lethal to the organism. Not all mutations, however, have such drastic effects. Some cause little or no disadvantage to their carriers, and some may even be advantageous.

Mutation rates in humans, as in other animals and plants, can be estimated. For example, the dominant mutation leading to the type of dwarfism known as achondroplasia appears once in every 20,000 gametes. This is one of the highest mutation rates known. Another frequent mutation is that leading to hemophilia, the coagulation disease, which, as has been described, is sex-linked. The rate of that mutation is about one in every 50,000 gametes. These two examples have been chosen from among the highest rates; the average mutation rate per gene in humans is likely to be lower than 1 in every 1,000,000 gametes. Then in every generation one gene of a given individual has a chance of about 1/1,000,000 of being mutated. But since there are many diff-erent genes in humans, perhaps as many as a million, every gamete may contain a new mutation of one of its genes. These estimates all refer to what are currently called spontaneous mutations, that is, mu-tations that occur in the absence of any specific (additional) outside agent.

Causes of Mutations

X rays and all types of ionizing radiation from cosmic rays and var-ious particles emitted in radioactive decay are known to produce mu-tations. The use of X rays for medical purposes has also contributed heavily to the increase. What fraction of the entire mutation rate can be attributed to this radioactive background? Exact figures cannot be given for humans, but it is estimated that of the total "spontaneous" mutation rate, the fraction that is due to the radioactive background is somewhere between 3 and 30 percent. One important consideration here is that there is no threshold below which radioactivity is com-pletely inactive in producing mutation. Therefore, any artificial addi-tion to the radioactive background will increase the rate of mutation.

It seems almost certain that radioactivity does not explain all of the mutation rate observed. What is the rest due to? Probably chemicals that enter into the composition of food or the atmosphere have a part. Clearly, study of the possible mutagenic action of substances that are widespread throughout the world, such as pesticides, is of considerable medical importance. Even some viruses have been changed with mu-tagenic action, especially chromosome changes. Many substances with mutagenic action are also endowed with carcinogenic (cancer-pro-ducing) action. Thus human genetics has a high relevance for human welfare and is coming to take a greater and greater place in our knowl-edge of heredity.

SUMMARY

Cells and organisms have mechanisms that make it possible to produce copies of themselves that are practically perfect. The necessary information is contained in the chromosomes; consequently chromosomes must be distributed equally when cells or organisms reproduce. Errors in distribution of chromosomes have serious consequences. The X chromosome exists in only one copy in males and in two copies in females. This situation gives us a good introduction to the laws of inheritance and to the mode of action of genes. Nonsexual chromosomes make up the great majority of genetic material. It is to chromosome behavior that Mendel's laws owe their validity. These laws help us predict the offspring expected in a particular mating.

VOCABULARY REVIEW

Match the statement in the left column with the correct word(s) in the right column. *Do not write in this book.*

1. A structural element on a chromosome that determines a specific characteristic.
2. A characteristic that is transmitted by the X chromosome.
3. An alternate form of a gene.
4. The genetic composition of a gamete.
5. The threadlike substance composing a chromosome.
6. A map of human chromosomes.
7. Chromosomal pairs that fail to separate.
8. A structural unit consisting of a base, a sugar, and a phosphate group.
9. A change in the DNA molecule.
10. Individuals who receive the same type of gene from both parents.

a. nucleotide
b. DNA
c. homozygous
d. heterozygous
e. homologous
f. phenotype
g. genotype
h. karyotype
i. nondisjunction
j. sex-linked
k. mutation
l. gene
m. allele

TEST YOUR KNOWLEDGE
Group A

Write the letter of the word(s) that correctly completes the statement. *Do not write in this book.*

1. The sex of an individual is determined at the time of   (a) spermatogenesis   (b) differentiation   (c) fertilization   (d) oogenesis.
2. The number of chromosomes in a gamete is called   (a) haploid   (b) diploid   (c) triploid   (d) tetraploid.
3. The material in the cell nucleus that determines hereditary traits is the   (a) chromosome   (b) centrosome   (c) mitochrondia   (d) Golgi body.
4. A strand of DNA is copied into a strand of mRNA by a process called   (a) mutation   (b) transcription   (c) translation   (d) nondisjunction.
5. In the replication of DNA, the base *guanine* pairs with the base   (a) thymine   (b) uracil   (c) cytosine   (d) adenine.
6. A cell that contains the haploid number of chromosomes is the   (a) spermatozoan   (b) oogonium   (c) spermatogonium   (d) ootid.
7. Sickle cell anemia is a disease of the   (a) brain   (b) muscles   (c) blood   (d) reproductive system.
8. In humans, one sex-linked characteristic is   (a) sickle-cell anemia   (b) Rh factor   (c) hemophilia   (d) hematin.

9. In a normal human, the number of chromosomes in a cell is (a) always an odd number   (b) usually 46   (c) twice as great in sperm as in a muscle cell   (d) half as great in a sperm as in an egg.
10. The genotype of an individual   (a) is always the same as the phenotype   (b) refers to the visible traits   (c) refers to the composition of the genes   (d) is indicated by the presence of either an X or Y chromosome.

Group B

Answer the following. Briefly explain your answers. *Do not write in this book.*

1. A woman heterozygous for the X chromosome recessive gene determining the disease ichthyosis (a skin disease) is married to a normal male. (a) Is the woman affected by the disease? (b) Will her daughters be affected? (c) May she have affected sons? (d) Will her normal sons transmit the disease? (e) Could her normal daughters transmit the disease?
2. A heterozygote for both the sickle-cell trait and for the Rh factor (Hb^SHb^ARr) marries another double heterozygote like himself. What is the expected proportion, in the progeny, of sickle-cell anemia Rh negative individuals, sickle-cell anemia Rh+, nonanemic Rh−, and nonanemic Rh+? (This problem is more advanced than given in the chapter. By noting what gametes are formed by each parent and in which proportions, and by forming all the possible combinations of the parental gametes, you should be able to come up with the right answer that was first given by Mendel.)
3. Why do most genetic diseases have a low incidence?
4. How are new genes formed?

GLOSSARY

A

abdomen The region of the lower trunk extending from the diaphragm to the upper part of the pelvis.

abduction The withdrawing of a part of the body from the body's axis.

absorption See *diffusion* and *osmosis*.

acetylcholine A substance produced by nerve fibers, causing muscles to contract.

actin One of the two proteins involved in muscle contraction. See *myosin*.

action current The electric current accompanying the activity of excitable tissue.

adduction Any movement resulting in one part of the body or a limb being brought toward another or toward the median line of the body.

adenosine diphosphate (ADP) A compound consisting of one molecule each of adenine and ribose and two molecules of phosphoric acid.

adenosine triphosphate (ATP) A compound consisting of one molecule each of adenine and ribose and three molecules of phosphoric acid.

adipose tissue A type of tissue in which each cell contains a single large vacuole filled with fat.

ADP See *adenosine diphosphate*.

adrenal gland An endocrine gland located above each kidney that produces the corticosteroids and adrenaline.

afferent Carrying toward, as in the direction of an afferent nerve impulse.

afterimage The visual image retained on the retina after the light stimulus ceases.

agglutination The clumping together of antigens, caused by the chemical action of agglutinins on the surface of the antigens.

agglutinin An antibody that causes antigens to stick together.

agglutinogen An antigen toward which a particular antibody is antagonistic.

albumin One of a group of relatively simple proteins.

albuminuria Presence of albumin in urine.

all-or-none law The response of a fiber of nerve or muscle to a stimulus occurs completely or not at all.

alveolus An air sac in the lungs through whose walls respiratory gases pass between the air and the blood.

amino acid One of a class of organic compounds of the general formula RCH $(NH_2)COOH$, containing the amine (NH_2) and the organic acid (COOH) radicals, R representing an organic group.

amphiarthrosis. A joint that permits only slight movement.

anabolism Those chemical processes occurring in cells that result in the formation or the replacement of protoplasm.

anaphase The stage in mitosis in which the daughter chromosomes migrate to opposite poles of the cell.

androgen A male sex hormone.

anemia A reduction below normal in erythrocytes, hemoglobin, or hematocrit.

anoxia Deficiency of oxygen in tissues.

antagonistic muscles The opposing members of a pair of skeletal muscles.

anterior Situated in front; at or toward the head end of an animal.

antibody A chemical substance capable of reacting with specific foreign substances not present normally in an organism.

antigen A substance, foreign to the body, that stimulates the formation of a specific antibody.

anvil See *incus*.

aponeurosis An apronlike tendon that serves for attachment for muscles.

appendage Any structure having one end attached to the main part of the body and the other end more or less free.

appendicular skeleton The skeletal structures composing and supporting the appendages. These include the bones of the shoulder and hip girdles as well as those of the arms and legs.

arachnoid The middle membrane covering the brain and spinal cord.

arteriole A fine branch of an artery.

artery A muscular blood vessel that carries blood away from the heart.

arthritis An inflammation of the joints, sometimes resulting in fusion of the bones.

articulation   1. The production of speech sounds. **2.** A joint; the junction of two or more bones.

asphyxia   A deficiency of oxygen and an accumulation of carbon dioxide in the body.

assimilation   The formation or the replacement of protoplasm within cells from non-living materials.

aster   A structure composed of cytoplasmic fibers that appears during mitosis.

astigmatism   A defect of vision caused by abnormal curvature of the surface of either the cornea or the lens.

ATP   See *adenosine triphosphate.*

atrioventricular (AV) bundle   A bundle of nerve fibers that transmits impulses from the atria to the ventricles.

atrium   The chamber on either side of the heart that receives blood from the veins.

atrophy   A decrease in size of some organ or structure of the body.

auditory canal   See *external acoustic meatus.*

auricle   1. The pinna of the ear; the flap of cartilage and skin on the outer ear. **2.** An appendage to an atrium of the heart.

Avogadro's principle   Equal volumes of gases at the same temperature and pressure contain the same number of molecules.

axial skeleton   The central, supporting portion of the skeleton, composed of the skull, vertebral column, ribs, breastbone, and pectoral and pelvic girdles.

axon   The portion of the neuron that carries impulses away from the cell body.

B

backcross   A cross between either a homozygous dominant or a heterozygous dominant of the F_2 generation and a homozygous recessive of the P generation.

barometer   A device for measuring atmospheric pressure.

basal metabolic rate (BMR)   The quantity of energy expended by the body per unit of time, usually measured when a fasting subject is in a warm atmosphere, is conscious, and is resting.

basophil   A type of white blood cell in which the cytoplasmic granules stain blue with the basic dye methylene blue.

benign   Not malignant; refers to neoplasms that remain at their site of origin.

bilateral symmetry   The property of having a body structure in which bisection through the vertebral column and the breast-bone results in two halves that are mirror images of each other.

binocular vision   The ability to see an object with both eyes simultaneously.

blind spot   The region of the retina that is insensitive to light because it has no rods and cones.

blood platelet   A formed element of the blood that aids in blood clotting.

blood pressure   The pressure exerted by circulating blood on the walls of the blood vessels and heart.

bolus   A ball of food prepared by the mouth for swallowing.

bone marrow   The tissue contained in the central cavity of long bones and in the interstices of cancellous bones. See *red bone marrow* and *yellow bone marrow.*

Boyle's law   At any given temperature the volume of any given mass of gas varies inversely with the amount of pressure exerted upon it.

brain stem   The lowest part of the brain, composed of the midbrain, pons, and medulla oblongata.

buffer   A substance that, when added to a solution, resists any change in the solution's acidity or basicity.

bursa   A sac of fluid within a joint.

C

callus   Any thickening of tissue resulting from the normal growth of cells.

calorie   The amount of heat required to raise the temperature of 1000 grams of water 1° Celsius.

cancer   A malignant neoplasm.

capillary   A minute blood vessel connecting arteries with veins.

cardiac muscle   A branching, lightly striated type of muscle found only in the heart and capable of contracting rhythmically.

cartilage A flexible supporting tissue composed of a nonliving matrix within which are living nucleated cells.

catabolism Chemical processes that result in the breaking down of protoplasm and in the release of energy.

catalyst A substance capable of speeding up a chemical reaction without itself being changed by the reaction.

centriole One of two small bodies lying within the centrosome that play a major role in cell division.

centromere The point on a chromosome to which the spindle is attached during mitosis.

centrosome The granular area near the nucleus of an animal cell that contains the centrioles.

cerebellum The region of the brain that coordinates muscular activities.

cerebral hemisphere One of the two principal divisions of the cerebrum.

cerebrum The largest of the three principal parts of the brain.

cerumen The waxy secretion of the glands in the external acoustic meatus.

Charles' law At any given pressure, the volume of a given mass of gas varies with the temperature.

chemical bond The link between atoms due to the transfer or sharing of electrons.

cholinesterase An enzyme that neutralizes the action of acetylcholine.

choroid membrane A pigmented, highly vascular layer of the eye lying between the retina and the sclera.

chromatic aberration The scattering of light rays at the edges of a lens so that rings of color appear to surround the object that is being viewed.

chromatid One of the pair of spiral threads that make up a chromosome.

chromatin The genetic material within the cell nucleus that is easily stained with basic dyes.

chyme The fluid form of food within the stomach and intestine.

cilium A whiplike projection of a cell.

circumduction The movement of a limb so that its distal part rotates in a circle and its proximal end remains fixed.

clavicle The collar bone, found in the shoulder girdle.

coccyx The end of the vertebral column beyond the sacrum.

cochlea The bony canal of the inner ear in which auditory receptors are located.

coenzyme A nonprotein component of an enzyme.

collagen The albumin-like substance in connective tissue, cartilage, and bone. It is the source of gelatin.

colon The part of the large intestine beginning at the cecum and terminating at the sigmoid flexure.

cone A receptor in the retina that is sensitive to high-intensity light and color.

conjunctiva The delicate layer of epithelium that covers the outer surface of the cornea and the lining of the eyelids.

connective tissue A group of specialized cells that supports and holds together various parts of the body.

contractility The property that enables voluntary muscles to become shorter and thicker when stimulated, thus producing movement.

cornea The transparent front portion of the eyeball, the outer surface of which is covered by the conjunctiva.

cranial nerve A nerve that arises from the brain and passes to other body parts.

crossing-over The exchange of genes that results from a breakage and recombination of chromosomes during meiosis.

cytoplasm The material lying within the cell membrane and outside the nucleus.

D

Dalton's law The pressure of a mixture of gases equals the sum of the pressures of the individual gases.

deamination The removal of the amino ($-NH_2$) group from an organic substance, especially an amino acid.

deciduous Falling off or shed periodically. Used to describe any structure, such as the primary teeth, that is lost during the normal growth period of an organism.

dendrite   The projection of a nerve cell that carries impulses toward the cell body.

dentine   The hard, bonelike material that forms the major part of a tooth.

deoxyribonucleic acid   The nucleic acid that bears coded genetic information.

dermis   The layer of skin below the epidermis, composed of connective tissue, sensory receptors, blood vessels, and glands.

diabetes mellitus   A disease resulting from the failure of the pancreas to produce an adequate supply of insulin.

dialysis   The separation of substances by using their different concentration gradients across a porous membrane.

diaphragm   The wall of muscle separating the thoracic cavity from the abdominal cavity. It is present only in mammals.

diaphysis   The central shaft of a bone.

diarthrosis   A joint allowing free movement.

diastole   The period when the atria and ventricles fill with blood.

differentiation   The process that takes place during embryonic development whereby cells become specialized in structure and physiological function.

diffusion   The movement of molecules from a region of relatively high concentration to one of lower concentration.

digestion   The process of altering the physical and chemical composition of foodstuff.

diploid number   The number of chromosomes present in the body cells of an organism, with the exception of the reproductive sex cells.

disaccharide   A carbohydrate formed by the chemical union of two monosaccharides.

dislocation   The forcing of a bone out of its normal position in a joint.

distal   Toward the free end of an appendage.

DNA   See *deoxyribonucleic acid.*

dorsal   Situated in back; the side along which the backbone passes.

dura mater   The outermost of the meninges.

E

eardrum   See *tympanum.*

edema   The excessive accumulation of fluids in the tissue spaces.

effector   A muscle, gland, or other type of tissue that responds to a definite stimulus.

efferent   Carrying away, as in the action of an efferent nerve impulse.

electrocardiogram   A record of the electric currents produced by the beating heart.

electroencephalogram   A record of the electric currents produced by the brain.

electrolyte   A compound that ionizes upon being dissolved.

electron   The negatively charged particle of an atom.

emulsion   The suspension of small droplets of an insoluble liquid in another liquid.

enamel   The hard outer layer of a tooth.

endocardium   A thin membrane of endothelial and connective-tissue cells that lines the inner surface of the heart.

endocrine gland   A gland that secretes hormones directly into the bloodstream.

endoplasmic reticulum   A network of channels passing through the cell on whose outer walls ribosomes are located.

endosteum   A thin membrane that lines the medullary canal.

endothelium   The squamous epithelial tissue lining the blood and lymph vessels and the heart.

enzyme   An organic catalyst.

eosinophil   A type of white blood cell in which the cytoplasmic granules are stained by eosin.

epicardium   The visceral layer of the membrane surrounding the heart.

epicranial muscle   A muscle that draws the scalp backward and raises the eyebrows.

epidermis   The outermost portion of the skin covering the body, consisting of several tissue layers.

epimysium   The sheath of fibrous connective tissue covering skeletal muscles.

epiphyseal plate   The cartilage mass between the epiphysis and the diaphysis.

epiphysis   A portion of bone attached to another bone by a layer of cartilage.

epithelium   The type of tissue that covers body surfaces and lines organs.

erythrocyte   See *red blood corpuscle.*

esophagus   The collapsible tube that extends from the pharynx to the stomach.

estrogen   A female sex hormone.

excretion The process of removing the waste products resulting from cellular activity.

exocrine gland A gland with a duct or ducts through which secretions pass.

expiration The act of expelling air from the lungs.

extension A stretching out, as in straightening a limb.

external acoustic meatus The slightly curved canal in the temporal bone that conducts sound waves from the exterior to the tympanum.

exteroceptor A sensory organ that receives stimuli from the external surroundings.

extrinsic eye muscles The muscles attached to the outer layer of the eyeball.

F

farsightedness See *hypermetropia*.

fasciculus **1.** A bundle of nerve, muscle, or tendon fibers separated by connective tissue. **2.** A bundle or tract of nerve fibers with common connections and functions.

fat An organic compound composed of glycerol and fatty acid.

fatigue The inability of nerves and tissue to respond to a stimulus following overactivity.

fatty acid An organic acid present in lipids.

feces Solid intestinal excretions.

fertilization The union of gametes.

fetus The unborn offspring of most mammals in the later stages of development.

fibrinogen A plasma protein required for the formation of a blood clot.

filtration The movement of fluid and solutes through a membrane where a mechanical-pressure gradient exists.

fissure Any groove occuring in an organ.

flagellum A small whiplike projection similar to a cilium, yet is longer.

flexion The bending of a joint, or of body parts having joints.

focal point The point at which the rays of light passing through a lens are focused.

foramen A normal opening in a tissue through which fluids, nerves, or blood vessels pass.

fovea centralis The most sensitive region in the retina. It contains only cones.

frontal (coronal) plane A plane that divides the body into anterior and posterior halves.

G

gamete An egg or sperm cell.

gametogenesis The origin and formation of gametes.

gastric juice The secretion of the glands of the stomach for the digestion of food.

gene A factor in determining inheritance of a trait.

genotype The genetic makeup of an organism.

gland A cell, tissue, or organ that manufactures and secretes a substance used elsewhere in the body.

globin One of a class of proteins.

glycosuria The presence of sugar in the urine.

goblet cell A columnar cell that secretes mucus that lubricates epithelial surfaces.

Golgi apparatus A small body in the cytoplasm of most animal cells that produces secretions.

gray matter The part of the CNS having nerve tissue lacking myelin.

H

hammer See *malleus*.

haploid number The number of chromosomes present in mature gametes.

haversian canal A small canal in bone tissue, occupied by blood vessels and connective tissue.

hematin The iron compound that is combined with the globin to form hemoglobin.

hematocrit The percentage (around 45 percent) of red cells in whole blood.

hematopoiesis The formation and maturation of blood cells.

hemodialysis The purification of the blood by dialysis.

hemoglobin The pigment in red cells capable of taking up and releasing oxygen.

hemophilia An inherited disorder in which the blood does not clot normally.

Henry's law The amount of gas dissolved in a liquid is proportional to the pressure of the gas.

heterozygous Possessing two different alleles for the same trait.

homeostasis The maintenance of a specific environment by the coordinated activities of various organ systems.

homozygous Possessing the same alleles for a specific trait.

hormone A chemical messenger produced by an endocrine gland.

hydrogen-ion concentration The number of ions of hydrogen per liter of solution.

hydrolysis Any reaction in which water combines with another compound, reducing it to a simpler form.

hypermetropia (farsightedness) A defect of vision in which the focal point of the lens falls behind the retina because of a shortening of the eyeball.

hypertrophy An increase in size of an organ or region of the body due to an enlargement of its cells.

hypothalamus A region of the brain that serves as the link between the nervous and the endocrine systems.

hypoxia The inadequate oxygenation of air, blood, or cells.

I

immunity The condition of a living organism whereby it resists infection and overcomes infection or disease.

incomplete dominance (codominance) A type of inheritance in which neither of a pair of genes for contrasting characteristics is completely dominant or recessive.

incus (anvil) One of the three small bones found in the middle ear.

infarction A localized area of dead tissue, caused by inadequate blood flow.

infrared radiation Heat in the form of rays beyond the red end of the visible spectrum, which is given off by an object when it is warmer than its surroundings.

ingestion The act of taking food into the body.

inspiration The act of taking air into the lungs.

integument The skin and its appendages.

interoceptor A sensory receptor situated in the viscera that receives stimuli connected with digestion, excretion, etc.

interphase The period in the life of a cell when it is not dividing.

involuntary muscle See *smooth muscle.*

iodopsin The visual pigment in cones.

ion An electrically charged atom.

iris The colored portion of the eye.

irritability The ability to respond to a stimulus.

islets of Langerhans Small, isolated groups of cells embedded in the pancreas that form insulin.

isotope An element that has the same number of protons as another but a different number of neutrons.

J

jaundice The yellow color of the skin, mucous membranes, and secretions due to an excess of bile pigment in the blood.

joint The point of union between two bones.

K

karyolymph The fluid within the nucleus of a cell.

karyotyping The arranging and mapping of chromosomes according to a standard classification.

kinetic energy The energy of motion.

L

labyrinth An intricate system of connecting passages; a maze; the interconnected canals of the inner ear.

lacrimal gland The gland that secretes tears.

lacteal A lymph duct in a villus.

lacuna A small depression or space.

lamella A thin scale or plate.

larynx The voicebox.

lateral Situated on the side.

lens An optical device that bends light so that its rays either diverge (biconcave lens) or converge (biconvex lens).

leukemia A disease of the blood characterized by an abnormally large number of white blood cells.

leukocyte See *white blood cell.*

leukopenia A decrease in the normal number of white blood cells, due usually to bone marrow damage or infections.

ligament   A tough band of connective tissue that connects bones to each other.

lipid   A group of organic compounds consisting of carbon, hydrogen, and oxygen; the simplest forms are fats and oils.

lumbar   That region of the back extending from the thorax to the sacrum.

lymph   A colorless fluid within the lymphatics, composed of the extracellular fluid.

lymphocyte   A type of white blood corpuscle formed in lymph tissues.

lysosome   A membrane-enclosed vesicle within the cytoplasm that disposes of unwanted foreign materials from a cell.

M

malleus   One of the small bones in the middle ear, one end of which is attached to the tympanum and the other to the incus.

marrow   The soft tissue contained in the medullary canals of long bones and in the interstices of cancellous bone.

matrix   That part of tissue into which any organ or process is set.

medial   Toward the midline of the body.

mediastinum   The space between the right and left pleural cavities.

medulla oblongata   The portion of the brain stem extending from the pons to the spinal cord.

meiosis   A process of cell division that occurs in the formation of gametes, resulting in a haploid number of chromosomes.

melanin   One of a group of pigments responsible for skin color.

Mendelian laws   The laws formulated by Gregor Mendel to explain the inheritance of characteristics.

meninges   The coverings of the brain and spinal cord.

mesentery   A fold of the peritoneum that supports parts of the digestive system.

metabolism   The sum total of all chemical processes occurring within an organism.

metaphase   The stage in mitosis in which the chromosomes separate.

micelle   A large chain of molecules in parallel arrangement.

microvillus   A structure made up of epithelial cells that lines the intestines.

midsagittal plane   An imaginary line that passes through the skull and spinal cord dividing the body into equal halves.

mitochondrion   A small living structure lying within the cytoplasm of a cell and associated with energy transformation.

mitosis   A process of cell division that results in two daughter cells containing the same number and kind of chromosomes.

monosaccharide   A carbohydrate that cannot be chemically broken down to a simpler carbohydrate; a simple sugar.

motor neuron   A neuron that stimulates a muscle fiber.

mucosa   The membrane lining those cavities and canals in the body that communicate with the outside. It secretes mucus.

mucus   A complex protein-carbohydrate compound that is secreted by mucous glands to lubricate internal body surfaces.

muscle tissue   A type of tissue that has the ability to contract.

mutation   A change in the characteristics of an organism that can be inherited.

myasthenia gravis   A disorder characterized by weakness of certain voluntary muscles, thought to be caused by an excess of a chemical that blocks acetylcholine at the myoneural junction.

myelin sheath   A covering of fatty material over a nerve fiber or process.

myocardium   The muscular tissue of the heart, consisting of interlacing cardiac muscle fibers. It is responsible for the contracting action of the heart.

myofibril   A threadlike subdivision of a striated muscle fiber.

myopia (nearsightedness)   A defect of vision, usually resulting from an elongation of the eyeball, thus causing the lens to focus the image in front of the retina.

myosin   One of the two proteins concerned in muscle contraction. See *actin*.

N

nearsightedness   See *myopia*.

negative afterimage   An afterimage seen in colors complementary to the initial stimulus.

nephron   The functional unit of the kidneys.

nerve impulse   A wavelike change passing over a nerve cell process that produces both electrical and chemical effects.

neurofibril   A strand in a nerve cell that crosses the nerve cell body to connect the dendrites and axons.

neuroglial cell   A nonconducting cell that is essential for the protection, structural support, and metabolism of the nerve cell.

neuron   The functional and structural unit of the nervous system; a nerve cell.

neuroplasm   The protoplasm in the neuron that fills the spaces between the fibrils.

neutron   An electrically neutral particle in the nucleus of an atom.

nuclear membrane   A semipermeable membrane surrounding the nucleus of a cell and separating it from the cytoplasm.

nucleic acid   One of a group of organic acids found in chromatin.

nucleolus   A body within the nucleus of a cell that appears to be associated with the formation of nucleoproteins.

nucleoprotein   A protein found in the nucleus of a cell consisting of a basic protein and nucleic acids.

nucleotide   A compound consisting of a purine or pyrimidine base combined with a sugar and phosphoric acid.

nucleus   **1.** The control center of a cell. **2.** The center of an atom, occupied by protons and neutrons. **3.** A group of specialized nerve cells lying within the white matter of the central nervous system.

O

oil   One of a group of relatively simple lipids that is fluid at ordinary temperatures.

olfaction   The process of smelling.

oligodendroglia   Small spheroidal cells of the nervous system that are found around nerve cells and between nerve fibers as supporting structures.

optic nerve   The nerve passing from the occipital lobe of the cerebrum to the eye.

orbit   The cavity of the anterior surface of the skull that contains the eye.

organ   A part of an organism that performs a definite body function.

organelle   A structure within a cell's protoplasm that has a specific function.

organ of Corti   The structure in the inner ear that contains the organs of hearing.

orifice   An opening to a cavity or tube.

osmosis   The passage of a solvent through a semipermeable membrane.

osmotic pressure   The pressure exerted on one side of a semipermeable membrane as a result of a difference in concentration of a substance found on both sides of the membrane.

ossicle   A small bone, especially one of the three bones of the middle ear.

ossification   The process of bone formation.

osteoblast   A cell concerned in the formation of bone.

osteoclast   A cell that destroys bone.

osteocyte   A bone cell.

osteomyelitis   An infection of bone tissue.

ovary   The egg-producing organ in the female.

ovulation   The growth and discharge of an ovum.

ovum   A female reproductive cell.

oxidation   A chemical reaction in which electrons are removed from one molecule or atom, usually releasing energy.

oxyhemoglobin   Iron compound containing oxygen. It is bright red in color.

P

pacemaker   The sinoatrial node; initiates each heart beat and regulates heart rate.

papilla   Any projection of tissue above a normal surface.

parietal   Situated in the outer layer or wall.

pentose   A sugar containing five carbon atoms, such as ribose.

pericardium   A double-walled sac of endothelial tissue surrounding the heart.

periosteum   The tissue that covers the outside of a bone.

peristalsis   A rhythmic, wavelike contraction moving over the walls of a tube that causes the passage of materials.

peritoneum   A layer of tissue that lines the abdominal cavity and many of its organs.

pH   The symbol used for expressing hydrogen-ion concentration.

phagocyte   A type of white blood cell that devours bacteria and other foreign bodies.

pharynx   The tube in the back of the mouth that is a common passageway to both the respiratory and digestive systems.

phenotype   The visible characteristics of an organism.

phonation   The production of sounds by means of the vocal cords.

photopic vision   Vision in a bright light resulting from the activities of the cones.

pia mater   The innermost of the meninges.

pinna   See *auricle*.

placenta   The organ on the wall of the uterus to which the embryo is attached by the umbilical cord and through which it receives its nourishment.

plasma   The fluid part of the blood.

plasma membrane   The external covering of an animal cell.

pleura   The membrane surrounding the lungs and lining the surface of the thoracic cavity.

plexus   A network of interlacing nerves, blood vessels, or lymphatics.

polysaccharide   One of a group of carbohydrates composed of monosaccharides.

portal vein   A large vein that passes from the digestive organs to the liver.

positive afterimage   An afterimage seen in the original color of the stimulus.

posterior   Situated in back; at or toward the tail end of an animal.

potential energy   The energy possessed by a body as stored energy.

prone   Lying face downward.

prophase   The first stage in the mitotic division of a cell, during which the chromosomes appear.

proprioceptor   A receptor whose function is connected with locomotion or posture.

prosthesis   An artificial body part.

protein   One of a group of complex chemical compounds composed of carbon, hydrogen, oxygen, nitrogen and other elements and characteristic of living matter.

protein-bound iodine (PBI)   Iodine present in protein molecules of the blood, usually reflecting the level of thyroid hormone.

proton   A positively charged particle in the nucleus of an atom.

proximal   Toward the main part of the body.

pulmonary circulation   The circulation of the blood through the lungs.

pulse   The blood flow along an artery.

pulse wave   A wave of muscular contraction that passes along the walls of the arteries.

pupil   The opening in the iris of the eye.

R

receptor   The end organ of a sensory nerve that is sensitive to a particular stimulus.

recessive characteristic   The member of a pair of opposing characteristics that does not appear in the phenotype when the dominant is present.

reciprocal innervation   The action in which the members of a pair of antagonistic muscles are stimulated and inhibited simultaneously by nerve impulses.

red blood corpuscle   A nonnucleated, biconcave disk found in blood that contains hemoglobin.

red bone marrow   The principal site of the formation of blood cells.

reduction   A chemical reaction in which electrons are gained by atoms.

reflex   A coordinated involuntary response of an organ to an appropriate stimulus.

refraction   The bending of a beam of light as it passes from one medium into another.

respiration   The exchange of gases between cells and their surroundings.

retina   The innermost layer of the eye, composed of nerve endings that are sensitive to light.

Rh factor   A hereditary blood factor present in the blood serum.

rhodopsin   A reddish blue pigment present in the choroid layer of the eye and in the rods in the retina.

ribonucleic acid (RNA)   One of a group of nucleic acids found principally in the cytoplasm of a cell.

ribose   A five-carbon sugar found in ribonucleic acid.

ribosome   A granule found in cytoplasm containing RNA.

rod   A receptor in the retina that is sensitive to low intensities of white light.

S

sacrum   The region of the vertebral column consisting of fused vertebrae that forms part of the hip girdle.

sagittal plane   A vertical line that divides the body into right and left segments.

sarcolemma   The delicate sheath surrounding a muscle fiber.

saturation   The quality of color that depends on the amount of white light present.

scapula   The shoulder blade, found in the upper skeletal region.

sclera   The tough outer layer of the eye; the white of the eye.

scotopic vision   Ability to see in a dim light due to the activities of the rods.

secretion   The act of releasing a substance that is produced by an organ or group of tissues; the substance itself.

semicircular canals   Three bony canals in the inner ear concerned with the maintenance of balance.

semipermeable membrane   A type of membrane that will permit the passage of certain materials but prevent the movement of others through it.

septum   A wall of tissue dividing a cavity.

serum   Plasma that lacks fibrinogen.

serum albumin   A type of plasma protein concerned with the maintenance of osmotic pressure.

serum globulin   A type of plasma protein that contains antibodies.

sesamoid bone   A small bone developed in the tendon of a muscle.

sex-limited characteristic   A characteristic that is determined by a gene that is present on the nonhomologous portion of the Y chromosome.

sex linkage   The type of inheritance in which the gene is carried on a sex chromosome.

sinus   A cavity within a bone or organ.

skeletal muscle   See *striated muscle.*

skull   The skeletal framework of the head.

smooth muscle   Elongated, thin, spindle-shaped involuntary muscle tissue.

solute   The dissolved substance in a liquid.

solution   A uniform mixture of a liquid and another substance.

solvent   A liquid substance capable of dissolving one or more other substances.

sperm   A male reproductive cell.

sphincter   A muscle arranged in a circular manner around a tube or opening. It behaves like a drawstring in controlling the size of the opening.

sphygmomanometer   A device used for measuring blood pressure.

spirometer   A mechanical device used to measure the capacity of the lungs.

sprain   A tearing or straining of tendons and ligaments at a joint.

stapes   The innermost of the three small bones of the middle ear.

stereoscopic vision   The ability to see in three dimensions.

stimulus   A change in the environment that brings about a response.

stirrup   See *stapes.*

stomach   An elongated pouch that lies in the left upper quandrant of the abdomen.

striated muscle   Muscle composed of spindle-shaped fibers with striations. Its action is voluntary.

stroke volume   The output of the left ventricle of the heart in a single systole.

subarachnoid space   The space between the arachnoid and the pia mater containing the cerebrospinal fluid.

subcutaneous tissue   A group of specialized connective cells, containing fat, that comprises one of the layers under the skin.

substrate   The substance acted upon by an enzyme.

summation   The effect produced by the accumulation of subthreshold stimuli.

superficial muscle   A muscle found near the surface of the body.

supine   Lying on the back with the face upward, or extending the arm with the palm upward; opposite of prone.

synapse   The region between neurons.

synarthrosis   A nonmovable joint.

synovial fluid   The clear fluid that is normally present in joint cavities.

synthesis   The process occurring when two or more substrates combine to form a new, more complex substance.

systole   The period of active contraction of the atria and ventricles.

T

tachycardia   A very rapid heart beat.

telophase   The final stage in mitosis in which a new cell membrane appears and the parent cell divides into two halves.

tendon   A tough band of connective tissue that attaches a muscle to a bone.

testis   The sperm-producing organ in the male.

tetanus   Sustained contraction of a muscle caused by the repeated flow of nerve impulses to the muscle.

thalamus   Either one of two large masses of gray matter located above the midbrain that serves as a relay station for incoming sensory impulses.

thoracic   Pertaining to the thorax, or the chest cavity.

threshold   The minimum level of stimulus required to bring about a response.

thrombus   A clot of blood that forms within a blood vessel.

tissue   A group of similar cells performing a specific function.

tonus   The partial contraction present in relaxed skeletal muscles.

tourniquet   A mechanical device used to prevent excessive blood loss.

transverse plane   An imaginary line passing at right angles to both the front and midsection; a cross section.

tumor   Any abnormal mass resulting from a growth disturbance of cells.

tympanum   A thin membrane covering the internal end of the external meatus. It vibrates when struck by sound waves.

U

ultraviolet radiation   Rays in that region of the spectrum lying just beyond the shortest visible violet rays.

uremia   A disorder due to kidney failure.

uterus   The cavity that receives the fertilized egg and in which the fetus develops.

V

vacuole   A space within a cell.

vein   A blood vessel that carries blood toward the heart.

ventral   Situated on the side opposite to that along which the backbone passes; sometimes called the anterior side in human beings.

ventricle   **1.** A chamber of the heart from which blood flows to parts of the body. **2.** One of several cavities in the brain.

venule   A small vein.

vestibule   The oval cavity of the inner ear that acts as an entrance to the semicircular canals.

villus   A small, fingerlike projection from the mucosa of the small intestine through the walls of which digested foods are absorbed by the blood.

viscera   The organs contained within the cranial, thoracic, and abdominal cavities.

visceral muscle   Muscle found in the walls of blood vessels, intestines, and other internal organs. Also called involuntary muscle.

visual acuity   The ability to see objects clearly with the naked eye.

visual purple   See *rhodopsin*

vitamin   Any of a group of organic compounds essential for the maintenance and growth of animals, who are unable to synthesize them.

voluntary muscle   See *striated muscle.*

W

white blood cell   Any of a large group of relatively large, nucleated blood cells having the power of independent movement.

white matter   The part of the CNS characterized by neurons having myelin.

Y

yellow bone marrow   Marrow in the medullary cavities of long bones, containing many fat cells.

yolk   Nonliving material in the cytoplasm of an ovum. It serves as food for the developing embryo.

Z

zygote   A fertilized egg cell.

INDEX

Page references for illustrations are in **boldface** type.

A

abdominal cavity, 7–8
abduction, 86, **114**–115
absorption, 40–42, 228–229, 262
accommodation, 177–178
acetabulum, 81
acetylcholine, 103–104, 141, 359
acid, 24–26
acid-base equilibrium, 305, 322–324
acromion process, 78
actin, 99–100
action potential, **137**–141
active site, 223
active transport, 39–40, 413
acuity, 183–184
Adam's apple, 274–275
adduction, 86, **114**–115
adenine, 471–473
adenoid, 234–**235**, 366
adenosine diphosphate (ADP), **41**, 43, 104–105
adenosine triphosphate (ATP), **41**, 43, 104–105
adipose tissue, 56–57
ADP, *see* adenosine diphosphate
adrenal gland, 422–**423**, 433
afferent process, 131–133
agglutination, 324
air, 299–307; composition of, 288–289
albuminuria, 412–413
aldosterone, 415, 439–440
all-or-none law, 106, 138–141
allele, 474
alveoli, 275–277, 281–282
amino acids, 43–44, 211, 215, 266–267; essential, 213
amniotic sac, 463–464
amphiarthroses, **84**–85
amplitude, 197–199
ampulla, 196
anabolism 22–23, 40–41, 395
anaphase, 47–**48**
androgens, 439, 441–442
anemia, 326; types of, 327–328
angular joint, **84**–85
antagonists, 114–115
anterior lobe, **430**–431, 436–437
antibody, 33, 324–325, 369–370
anticoagulants, 321
antigen, 33, 324–325, 369–370
aorta, 350–**351**–353
apocrine gland, **379**–380
aponeurosis, 113–115
appendicular skeleton, 76–**77**

appendix, 253–255
arachnoid, 146–**147**–152
arteriole, 346–**348**
artery, 333–**334**, 346–**347**
arthritis, 91–92
artificial respiration, 292–**293**–294
asphyxiation, 280–281
assimilation, 45–47
aster, **47**
asthma, 296
astigmatism 184–185
atherosclerosis, 360–361
atom, 13; structure of, 14–18
atomic energy levels, 15–16
ATP, *see* adenosine triphosphate
atria, 333–**334**
atrioventricular bundle, 335–336
autonomic nervous system, 155–**156**–157; functions of, 163–164
Avogadro's principle, 287
axis cylinder, 132–133
axon, 131–**132**

B

Bainbridge reflex, 338–339
ball-and-socket joint, 81, **84**–85
barometric physiology, 305–307
basal ganglia, 152–153
basal metabolism, 398–399
bel, 198–199
beta cells, 258–259
biceps, **113**–115
biceps femoris, 122–**125**
bile, ducts, 256–259; salts, 262–263, 265; secretion, 260, 265–267
binding site, 223
biological oxidation, 395–397
biological sciences, 2–4
bipolar neuron, **132**–134
blind spot, 176
blood, 57, 310–329; cells, 311–318; clots, 320–321; composition of, **58**, 310–318; disorders of, 326–329; flow, 348–**349**; groups, 324; physiology, 318–326; pressure, 354–360; types, 324–326, 479–480; viscosity, 354
blood vessels, anatomy of, 346–355; physiology of, 355–360
body temperature, dysfunctions of, 392–393; maintaining, 389–391; normal, 391–392; principles of, 386–389;

variations in, 391–392
bonds, covalent, 18–19; and energy exchange, 21
bone, compact, 69–**70**; development of, 87–91; marrow, 69; sesamoid, 69; spongy, 69–**70**, 87–88; structure of, 69–**70**, 71; tissue, 69; types of, **69**
Boyle's law, 287
brachioradialis, 121–123
brain, development of, **148**–**149**–150; structure of, 148
brainstem, **149**–152, 159–160
breast, 455, 465–466
breathing, 285–296
bronchi, 275–276
bronchitis, 282–283
burns, 384
bursa, 86

C

calcaneus, 82–83
caloric, 396–**397**
canal of Schlemm, 172–**173**
cancer, 47–48
capillary, 346–**349**
carbohydrate, 20, 206–208; metabolism, 400–402
carboxyl group, 211–215
cardiac cycle, 336–338
cardiac muscles, 61, 96–97
cardiac output, 337–338, 355–356
cardiovascular system, 310–329, 331–343
carotid sinus, 338–339
cartilage, 57–58, 76; hyaline, 87
catabolism, 22–23, 40–42, 395
cataract, 186
cavities, abdominal, 7–**9**; body, 7–**9**; cranial, 7–**9**; dorsal, 7–**9**; pelvic, **8**–**9**; spinal, 7–**9**; thoracic, 7–**9**; ventral, 7–**9**
cecum, 253–255
cell, 29–48; blood, 31; defined, 13; division, 35, 46–47; function of, 36–47; metabolic functions in, 40–44; muscle, 31; nerve, 31; pathology, 47–48; physical processes in, 37–39; physiological processes in, 39–40; reproduction of, 45–47; structure of, 29–36; transport mechanisms, 37–44
cell body, 131–**132**

cell membrane, **31**–33
central nervous system, 130–141,
    145–157
centriole, **31**, 34, **46–48**
cephalic phase, 244
cerebellum, **149**–152, 159–160
cerebral cortex, 151–152
cerebral palsy, 164
cerebrum, **149**–152, 160–161
cervical, 74–**75**–76
cervix, 453–455
Charles' law, 287
chemistry, principles of, 13–26
chemoreceptors, 358–359
cholesterol, 265
chondriocytes, 88
choroid coat, **173**
chromatin, 35, **46**–47
chromosome, 35, **48**, 469–483;
    nondisjunction of, 470–471;
    structure of, 469–470
chyme, 245–247
cilia, 36, 56; function of, 36
ciliated epithelium, 56
circulatory system, 10, 310–371
circumduction, **115**
citric acid cycle, 396, 402
clavicle, 78
coagulation, 320–321
coccyx, 74–**75**–76
cochlea, **193**–**194**–195
coenzymes, 42
cofactor, 221–222
collagen, 56
colon, 254–255, 263–264
color blindness, 186
conducting system, 334–335
conduction, 135–141, 389
condyle, 70
cones, **174**–176
conjunctiva, 172–**173**
conjunctivitis, 187
connective tissue, 52, 56–59, 229
contractility, 51, 101–**102**
contusion, 125–126
convection, 389
convergence, 177
convolution, 152
coracoid process, **77**–78
cornea, 172–**173**
coronary circulation, **334**
corpuscle, 57–**58**
cortex, **149**–152
corticoids, 433, 439–440
costal cartilage, 76–**77**
cranium, 71–**72**
cricoid, 274–275

cupula, 196
cyanosis, 307
cyanotic, 343
cyclic AMP, 426
cytoplasm, **31**, 33–36
cytoplasmic membrane, 32
cytosine, 471–473

D

Dalton's law, 288
dead space, 300
deamination, 266–267
decibel, 198–199
deciduous teeth, **231**–232
deglutition, 237–**238**
deltoid, 120–**121**–122
denature, 223–224
dendrite, 131–**132**
dentin, 232–233
deoxyribonucleic acid, *see* DNA
dermis, **375**, 377–378
diabetes mellitus, 401, 413,
    446–447
dialysis, 37, 39
diaphysis, 69
diarthroses, **84**–85
diastole, **337**–338, 354, 356
diencephalon, 149–152
differentiation, 463–464
diffusion, **37**–38, 262, 303–305,
    413; facilitated, 39
digestive system, 9, 206–268
digestive tract, 228–239; anatomy
    of, 228–236; functions of,
    236–238
diglycerides, 208–211
diploid, 462, 470–471
disaccharide, 207–208
dislocation, 92–93
diuresis, 415–416
DNA, 21, 35, 43–44, 46, 469,
    471–473
double helix, 43–44
Down's syndrome, 470–471
duodenum, 251–255
dura mater, 136–**147**–152
dyspnea, 241

E

ear, anatomy of, 190–**191**–196;
    disorders of, 201–202;
    physiology of, 197–201
eccrine gland, 379–380
edema, 212–215, 370–371
elastin, 56
electric potential difference,
    136–141

electrocardiogram (EKG), 341
electrolytes, 26
electron, 14–21
electron-transport system, 402
elements, 13–21; mineral,
    215–218
embryo, 461–464
emphysema, 302
encephalon, 149–152
endocardium, **332**–333
endocrine system, 10,
    421–**422**–**423**–427; anatomy of,
    429–435; characteristics of,
    425–426; dysfunctions of, 427,
    442–447; physiology of,
    435–442; types of, 420–424
endolymph, 193
endomysium, 97
endoplasmic reticulum, **31**–35
endosteum, 69
endothelium, 54
energy, in cell movement, 37–44;
    defined, 21; exchange 22–24;
    kinetic, 21; in metabolism,
    40–42; for muscle contraction,
    104–105; potential, 21
energy of activation, 222–223
enzyme, 21, 33, 35, 221–224, 396,
    474; activity of, **222**–224;
    composition of, 221–224; in
    metabolic processes, **41**, 43;
    specificity of, 223–224
epicardium, 332
epicranius, 116–**117**–118
epidermis, **375**–377
epinephrine, 266, 359
epiphyseal plate, 89
epiphysis, 69
epithelial tissue, 36, 52, 54–**55**–**56**
equilibrium, 200–201
erythrocytes, 310–329
erythropoiesis, 312–**313**
erythropoietin, 306
esophagus, 228–239, 275
estrogens, 439, 442
Eustachian tube, **191**–193
excitability, 135–141
exocrine gland, 421–**422**–**423**
extension, 85, **114**–115
extensor carpi, 122–123
external oblique, 119–**120**
external respiration, 301–302
exteroceptors, 135
eye, architecture of, 169–176;
    components of, 172–176;
    disorders of, 184–187; organs of,
    170–172; physiology of, 176–184

F

facial bones, 71, **73–74**
Fallopian tubes, 453–454
fat, 208–211
fatty acid, 403–404
feces, 264
fertilization, 452–453, 462
fetus, 461–463
fibrillation, 342–343
fibrin, 320
fibroserous layer, 229
fibrous membranes, 87
fibula, 81–**82**
filtration, 37, **39**
fissure of Rolando, **151**–152
fissure of Sylvius, **151**–152
flagella, 36
flexion, 85, **114**–115
flexor, 122–123
follicular cells, 431–432
fontanels, **88**
foramen, 71, 76
foramen of Monro, 149–150
forearm, 78–79
fossa, 71
fracture, 93
frontal, 71, **73–74**, **151**–152
fundus, 241–242, 453

G

galactosemia, 477–478
gallbladder, 257
gamete, 461–463
gamma globulin, 322
gas(es), 13; diffusion of, 288–289,
    303–305; exchange, 285–290,
    299–300; laws, 287–288; partial
    pressures of, 301–302;
    properties of, 287–289;
    transport, 299–300
gastric glands, **243**–244
gastric juice, 244, 247
gastric motility, 245–247
gastritis, 248–249
gastrocnemius, 124–**125**
gastroenteritis, 268
gastrointestinal tract, 229–234
genetics, 4, 468–483; disorders,
    482–483
genitalia, 453–455
glaucoma, 187
gliding joint, **84**–85
glomerular filtration, 412–413
gluconeogenesis, 266
glucose, 20–21
gluteus maximus, 122–**125**
glycerol, 208–211

glycogen, 20
glycogenolysis, 266
glycolysis, 401–402
glycoproteins, 34, 208
glycosuria, 400–401, 413
goblet cell, 56, **243**
goiter, 444
Golgi apparatus, **31**–34
gonad, **423**–424, 434–435, 453
gonadotropic cycle, 460–461
granulocytes, 316
gray matter, 133
guanine, 471–473
gyri, 150–**151**–152

H

haploid, 462, 470–471
harmonics, 199
Haversian system, **90**–91
hearing, 190–202
heart, 331–343; beat, 341–342;
    chambers, 333; control of,
    338–340; disorders, 342–343;
    effects of drugs on, 339–340;
    murmurs, 343; physiology of,
    334–342; structure of,
    331–**332**–334; valves, 333
heat, gain, 390–391; loss,
    386–**389**, 390–391; production,
    387–388
hematocrit, 311
hemodialysis, 416
hemoglobin, 304–305, 314–315
hemolysis, 326
hemophilia, 328, 475–476
hemopoiesis, 311–312
hemorrhage, 329
hemostasis, 319–320
Henry's law, 288–289
heparin, 321
hepatic sinusoid, 256–257
hepatitis, 268
heredity, 35, 44, 468–483; basis
    of, 468–476; disorders,
    482–483; principles of, 476–482
hernia, 125–126
heterocrine gland, 421–**422**–**423**
heterozygotes, 474, 477
hilum, 364–365, 409
hinge joint, **84**–85
Hodgkin's disease, 370–371
homeostasis, 25, 305–307, 386
heterozygotes, 474, 477
hormone, 33, 244–247, 420–427;
    balance, 426–427;
    characteristics of, **425**;
    gastrointestinal, 260–261;

physiological effects of, 425–426
human body, 2; cavities in, **7–8**;
    organization of, 2–10; planes in,
    6–7
humerus, **70**, **77**–78
hydrogen ions, 24–26
hydrolysis, 22–23, 211–215
hydroxyl ions, 24–26
hyoid bone, 73–**75**
hyperglycemia, 266
hypermetropia, 185–**186**
hypertension, 360–361
hyperthyroidism, 445
hypertonic, 39
hyperventilation, 305–306
hypoglycemia, 266
hypophysis, **430**–431, 442–444
hypothalamus, 151–152
hypothyroidism, 444–445
hypotonic, 39, 263
hypoxia, 291–292, 305, 307

I

ideal gas law, 287–288
ileum, 251–255
ilium, 79–81
immunity, 369–370
incisors, **231**–232
incus, **191**–192
independent assortment, 479–481
infarction, 342–343
insertion, **113**–115
inspiration, 290–291
insulin, 258–259, 441
integration, 135–141
internal respiration, 302
interoceptor, 135
interphase, 46
interstitial fluid, 368–369
intestines, anatomy of, 251–259;
    disorders of, 268; function of,
    259–268, large, 253–259; small,
    251–252, 259–263
iris, **173**
irritability, 45, 101
ischemia, 342–343
ischium, 79–81
islet of Langerhans, **423**–424,
    433–434
isometric, 107–108
isotonic, 39

J

jaundice, 268
jejunum, 251–258
joints, different types of, **84**–86;
    movement of, 85–86
jugular vein, 352–353

K

karyotyping, 470
keratin, 376
ketosis, 405–406
kidney, 408–416; anatomy of, 408–411; artificial, 416; diagram of, **410;** disorders of, 416; physiology of, 411–416
Klinefelter's syndrome, 470
Kreb's cycle, 396
Kupffer cells, 256–257

L

labor, 464–**465**
lacrimal apparatus, 170–**171**
lactation, 465–466
lacteal, 364, 370
lacunae, 88
lamellae, 90–91
large intestine, 253–255, 263–268
larynx, 274–275, 278
latent period, 105–106
latissimus dorsi, 120–**121**–122
learning, 161
lens, 169–187
leukocytes, 310–329
levers, classes of, 86–**87**
ligament, 91
lipid, 20–21, 208–211; metabolism, 402–404
liver, 255–257, 265–267
loop of Henle, 410
lungs, 276–277, 280–283
lymph node, 363–**365**–**367**
lymph system, 310–329, 363–371; circulation of, 363–366; disorders of, 370–371; function of, 367–370; organs in, 366–367
lymphocytes, 369–370
lysosome, **31**–35

M

macronutrients, 266
macula lutea, 176
malleus, **191**–192
mammary gland, 453–455
mandible, 73–**74**, 236–237
marrow, bone, 69
masseter, **117**–118, 236–237
mastication, 231, 236–237
mastoid process, 71, **73**
matrix, 52
maxillae, 73–**74**
mediastinal surface, 277
mediastinum, 286, 331
medulla, 150–152, 159–160
medullary canal, 69, 90
meiosis, 461–463

melanin, 174, 376–377
membranes, metabolic processes in, 40–44; nuclear, **32,** 35; physical processes in, 37–39; physiological processes in, 39–40
Mendel, Gregor, 468; laws of, 476–478
meninges, 146–**147**–152
menstruation, 458–461
metabolism, 40–44, 395–406; basal, 398–399; carbohydrate, 400–402; disorders in, 405–406; energy of, 395–397; lipid, 402–404; protein, 404–405
metacarpal bones, 79
metaphase, 46–**48**
metatarsal bones, 82–83
micelles, 262–263
micronutrients, 215–221
microtubules, 36
microvilli, 36
micturition, 415–416
midbrain, 150–152
mitochondria, **31**–32–**33**
mitosis, 46–47, 409
mitral valve, 333–**334**
monoglycerides, 208–211
monosaccharide, 207–208
morphology, 2
motor neuron, 103–104, 135
mucous membrane, 273–274
multiple sclerosis, 141–142
muscle tissue, 53, 60–61, 96–109; anatomy of, 96–100; disorders of, 108; physiology of, 101–108
muscles, axial skeleton, 116–**117**–121; cardiac, 61, 96–97; lower extremity, 122–124; smooth, 61, 96–97; striated, 60, 96–97; upper extremity, 120–**121**–122
mutation, 471–473; causes of, 483
myasthenia gravis, 109
myelin sheath, 132–**133**
myocardium, 332–333
myofibrils, 60, 98–99, **100**–**101**
myogram, 105–108
myopia, 184–**185**
myosin, 99–100

N

nasal bones, 73–**74**
nasal cavity, 273–274
nasopharyngitis, 282
nephritis, 412–413
nephron, 410–411
nerve, cranial, 154–155; mixed,

153–155; motor, 153–155; sensory, 153–155; spinal, 153–155; vagus, 154–155
nerve fiber, 131–132; regeneration of, **134**
nerve impulse, **136**–**137**–**138**–**139**–**141**
nervous system, 9, 145–165; anatomy of, 145–157; disorders of, 164–165; peripheral, 153–157; physiology of, 157–164
nervous tissue, 52, 59, 130–142; disorders of, 141–142; organization of, 130–135; physiology of, 135–141
neuron, 59–**62**, 130–**132**; classification of, **132**–135; motor, 153–**154**; physiology of, 135–141; sensory, 153–**154**; structure of, 131–135
neurotransmitter, 33, 208
neutron, 14–18
Nissl bodies, 131–**132**
nitrogen narcosis, 306–307
nitrogenous waste, 408
nodes of Ranvier, 132–133
norepinephrine, 141
nucleic acid, 20–21
nucleolus, 36, **46**–47, 131–135
nucleoplasm, 35
nucleotide, 471–473
nucleus, 14–17, **31**, 35–36; function of, 35
nutrition, 206–225

O

occipital, 71, **73**–**74**, **151**–152
occipitalis, 116–**117**–118
odontoclast, 231–233
olecranon process, 79
olfaction, 277–278
olfactory bulb, 152
oligodendroglia, 130–141
oliguria, 415–416
optic disc, 176
optic nerve, 170, 175
oral cavity, 229–234; parts of, 229–234
orbicularis oculi, **117**–118
orbital cavities, 170–172
organ of Corti, **195**
organogenesis, 464
osmosis, 37–**38**–39, 262, 413
osmotic pressure, 39, 368
osseous tissue, 58–59, 69
ossicles, 73–**74**, **191**–192
ossification, 87–91

osteoblasts, 59, 87–88, 90
osteomyelitis, 91–93
osteosclerosis, 202
oval window, **191**–192
ovarian cycle, 458–461
ovary, **423**–424, 453–455;
    function of, 458–461
ovulation, 459–461
oxidative phosphorylation, 401–402
oxygen, 299–307

P

pacemaker, 335
pancreas, 258–259, 267–268, 441
pancreatic juice, 258–260
papillae, **230**–231
parasympathetic, 153–**156**–157
parasympathetic stimulation, 260
parathyroid gland, 422–**423,**
    432–433, 445–446
parietal, 71–**73**–**74, 151**–152; cell,
    243
Parkinson's disease, 165
parotid, **233**–234
patella, 81–**82**
pathogens, 369–370
pectoral girdle, **77**–**78**–**80**
pectoralis major, 120–**121**–122
peduncles, 150–152, 159–160
pelvic girdle, **8, 77**–79–**81**
penis, 453, 457
peptide bond, 221–222
pericardial fluid, 54
pericardial sac, 332
perilymph, 193
perimysium, 97
periosteum, 88
peripheral nervous system,
    131–**133,** 153–157
peristalsis, 238
peritoneum, 242–243, 252
permanent teeth, **231**–232
phagocytosis, 40, 315
phalanges, 79, 83
pharynx, 228–239, 274, 278
phonation, 277–278
phospholipids, 210–211
pia mater, 146–**147**–152
pineal body, 435, 442
pineal gland, **423**–425
pinna, **191**
pinocytosis, 40–41
pitch, 197–199, 279–280
pituitary gland, 421–**423, 430**–431
pivot joint, **84**–85
placenta, 424–425, 435
plasma, 310–329

plasma protein, 318, 322
platelets, 57–**58**, 317–318
platysma, **117**–119
pleural cavity, 276–277
pleural fluid, 54, 286
pneumonia, 282–283
pneumothorax, 296
poliomyelitis, 142
polycythemia, 307
polysaccharide, 207–208
pons, **149**–152
portal circulation, **353**–354
posterior lobe, **430**–431, 437–438
presbyopia, 186
pressoreceptor, 338–339
pronation, **114**–115
prophase, 46–**47**
proprioceptor, 135
protein, 20–21, 211–215; in meta-
    bolic processes, **41;** molecules,
    32, 43; synthesis, 471–473
protein-bound iodine, 398–399
protein synthesis, 43–44, 471–473
proton, 14–17
protoplasm, 13
pterygoid, 236–237
pubis, 79–81
pulmonary pressure, changes in,
    290
pulmonary ventilation, 287–288
pulse, arterial, 356–357; venous,
    357
pupil, **173**–174
pylorus, 241–**242**

Q

quadriceps, 123–**125**
quality, 197–199, 279–280

R

radiation, **388**
radioactivity, 17
radius, 78–79
reabsorption, tubular, 412–413
reaction time, 135–141
receptors, 153–155, 169
rectum, 255
rectus abdominis, 119–**120**
red marrow, 311–312
reflex, 158, 338–339; arc,
    158–**159;** conditioned, 161–162
refraction, 177, 184, **185, 186,** 187
refractory period, 137–141
renal, 409–410
reproduction, anatomy of, 452–453;
    of cells, 45–47; female, 453–455;
    male, 455–458; pathology of,
    466; physiology of, 458–466

respiration, 285–296; cellular, 24
respiratory cycle, 10, 290–296;
    control of, 295–296
respiratory gases, 299–307
respiratory tract, physiology of,
    277–282; problems of, 282–283;
    structure of, 272–277
retina, **174**–176
Rh factor, 326
rhodopsin, 175–176, 181
ribonucleic acid, *see* RNA
ribosomes, 34, 44
ringworm, 380–381
RNA, 21, 34, 44, 472–473;
    messenger, 44; ribosomal, 44;
    transfer, 44
rods, **174**–176

S

sacroiliac joint, 79–81
sacrum, 74–**75**–76
saddle joint, **84**–85
saliva, 237–238
salivary glands, **233**–234
saltatory conduction, 133,
    138–139
sarcolemma, 60, 98–100
sarcomere, 100–105
sarcoplasm, 98–**100**–101
saturation, 183–184, 209–211
scapula, **77**–78
Schwann cells, 131–**133**
sclera, 172–**173**
scrotum, **423**–424
sebaceous gland, 378–**379**
semen, 455–458
semicircular canals, **193**–196
semilunar valve, 333–**334**
sense receptor, **381**–382
sensory neuron, **132**–135
serotonin, 208
serum albumin, 322
sesamoid, 69
sex determination, 469–470
sex hormone, 435
sex-linked inheritance, 474
shoulder girdle, **77**–**78**–**80**
sickle-cell anemia, 479
sinoatrial node, 335–336
sinus, 274
skeletal muscles, 96–**97**–100,
    112–126; in action,
    112–**114**–115; disorders of,
    125–126; superficial, 115–125
skeleton, 8, 68–93; anatomy of,
    68–71; appendicular, 76–83;
    axial, 71–**72**–**73**–74; disorders
    of, 91–93; divisions of,

71–**72**–**73**–76; physiology of, 84–91

skin, anatomy of 374–375; appendages of, 378–382; glands of, 378–**379**–380; layers of, **375**–378; pathology of, 383–384; physiology of, 382–383

skull, 71–**72**–**73**–**74**, 88

sliding-filament theory, **102**–104

small intestine, 251–252, 259–263

smooth muscles, 61, 97

sodium pump, **136**–**137**–141

solvent, 25–26

somatic division, 153–155

somatostatin, 259

sound, characteristics of, 197–199; formation of, 278–279; nature of, 197; properties of, 279–280

sperm, 455–**457**–458

sphenoid, 71–**73**–74

sphincter muscle, 242–243

sphygmomanometer, 356

spinal cord, 76, 146–**147**–**148**; conduction pathways, 157–160

spinal nerves, 148

spindle, **47**–**48**

spirogram, 294–295

spleen, 366; function of, 370

stapes, **191**–192

stenosis, 343

sternomastoid, **117**–119

sternum, 71–**72**, 76–**77**

steroids, 433

sterols, 210–211

stimulus, 135–141; liminal, 106

stomach, 241–249; cancer of, 249, disorders of, 248–249; muscle coats in, **242**; organs of, 241–243; physiology of, 244–247

striated muscles, 60, 96–97

stroke volume, 337–338

subarachnoid space, 146–**147**

subcutaneous tissue, **375**–378

sublingual, **233**–234

submandibular, **233**–234

subthreshold, 106–107

sulci, 150–**151**–152

summation, 106–107, 138–141

superficial muscles, 115–**116**–**117**, **121**, **125**

supination, **114**–115

suprarenal gland, 433

swallowing, 237–**238**

sweat gland, 378–**379**–380

symmetry, bilateral, 7

sympathetic, 153–**156**–157

synapse, 139–141

synaptic cleft, 139–**141**

synaptic knob, 132–133

synarthroses, 84–85

synovial fluid, 86

systemic circulation, 350–**351**–353

systems, circulatory, 10, 310–371; digestive, 9, 206–268; endocrine, 10, 420–447; muscular, 9, 96–109; nervous, 9, 145–165; reproductive, 10, 452–483; respiratory, 10, 290–296; skeletal, 8, 68–196; urinary, 10, 408–416; vascular, 346–361;

systole, **337**–338, 355–356

T

tarsal bone, 82–**83**

taste buds, 230, **236**

telophase, 47–**48**

temporal, 71, **73**–**74**, **151**–152

temporal bone, 236–237

tendon, 91, **113**–115, 125–126

tendon of Achilles, 124–**125**

terminal filament, 132–133

testes, **423**–424, 455–457

tetanus, 107–**108**

thalamus, 151–152

thermometer, 392

thigh, 81–**82**

thoracic cavity, 7–**8**, 74, 286

thorax, 71–**72**, 76

threshold, 106–107, 138–141

thrombocytes, 310–329

thrombosis, 328

thymine, 471–473

thymus, 367

thyroid, 274–275, 399, 444

thyroid gland, **421**–**423**, 431–**432**; function of, 438–439

tibia, 81–**82**

tissue, 47–48; classification of, 54–61; connective, 52, 56–59; epithelial, 52, 54–**55**–**56**; muscle, 60–61, 96–109; nervous, 52, 59; pathology of, 61

tongue, **230**, 236–237

tonsil, 234–**235**, 274, 366

tonsillitis, 370–371

trachea, 275, 280–281

transamination, 404–405

transfusion, 324–325

transudation, 368

transverse plane, 7

trapezius, 120–**121**–122

triceps, **113**–115

triglycerides, 208–211, 402–404

tumor, 61

Turner's syndrome, 470

twitch, 105–**106**

tympanic cavity, 191–192

U

ulna, 78–79

unsaturated, 209–211

urinary system, 10, 408–416

urine, 410–415; composition of, 415–416

uterus, 453–455

utricle, 194

uvea, **173**

uvula, 234–235

V

vagina, 454

valve, 333, 350

vaporization, 388–389

vascular system, 346–361; disorders of, 360–361

vasoconstriction, 348, 358

vasodilation, 348, 358, 390–391

vein, 346, 349–350

ventricles, **149**–152

venule, 349–350

vertebral column, 71–**72**, 74–**75**

vestibule, **193**–194

villi, 252

visceral muscles, 96–97

vision, 170–184; binocular, 179; chemistry of, 180–183; color, 181–182; peripheral, 179

visual purple, 175

vitamin, fat-soluble, 218–221; water-soluble, 218–221

vocal cords, 274–275

voicebox, 278–279

voluntary muscles, 60, 96–97

vomer, **73**–74

W

water, 214–215

wavelengths of the spectrum, 176–177, **181**–182

white blood cell, 35, 40, 315–317

white matter, 133

Wright's stain, 316

wrist, 79

Y

yellow marrow, 311–312

Z

zygomatic arch, 71, **73**–**74**

zygote, 462